Robert H. Mohlenbrock
and
Douglas M. Ladd

Distribution of Illinois Vascular Plants

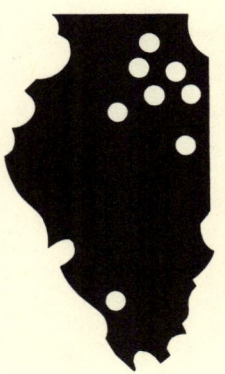

Southern Illinois
University Press
Carbondale
and
Edwardsville

Library of Congress Cataloging in Publication Data

Mohlenbrock, Robert H
    Distribution of Illinois vascular plants.

    Bibliography: p.
1. Botany—Illinois—Maps. I. Ladd, Douglas M.,
joint author. II. Title.
G1406.D2M6   1978   912′.1′5819773   77-15987

ISBN 0–8093–0848–7

Contents

Preface

This book is intended to bring up-to-date the distribution of every vascular plant known to occur in Illinois as a native, naturalized, or escaped species. All species and recognizable varieties are included in this work, but forms have been omitted, principally because of their sporadic occurrence. All named hybrids are included, but unnamed hybrids are not. Keys, descriptions, and ecological notes are not included, but these may be found in the companion volume, *Guide to the Vascular Flora of Illinois* (1975). This is the first attempt to update the distribution of all Illinois vascular plants since Winterringer and Evers (1960).

A voluminous amount of knowledge about the Illinois flora has been gained since Michaux's visit in 1795. In those eighteen decades, 3,001 taxa have been found within the boundaries of the state. Some plants which may have been attributed to Illinois in the past have been omitted in this work either because the earlier reports were based on misidentifications, or because substantiating specimens could not be found, or because they are no longer recognized as distinct taxonomic entities.

The distribution of the 3,001 taxa of vascular plants which follow is based on studies of collections at a number of public and private institutions. Several personal collections have been examined. With the exception of a few sight records contributed by John Schwegman for Pope County, each dot represents at least one specimen collected from that county.

The plants are arranged alphabetically by genus and, under each genus, alphabetically by species. The nomenclature follows Mohlenbrock's *Guide to the Vascular Flora of Illinois* (1975).

A list of synonyms applied to our taxa by Fernald (1950), Gleason (1952), and Jones (1963) follows the maps.

In order to gain an understanding of relationships of the plants in the Illinois flora, all 3,001 taxa are arranged in a phylogenetic sequence at the end of this book. The entries in the phylogenetic listing bear the proper author citation for each taxon.

The authors urge all Illinois botanists to report their finds to them so that the distribution of Illinois vascular plants can be updated continuously.

A WORK OF THIS NATURE represents the cumulative efforts of numerous individuals, only a few of whom can be mentioned here. The authors are most grateful to the generous support of the J. Sterling Morton Foundation which supplied funds for some of the travel necessary to compile the data for this book.

For providing distribution data and assistance in the field, the authors are grateful to Marion Cole, Mike Homoya, John Schwegman, Paul Shildneck, Robert Tatina, W. Carl Taylor, and Randy Vogel.

Several curators of major herbaria in the state have generously provided assistance while the authors were visiting their institutions. We would like to thank John Ebinger, Robert Evers, Alfred Koelling, and Kenneth Robertson for their aid.

Ray Schulenberg, Floyd Swink, and Gerould Wilhelm of the Morton Arboretum have freely provided distributional information for the Illinois counties included in their thorough documentation of the flora of the Chicago region.

In addition, the authors are indebted to Deborah Bowen and Paul Nelson for their continual assistance in the field and during manuscript preparation, and to Beverly Mohlenbrock for typing the text material. Without their help, this work would not have been possible.

ROBERT H. MOHLENBROCK
DOUGLAS M. LADD
Southern Illinois University
March 17, 1977

**Distribution
of Illinois
Vascular Plants**

WISCONSIN

IOWA

INDIANA

MISSOURI

KENTUCKY

JO DAVIESS | STEPHENSON | WINNEBAGO | BOONE | McHENRY | LAKE

CARROLL | OGLE | DeKALB | KANE | COOK

WHITESIDE | LEE | DuPAGE

ROCK ISLAND | HENRY | BUREAU | La SALLE | KENDALL | WILL

MERCER | PUTNAM | GRUNDY | KANKAKEE

STARK | MARSHALL | LIVINGSTON

WARREN | KNOX | PEORIA | WOODFORD | IROQUOIS

HENDERSON | FULTON | TAZEWELL | McLEAN | FORD

HANCOCK | McDONOUGH | MASON | LOGAN | DeWITT | CHAMPAIGN | VERMILION

SCHUYLER | MENARD | PIATT

ADAMS | BROWN | CASS | SANGAMON | MACON | DOUGLAS | EDGAR

MORGAN | MOULTRIE | COLES

PIKE | SCOTT | CHRISTIAN | SHELBY | CLARK

GREENE | MACOUPIN | MONTGOMERY | CUMBERLAND

CALHOUN | JERSEY | FAYETTE | EFFINGHAM | JASPER | CRAWFORD

MADISON | BOND | CLAY | RICHLAND | LAWRENCE

CLINTON | MARION | WABASH

ST. CLAIR | WASHINGTON | JEFFERSON | WAYNE | EDWARDS

MONROE | RANDOLPH | PERRY | HAMILTON | WHITE

JACKSON | FRANKLIN | SALINE | GALLATIN

WILLIAMSON | UNION | JOHNSON | POPE | HARDIN

ALEXANDER | PULASKI | MASSAC

County Map of Illinois

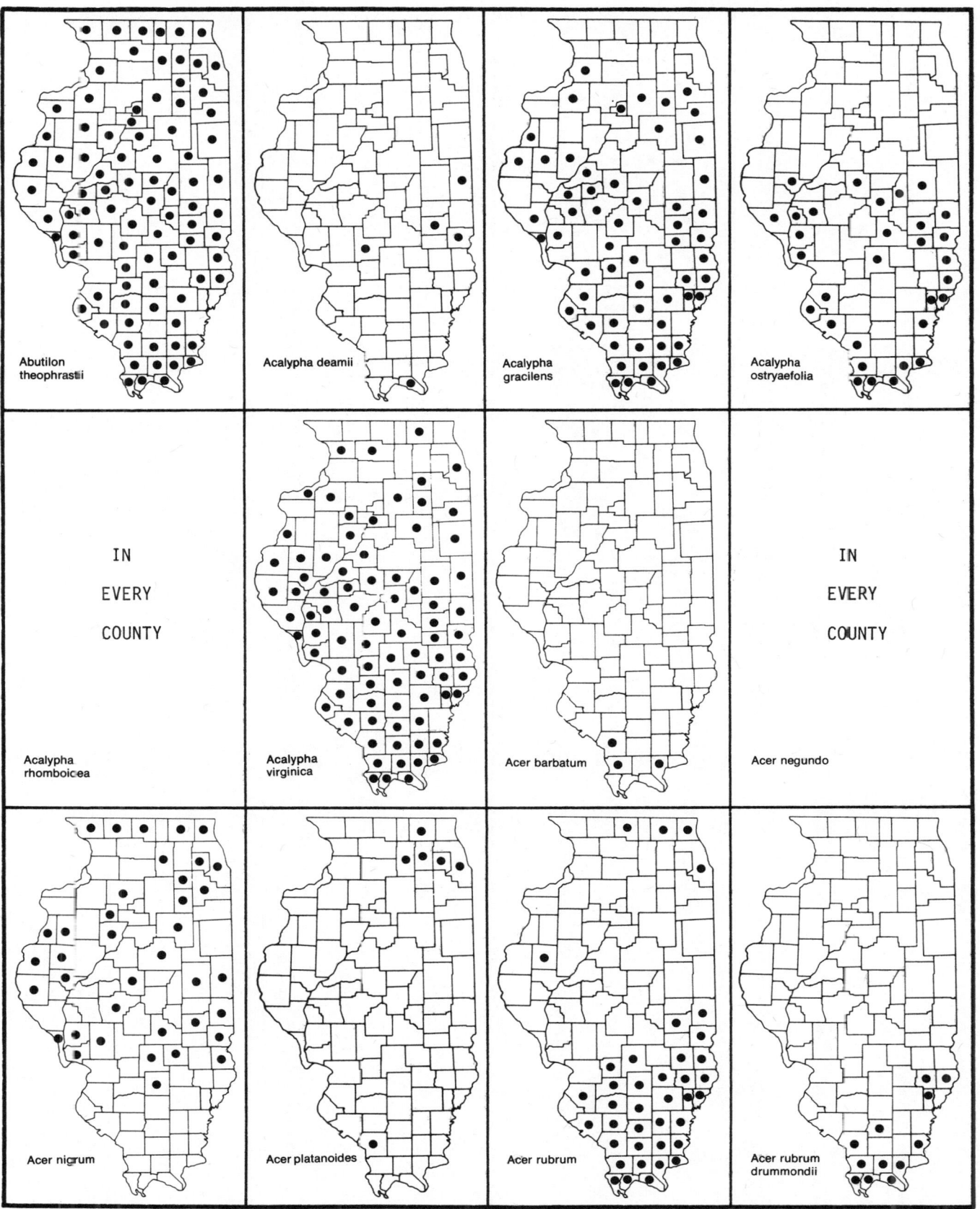

Abutilon
theophrasti

Acalypha deamii

Acalypha
gracilens

Acalypha
ostryaefolia

IN

EVERY

COUNTY

Acalypha
rhomboicea

Acalypha
virginica

Acer barbatum

IN

EVERY

COUNTY

Acer negundo

Acer nigrum

Acer platanoides

Acer rubrum

Acer rubrum
drummondii

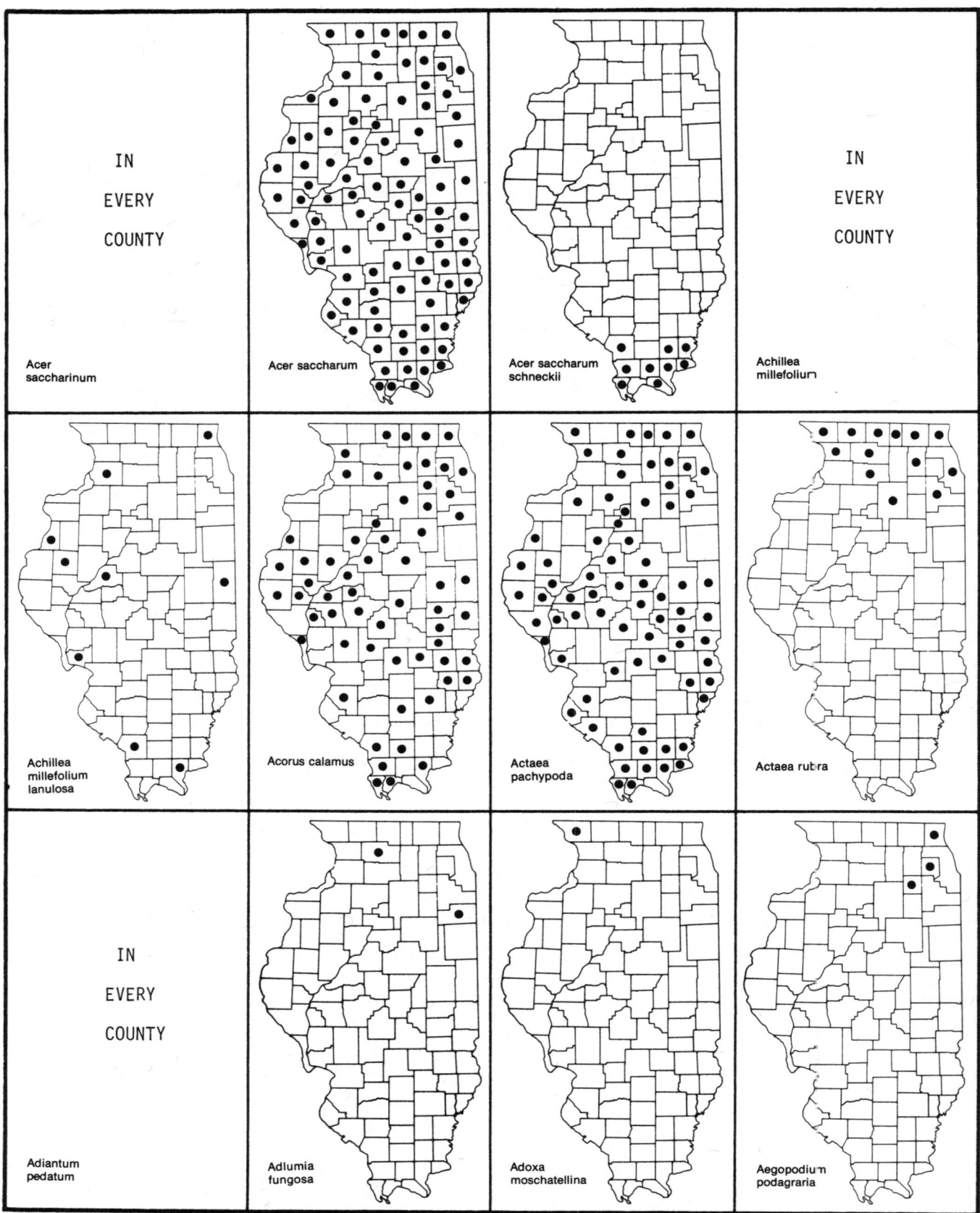

IN EVERY COUNTY

Acer saccharinum

Acer saccharum

Acer saccharum schneckii

IN EVERY COUNTY

Achillea millefolium

Achillea millefolium lanulosa

Acorus calamus

Actaea pachypoda

Actaea rubra

IN EVERY COUNTY

Adiantum pedatum

Adlumia fungosa

Adoxa moschatellina

Aegopodium podagraria

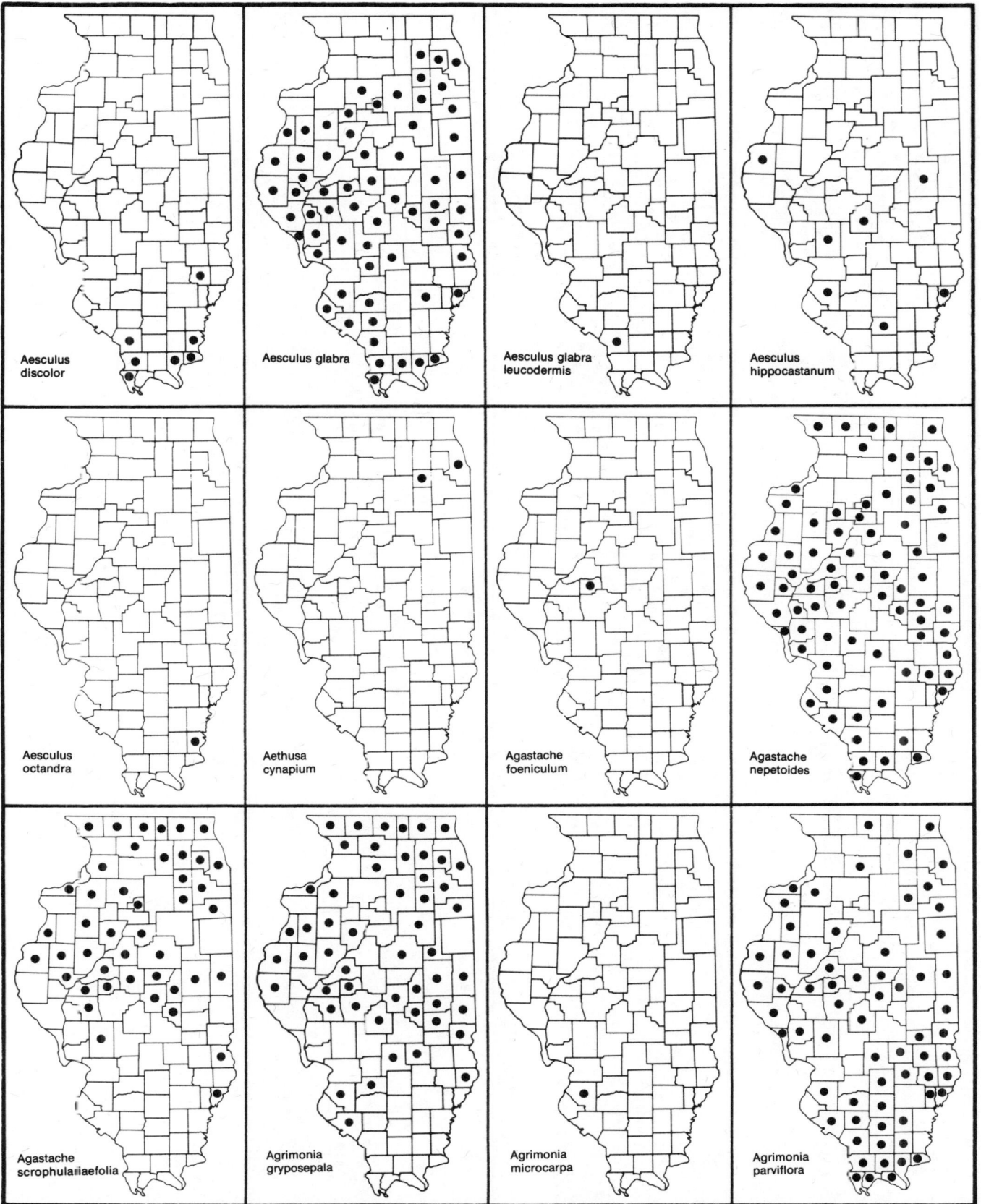

Aesculus
discolor

Aesculus glabra

Aesculus glabra
leucodermis

Aesculus
hippocastanum

Aesculus
octandra

Aethusa
cynapium

Agastache
foeniculum

Agastache
nepetoides

Agastache
scrophulariaefolia

Agrimonia
gryposepala

Agrimonia
microcarpa

Agrimonia
parviflora

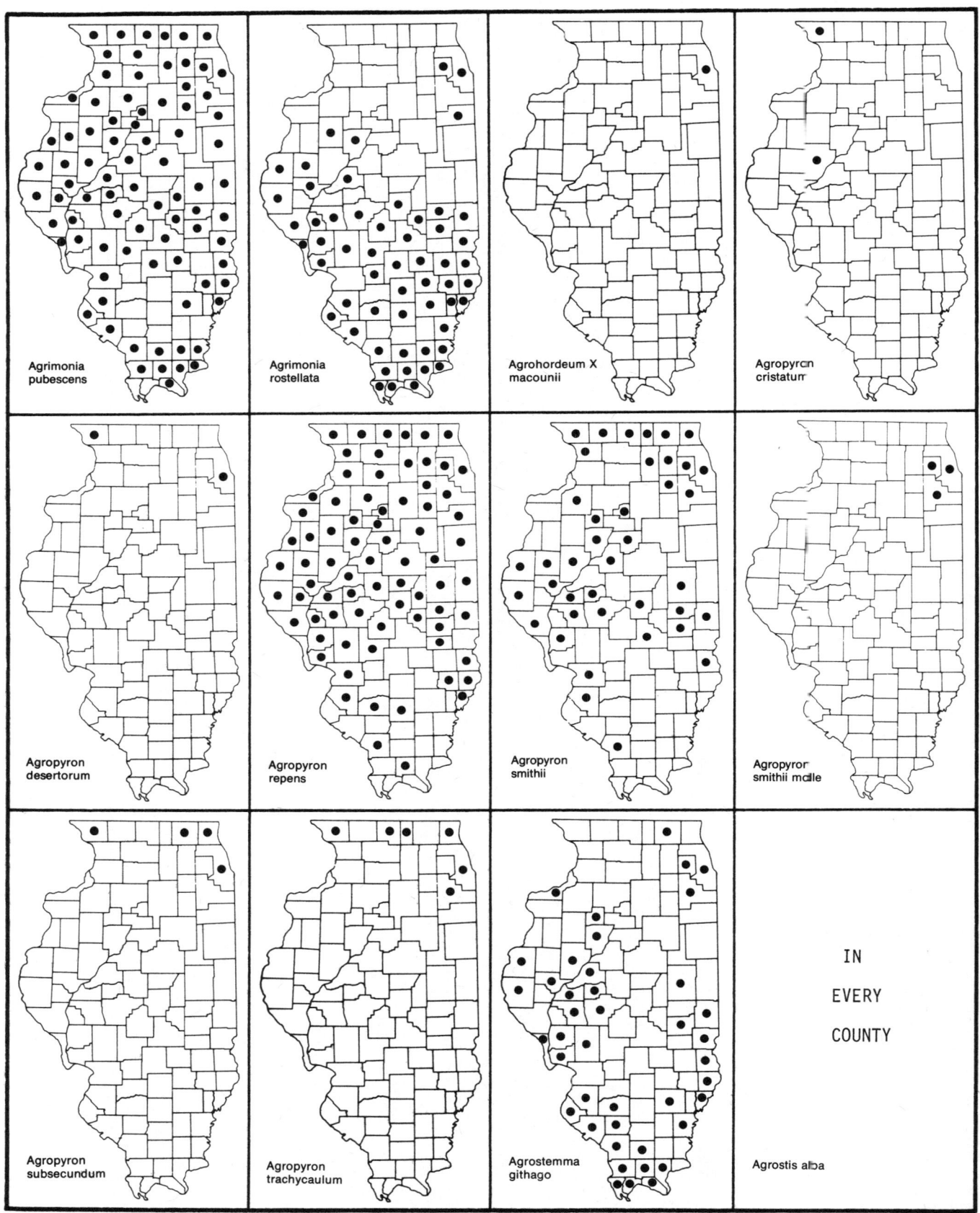

Agrimonia
pubescens

Agrimonia
rostellata

Agrohordeum X
macounii

Agropyron
cristatum

Agropyron
desertorum

Agropyron
repens

Agropyron
smithii

Agropyron
smithii molle

Agropyron
subsecundum

Agropyron
trachycaulum

Agrostemma
githago

IN

EVERY

COUNTY

Agrostis alba

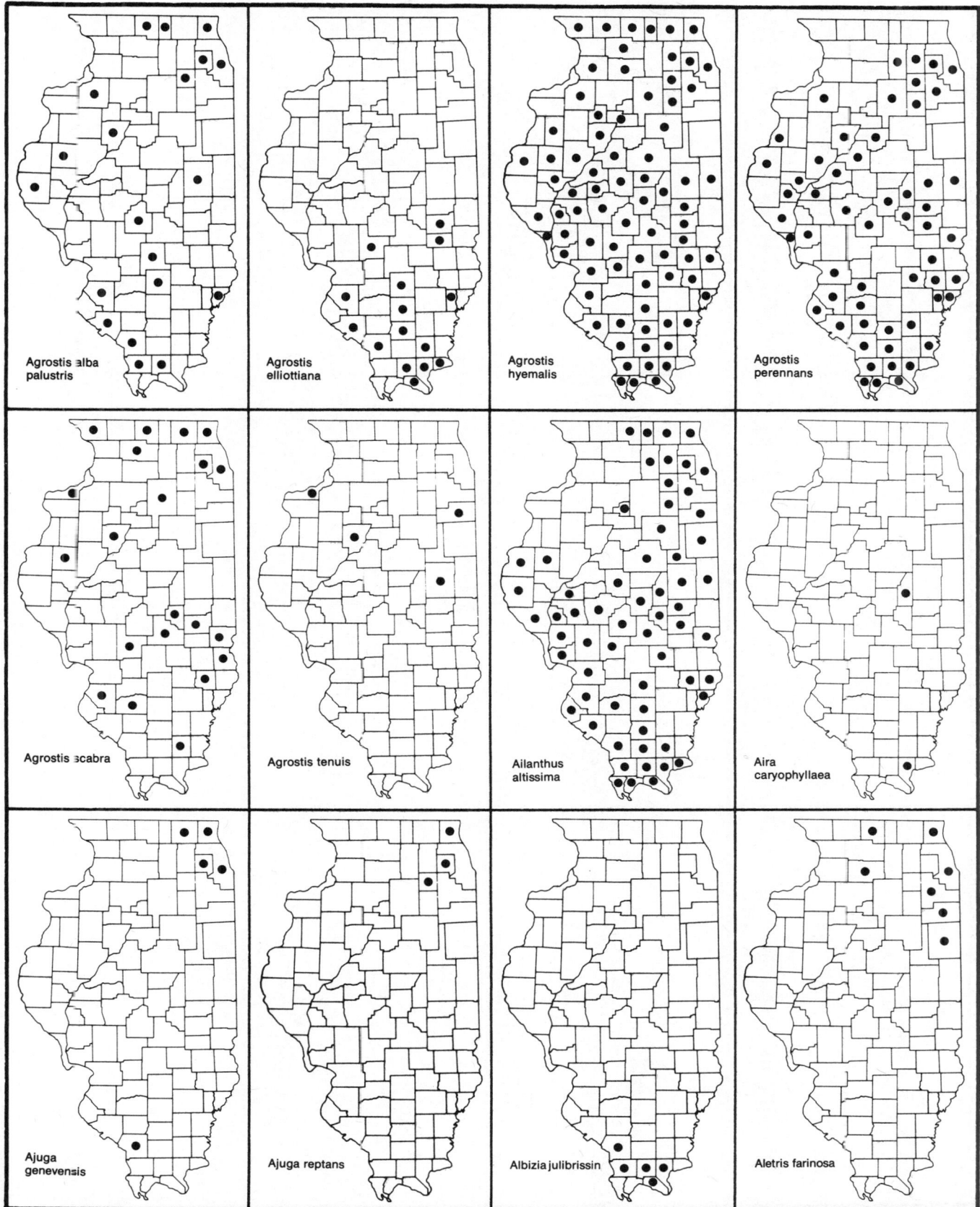

Agrostis alba
palustris

Agrostis
elliottiana

Agrostis
hyemalis

Agrostis
perennans

Agrostis scabra

Agrostis tenuis

Ailanthus
altissima

Aira
caryophyllaea

Ajuga
genevensis

Ajuga reptans

Albizia julibrissin

Aletris farinosa

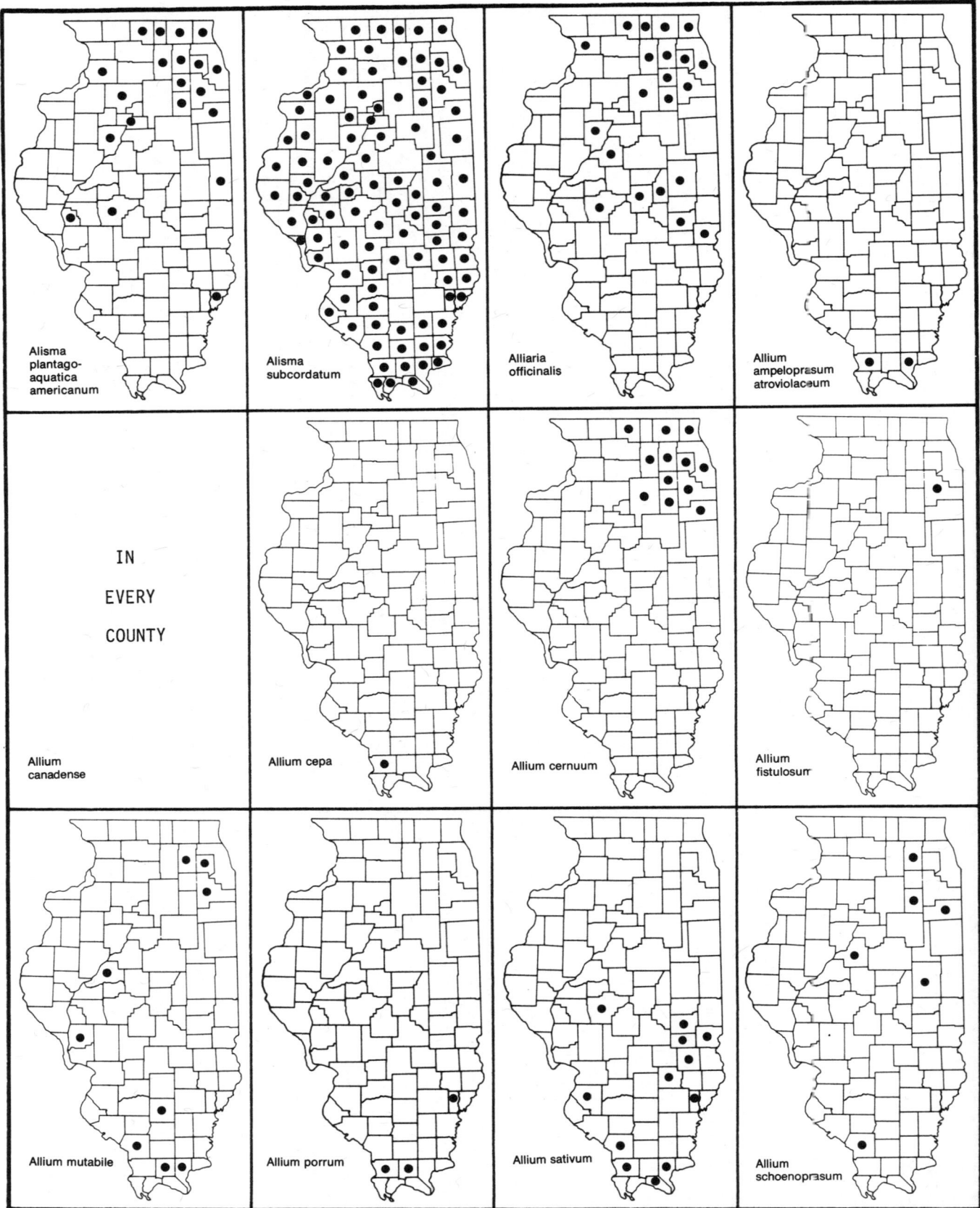

Alisma
plantago-
aquatica
americanum

Alisma
subcordatum

Alliaria
officinalis

Allium
ampeloprasum
atroviolaceum

IN

EVERY

COUNTY

Allium
canadense

Allium cepa

Allium cernuum

Allium
fistulosum

Allium mutabile

Allium porrum

Allium sativum

Allium
schoenoprasum

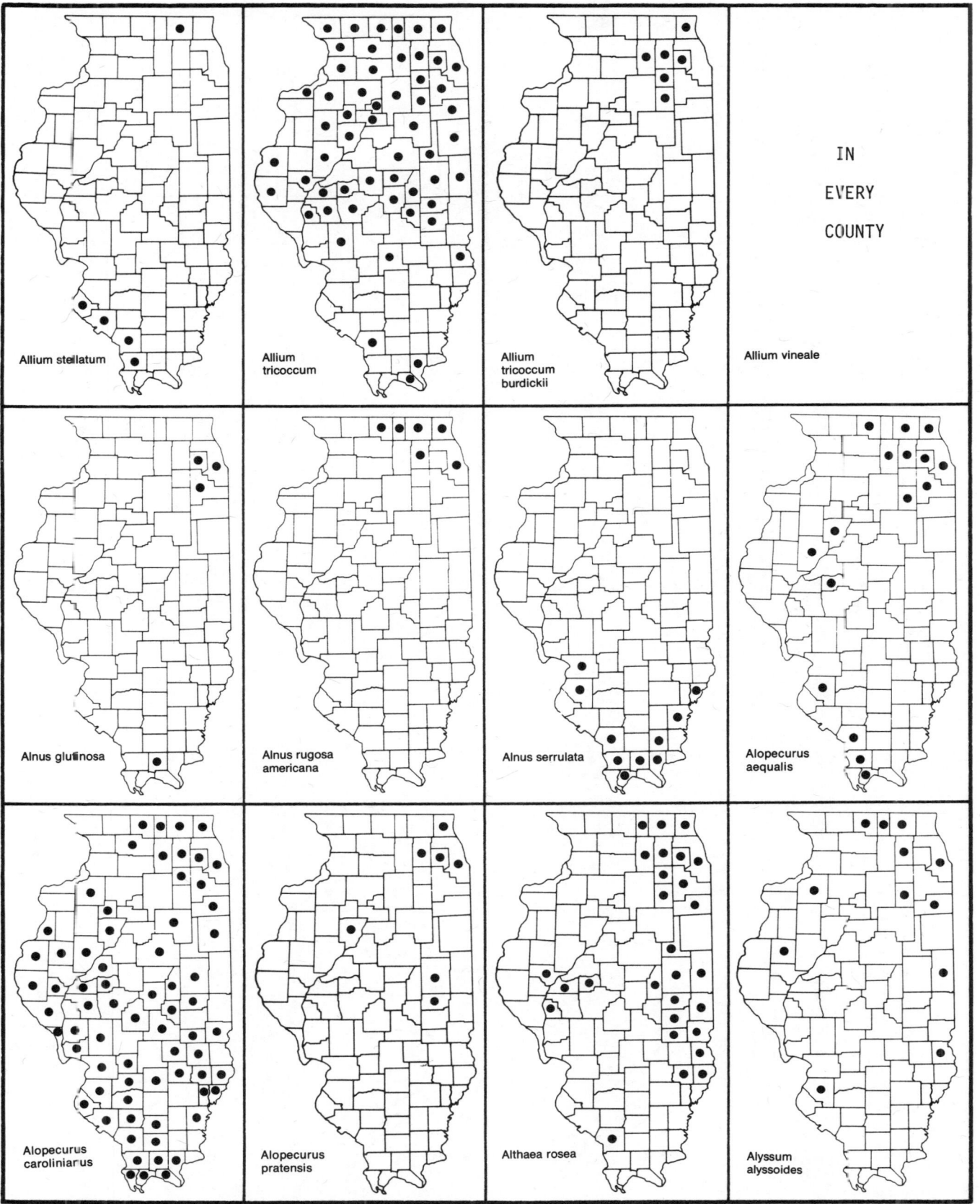

Allium stellatum

Allium tricoccum

Allium tricoccum burdickii

Allium vineale

IN

EVERY

COUNTY

Alnus glutinosa

Alnus rugosa americana

Alnus serrulata

Alopecurus aequalis

Alopecurus carolinianus

Alopecurus pratensis

Althaea rosea

Alyssum alyssoides

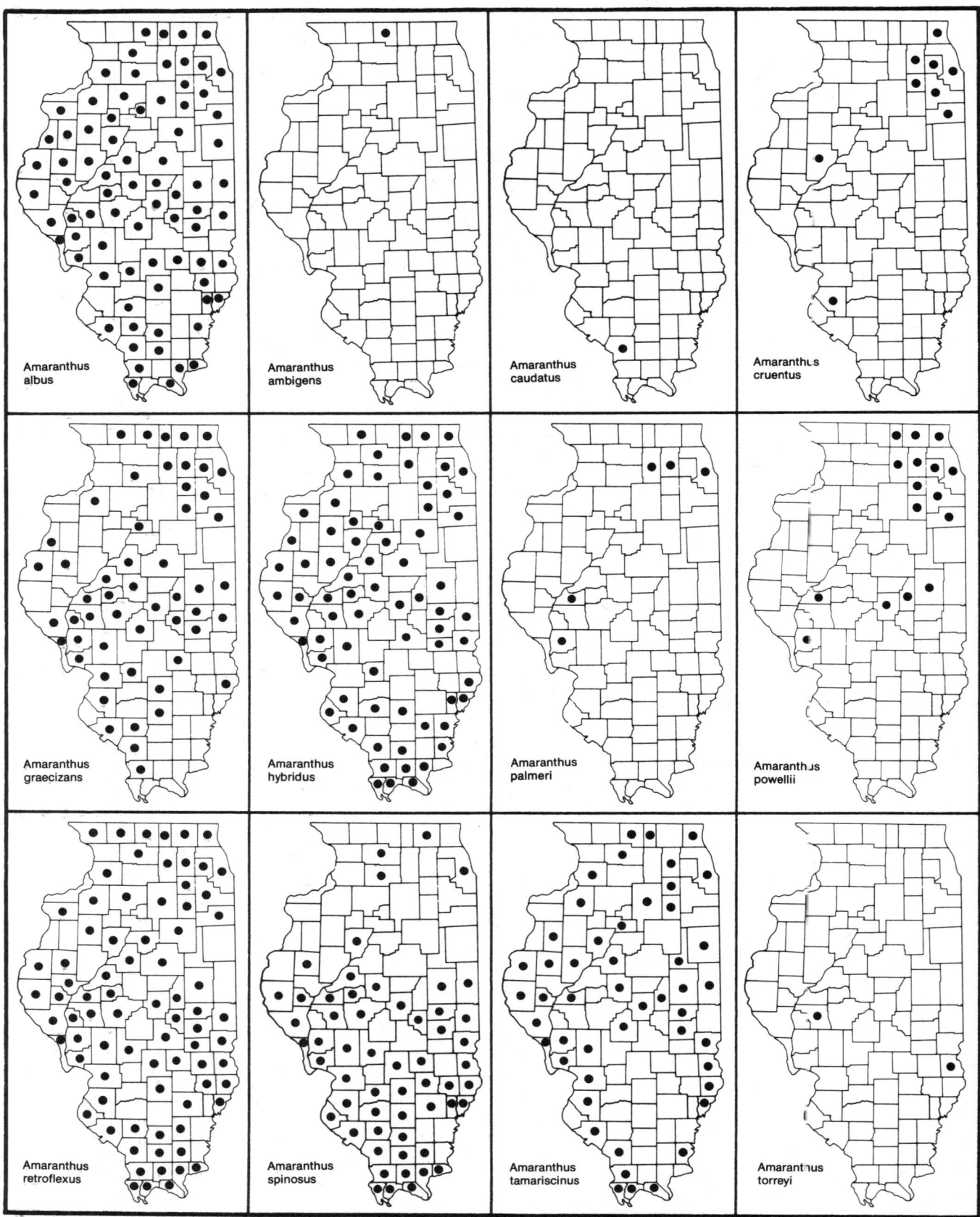

Amaranthus
albus

Amaranthus
ambigens

Amaranthus
caudatus

Amaranthus
cruentus

Amaranthus
graecizans

Amaranthus
hybridus

Amaranthus
palmeri

Amaranthus
powellii

Amaranthus
retroflexus

Amaranthus
spinosus

Amaranthus
tamariscinus

Amaranthus
torreyi

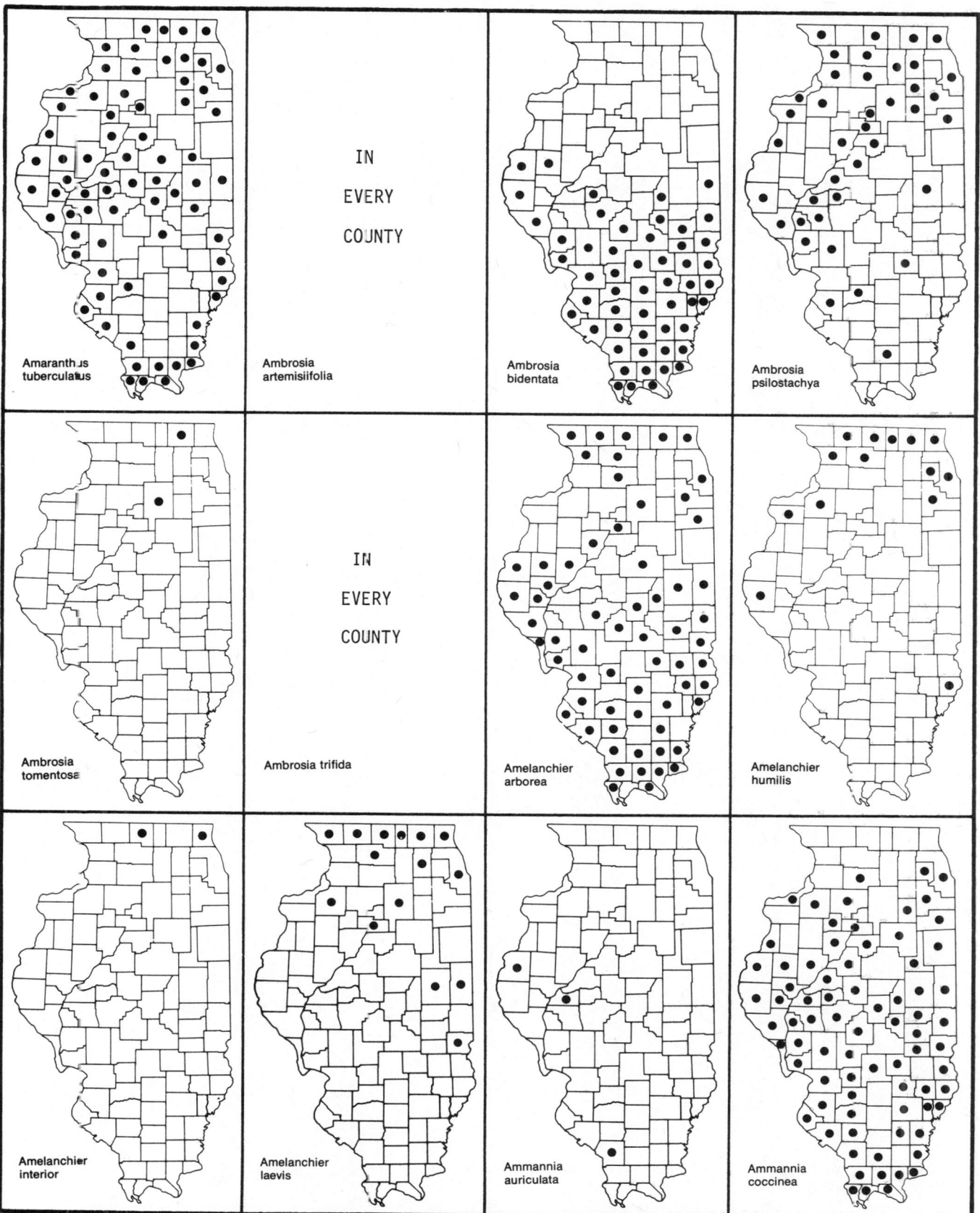

Amaranthus
tuberculatus

Ambrosia
artemisiifolia

IN

EVERY

COUNTY

Ambrosia
bidentata

Ambrosia
psilostachya

Ambrosia
tomentosa

Ambrosia trifida

IN

EVERY

COUNTY

Amelanchier
arborea

Amelanchier
humilis

Amelanchier
interior

Amelanchier
laevis

Ammannia
auriculata

Ammannia
coccinea

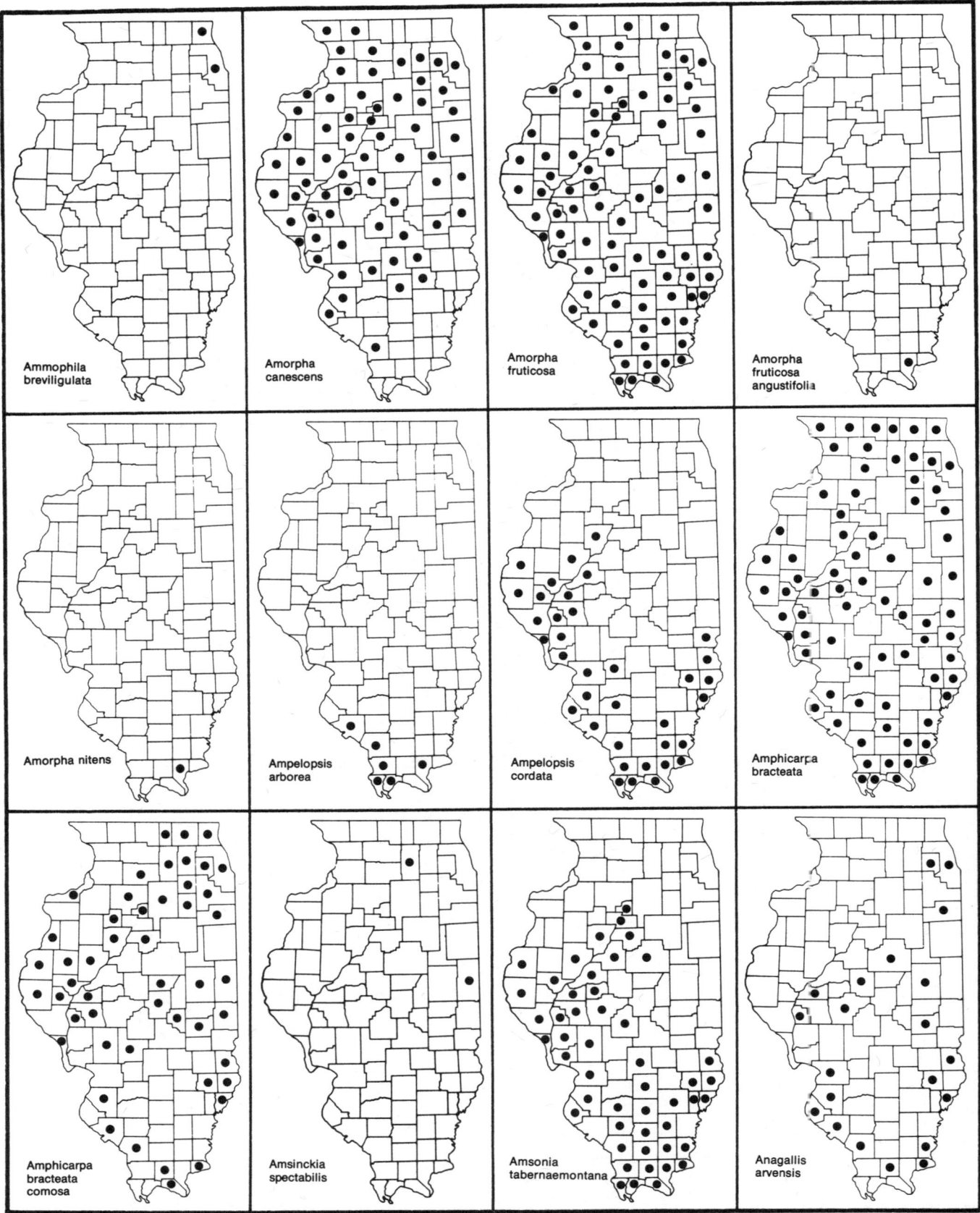

Ammophila
breviligulata

Amorpha
canescens

Amorpha
fruticosa

Amorpha
fruticosa
angustifolia

Amorpha nitens

Ampelopsis
arborea

Ampelopsis
cordata

Amphicarpa
bracteata

Amphicarpa
bracteata
comosa

Amsinckia
spectabilis

Amsonia
tabernaemontana

Anagallis
arvensis

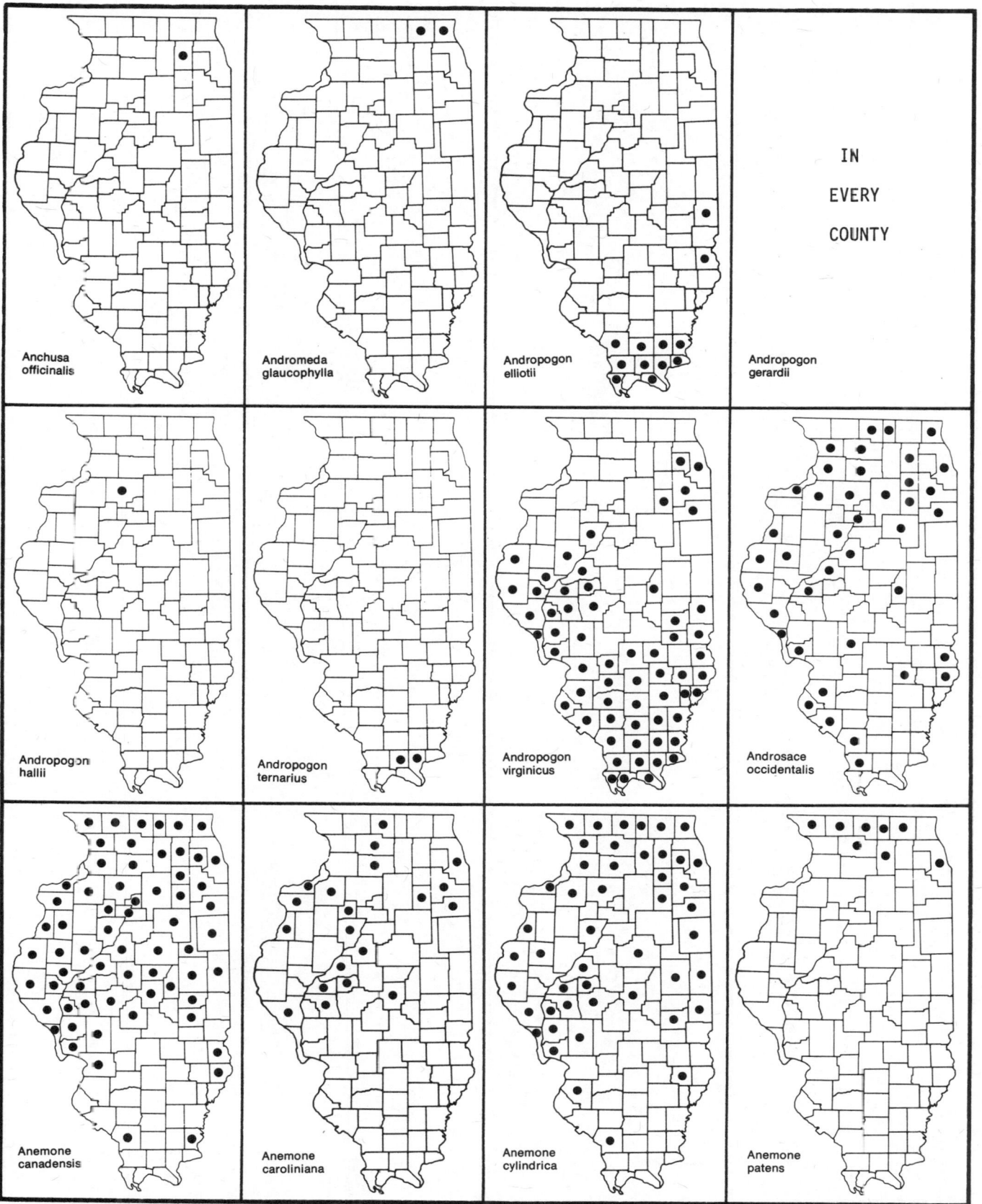

Anchusa
officinalis

Andromeda
glaucophylla

Andropogon
elliotii

Andropogon
gerardii

IN

EVERY

COUNTY

Andropogon
hallii

Andropogon
ternarius

Andropogon
virginicus

Androsace
occidentalis

Anemone
canadensis

Anemone
caroliniana

Anemone
cylindrica

Anemone
patens

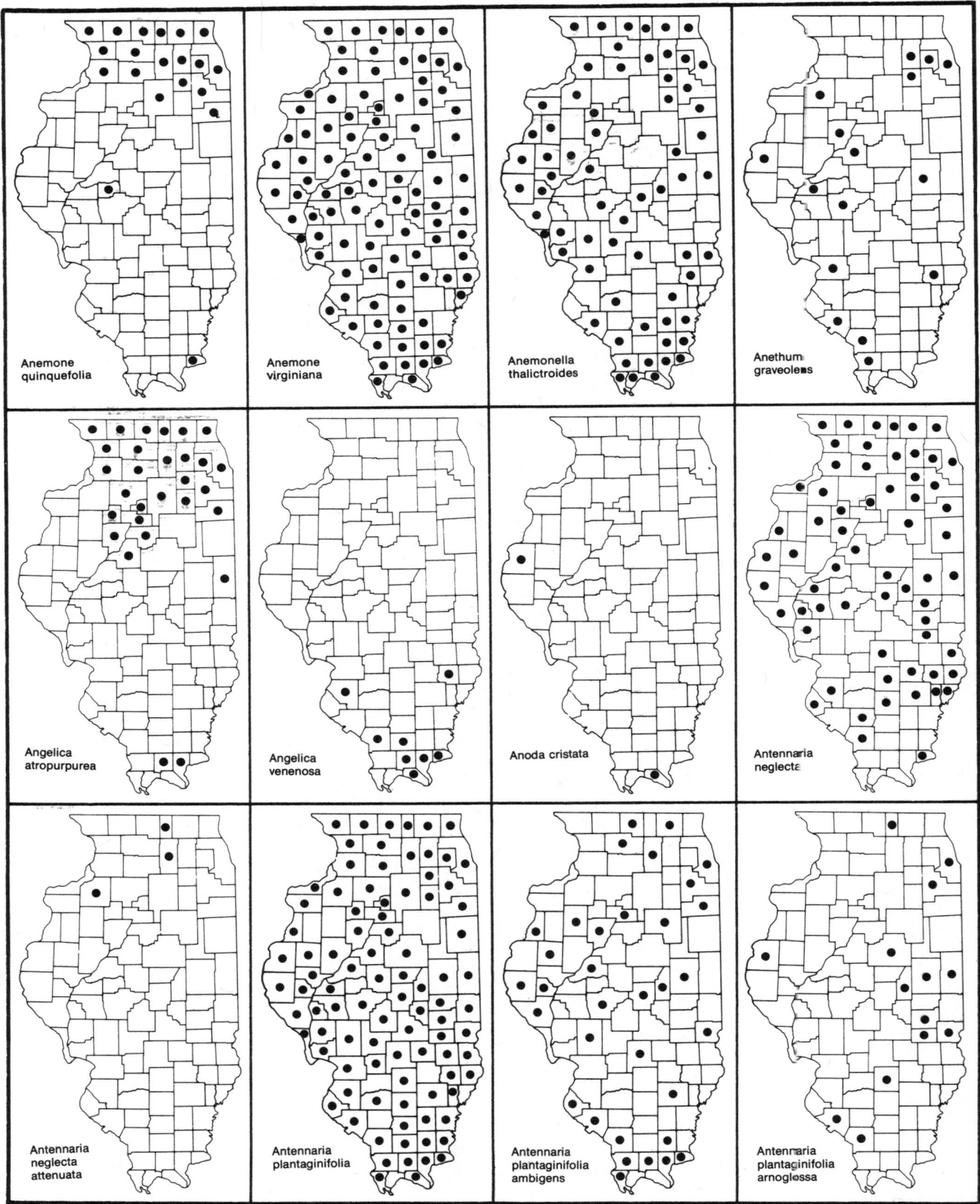

Anemone
quinquefolia

Anemone
virginiana

Anemonella
thalictroides

Anethum
graveolens

Angelica
atropurpurea

Angelica
venenosa

Anoda cristata

Antennaria
neglecta

Antennaria
neglecta
attenuata

Antennaria
plantaginifolia

Antennaria
plantaginifolia
ambigens

Antennaria
plantaginifolia
arnoglossa

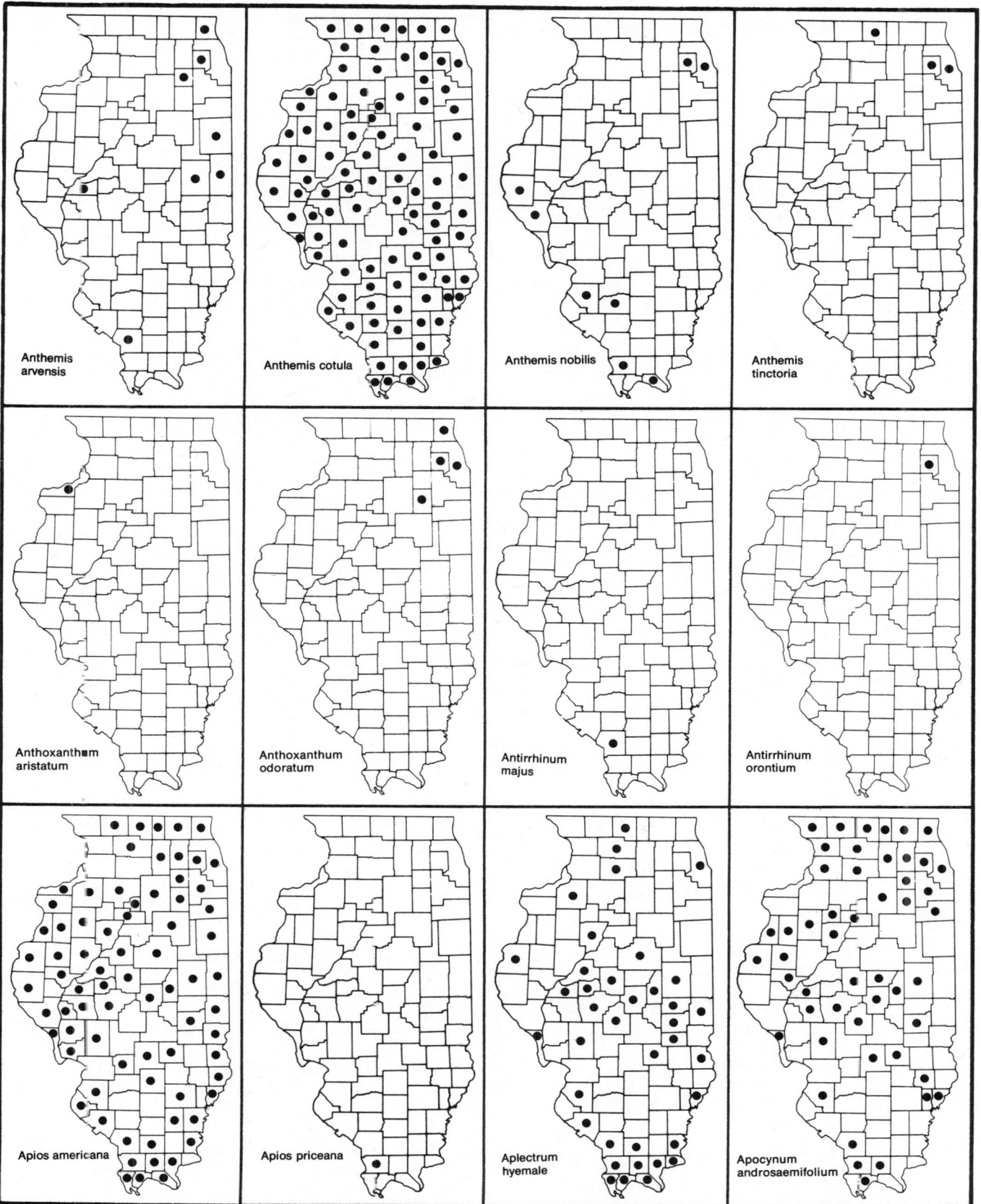

Anthemis
arvensis

Anthemis cotula

Anthemis nobilis

Anthemis
tinctoria

Anthoxanthum
aristatum

Anthoxanthum
odoratum

Antirrhinum
majus

Antirrhinum
orontium

Apios americana

Apios priceana

Aplectrum
hyemale

Apocynum
androsaemifolium

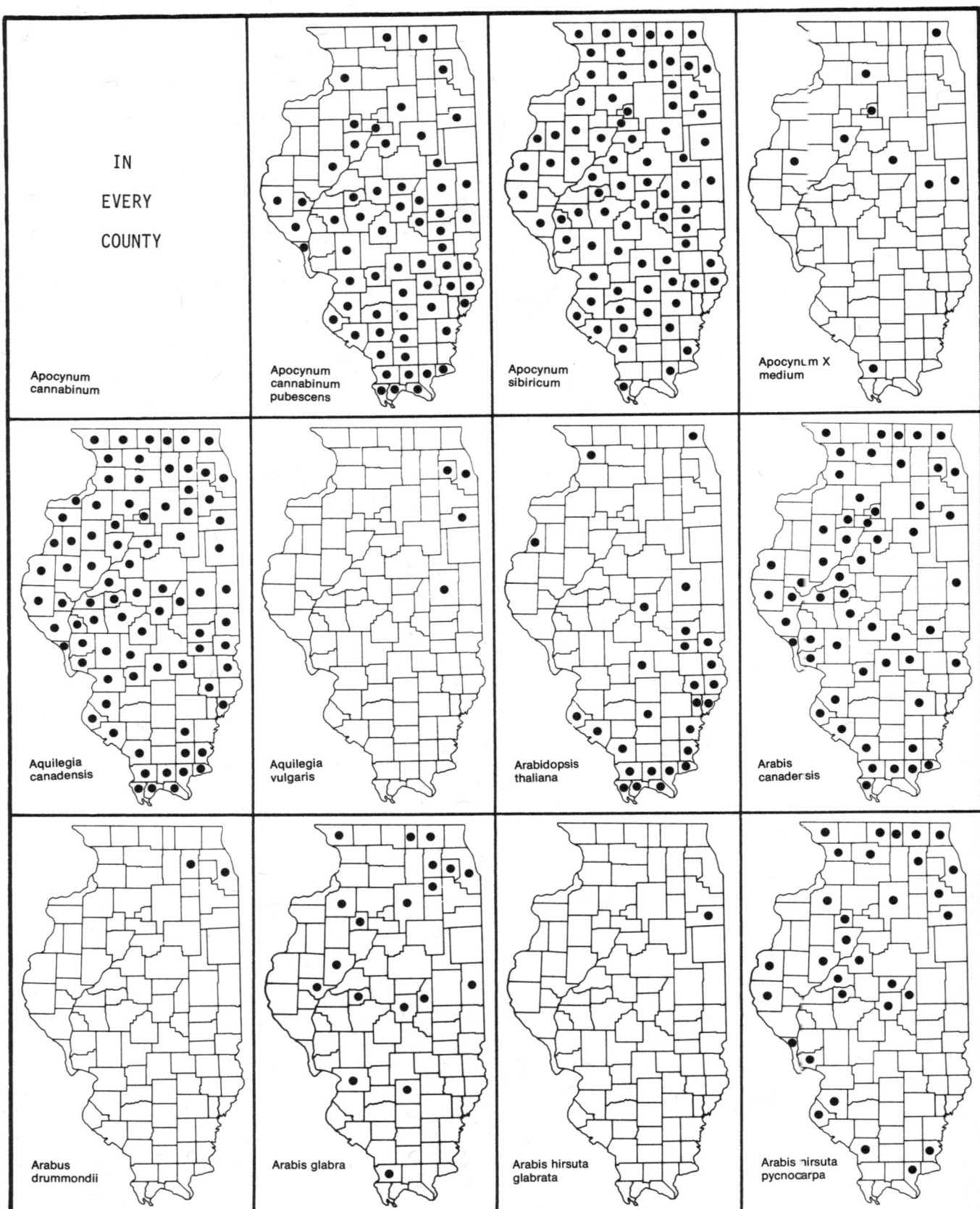

IN
EVERY
COUNTY

Apocynum
cannabinum

Apocynum
cannabinum
pubescens

Apocynum
sibiricum

Apocynum X
medium

Aquilegia
canadensis

Aquilegia
vulgaris

Arabidopsis
thaliana

Arabis
canadensis

Arabus
drummondii

Arabis glabra

Arabis hirsuta
glabrata

Arabis hirsuta
pycnocarpa

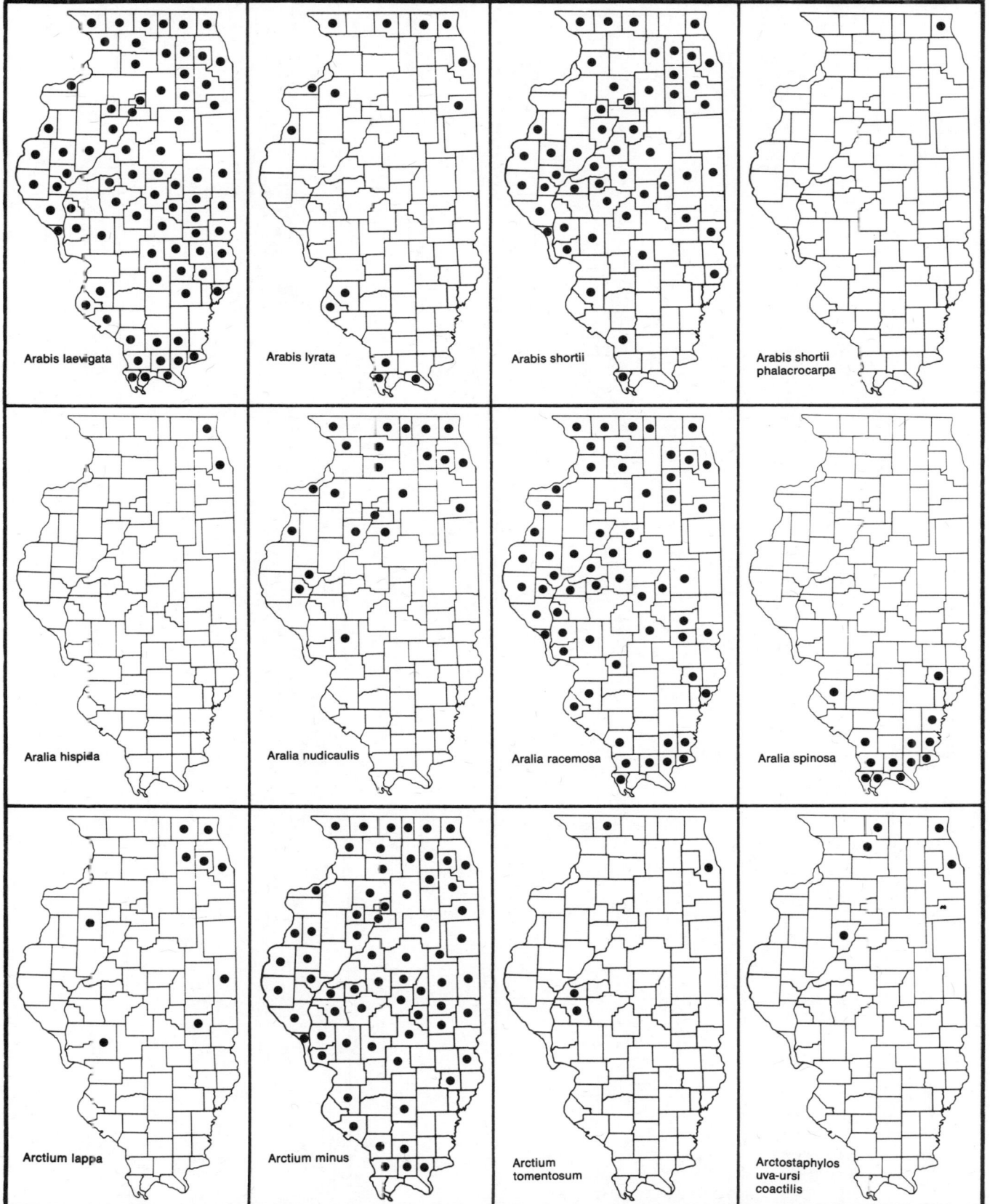

Arabis laevigata

Arabis lyrata

Arabis shortii

Arabis shortii
phalacrocarpa

Aralia hispida

Aralia nudicaulis

Aralia racemosa

Aralia spinosa

Arctium lappa

Arctium minus

Arctium
tomentosum

Arctostaphylos
uva-ursi
coactilis

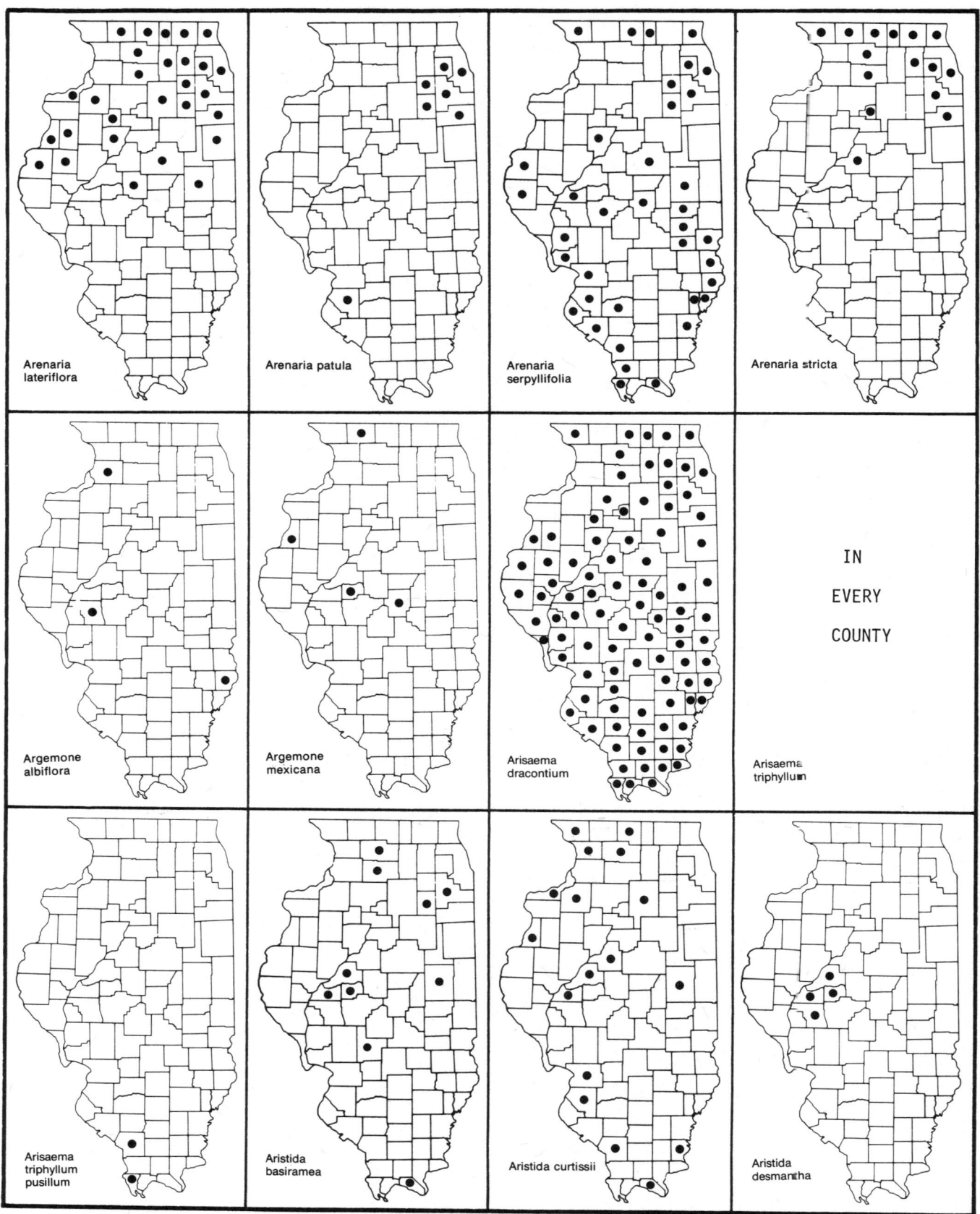

Arenaria
lateriflora

Arenaria patula

Arenaria
serpyllifolia

Arenaria stricta

Argemone
albiflora

Argemone
mexicana

Arisaema
dracontium

Arisaema
triphyllum

IN

EVERY

COUNTY

Arisaema
triphyllum
pusillum

Aristida
basiramea

Aristida curtissii

Aristida
desmantha

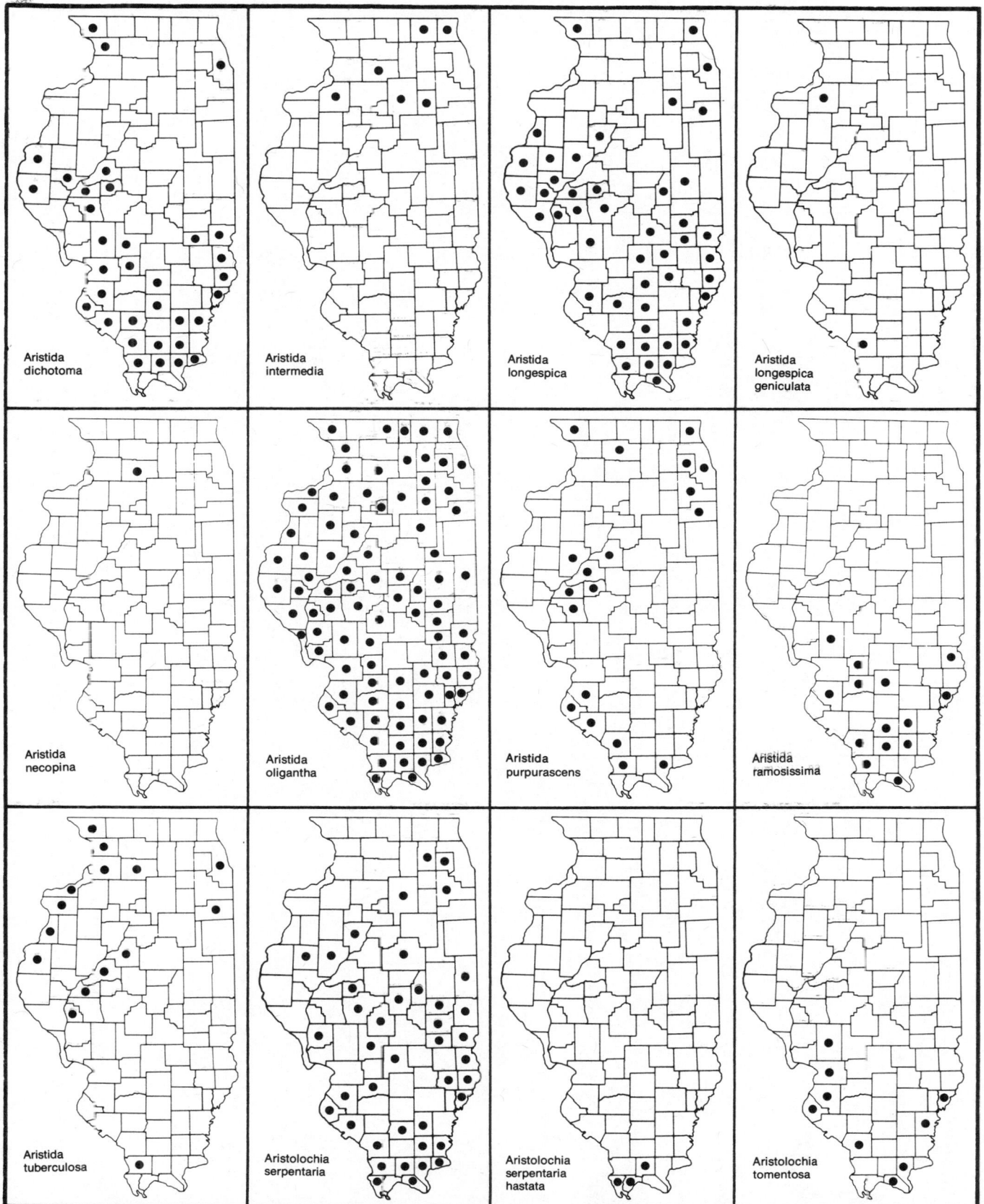

Aristida
dichotoma

Aristida
intermedia

Aristida
longespica

Aristida
longespica
geniculata

Aristida
necopina

Aristida
oligantha

Aristida
purpurascens

Aristida
ramosissima

Aristida
tuberculosa

Aristolochia
serpentaria

Aristolochia
serpentaria
hastata

Aristolochia
tomentosa

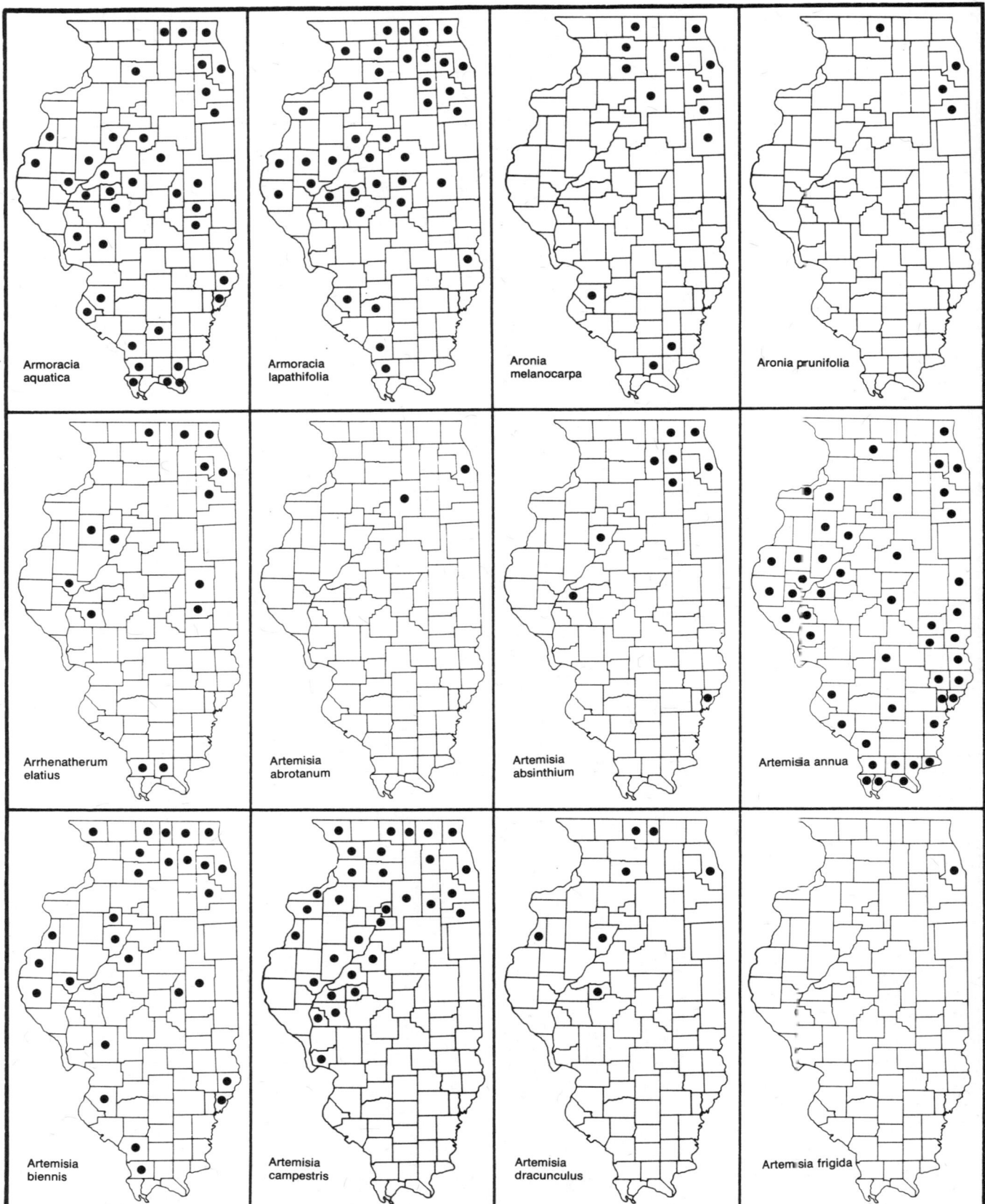

Armoracia
aquatica

Armoracia
lapathifolia

Aronia
melanocarpa

Aronia prunifolia

Arrhenatherum
elatius

Artemisia
abrotanum

Artemisia
absinthium

Artemisia annua

Artemisia
biennis

Artemisia
campestris

Artemisia
dracunculus

Artemisia frigida

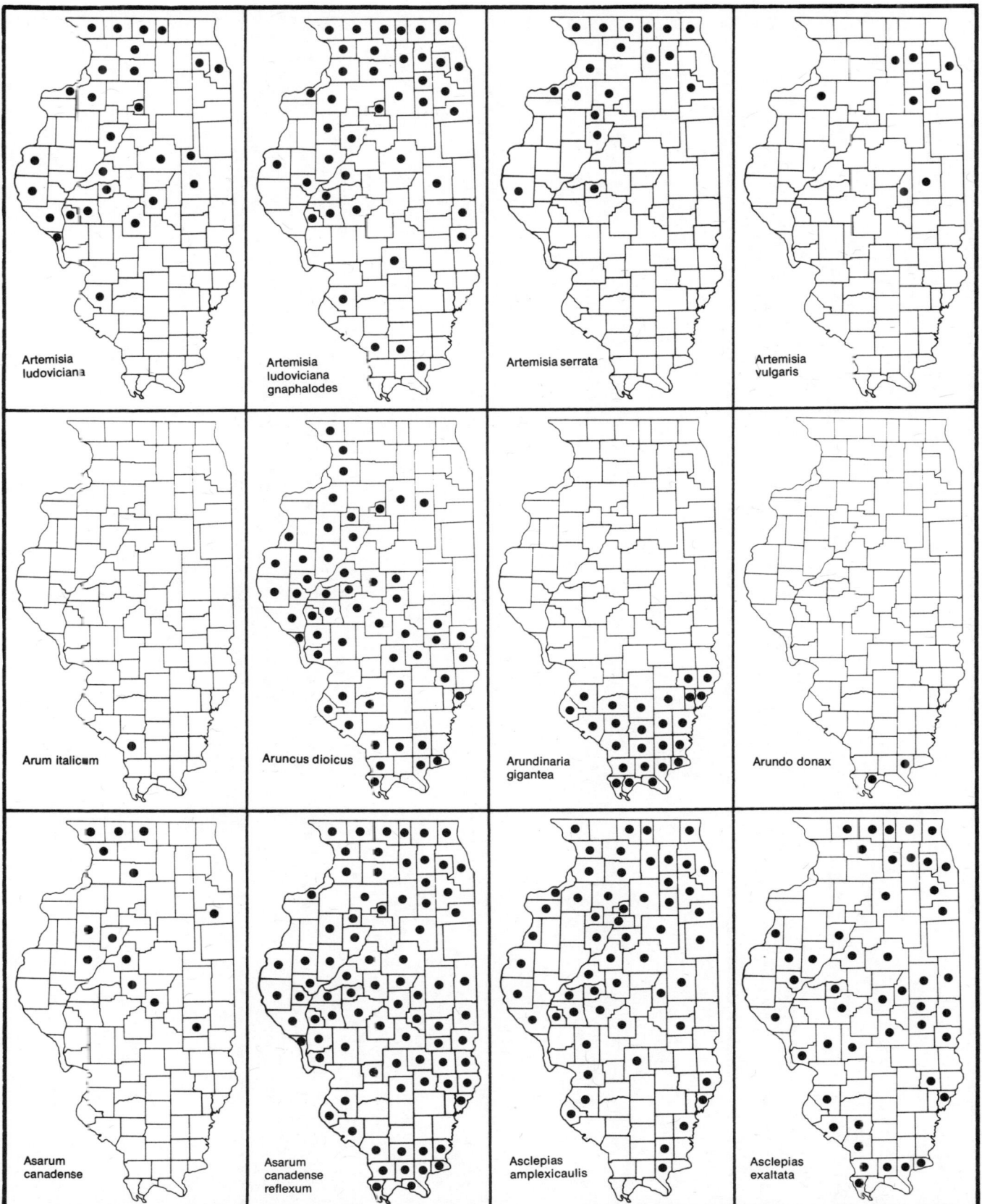

Artemisia
ludoviciana

Artemisia
ludoviciana
gnaphalodes

Artemisia serrata

Artemisia
vulgaris

Arum italicum

Aruncus dioicus

Arundinaria
gigantea

Arundo donax

Asarum
canadense

Asarum
canadense
reflexum

Asclepias
amplexicaulis

Asclepias
exaltata

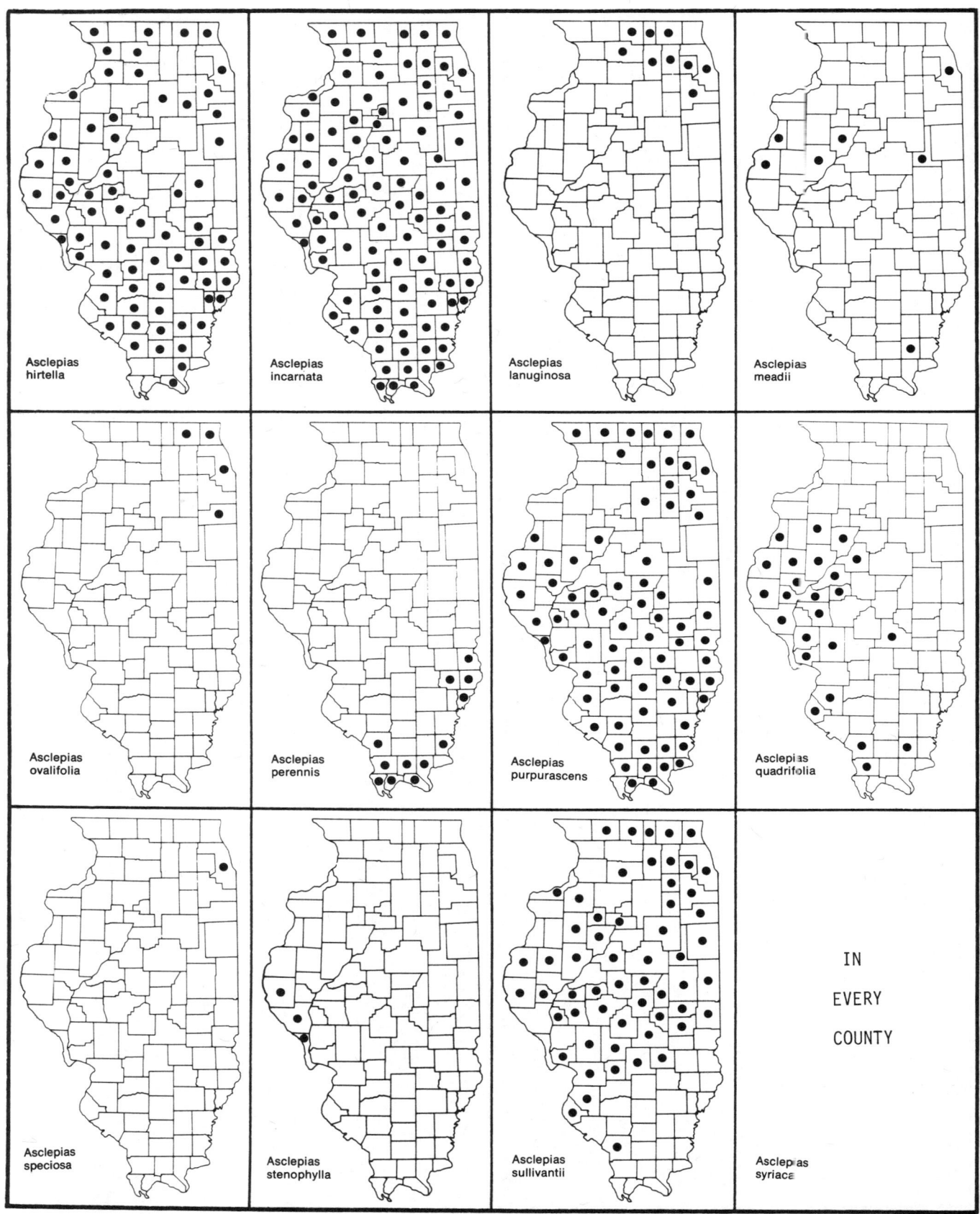

Asclepias
hirtella

Asclepias
incarnata

Asclepias
lanuginosa

Asclepias
meadii

Asclepias
ovalifolia

Asclepias
perennis

Asclepias
purpurascens

Asclepias
quadrifolia

Asclepias
speciosa

Asclepias
stenophylla

Asclepias
sullivantii

Asclepias
syriaca

IN

EVERY

COUNTY

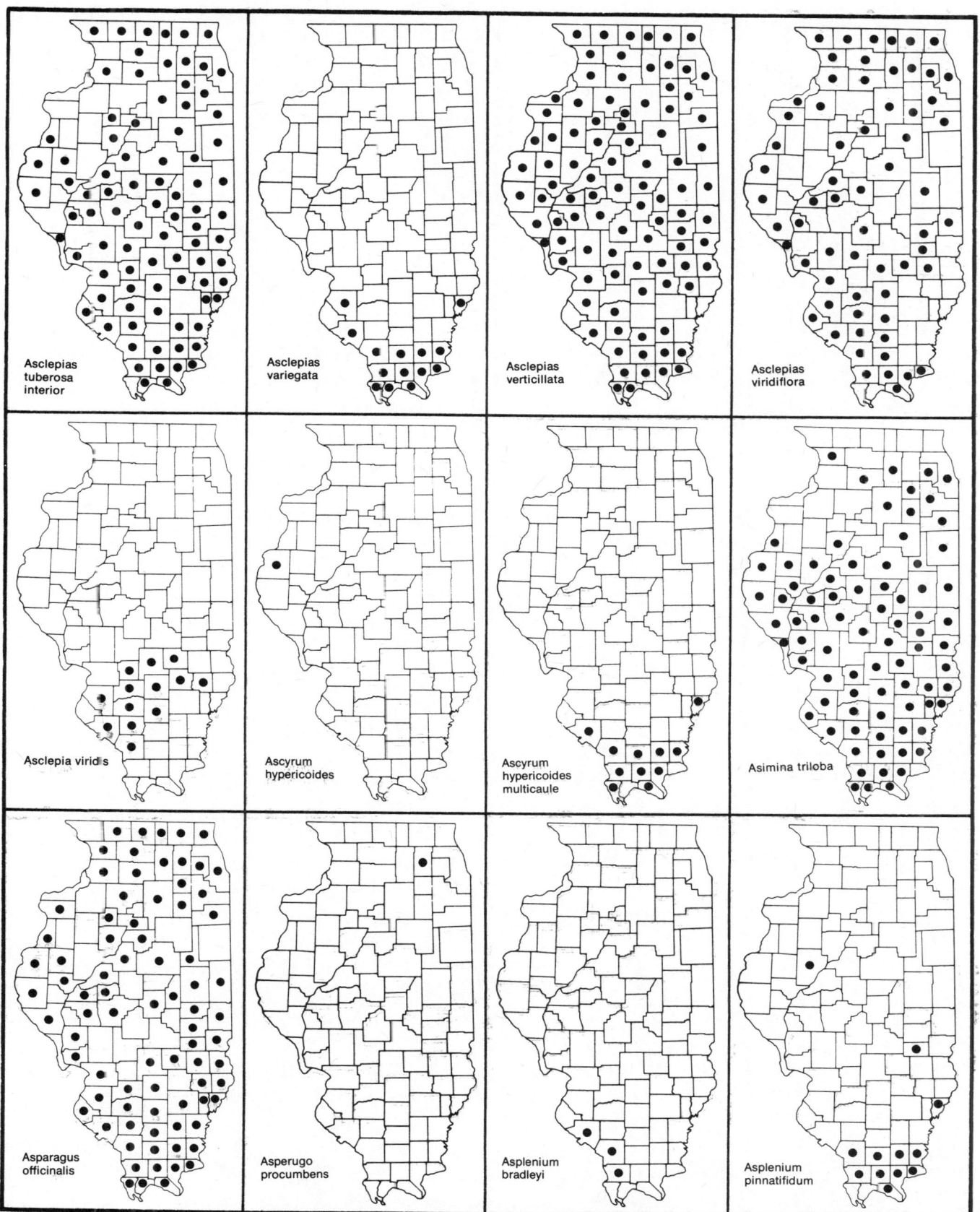

Asclepias tuberosa interior

Asclepias variegata

Asclepias verticillata

Asclepias viridiflora

Asclepia viridis

Ascyrum hypericoides

Ascyrum hypericoides multicaule

Asimina triloba

Asparagus officinalis

Asperugo procumbens

Asplenium bradleyi

Asplenium pinnatifidum

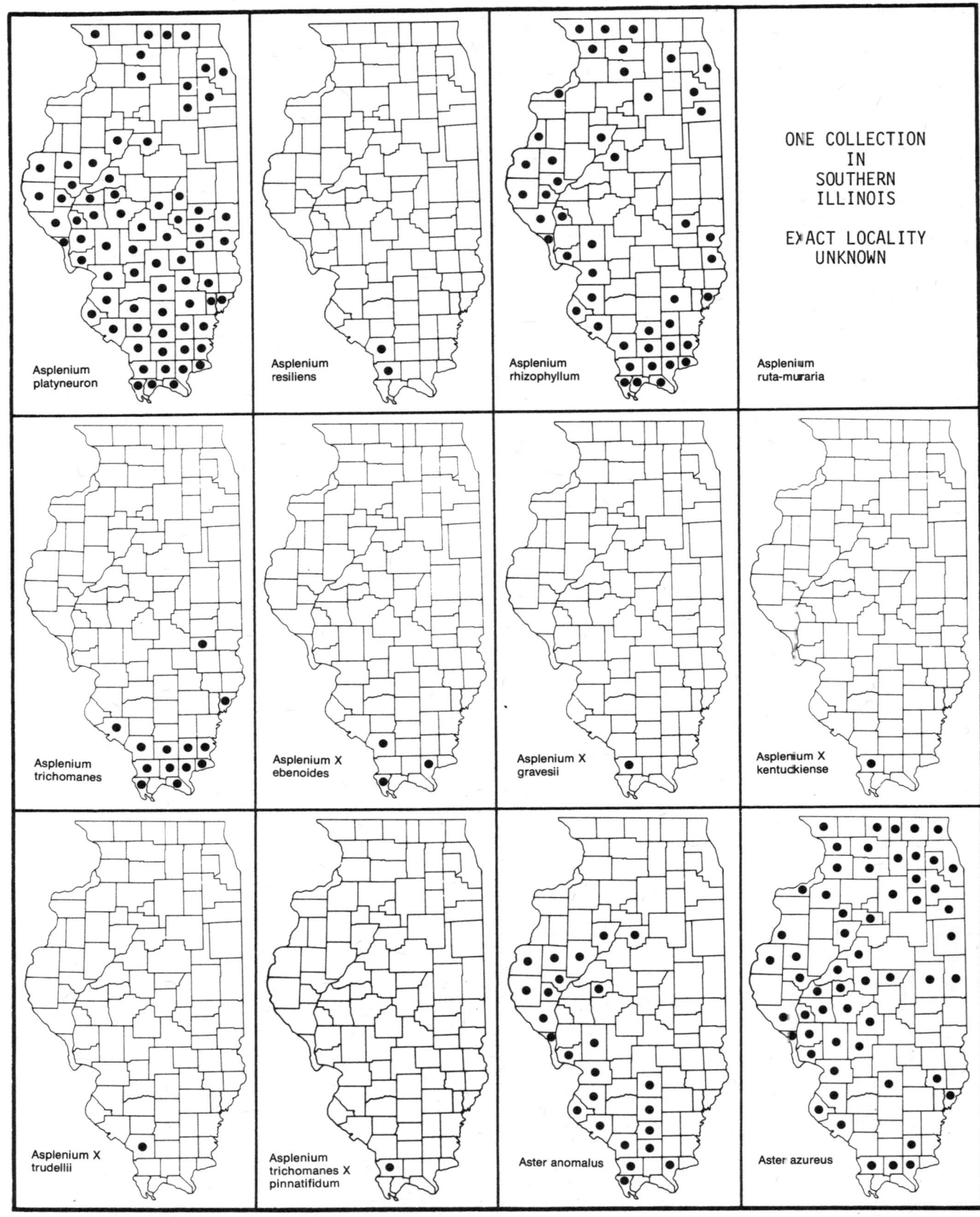

ONE COLLECTION
IN
SOUTHERN
ILLINOIS

EXACT LOCALITY
UNKNOWN

Asplenium
platyneuron

Asplenium
resiliens

Asplenium
rhizophyllum

Asplenium
ruta-muraria

Asplenium
trichomanes

Asplenium X
ebenoides

Asplenium X
gravesii

Asplenium X
kentuckiense

Asplenium X
trudellii

Asplenium
trichomanes X
pinnatifidum

Aster anomalus

Aster azureus

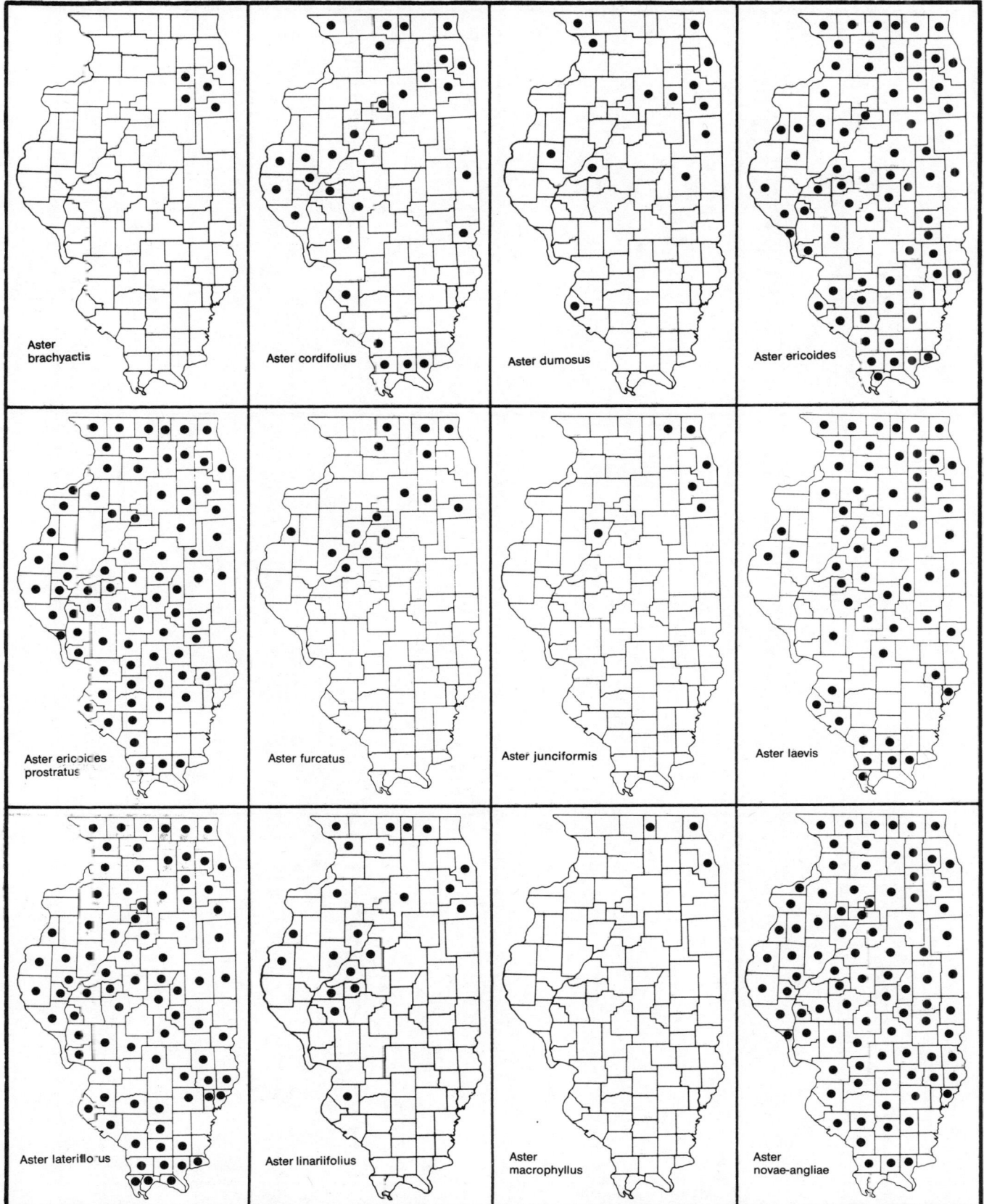

Aster
brachyactis

Aster cordifolius

Aster dumosus

Aster ericoides

Aster ericoides
prostratus

Aster furcatus

Aster junciformis

Aster laevis

Aster lateriflorus

Aster linariifolius

Aster
macrophyllus

Aster
novae-angliae

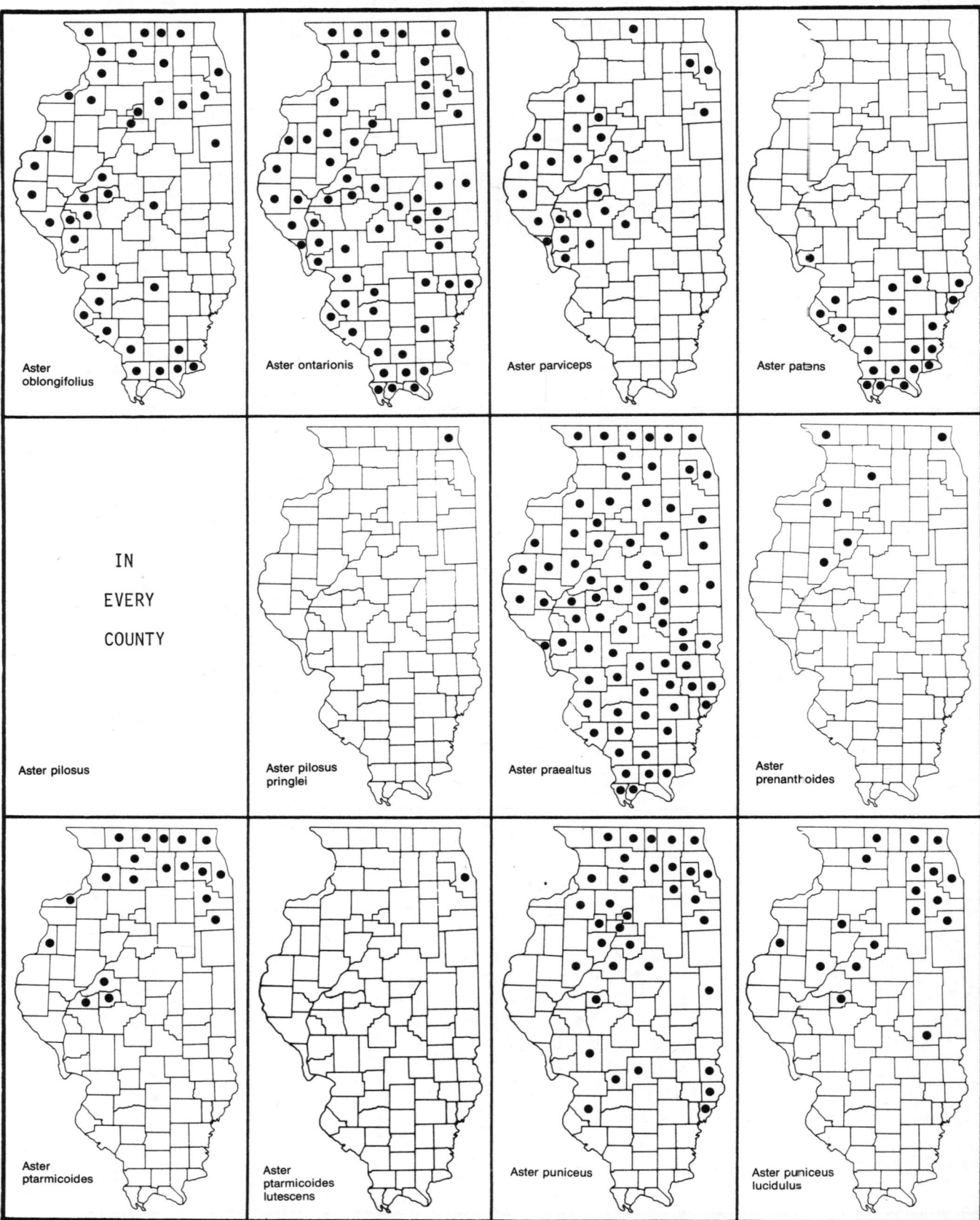

Aster
oblongifolius

Aster ontarionis

Aster parviceps

Aster patens

IN

EVERY

COUNTY

Aster pilosus

Aster pilosus
pringlei

Aster praealtus

Aster
prenanthoides

Aster
ptarmicoides

Aster
ptarmicoides
lutescens

Aster puniceus

Aster puniceus
lucidulus

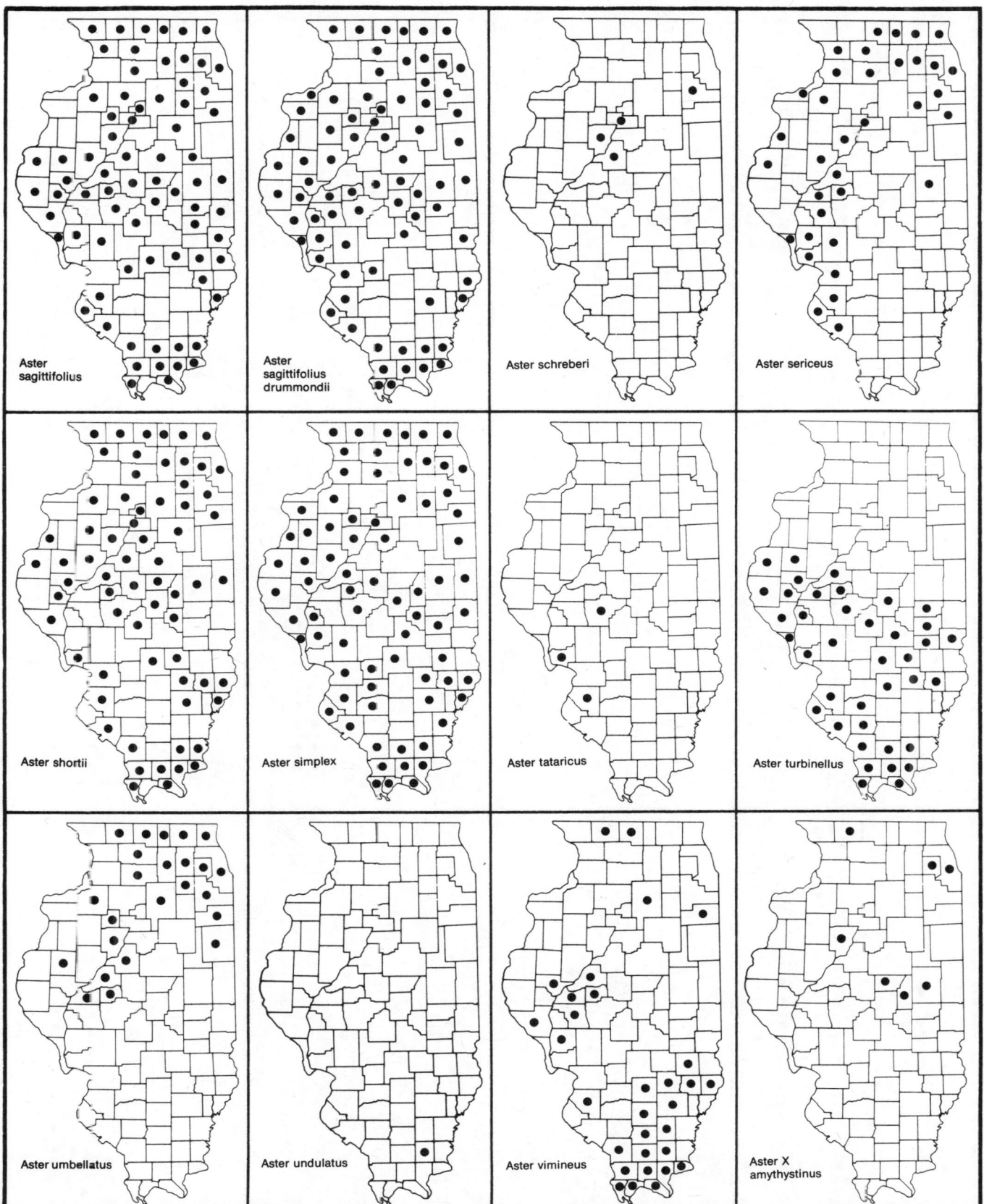

Aster
sagittifolius

Aster
sagittifolius
drummondii

Aster schreberi

Aster sericeus

Aster shortii

Aster simplex

Aster tataricus

Aster turbinellus

Aster umbellatus

Aster undulatus

Aster vimineus

Aster X
amythystinus

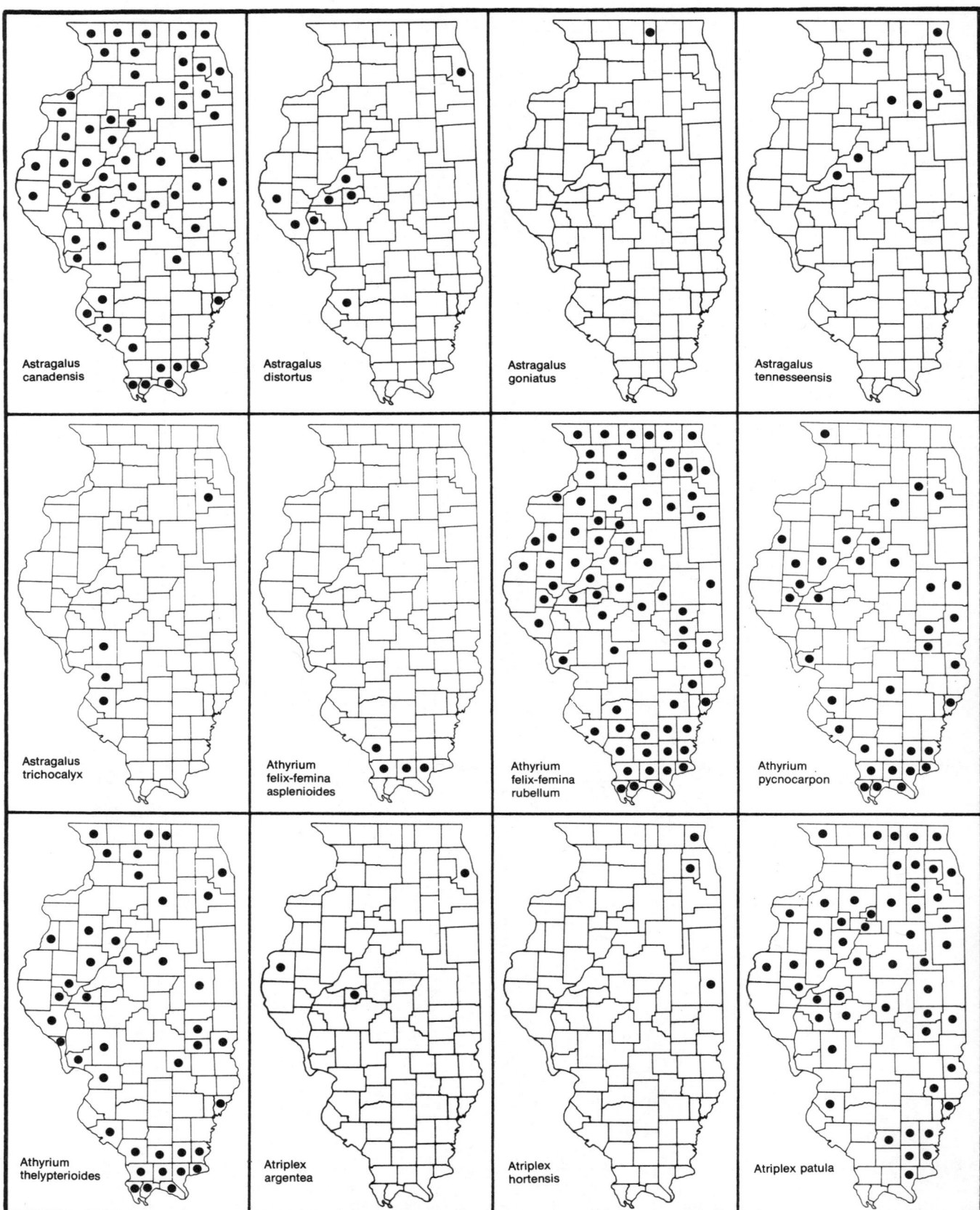

Astragalus
canadensis

Astragalus
distortus

Astragalus
goniatus

Astragalus
tennesseensis

Astragalus
trichocalyx

Athyrium
felix-femina
asplenioides

Athyrium
felix-femina
rubellum

Athyrium
pycnocarpon

Athyrium
thelypterioides

Atriplex
argentea

Atriplex
hortensis

Atriplex patula

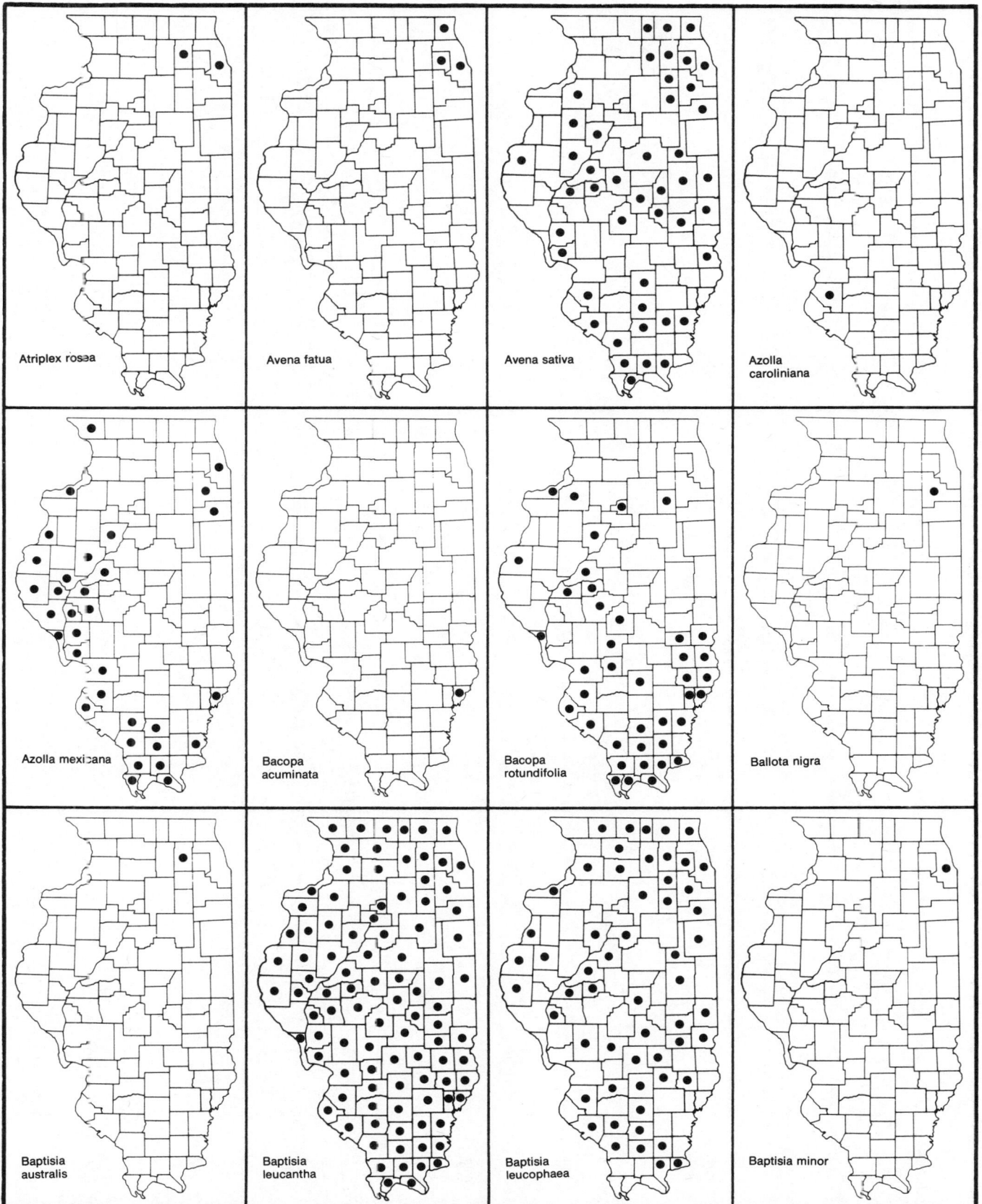

Atriplex rosea

Avena fatua

Avena sativa

Azolla
caroliniana

Azolla mexicana

Bacopa
acuminata

Bacopa
rotundifolia

Ballota nigra

Baptisia
australis

Baptisia
leucantha

Baptisia
leucophaea

Baptisia minor

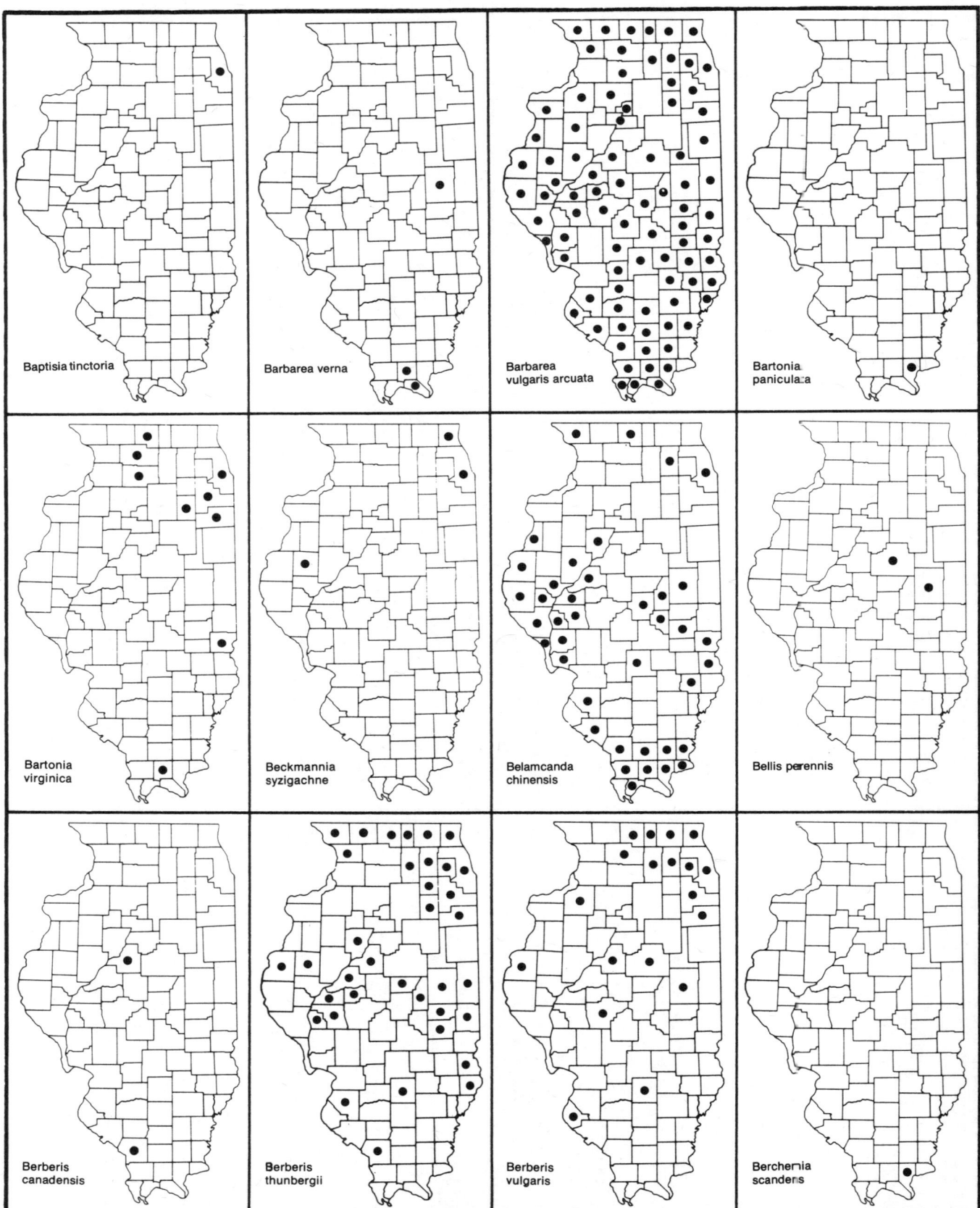

Baptisia tinctoria

Barbarea verna

Barbarea
vulgaris arcuata

Bartonia
paniculata

Bartonia
virginica

Beckmannia
syzigachne

Belamcanda
chinensis

Bellis perennis

Berberis
canadensis

Berberis
thunbergii

Berberis
vulgaris

Berchemia
scandens

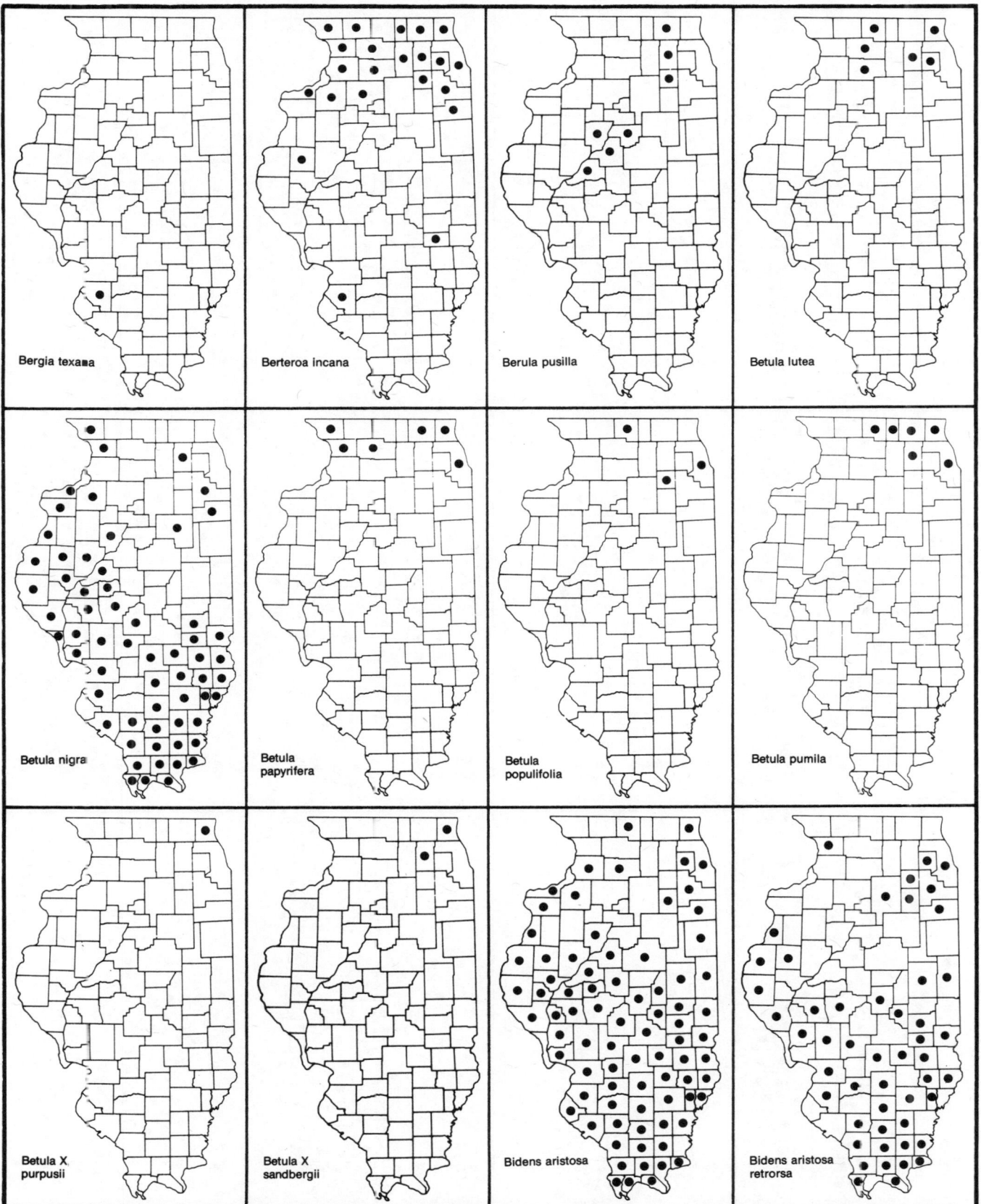

Bergia texana

Berteroa incana

Berula pusilla

Betula lutea

Betula nigra

Betula
papyrifera

Betula
populifolia

Betula pumila

Betula X
purpusii

Betula X
sandbergii

Bidens aristosa

Bidens aristosa
retrorsa

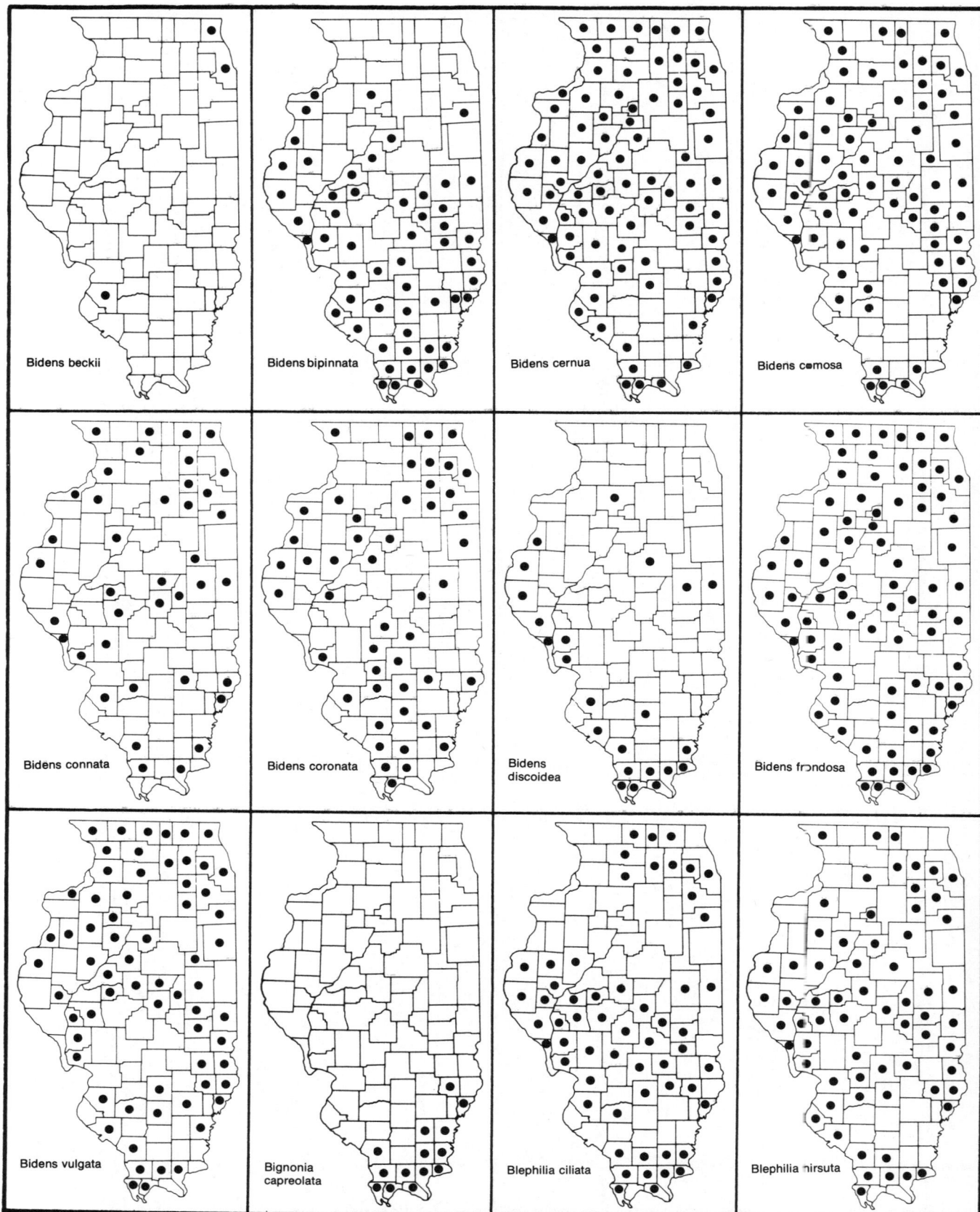

Bidens beckii

Bidens bipinnata

Bidens cernua

Bidens comosa

Bidens connata

Bidens coronata

Bidens discoidea

Bidens frondosa

Bidens vulgata

Bignonia capreolata

Blephilia ciliata

Blephilia hirsuta

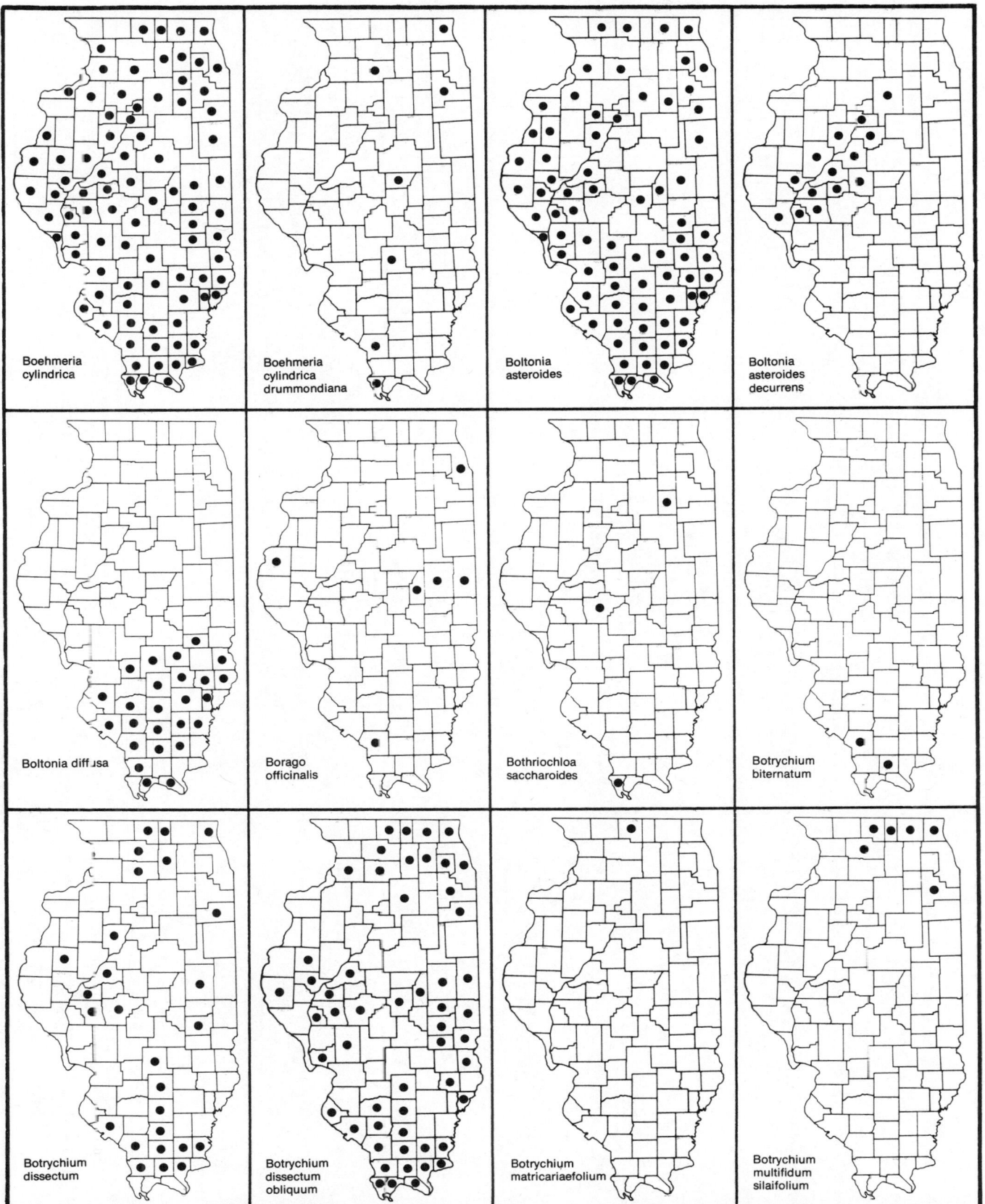

Boehmeria
cylindrica

Boehmeria
cylindrica
drummondiana

Boltonia
asteroides

Boltonia
asteroides
decurrens

Boltonia diffusa

Borago
officinalis

Bothriochloa
saccharoides

Botrychium
biternatum

Botrychium
dissectum

Botrychium
dissectum
obliquum

Botrychium
matricariaefolium

Botrychium
multifidum
silaifolium

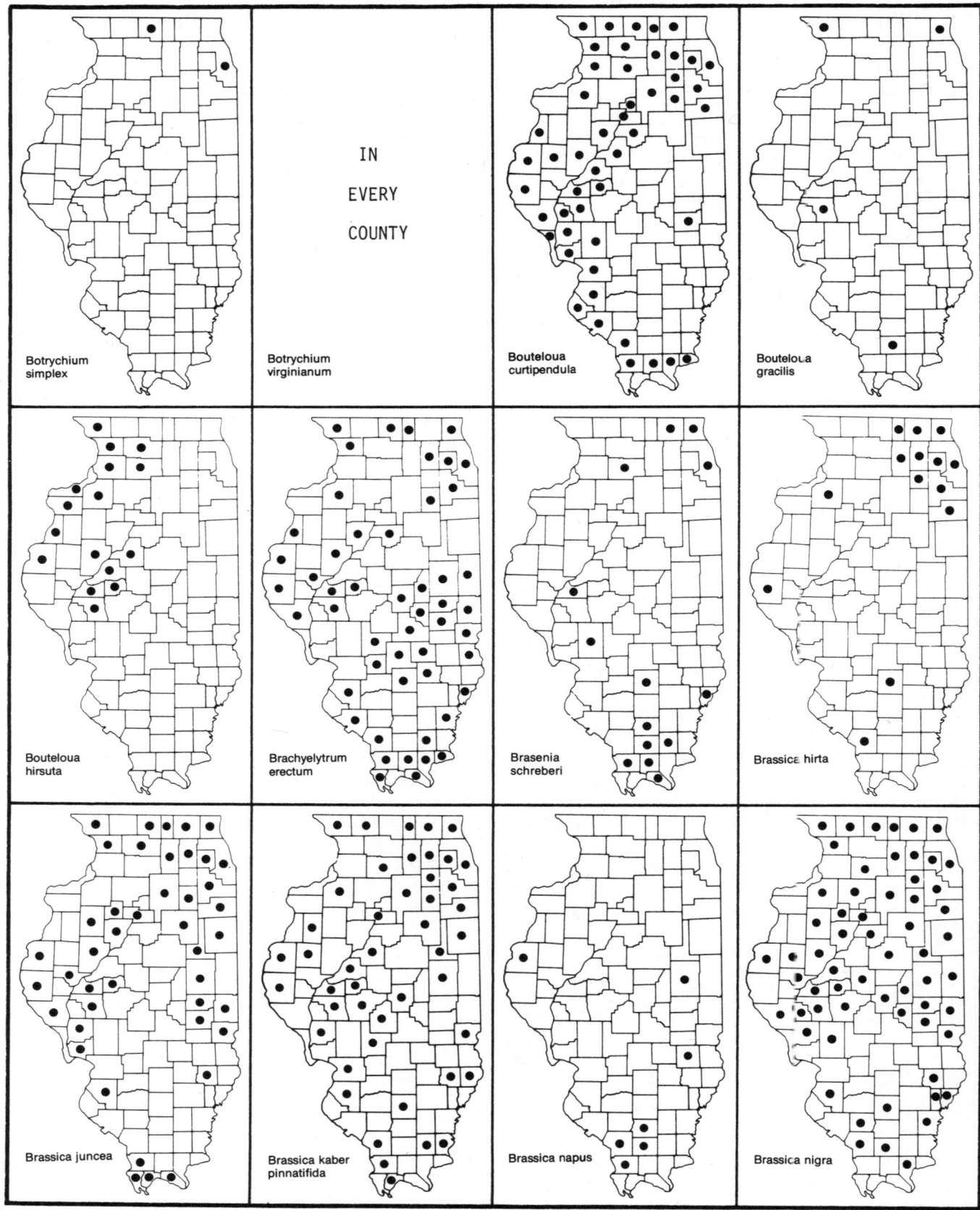

Botrychium
simplex

Botrychium
virginianum

IN

EVERY

COUNTY

Bouteloua
curtipendula

Bouteloua
gracilis

Bouteloua
hirsuta

Brachyelytrum
erectum

Brasenia
schreberi

Brassica hirta

Brassica juncea

Brassica kaber
pinnatifida

Brassica napus

Brassica nigra

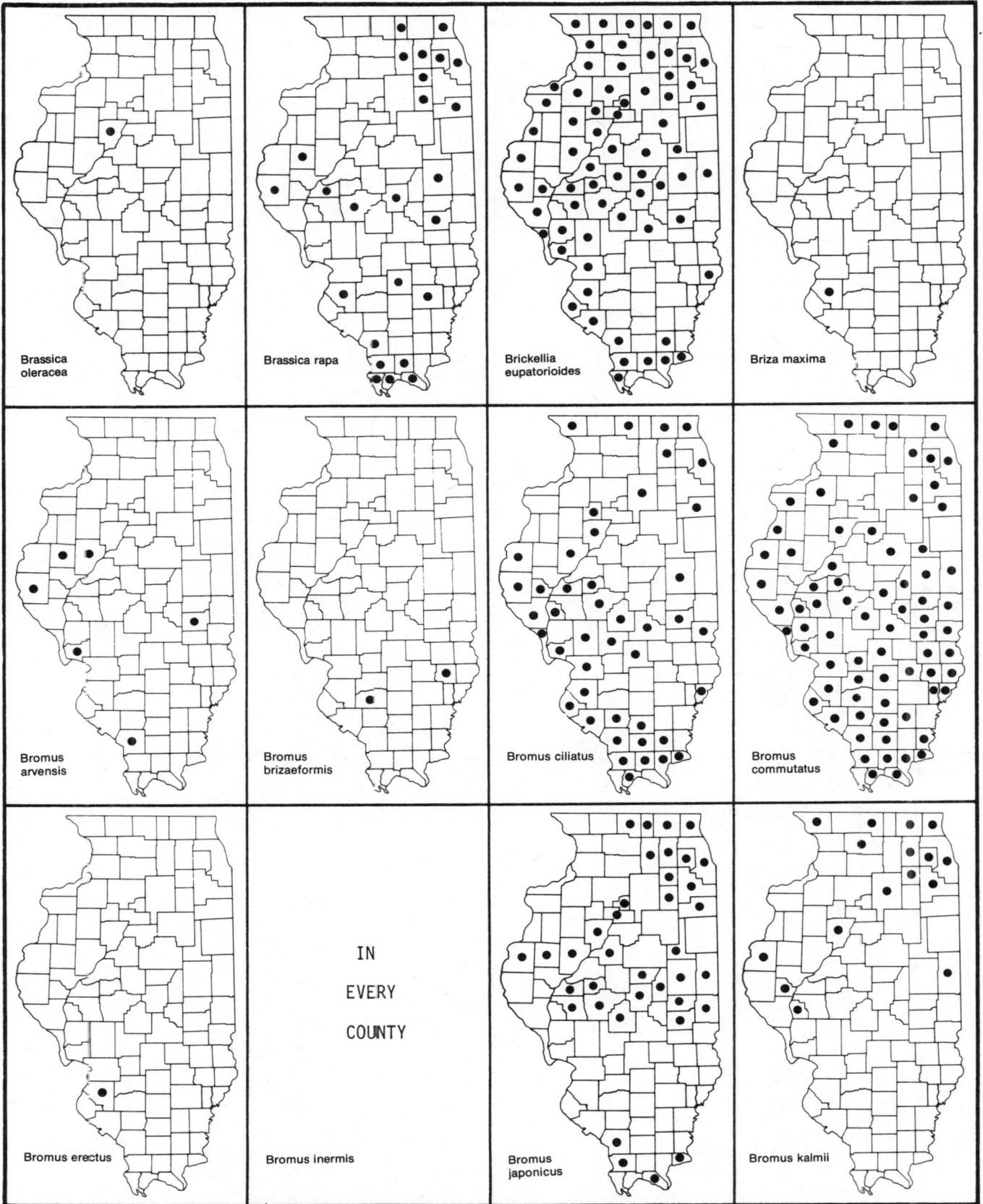

Brassica
oleracea

Brassica rapa

Brickellia
eupatorioides

Briza maxima

Bromus
arvensis

Bromus
brizaeformis

Bromus ciliatus

Bromus
commutatus

Bromus erectus

IN

EVERY

COUNTY

Bromus inermis

Bromus
japonicus

Bromus kalmii

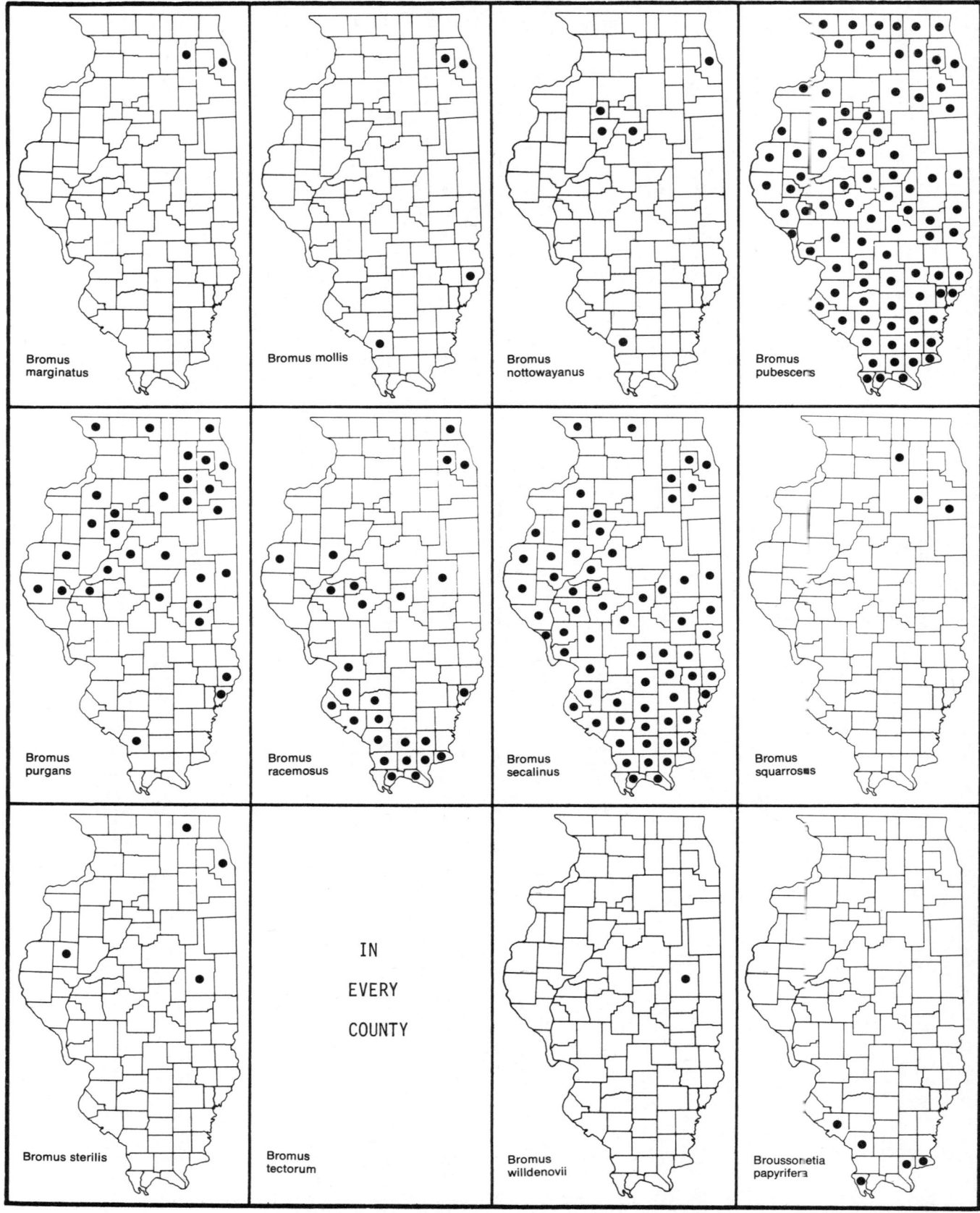

Bromus
marginatus

Bromus mollis

Bromus
nottowayanus

Bromus
pubescens

Bromus
purgans

Bromus
racemosus

Bromus
secalinus

Bromus
squarrosus

Bromus sterilis

Bromus
tectorum

IN

EVERY

COUNTY

Bromus
willdenovii

Broussonetia
papyrifera

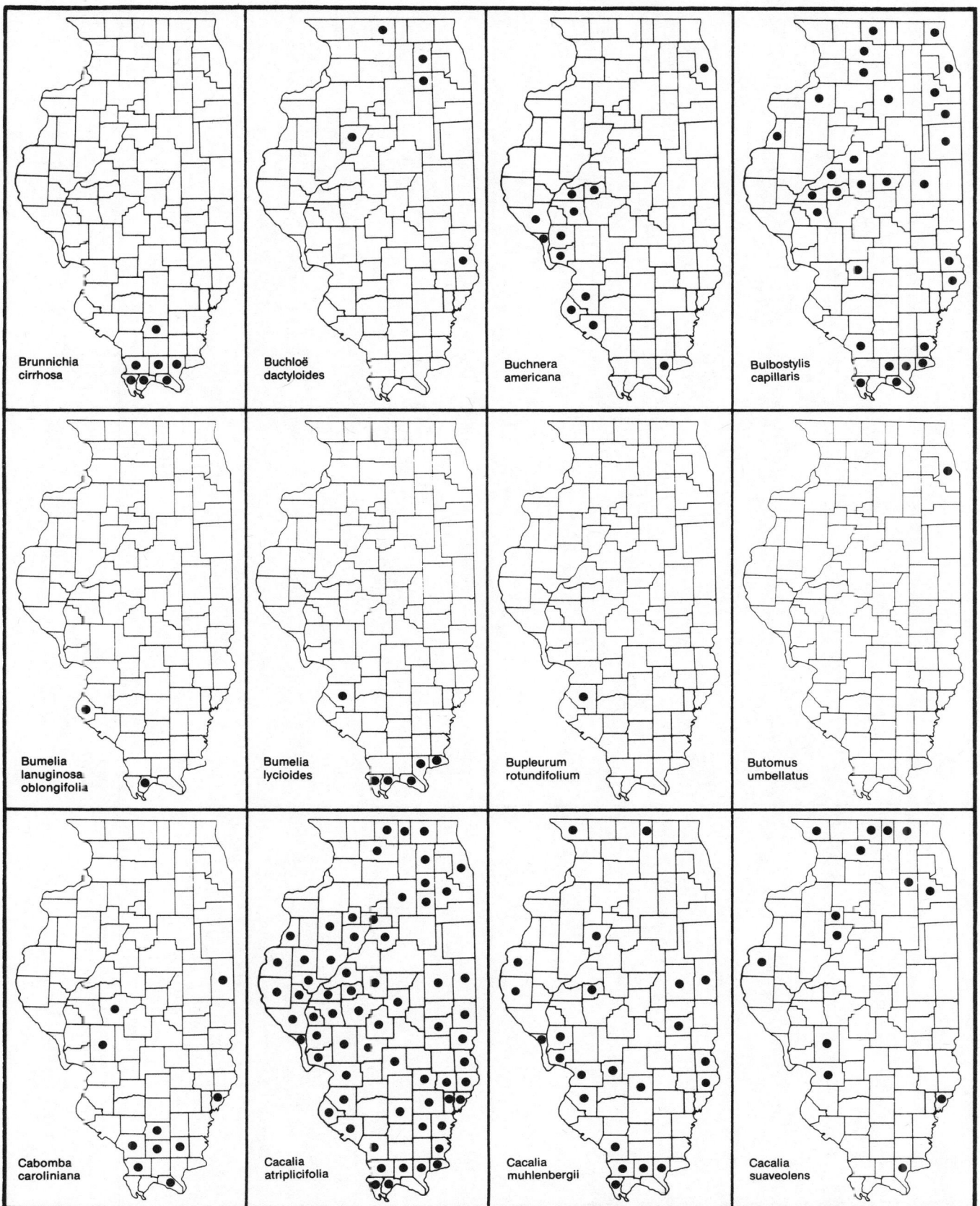

Brunnichia
cirrhosa

Buchloë
dactyloides

Buchnera
americana

Bulbostylis
capillaris

Bumelia
lanuginosa
oblongifolia

Bumelia
lycioides

Bupleurum
rotundifolium

Butomus
umbellatus

Cabomba
caroliniana

Cacalia
atriplicifolia

Cacalia
muhlenbergii

Cacalia
suaveolens

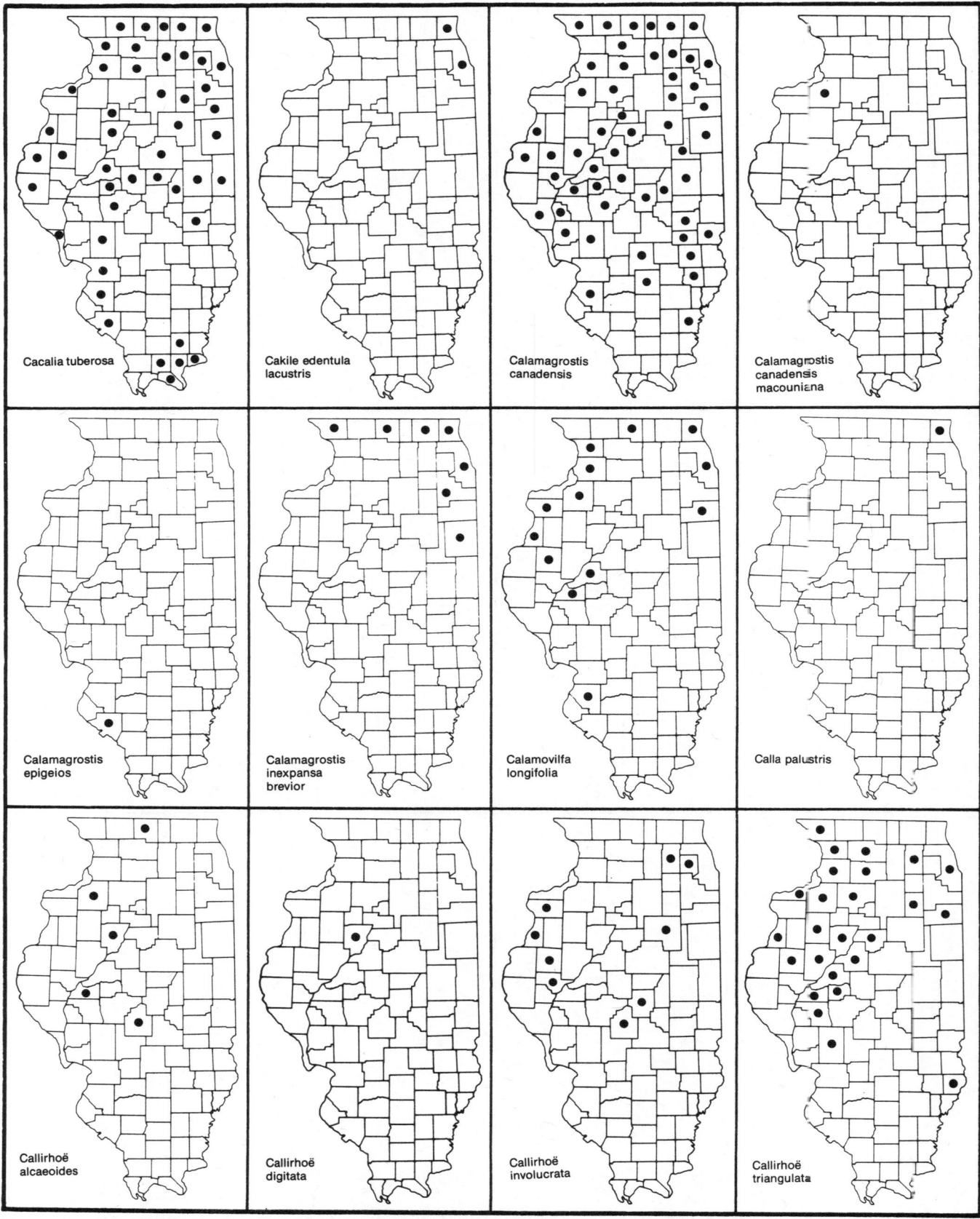

Cacalia tuberosa

Cakile edentula
lacustris

Calamagrostis
canadensis

Calamagrostis
canadensis
macouniana

Calamagrostis
epigeios

Calamagrostis
inexpansa
brevior

Calamovilfa
longifolia

Calla palustris

Callirhoë
alcaeoides

Callirhoë
digitata

Callirhoë
involucrata

Callirhoë
triangulata

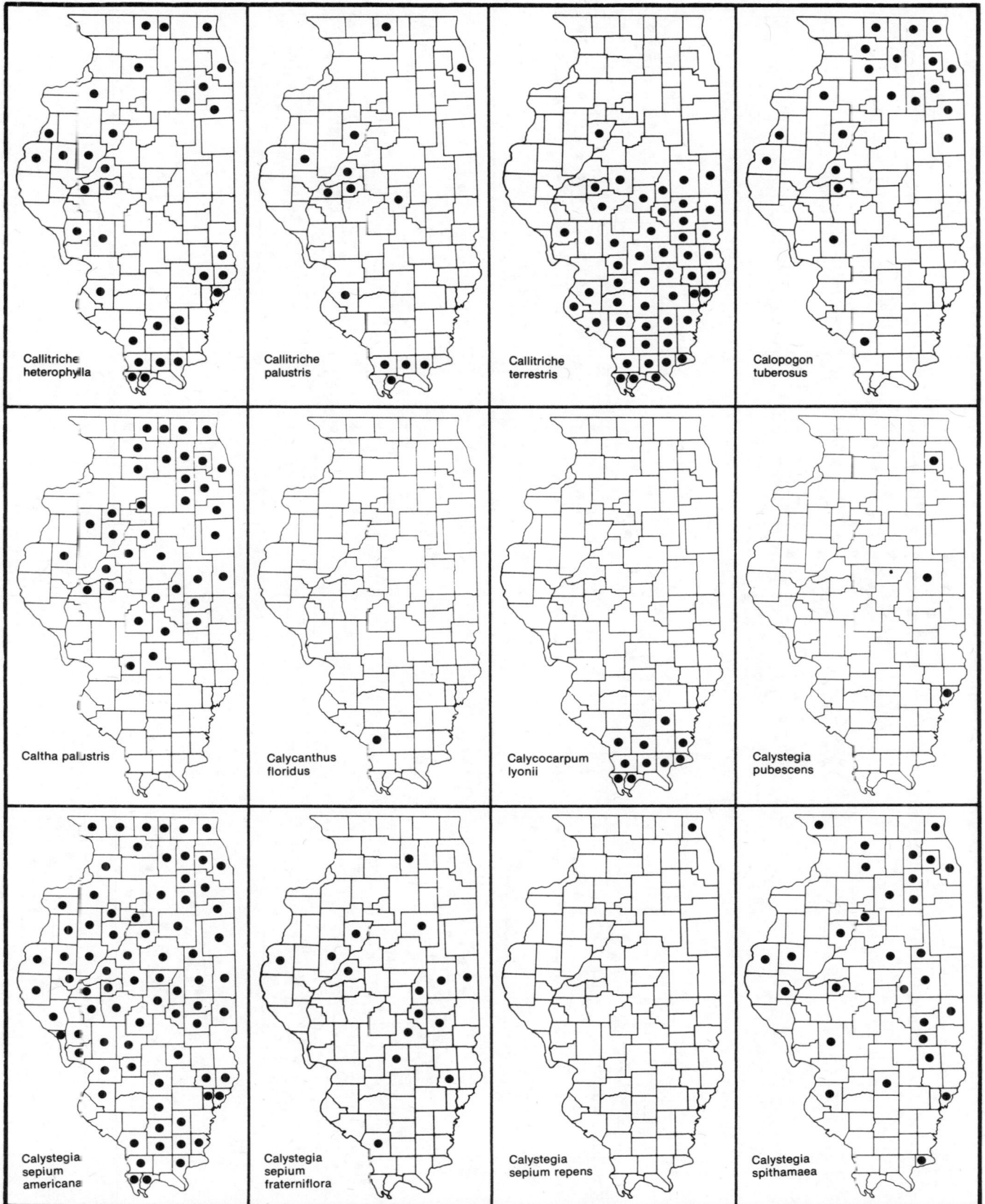

Callitriche
heterophylla

Callitriche
palustris

Callitriche
terrestris

Calopogon
tuberosus

Caltha palustris

Calycanthus
floridus

Calycocarpum
lyonii

Calystegia
pubescens

Calystegia
sepium
americana

Calystegia
sepium
fraterniflora

Calystegia
sepium repens

Calystegia
spithamaea

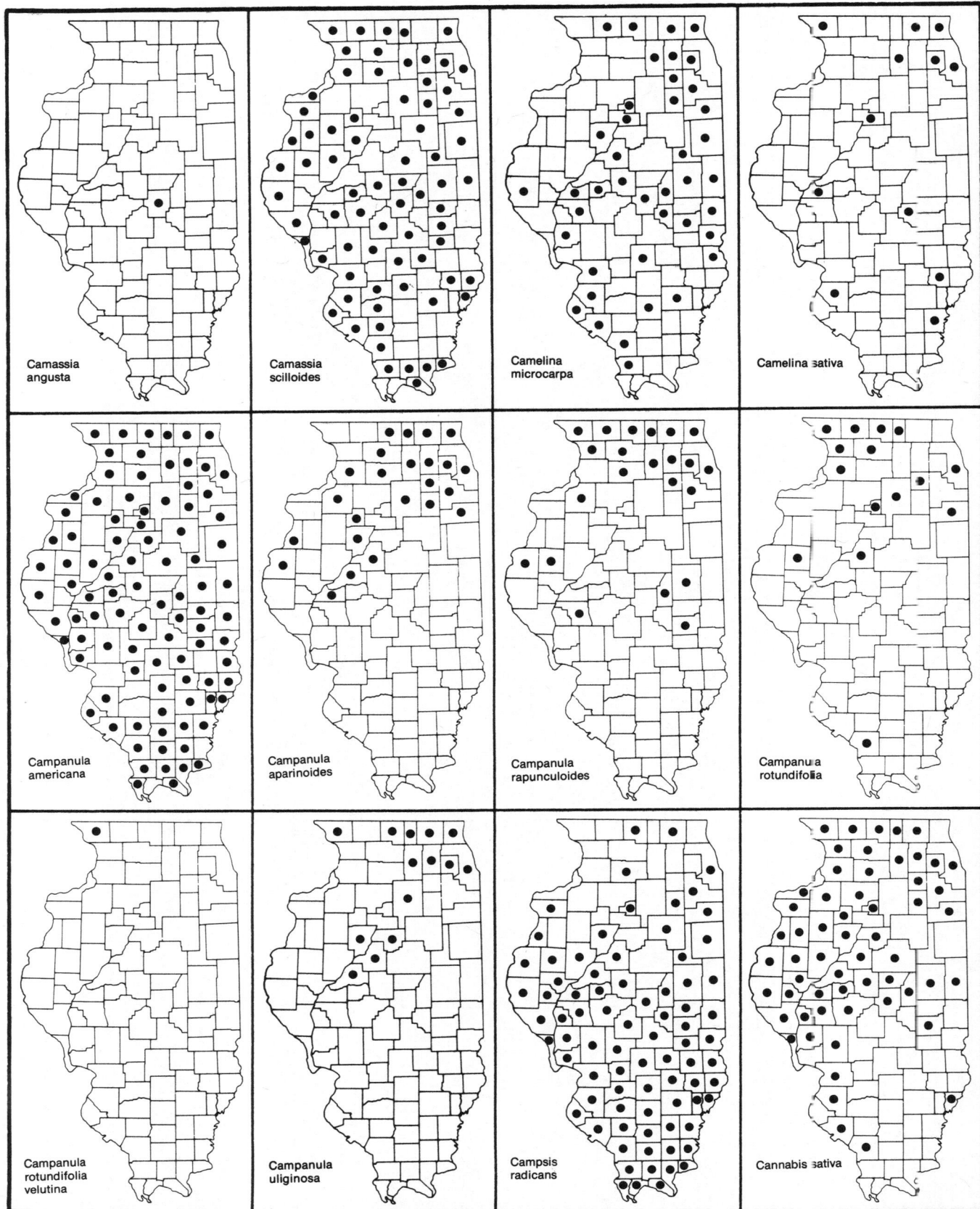

Camassia
angusta

Camassia
scilloides

Camelina
microcarpa

Camelina sativa

Campanula
americana

Campanula
aparinoides

Campanula
rapunculoides

Campanula
rotundifolia

Campanula
rotundifolia
velutina

Campanula
uliginosa

Campsis
radicans

Cannabis sativa

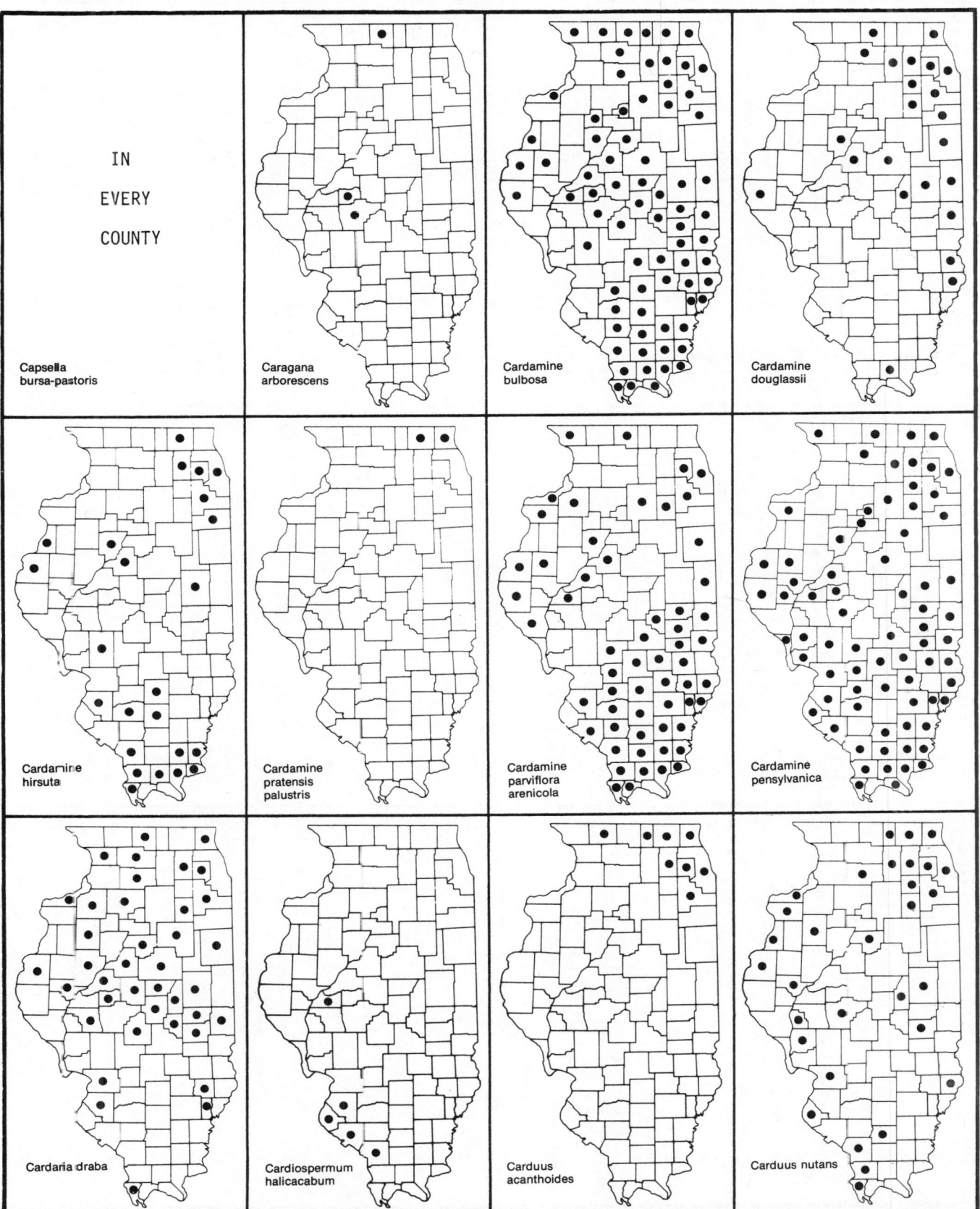

IN
EVERY
COUNTY

Capsella
bursa-pastoris

Caragana
arborescens

Cardamine
bulbosa

Cardamine
douglassii

Cardamine
hirsuta

Cardamine
pratensis
palustris

Cardamine
parviflora
arenicola

Cardamine
pensylvanica

Cardaria draba

Cardiospermum
halicacabum

Carduus
acanthoides

Carduus nutans

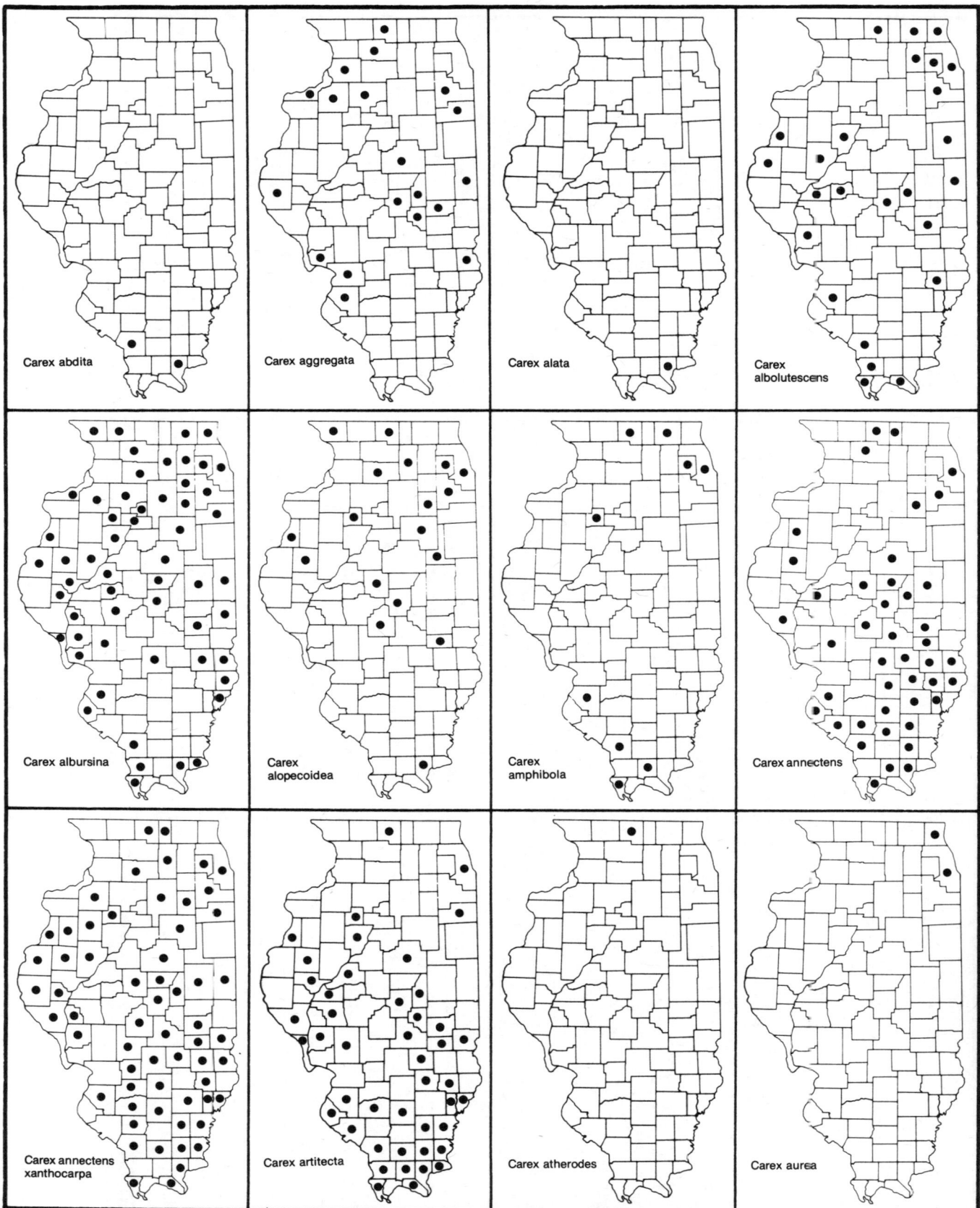

Carex abdita

Carex aggregata

Carex alata

Carex albolutescens

Carex albursina

Carex alopecoidea

Carex amphibola

Carex annectens

Carex annectens xanthocarpa

Carex artitecta

Carex atherodes

Carex aurea

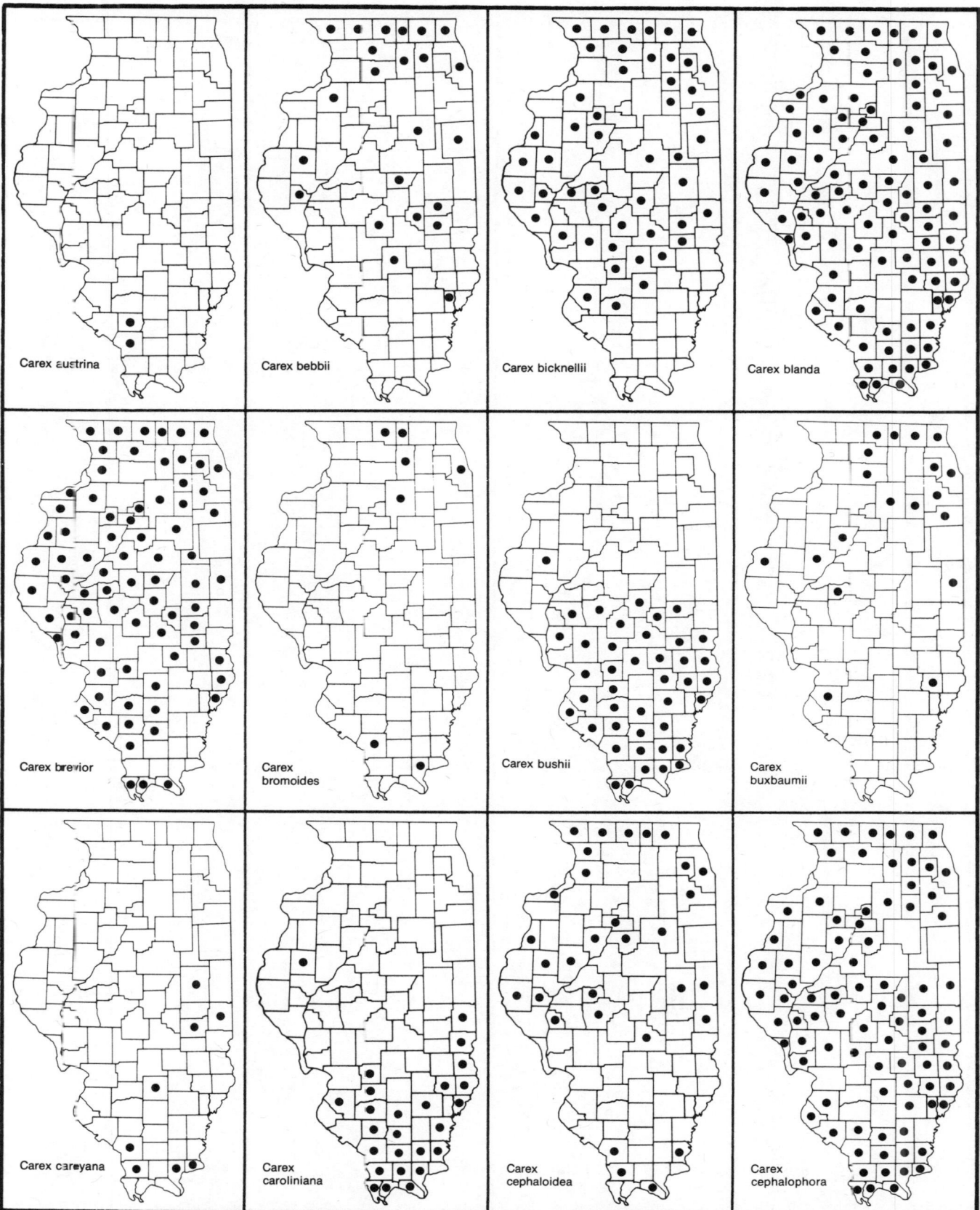

Carex austrina

Carex bebbii

Carex bicknellii

Carex blanda

Carex brevior

Carex bromoides

Carex bushii

Carex buxbaumii

Carex careyana

Carex caroliniana

Carex cephaloidea

Carex cephalophora

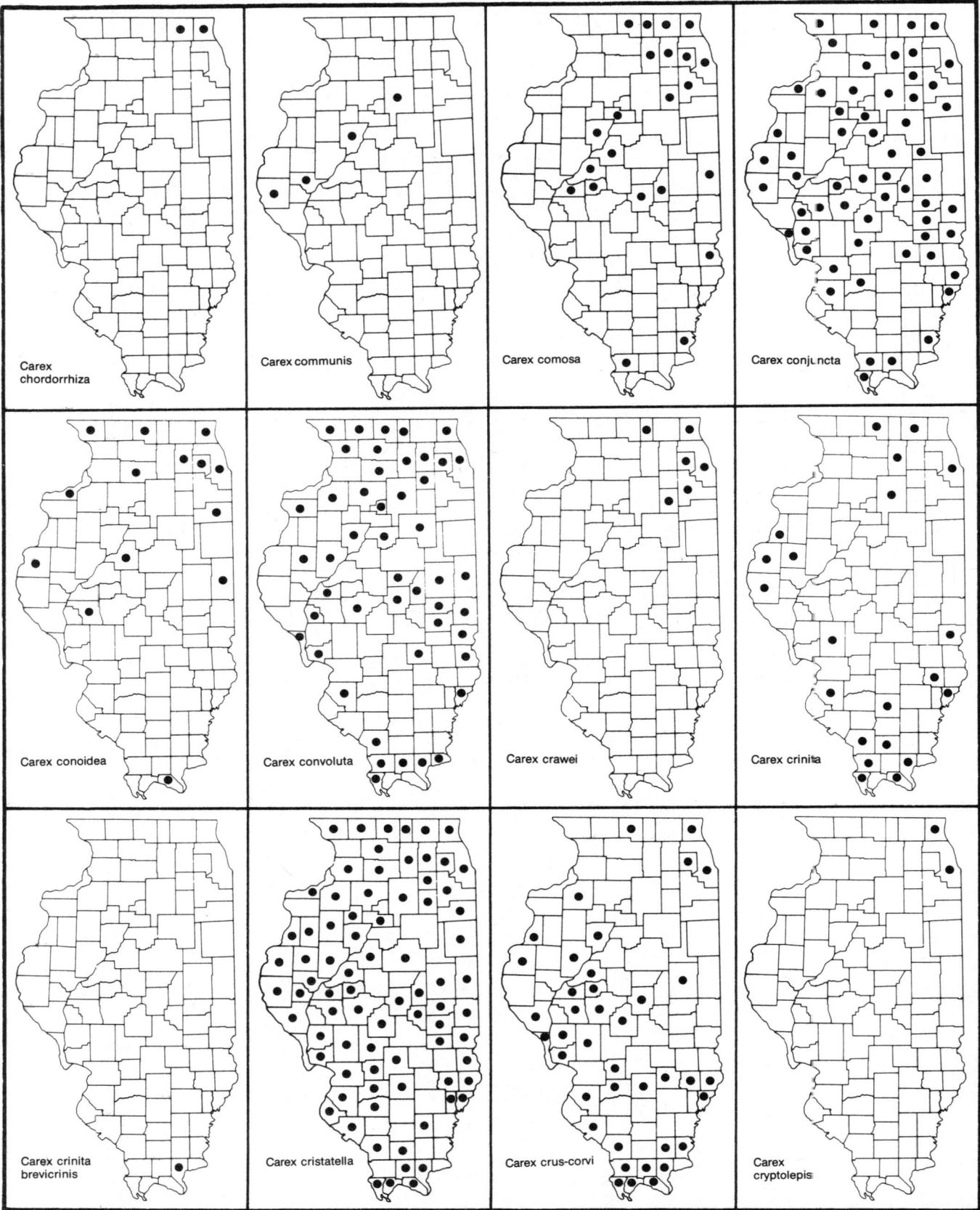

Carex
chordorrhiza

Carex communis

Carex comosa

Carex conjuncta

Carex conoidea

Carex convoluta

Carex crawei

Carex crinita

Carex crinita
brevicrinis

Carex cristatella

Carex crus-corvi

Carex
cryptolepis

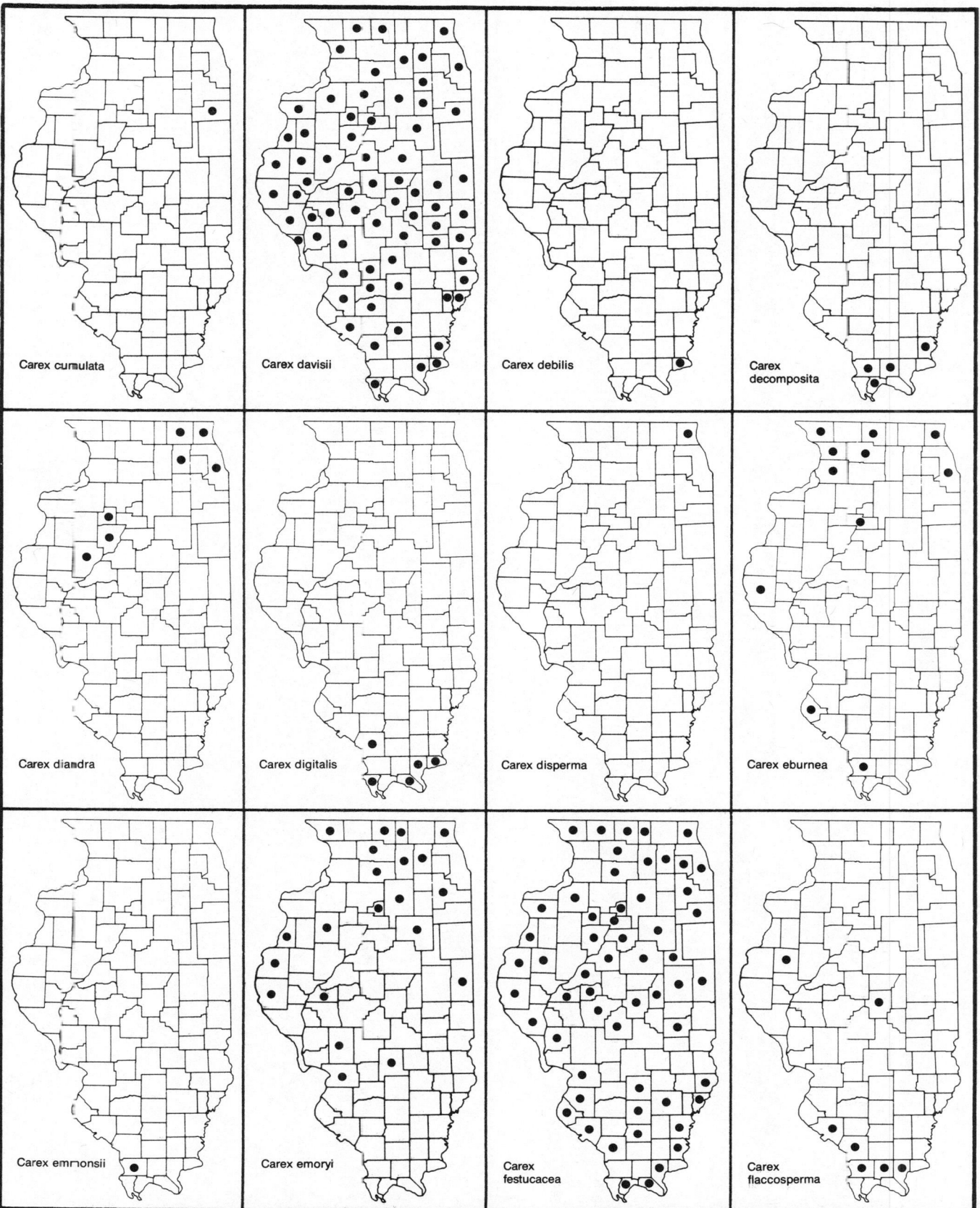

Carex cumulata

Carex davisii

Carex debilis

Carex decomposita

Carex diandra

Carex digitalis

Carex disperma

Carex eburnea

Carex emmonsii

Carex emoryi

Carex festucacea

Carex flaccosperma

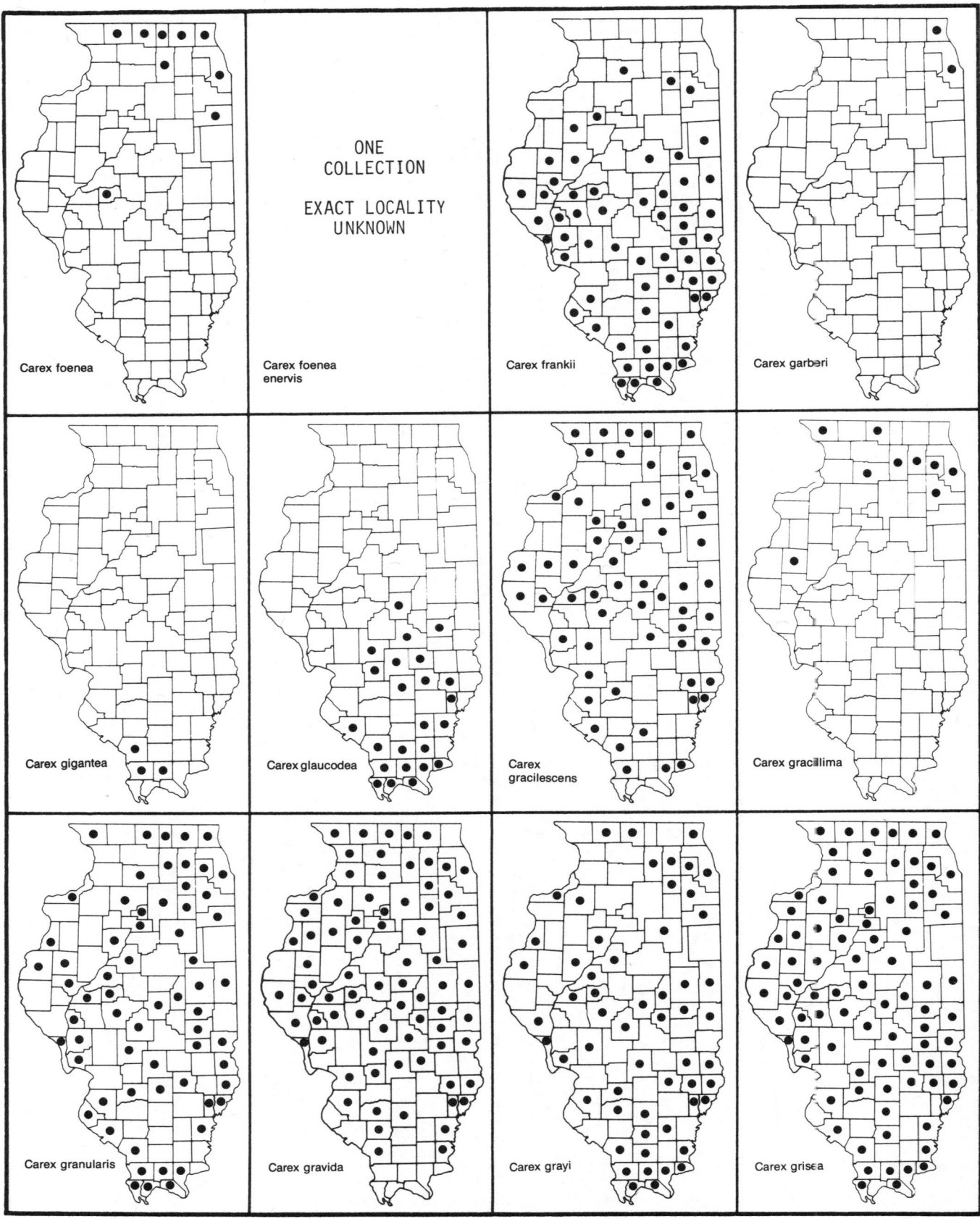

ONE
COLLECTION

EXACT LOCALITY
UNKNOWN

Carex foenea

Carex foenea
enervis

Carex frankii

Carex garberi

Carex gigantea

Carex glaucodea

Carex
gracilescens

Carex gracillima

Carex granularis

Carex gravida

Carex grayi

Carex grisea

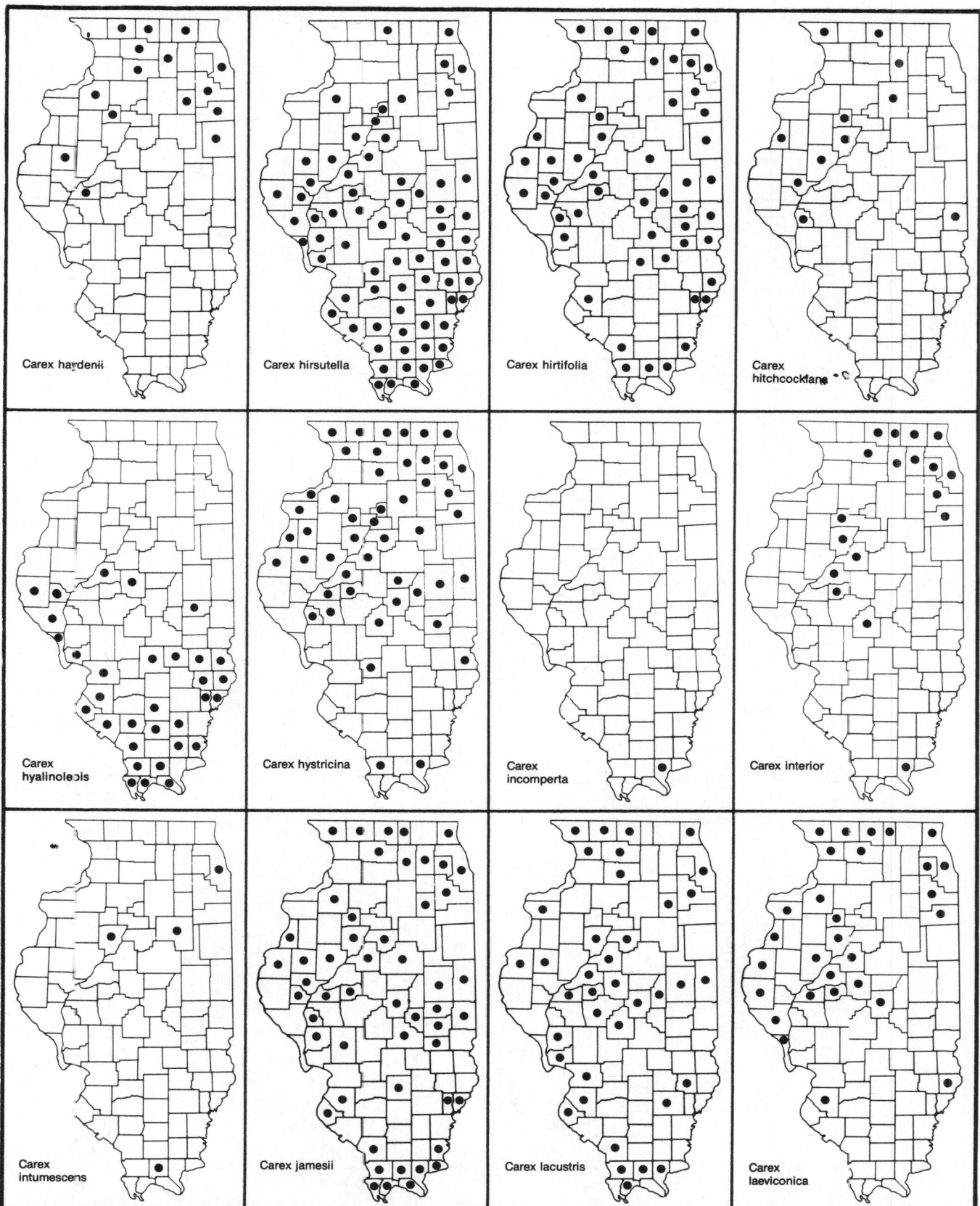

Carex hardenii

Carex hirsutella

Carex hirtifolia

Carex hitchcockiana

Carex hyalinolepis

Carex hystricina

Carex incomperta

Carex interior

Carex intumescens

Carex jamesii

Carex lacustris

Carex laeviconica

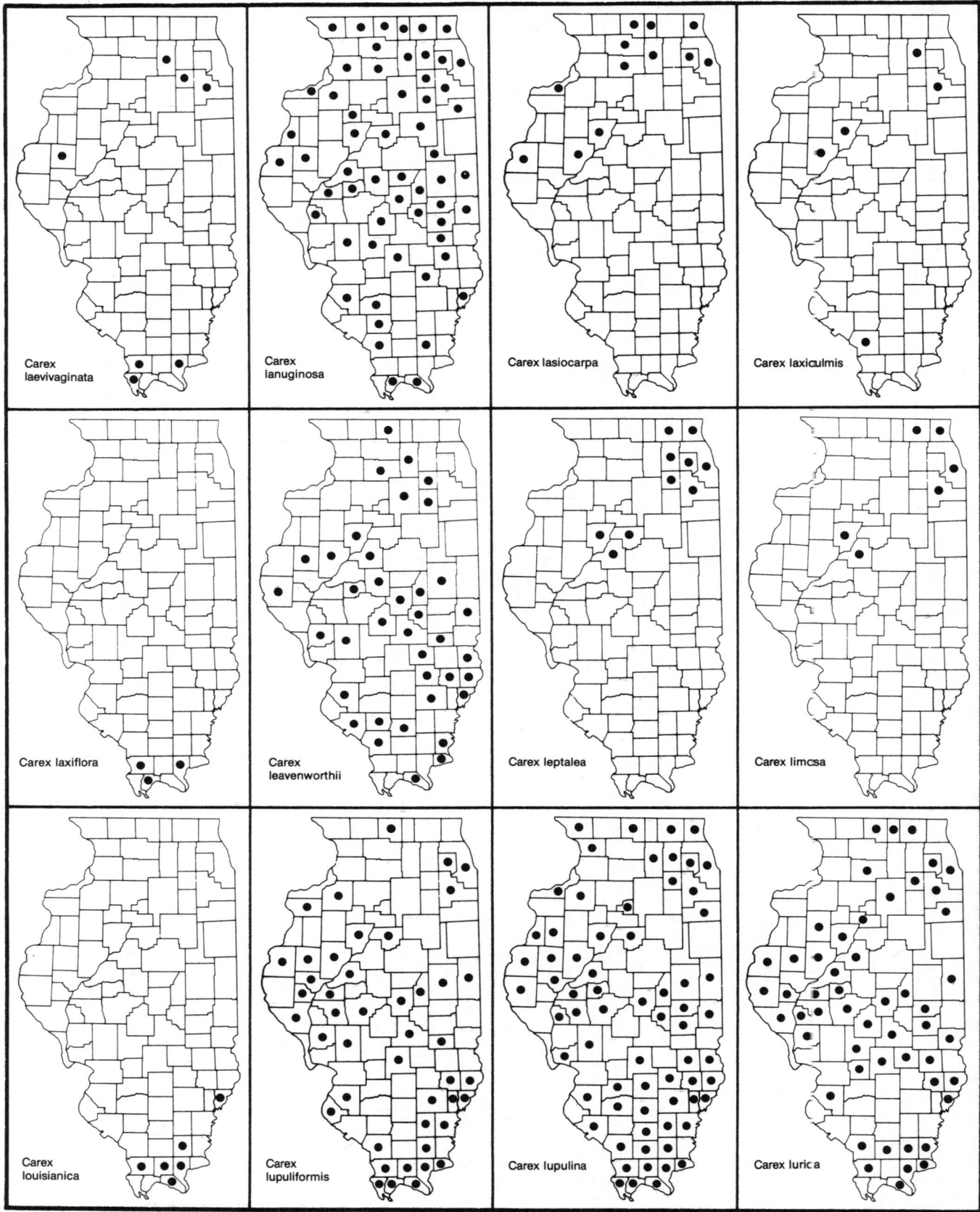

Carex
laevivaginata

Carex
lanuginosa

Carex lasiocarpa

Carex laxiculmis

Carex laxiflora

Carex
leavenworthii

Carex leptalea

Carex limosa

Carex
louisianica

Carex
lupuliformis

Carex lupulina

Carex lurida

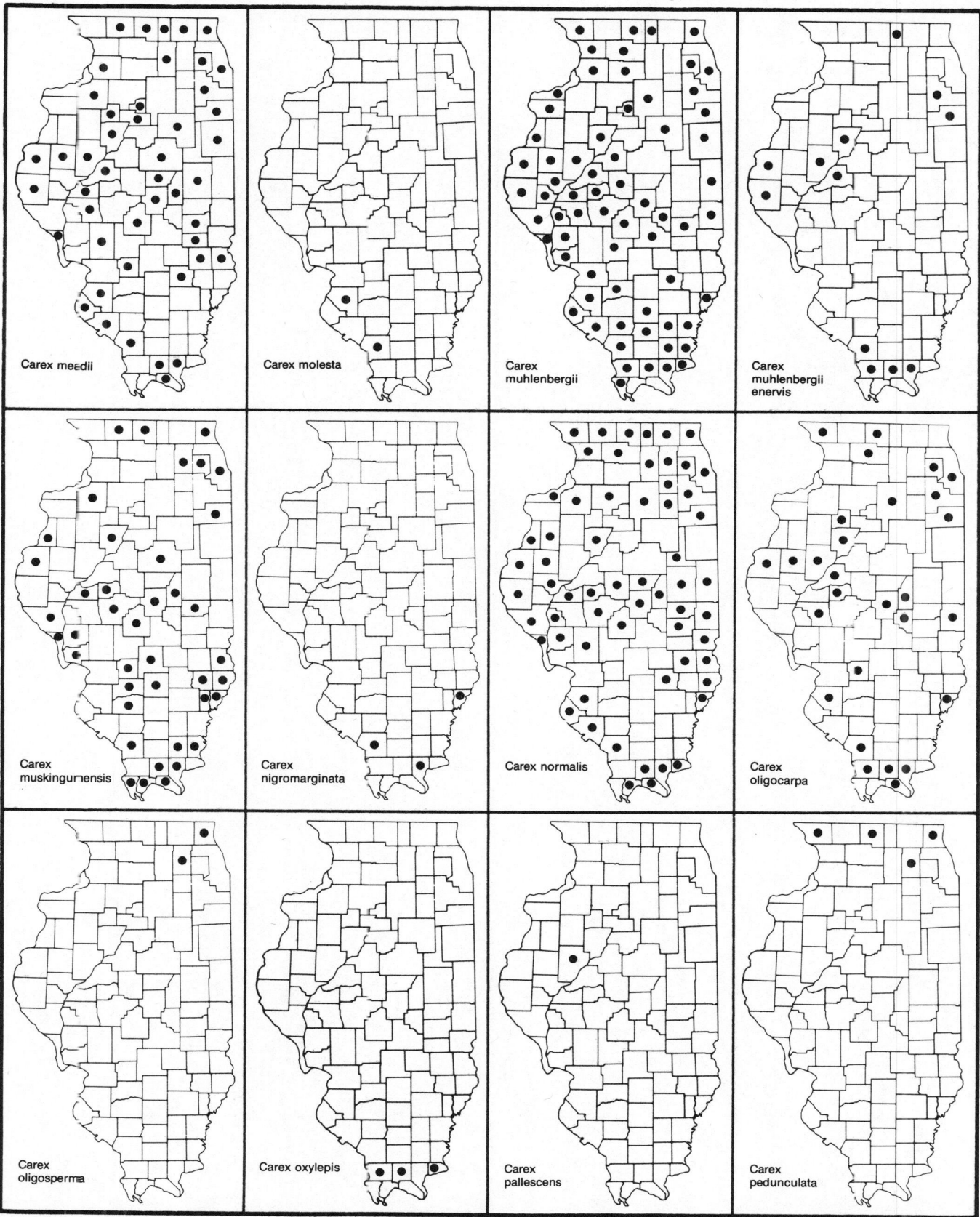

Carex meadii

Carex molesta

Carex
muhlenbergii

Carex
muhlenbergii
enervis

Carex
muskingumensis

Carex
nigromarginata

Carex normalis

Carex
oligocarpa

Carex
oligosperma

Carex oxylepis

Carex
pallescens

Carex
pedunculata

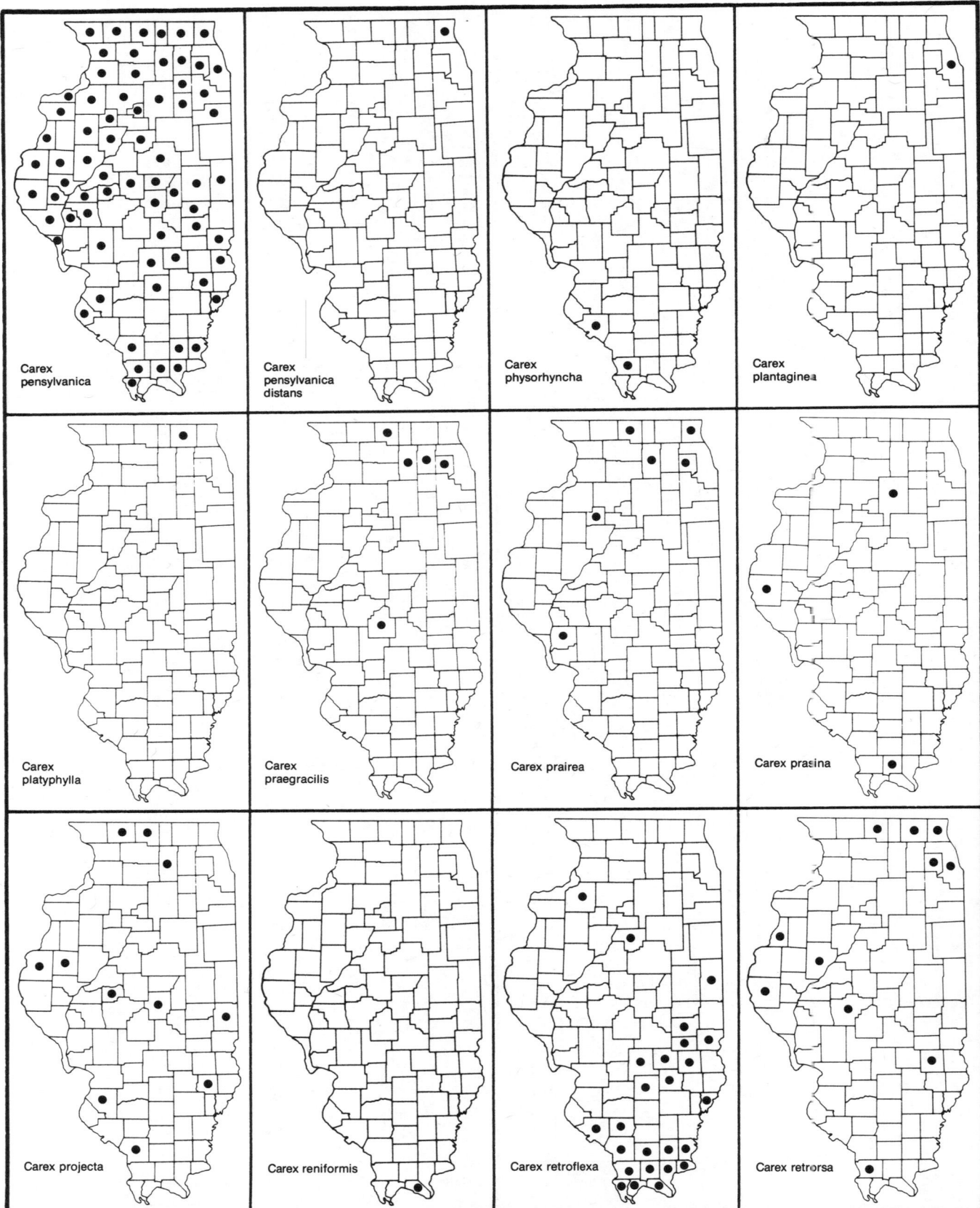

Carex
pensylvanica

Carex
pensylvanica
distans

Carex
physorhyncha

Carex
plantaginea

Carex
platyphylla

Carex
praegracilis

Carex prairea

Carex prasina

Carex projecta

Carex reniformis

Carex retroflexa

Carex retrorsa

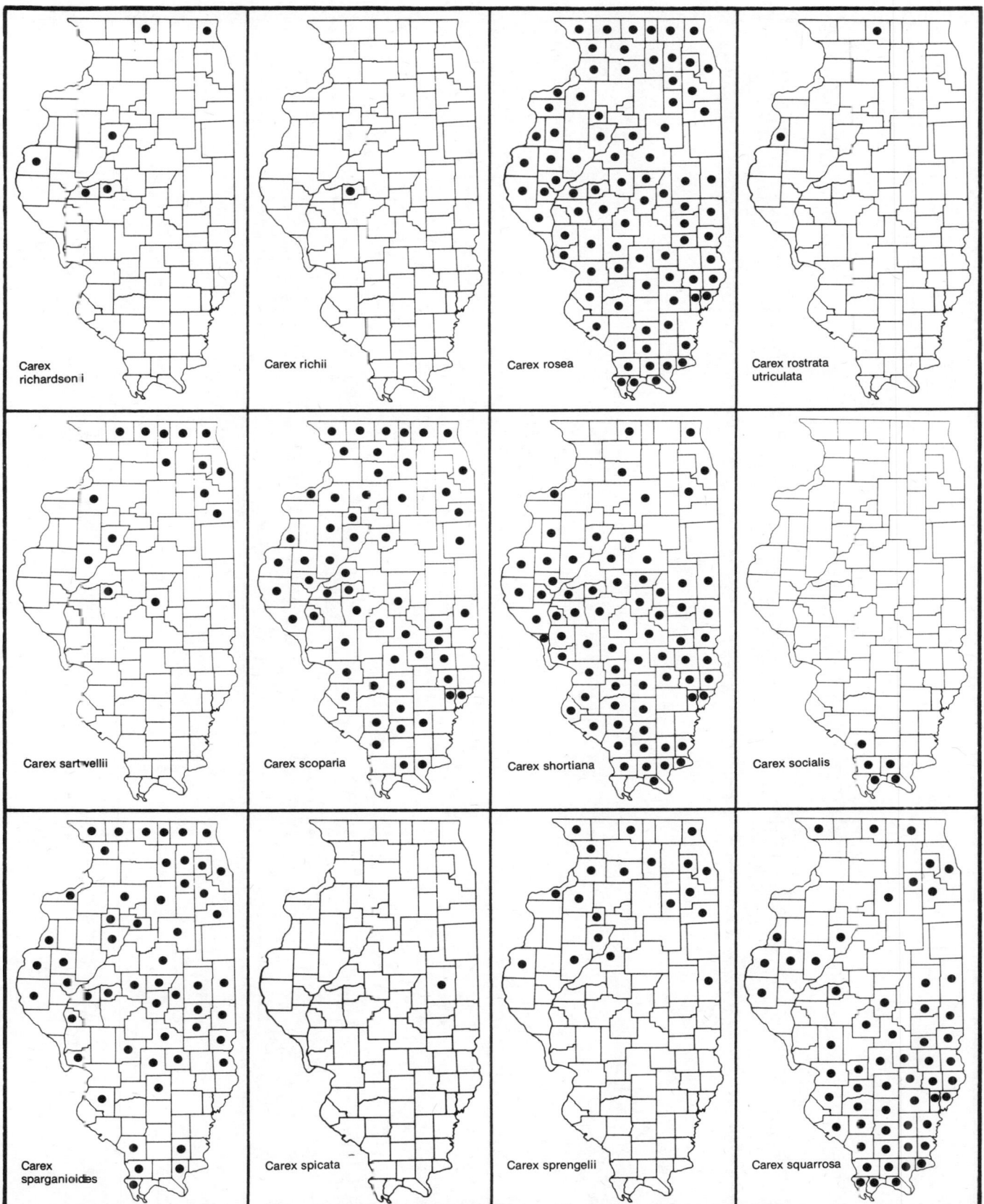

Carex
richardsoni

Carex richii

Carex rosea

Carex rostrata
utriculata

Carex sartwellii

Carex scoparia

Carex shortiana

Carex socialis

Carex
sparganioides

Carex spicata

Carex sprengelii

Carex squarrosa

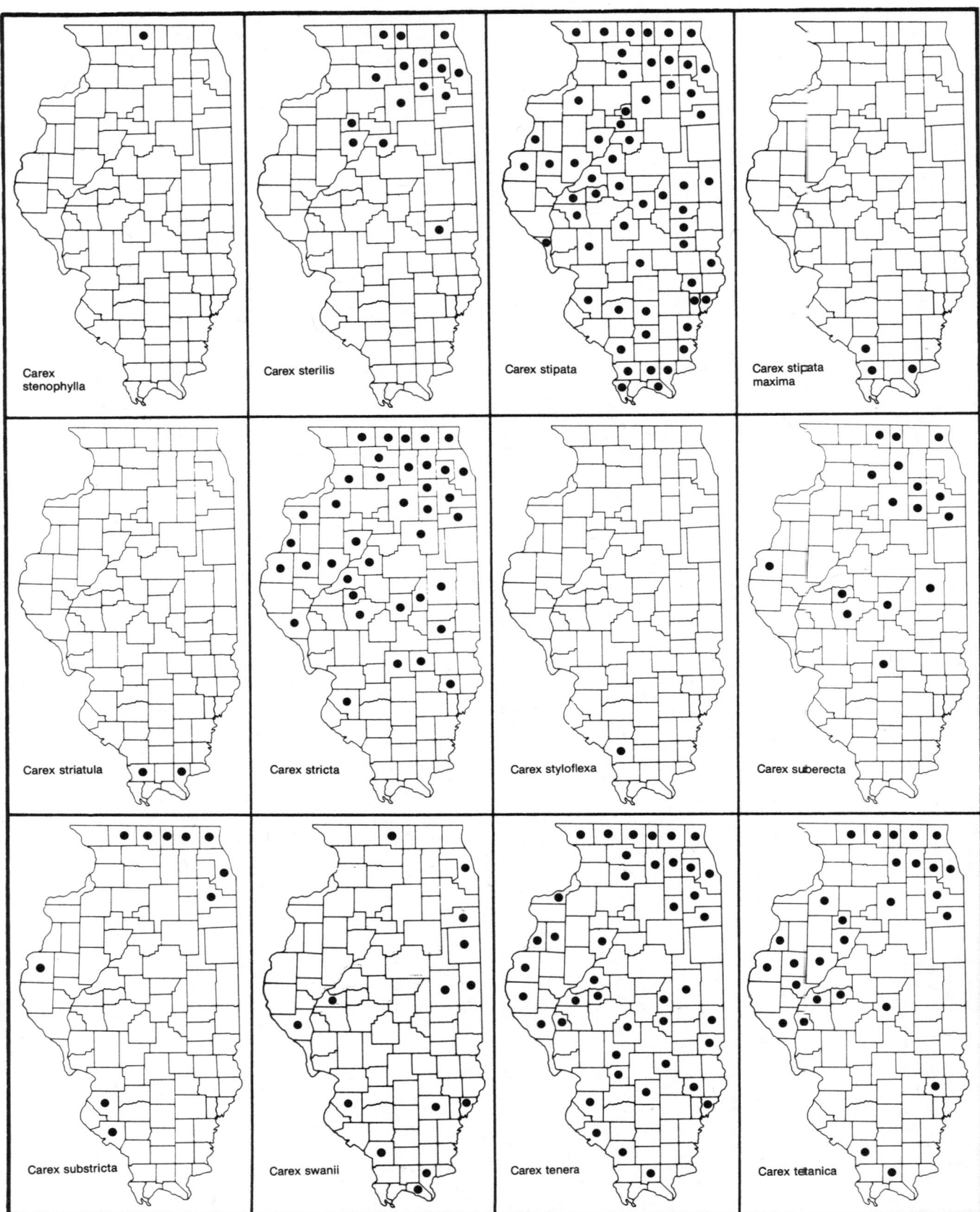

Carex
stenophylla

Carex sterilis

Carex stipata

Carex stipata
maxima

Carex striatula

Carex stricta

Carex styloflexa

Carex suberecta

Carex substricta

Carex swanii

Carex tenera

Carex tetanica

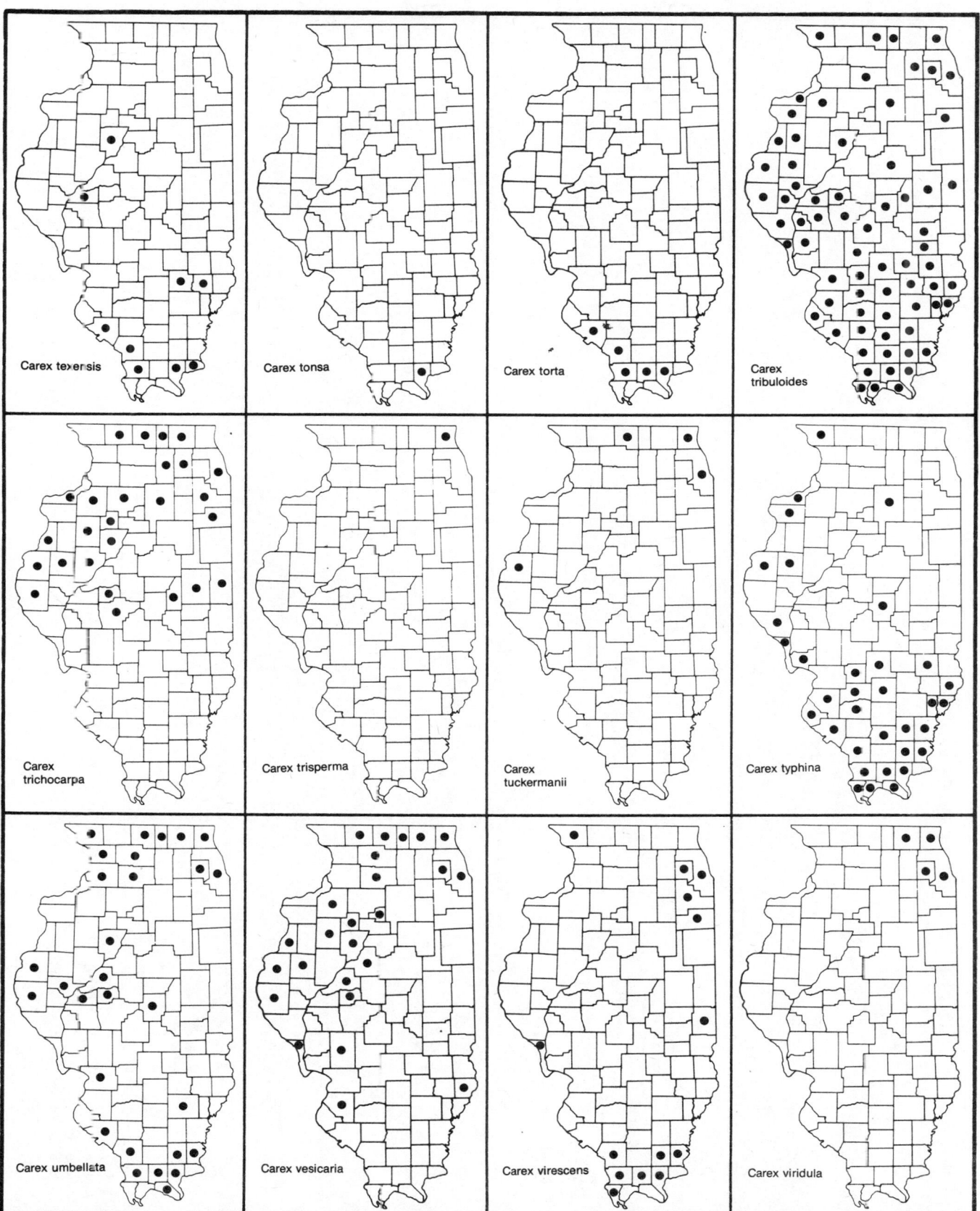

Carex texensis

Carex tonsa

Carex torta

Carex tribuloides

Carex trichocarpa

Carex trisperma

Carex tuckermanii

Carex typhina

Carex umbellata

Carex vesicaria

Carex virescens

Carex viridula

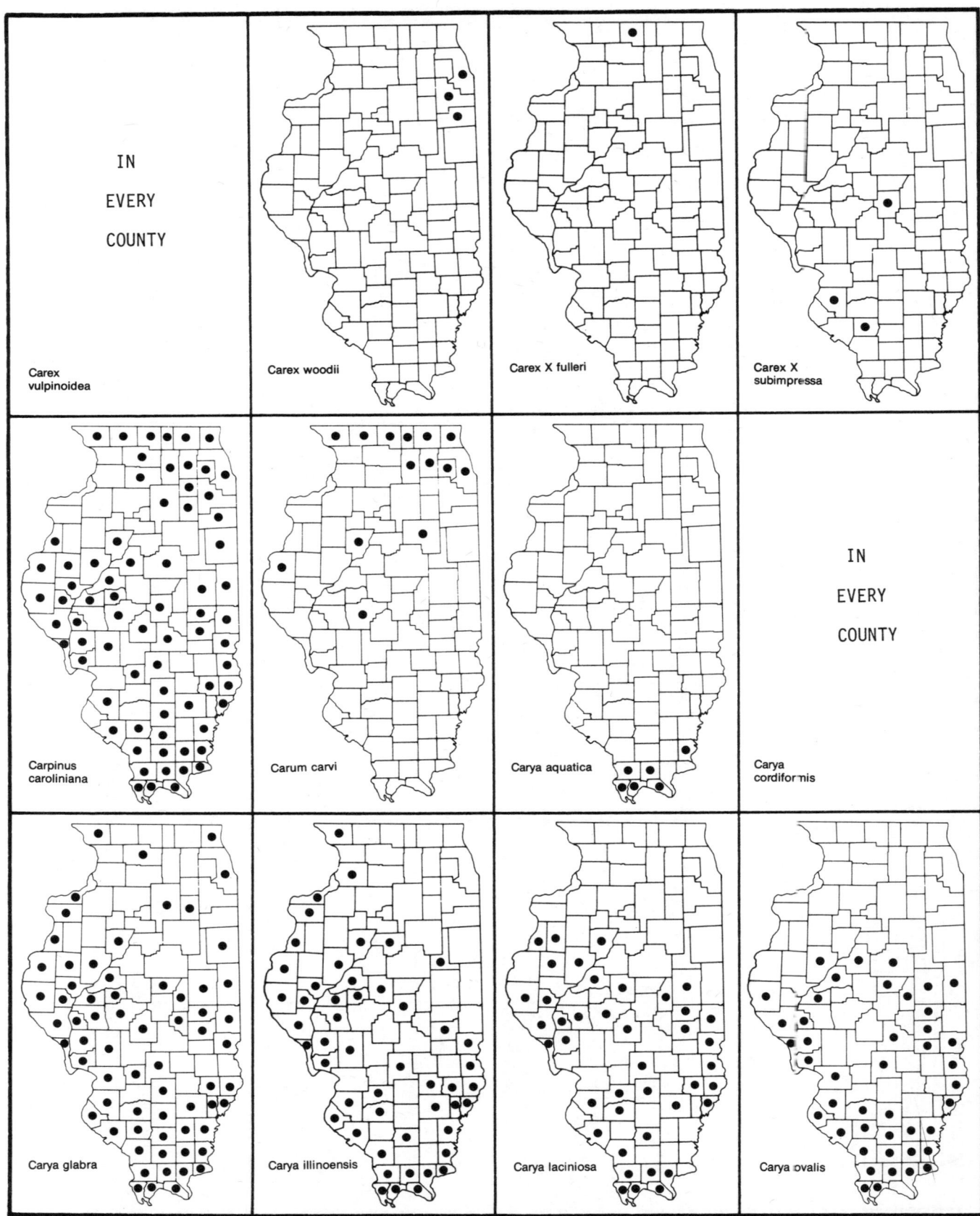

IN EVERY COUNTY

Carex vulpinoidea

Carex woodii

Carex X fulleri

Carex X subimpressa

Carpinus caroliniana

Carum carvi

Carya aquatica

IN EVERY COUNTY

Carya cordiformis

Carya glabra

Carya illinoensis

Carya laciniosa

Carya ovalis

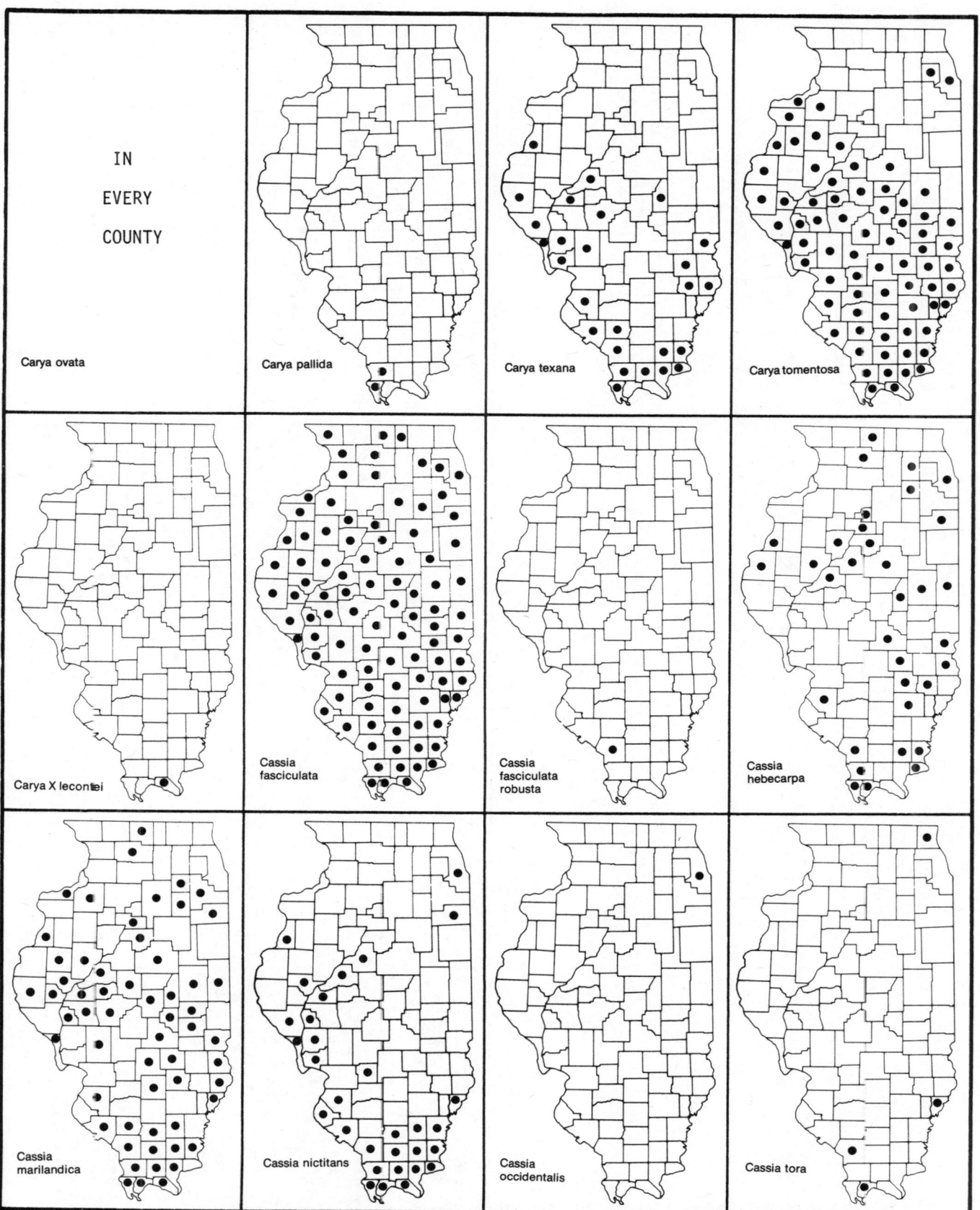

IN EVERY COUNTY

Carya ovata

Carya pallida

Carya texana

Carya tomentosa

Carya X lecontei

Cassia fasciculata

Cassia fasciculata robusta

Cassia hebecarpa

Cassia marilandica

Cassia nictitans

Cassia occidentalis

Cassia tora

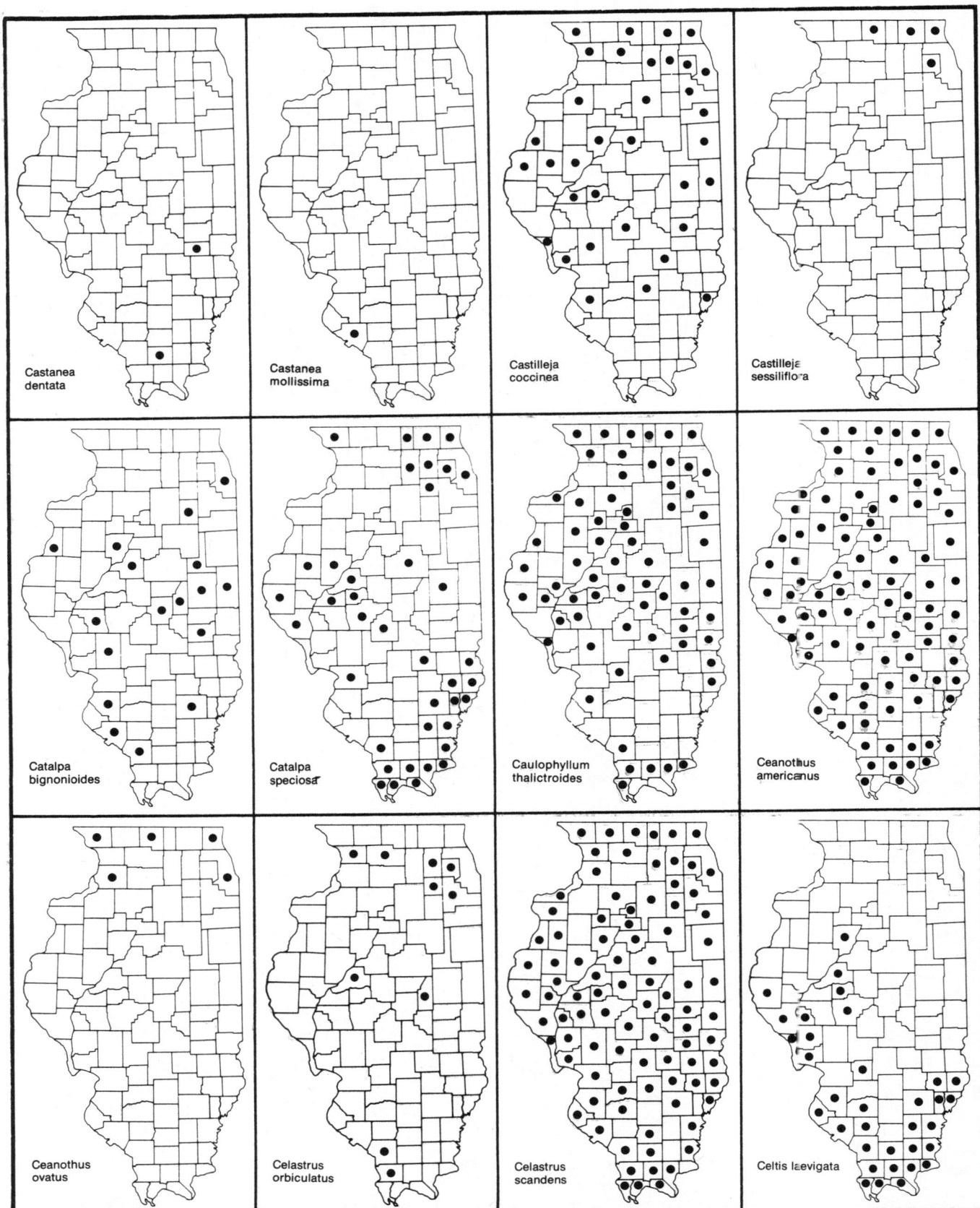

Castanea
dentata

Castanea
mollissima

Castilleja
coccinea

Castilleja
sessiliflora

Catalpa
bignonioides

Catalpa
speciosa

Caulophyllum
thalictroides

Ceanothus
americanus

Ceanothus
ovatus

Celastrus
orbiculatus

Celastrus
scandens

Celtis laevigata

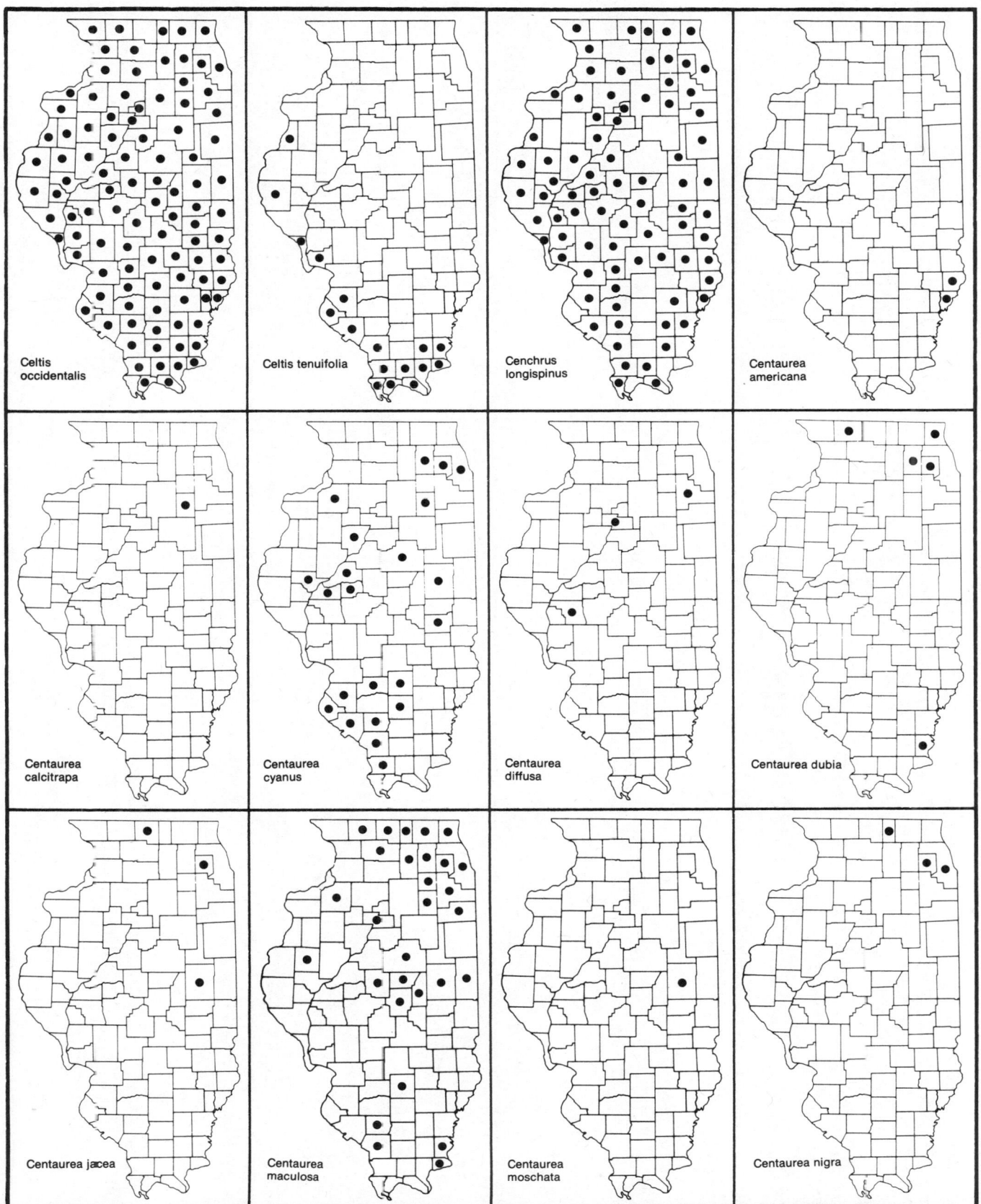

Celtis
occidentalis

Celtis tenuifolia

Cenchrus
longispinus

Centaurea
americana

Centaurea
calcitrapa

Centaurea
cyanus

Centaurea
diffusa

Centaurea dubia

Centaurea jacea

Centaurea
maculosa

Centaurea
moschata

Centaurea nigra

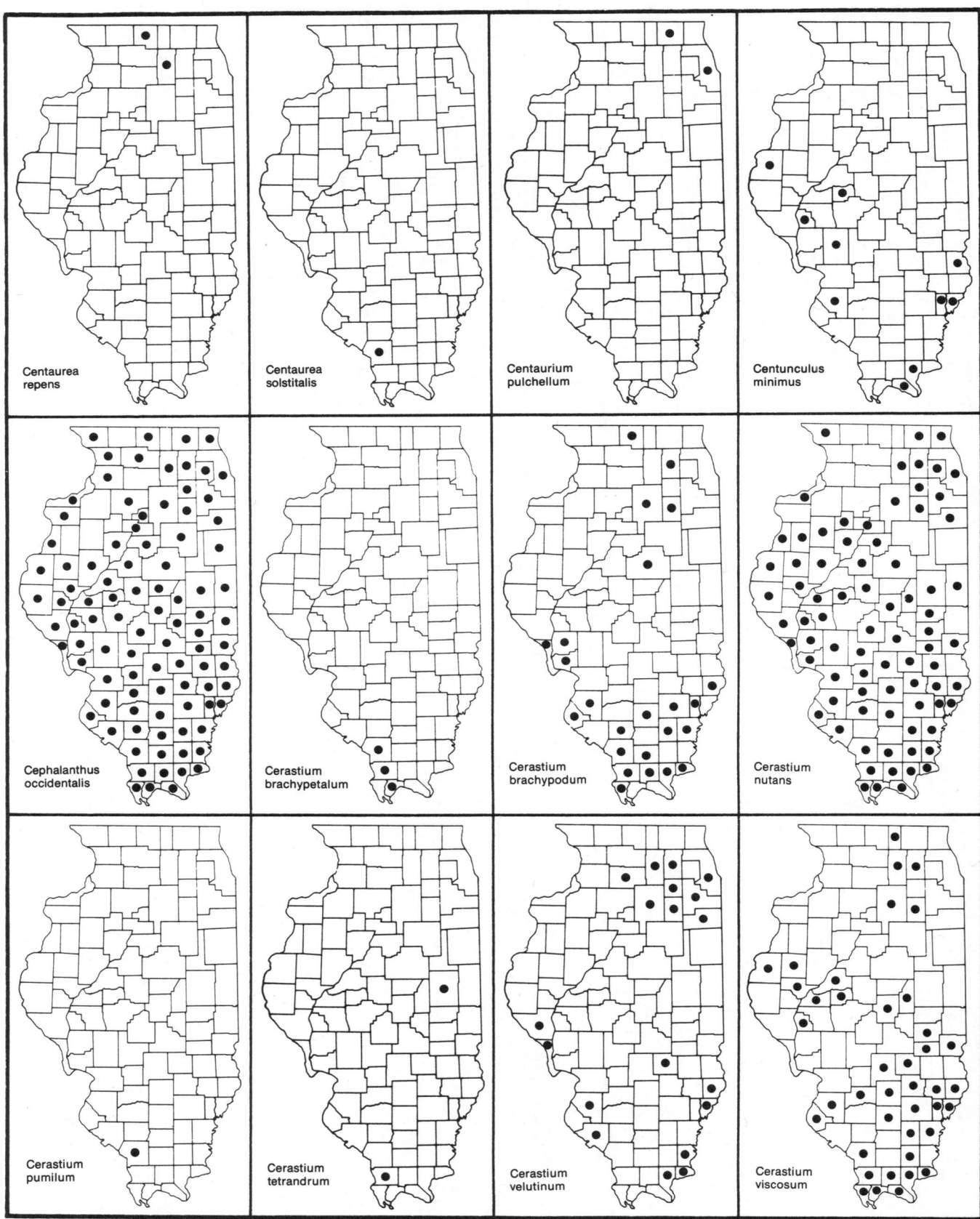

Centaurea
repens

Centaurea
solstitalis

Centaurium
pulchellum

Centunculus
minimus

Cephalanthus
occidentalis

Cerastium
brachypetalum

Cerastium
brachypodum

Cerastium
nutans

Cerastium
pumilum

Cerastium
tetrandrum

Cerastium
velutinum

Cerastium
viscosum

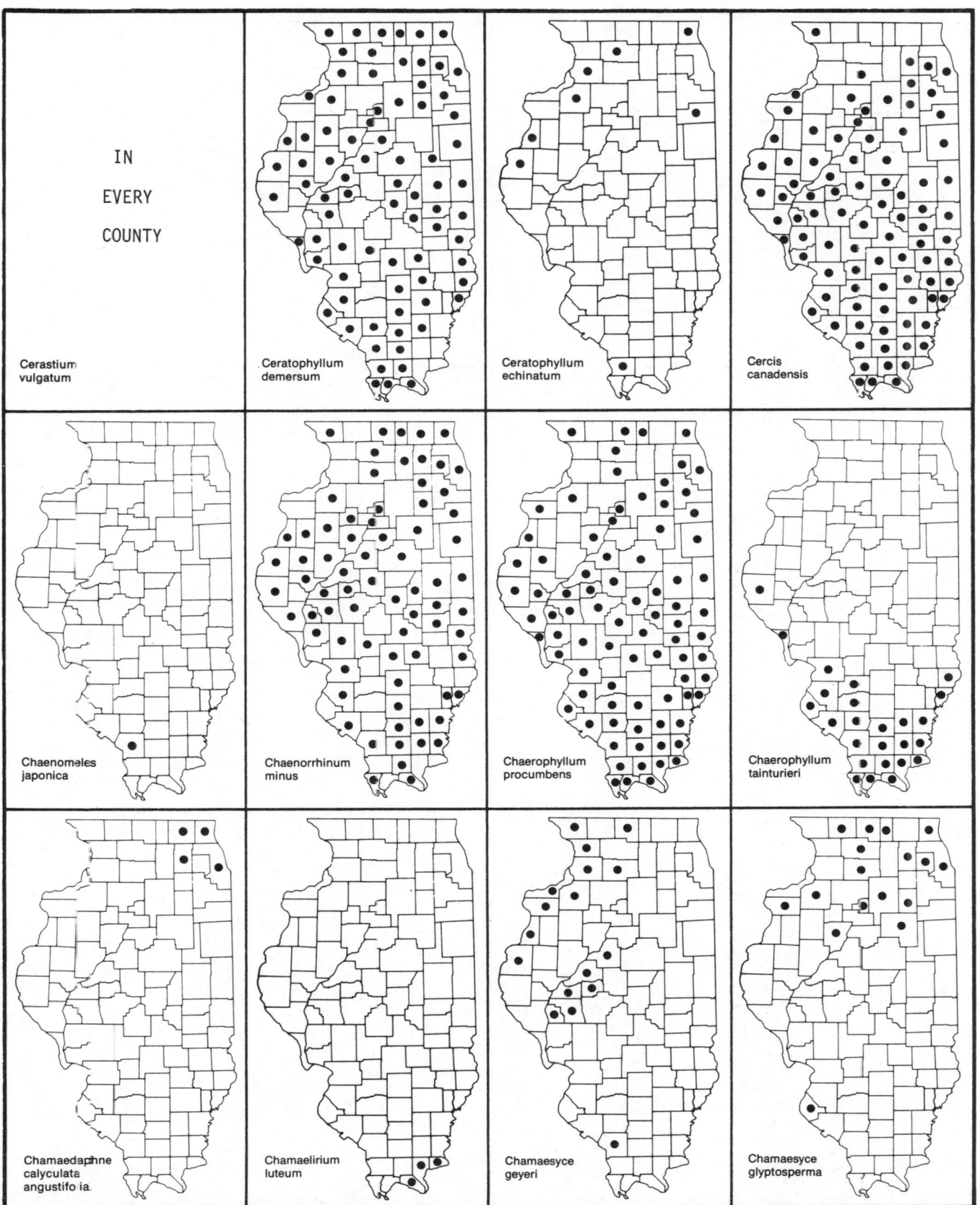

IN

EVERY

COUNTY

Cerastium
vulgatum

Ceratophyllum
demersum

Ceratophyllum
echinatum

Cercis
canadensis

Chaenomeles
japonica

Chaenorrhinum
minus

Chaerophyllum
procumbens

Chaerophyllum
tainturieri

Chamaedaphne
calyculata
angustifolia

Chamaelirium
luteum

Chamaesyce
geyeri

Chamaesyce
glyptosperma

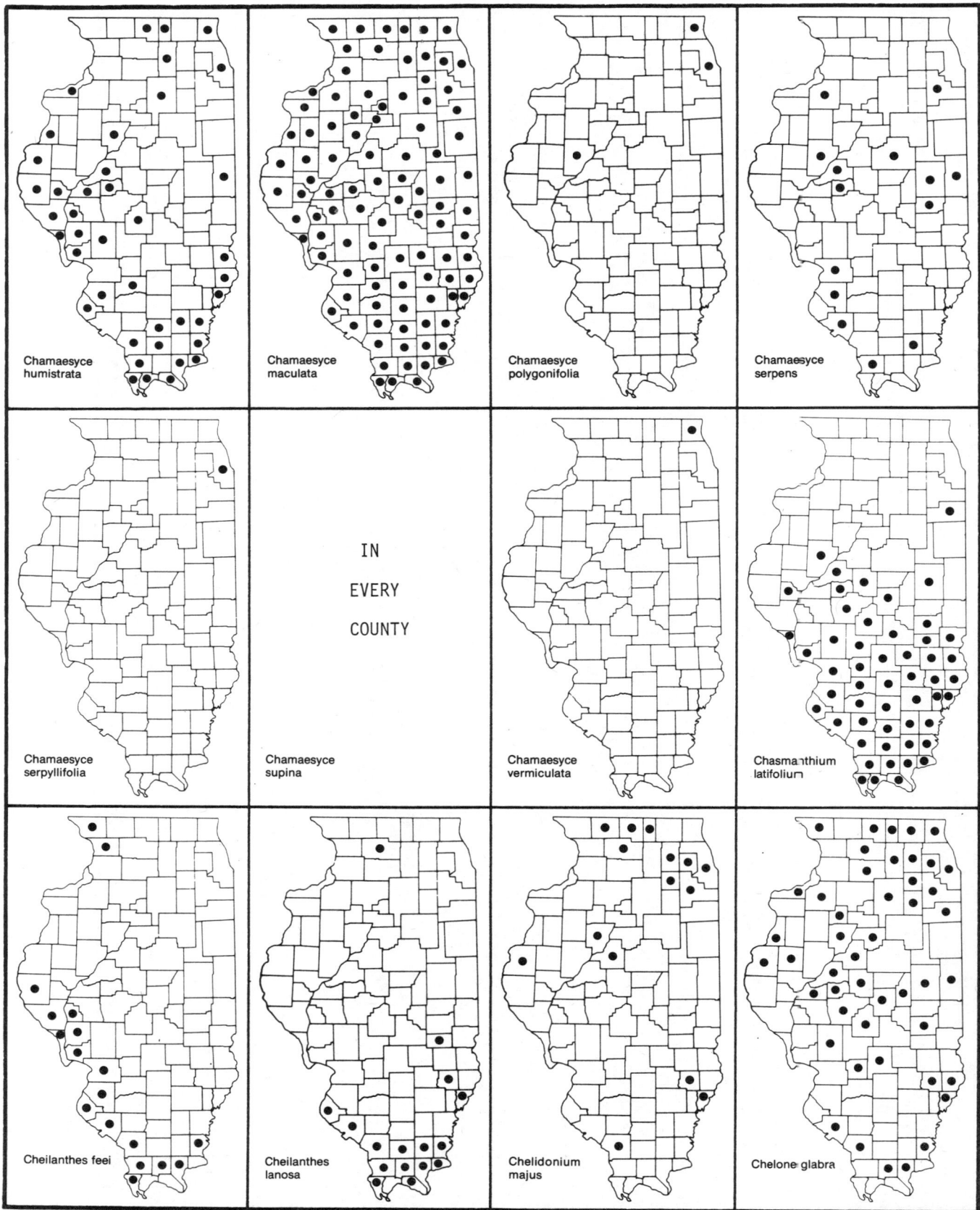

Chamaesyce
humistrata

Chamaesyce
maculata

Chamaesyce
polygonifolia

Chamaesyce
serpens

Chamaesyce
serpyllifolia

Chamaesyce
supina

IN

EVERY

COUNTY

Chamaesyce
vermiculata

Chasmanthium
latifolium

Cheilanthes feei

Cheilanthes
lanosa

Chelidonium
majus

Chelone glabra

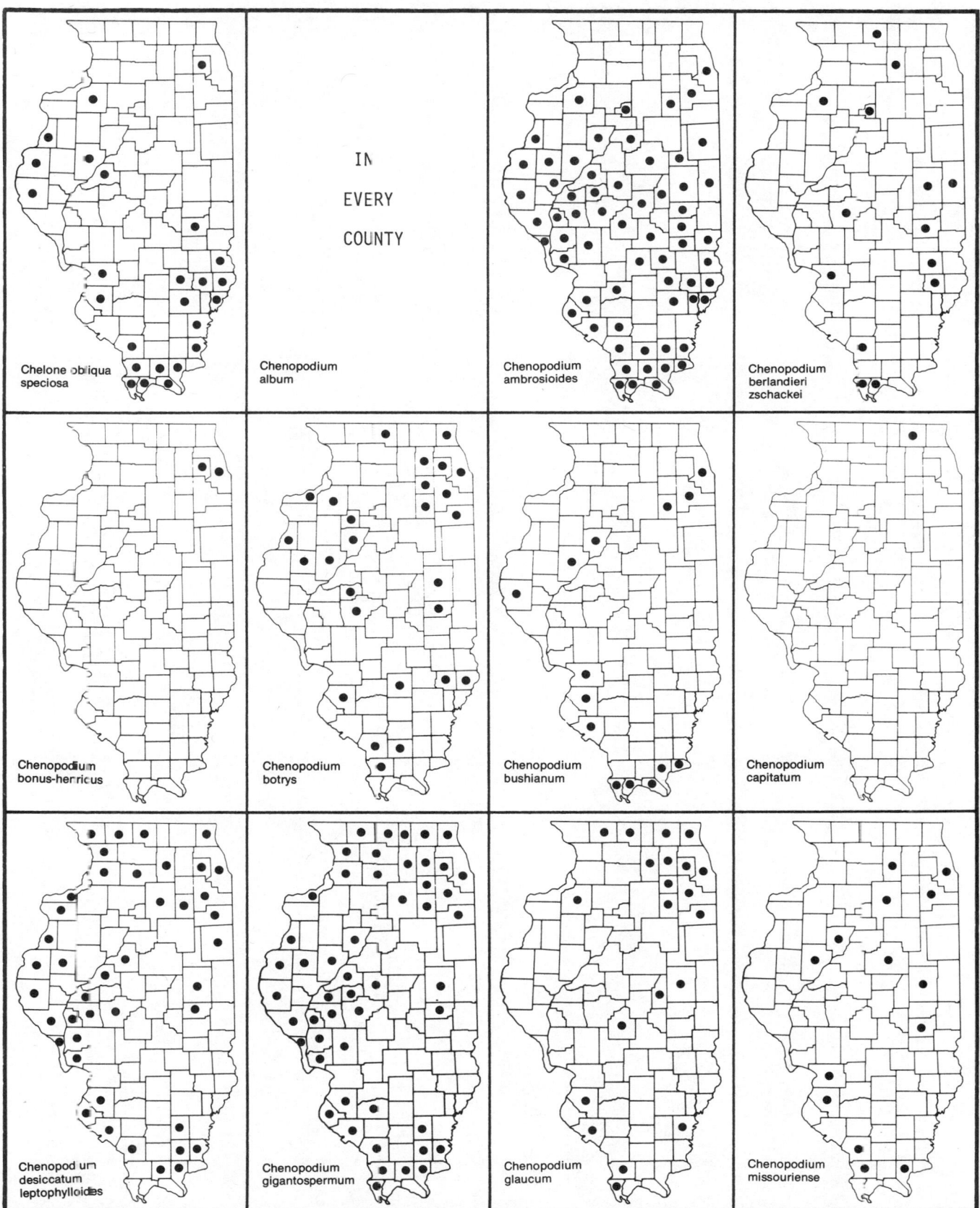

Chelone obliqua
speciosa

Chenopodium
album

IN

EVERY

COUNTY

Chenopodium
ambrosioides

Chenopodium
berlandieri
zschackei

Chenopodium
bonus-herricus

Chenopodium
botrys

Chenopodium
bushianum

Chenopodium
capitatum

Chenopodium
desiccatum
leptophylloides

Chenopodium
gigantospermum

Chenopodium
glaucum

Chenopodium
missouriense

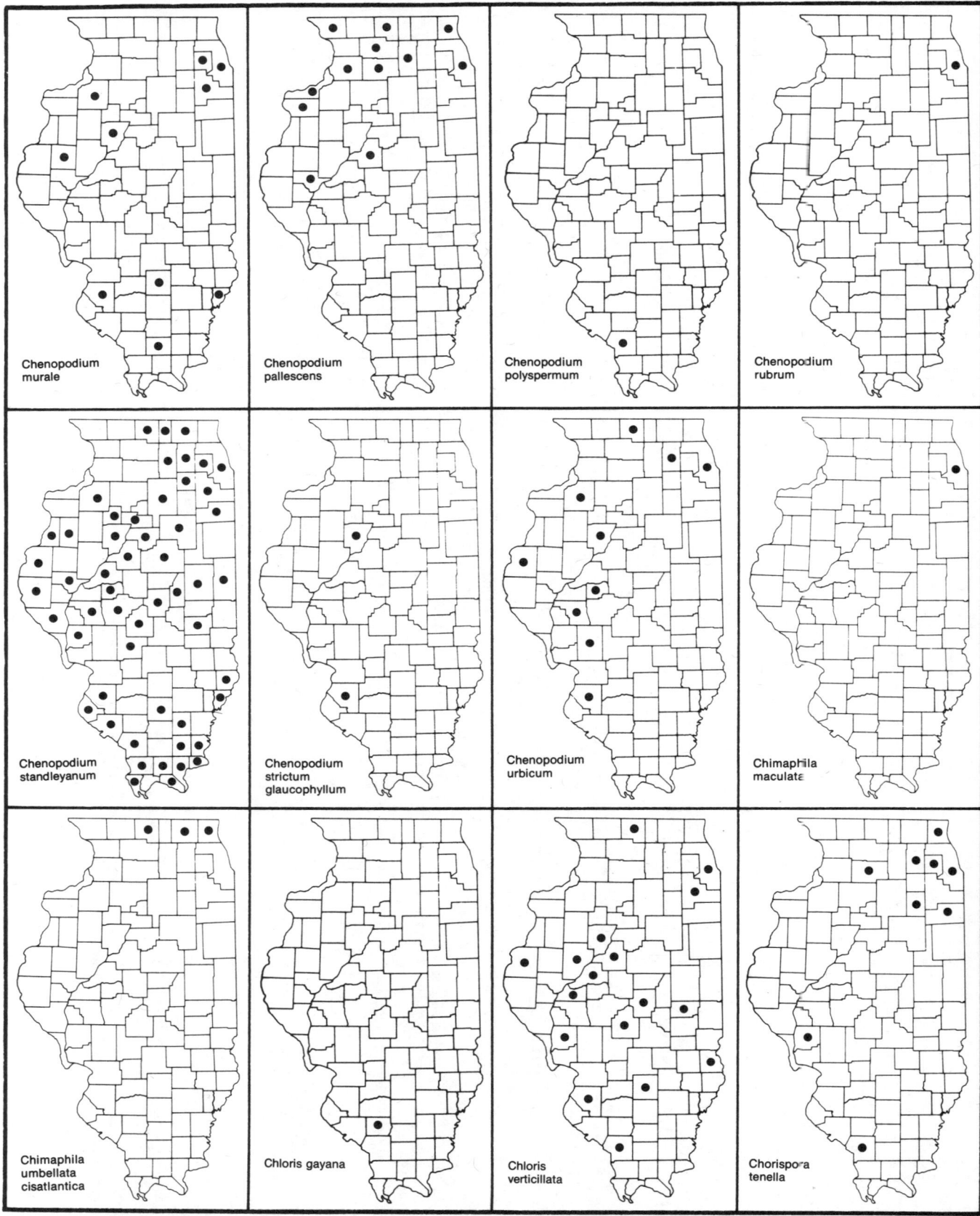

Chenopodium
murale

Chenopodium
pallescens

Chenopodium
polyspermum

Chenopodium
rubrum

Chenopodium
standleyanum

Chenopodium
strictum
glaucophyllum

Chenopodium
urbicum

Chimaphila
maculata

Chimaphila
umbellata
cisatlantica

Chloris gayana

Chloris
verticillata

Chorispora
tenella

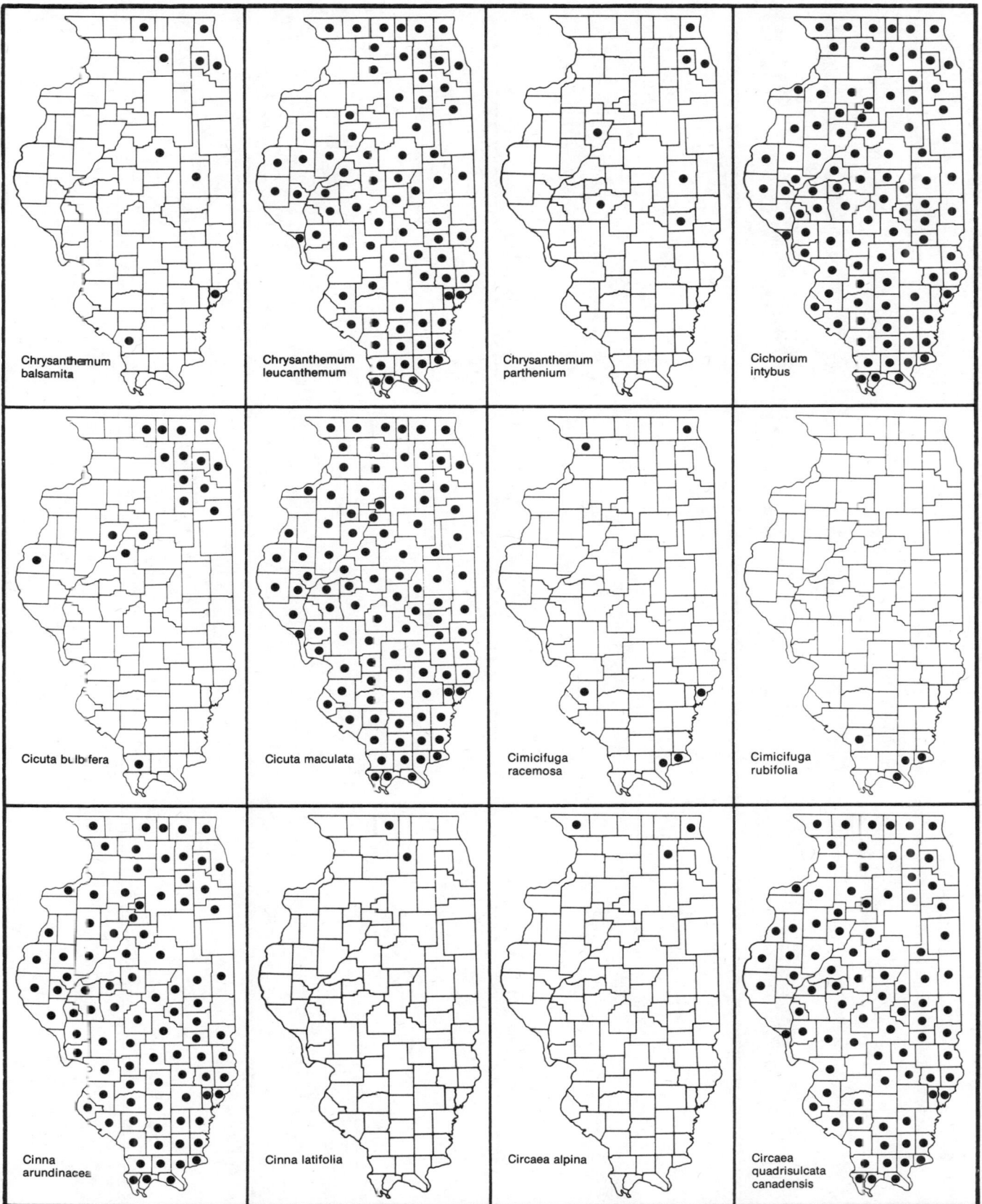

Chrysanthemum
balsamita

Chrysanthemum
leucanthemum

Chrysanthemum
parthenium

Cichorium
intybus

Cicuta bulbfera

Cicuta maculata

Cimicifuga
racemosa

Cimicifuga
rubifolia

Cinna
arundinacea

Cinna latifolia

Circaea alpina

Circaea
quadrisulcata
canadensis

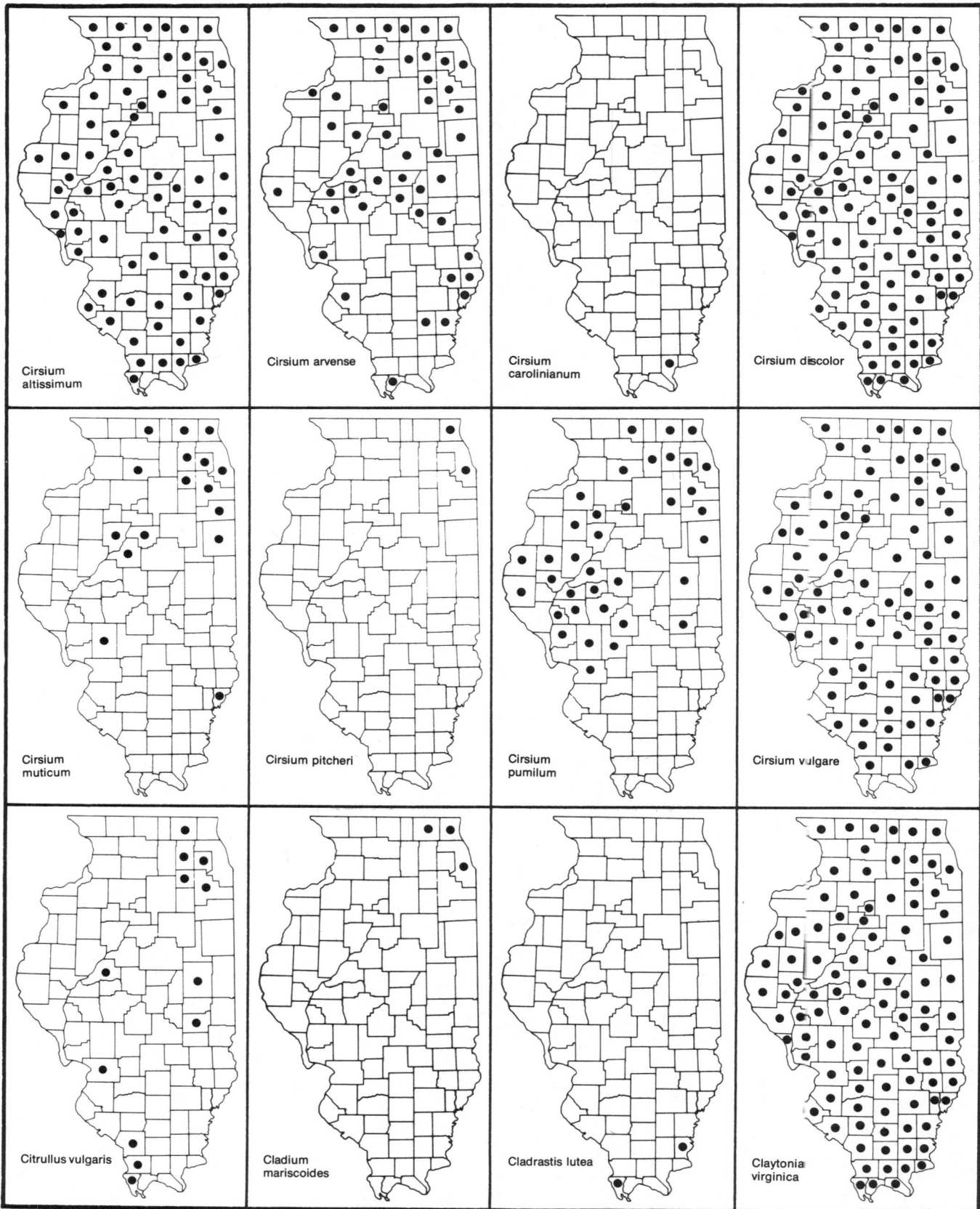

Cirsium
altissimum

Cirsium arvense

Cirsium
carolinianum

Cirsium discolor

Cirsium
muticum

Cirsium pitcheri

Cirsium
pumilum

Cirsium vulgare

Citrullus vulgaris

Cladium
mariscoides

Cladrastis lutea

Claytonia
virginica

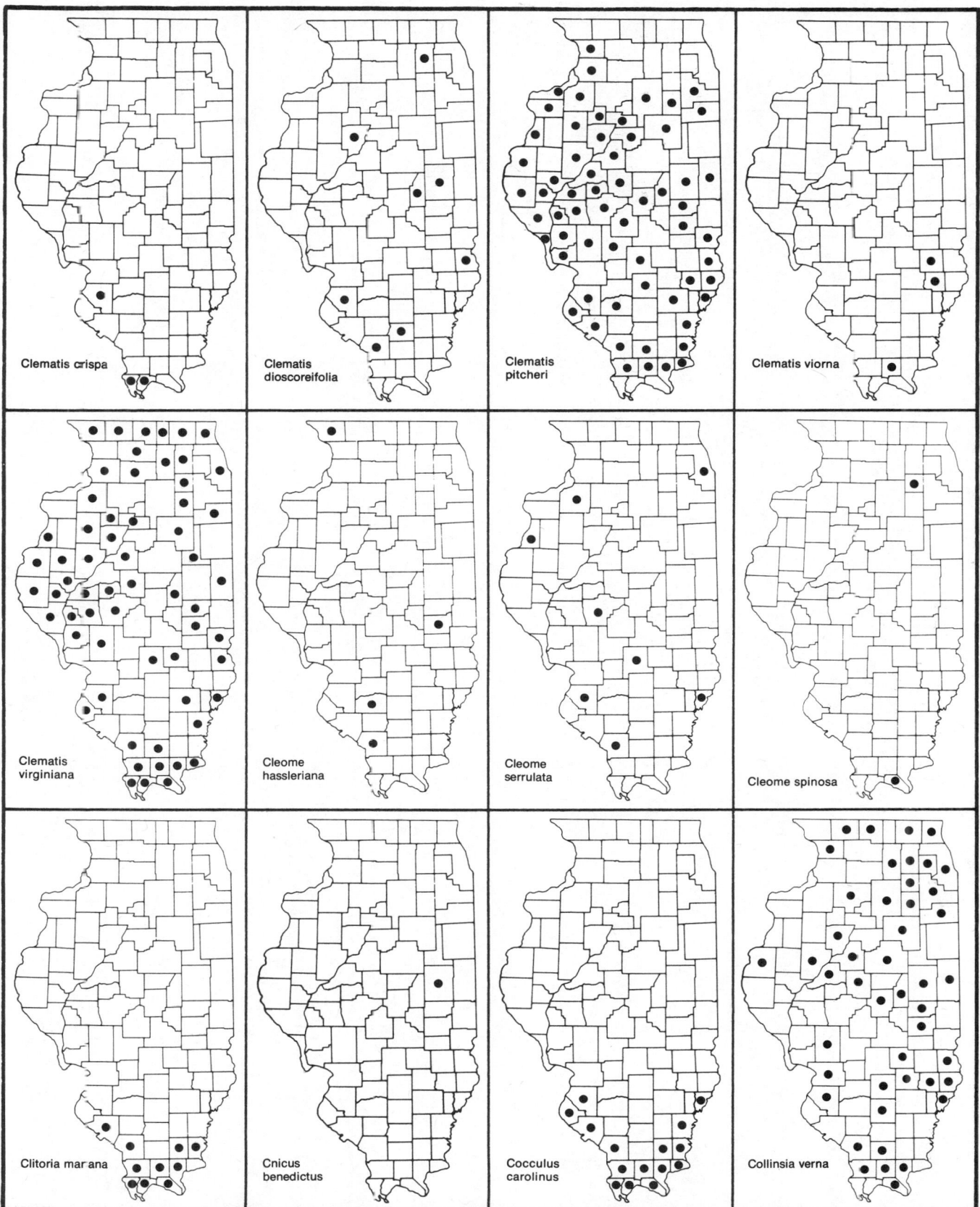

Clematis crispa

Clematis
dioscoreifolia

Clematis
pitcheri

Clematis viorna

Clematis
virginiana

Cleome
hassleriana

Cleome
serrulata

Cleome spinosa

Clitoria mariana

Cnicus
benedictus

Cocculus
carolinus

Collinsia verna

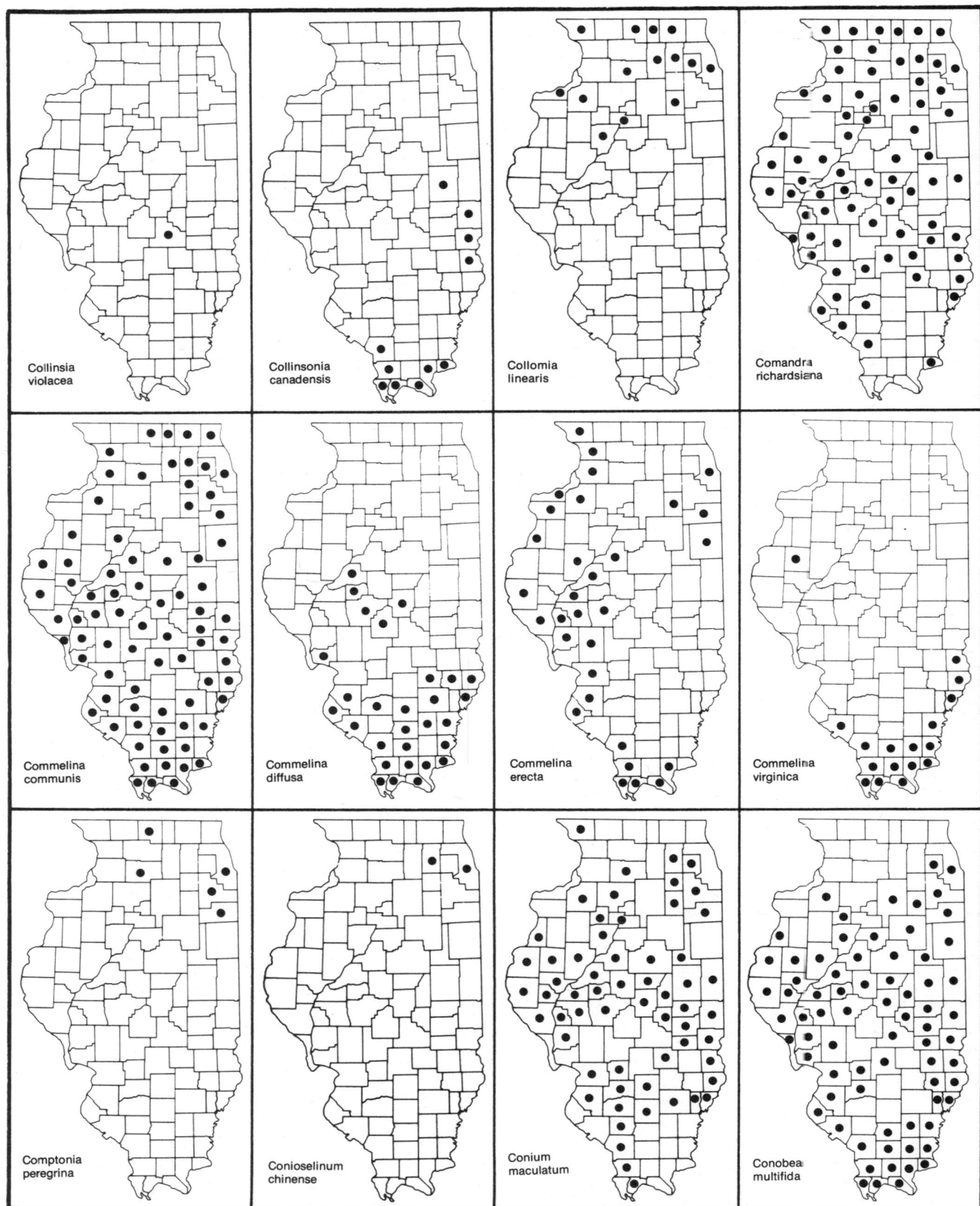

Collinsia
violacea

Collinsonia
canadensis

Collomia
linearis

Comandra
richardsiana

Commelina
communis

Commelina
diffusa

Commelina
erecta

Commelina
virginica

Comptonia
peregrina

Conioselinum
chinense

Conium
maculatum

Conobea
multifida

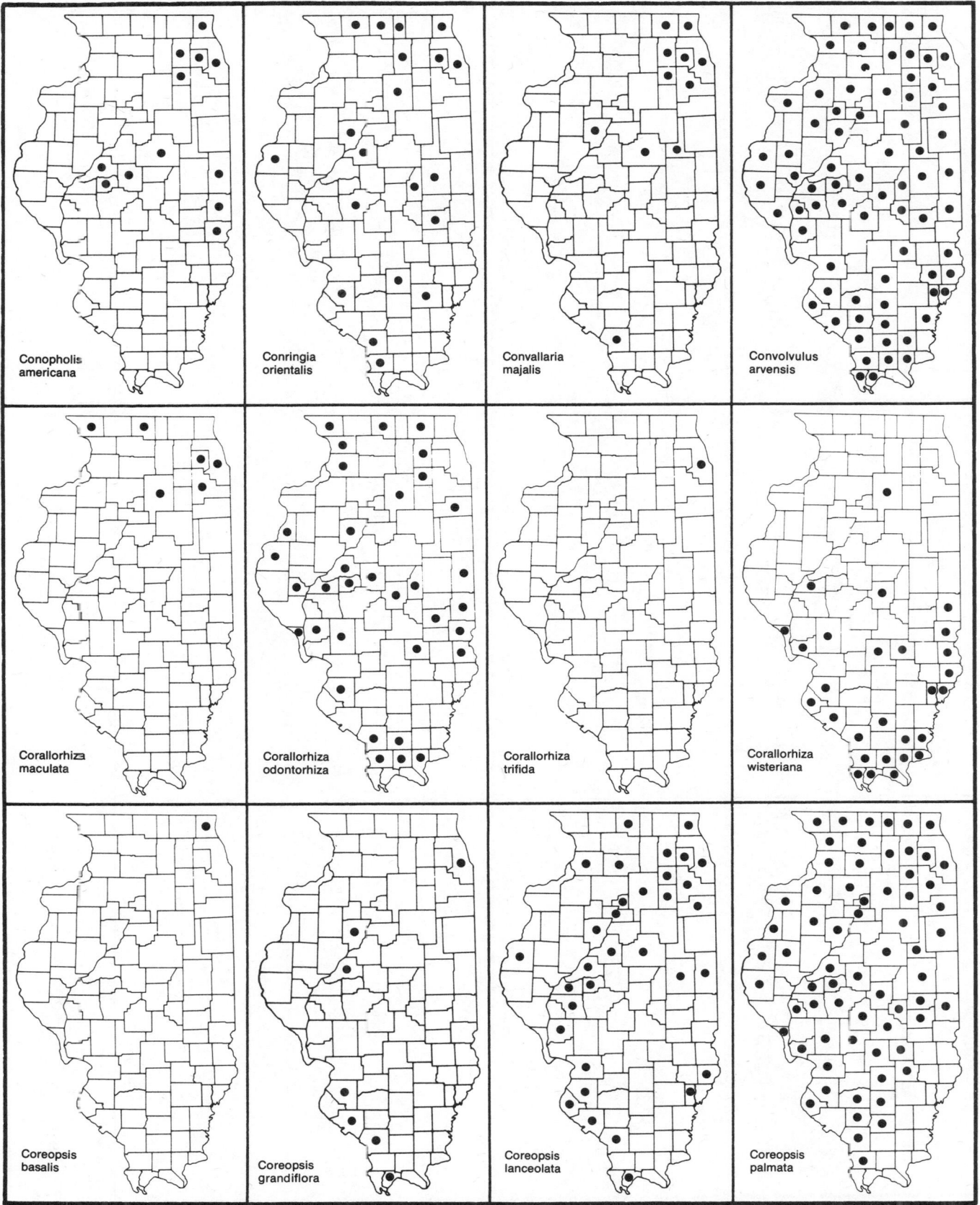

Conopholis
americana

Conringia
orientalis

Convallaria
majalis

Convolvulus
arvensis

Corallorhiza
maculata

Corallorhiza
odontorhiza

Corallorhiza
trifida

Corallorhiza
wisteriana

Coreopsis
basalis

Coreopsis
grandiflora

Coreopsis
lanceolata

Coreopsis
palmata

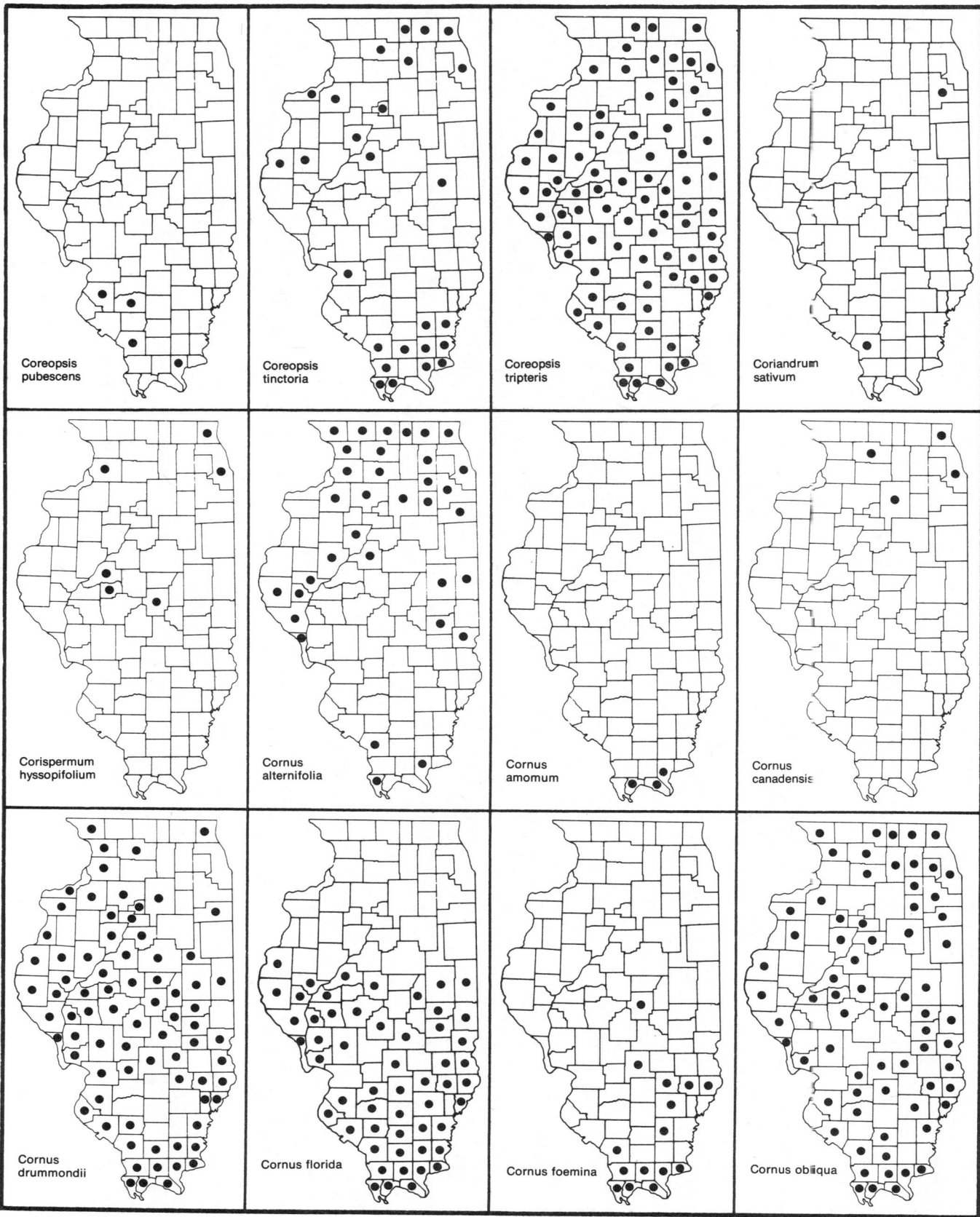

Coreopsis
pubescens

Coreopsis
tinctoria

Coreopsis
tripteris

Coriandrum
sativum

Corispermum
hyssopifolium

Cornus
alternifolia

Cornus
amomum

Cornus
canadensis

Cornus
drummondii

Cornus florida

Cornus foemina

Cornus obliqua

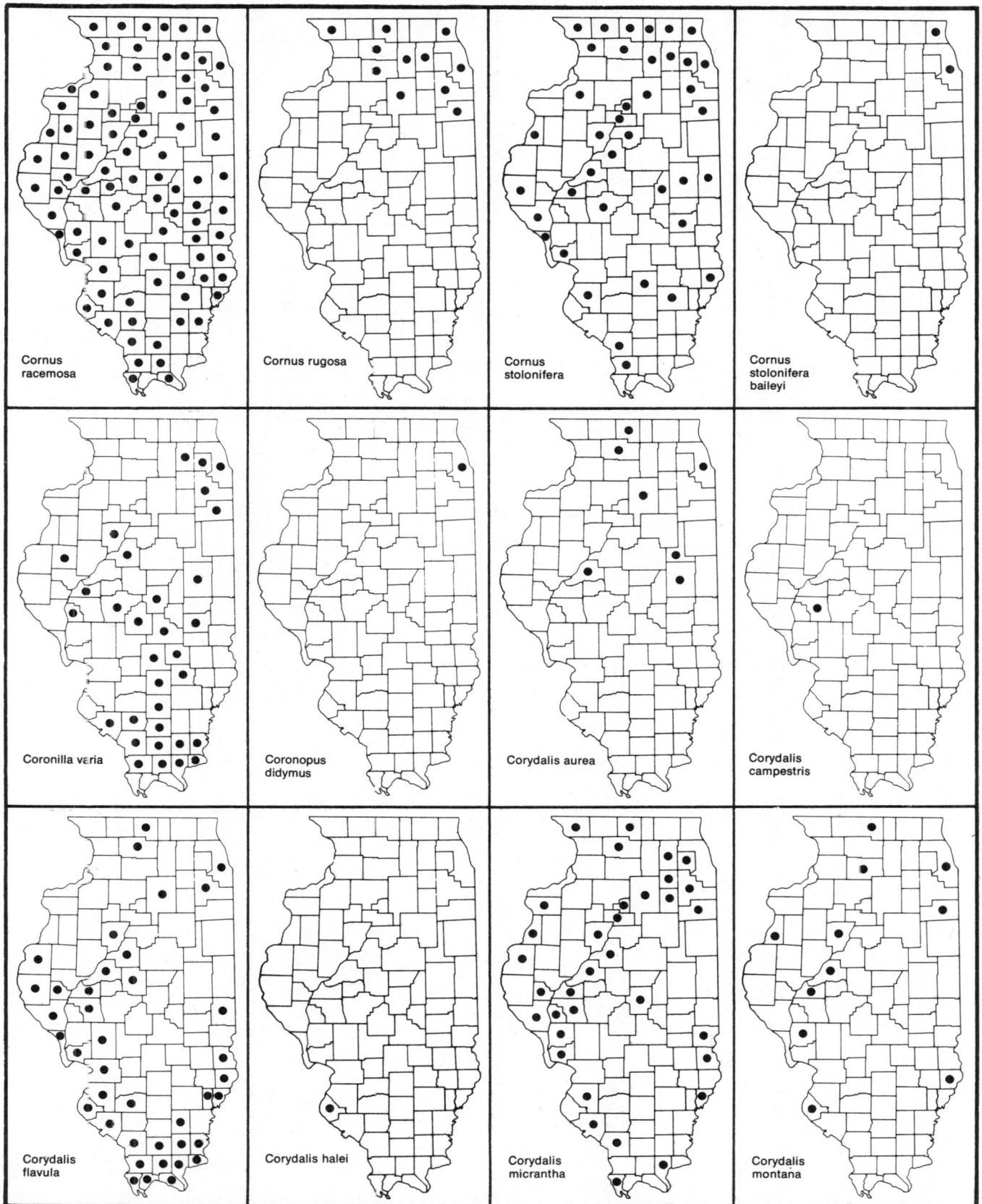

Cornus
racemosa

Cornus rugosa

Cornus
stolonifera

Cornus
stolonifera
baileyi

Coronilla varia

Coronopus
didymus

Corydalis aurea

Corydalis
campestris

Corydalis
flavula

Corydalis halei

Corydalis
micrantha

Corydalis
montana

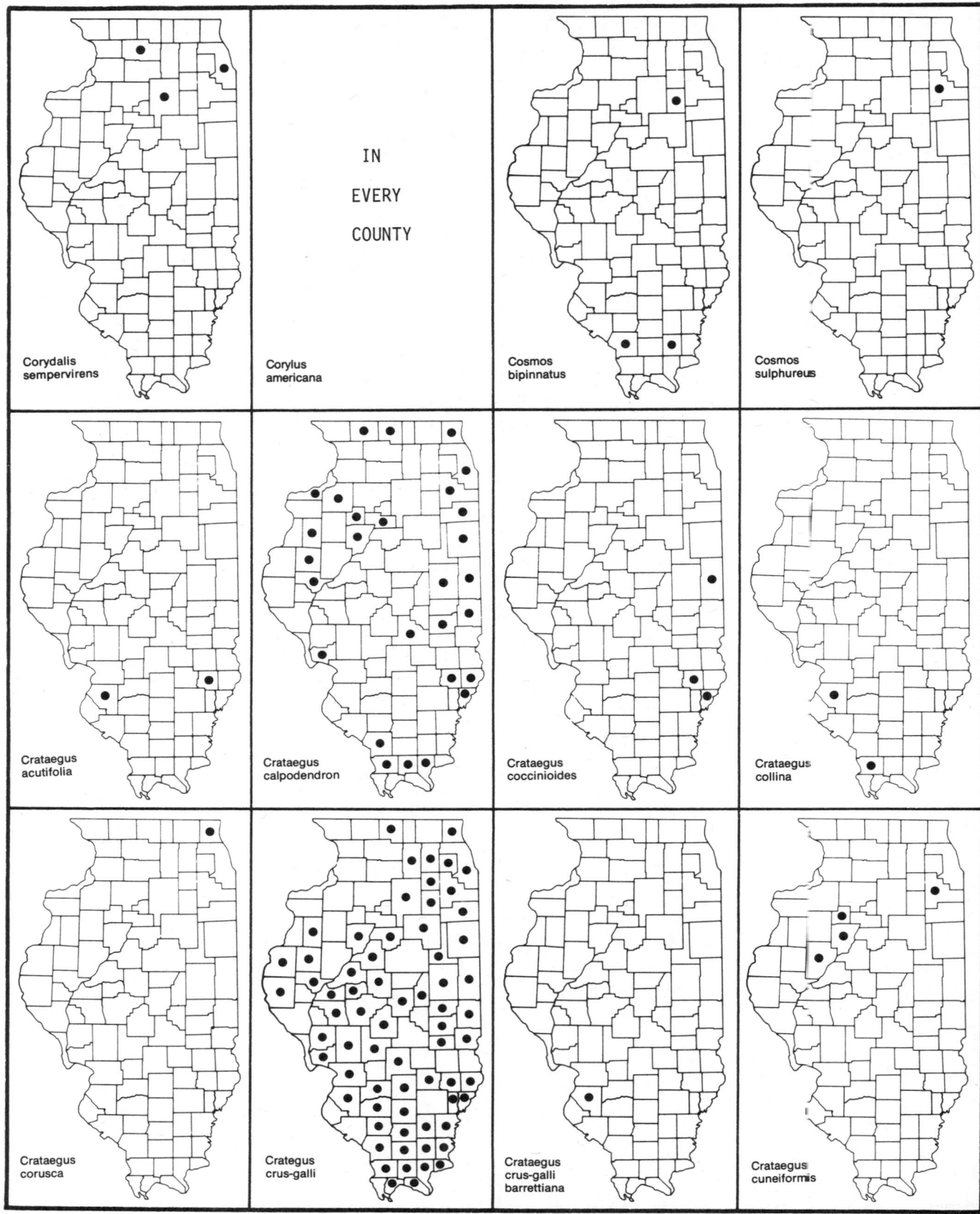

IN

EVERY

COUNTY

Corydalis
sempervirens

Corylus
americana

Cosmos
bipinnatus

Cosmos
sulphureus

Crataegus
acutifolia

Crataegus
calpodendron

Crataegus
coccinioides

Crataegus
collina

Crataegus
corusca

Crategus
crus-galli

Crataegus
crus-galli
barrettiana

Crataegus
cuneiformis

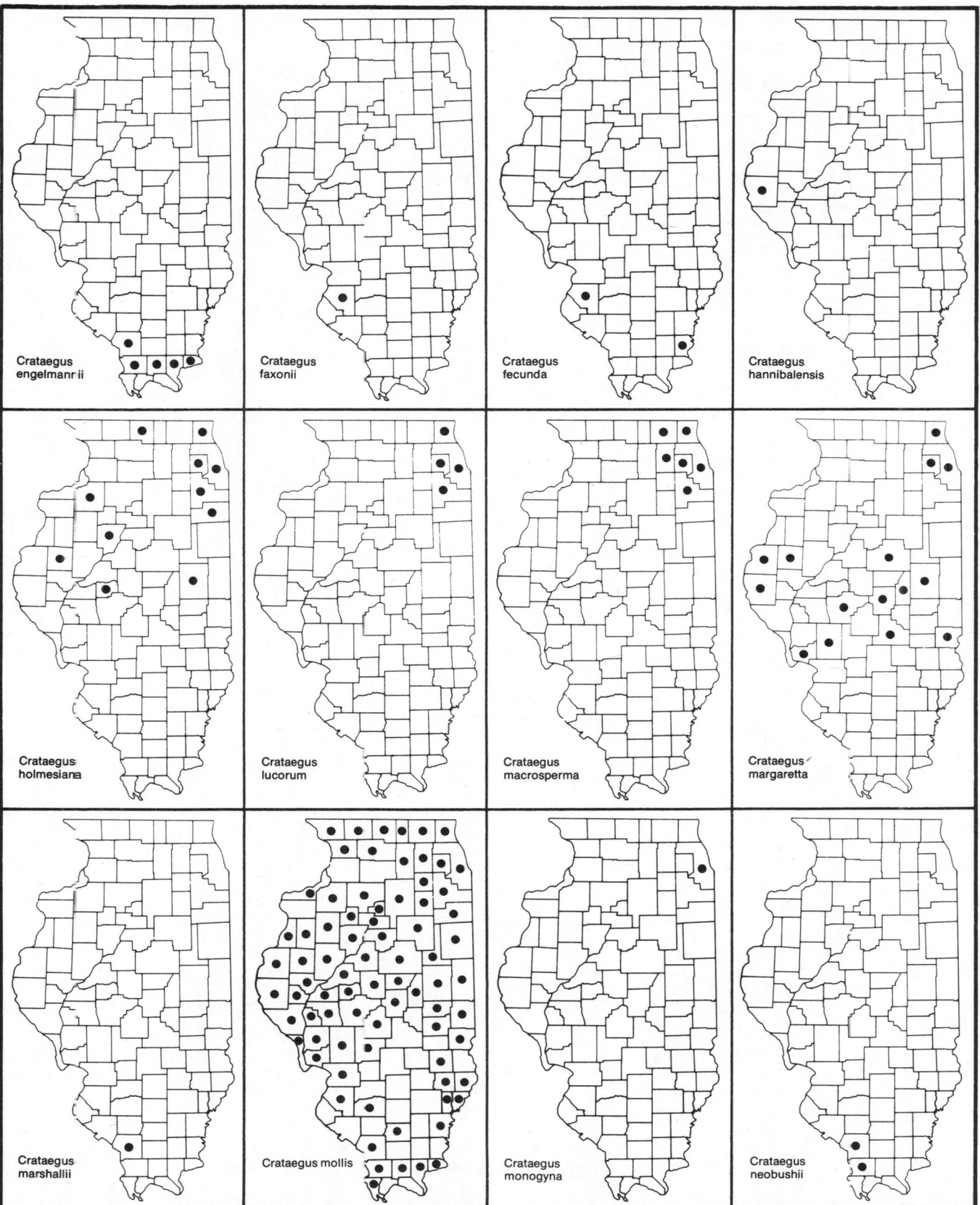

Crataegus
engelmann ii

Crataegus
faxonii

Crataegus
fecunda

Crataegus
hannibalensis

Crataegus
holmesiana

Crataegus
lucorum

Crataegus
macrosperma

Crataegus
margaretta

Crataegus
marshallii

Crataegus
mollis

Crataegus
monogyna

Crataegus
neobushii

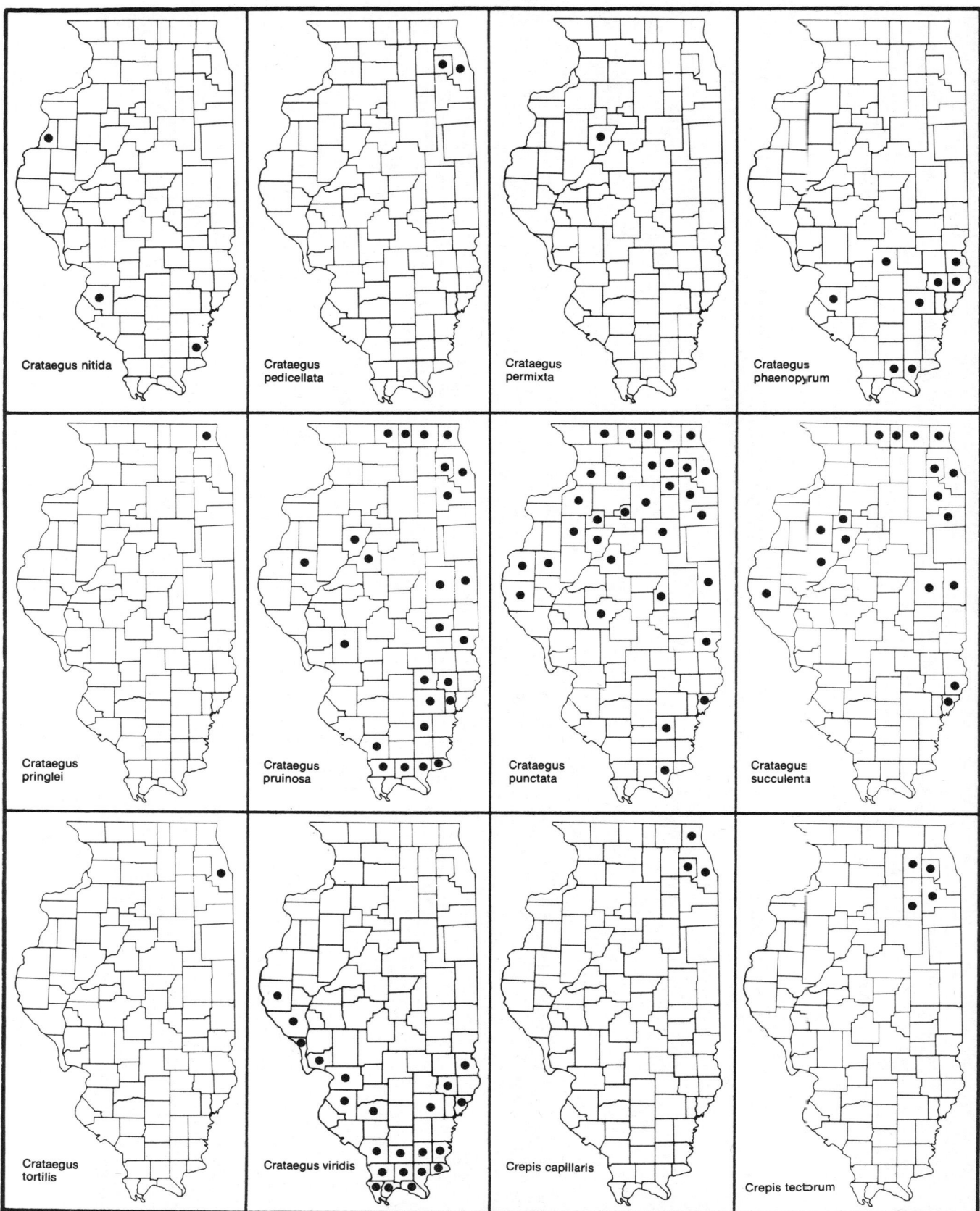

Crataegus
nitida

Crataegus
pedicellata

Crataegus
permixta

Crataegus
phaenopyrum

Crataegus
pringlei

Crataegus
pruinosa

Crataegus
punctata

Crataegus
succulenta

Crataegus
tortilis

Crataegus
viridis

Crepis
capillaris

Crepis
tectorum

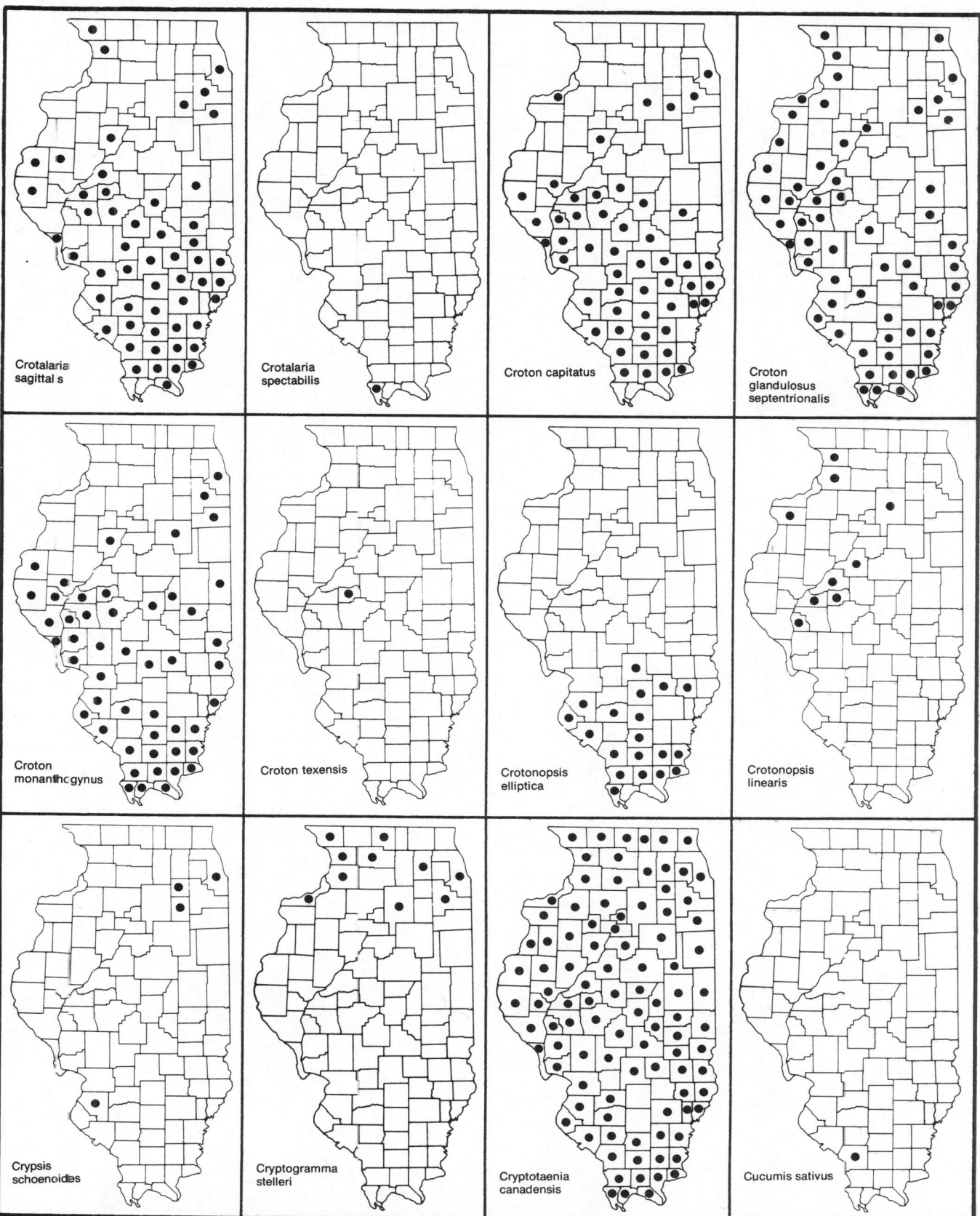

Crotalaria sagittal s

Crotalaria spectabilis

Croton capitatus

Croton glandulosus septentrionalis

Croton monanthogynus

Croton texensis

Crotonopsis elliptica

Crotonopsis linearis

Crypsis schoenoides

Cryptogramma stelleri

Cryptotaenia canadensis

Cucumis sativus

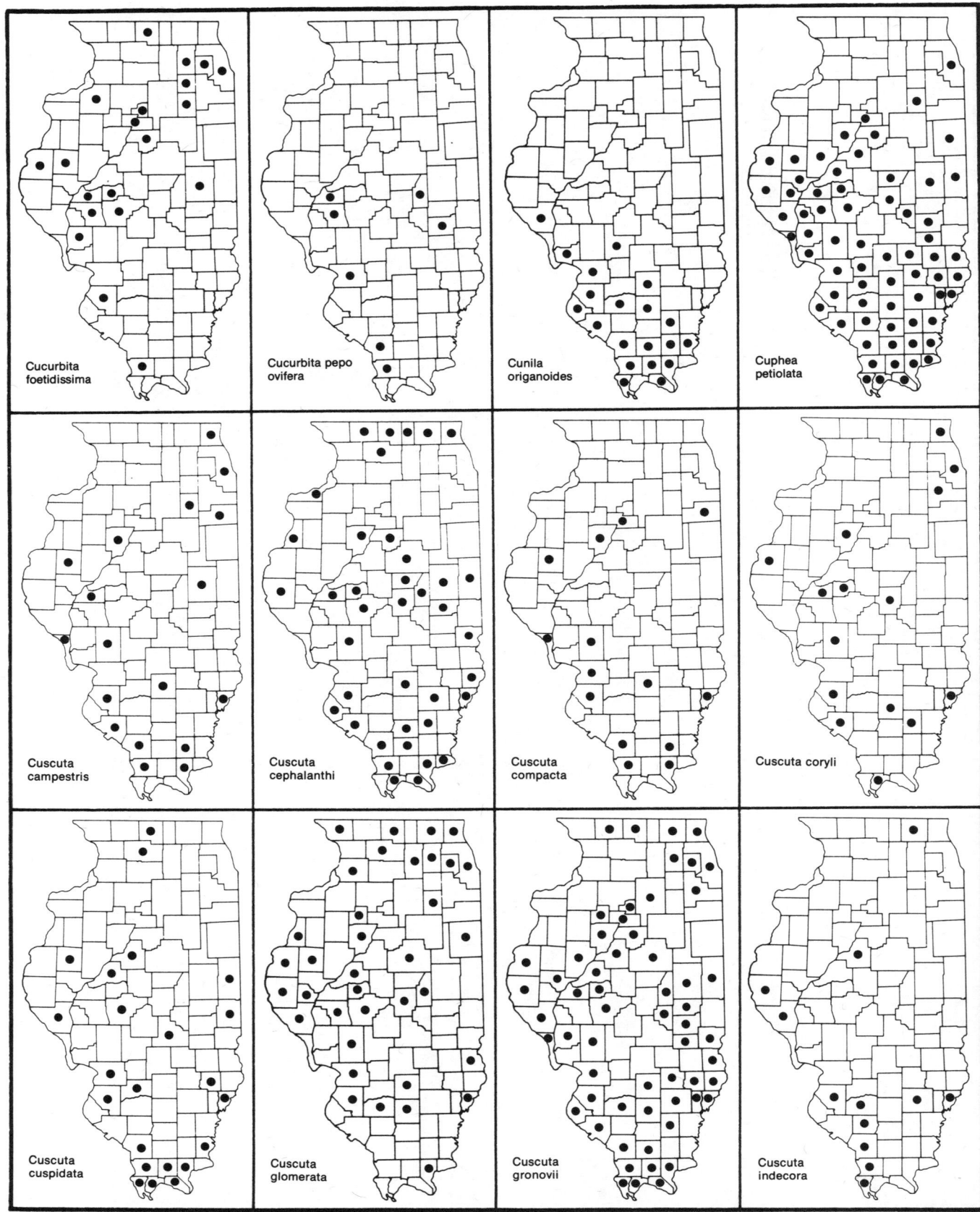

Cucurbita
foetidissima

Cucurbita pepo
ovifera

Cunila
origanoides

Cuphea
petiolata

Cuscuta
campestris

Cuscuta
cephalanthi

Cuscuta
compacta

Cuscuta coryli

Cuscuta
cuspidata

Cuscuta
glomerata

Cuscuta
gronovii

Cuscuta
indecora

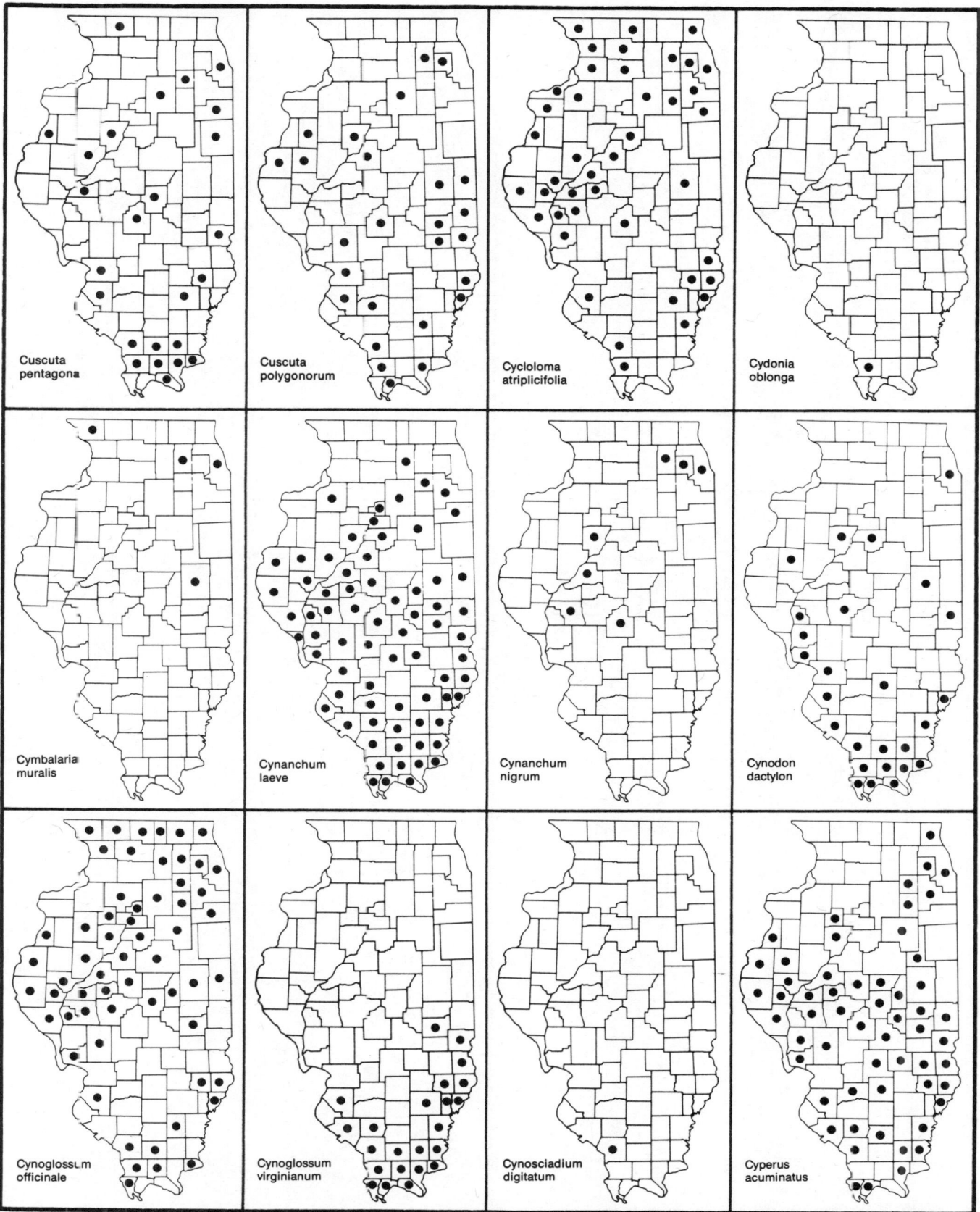

Cuscuta
pentagona

Cuscuta
polygonorum

Cycloloma
atriplicifolia

Cydonia
oblonga

Cymbalaria
muralis

Cynanchum
laeve

Cynanchum
nigrum

Cynodon
dactylon

Cynoglossum
officinale

Cynoglossum
virginianum

Cynosciadium
digitatum

Cyperus
acuminatus

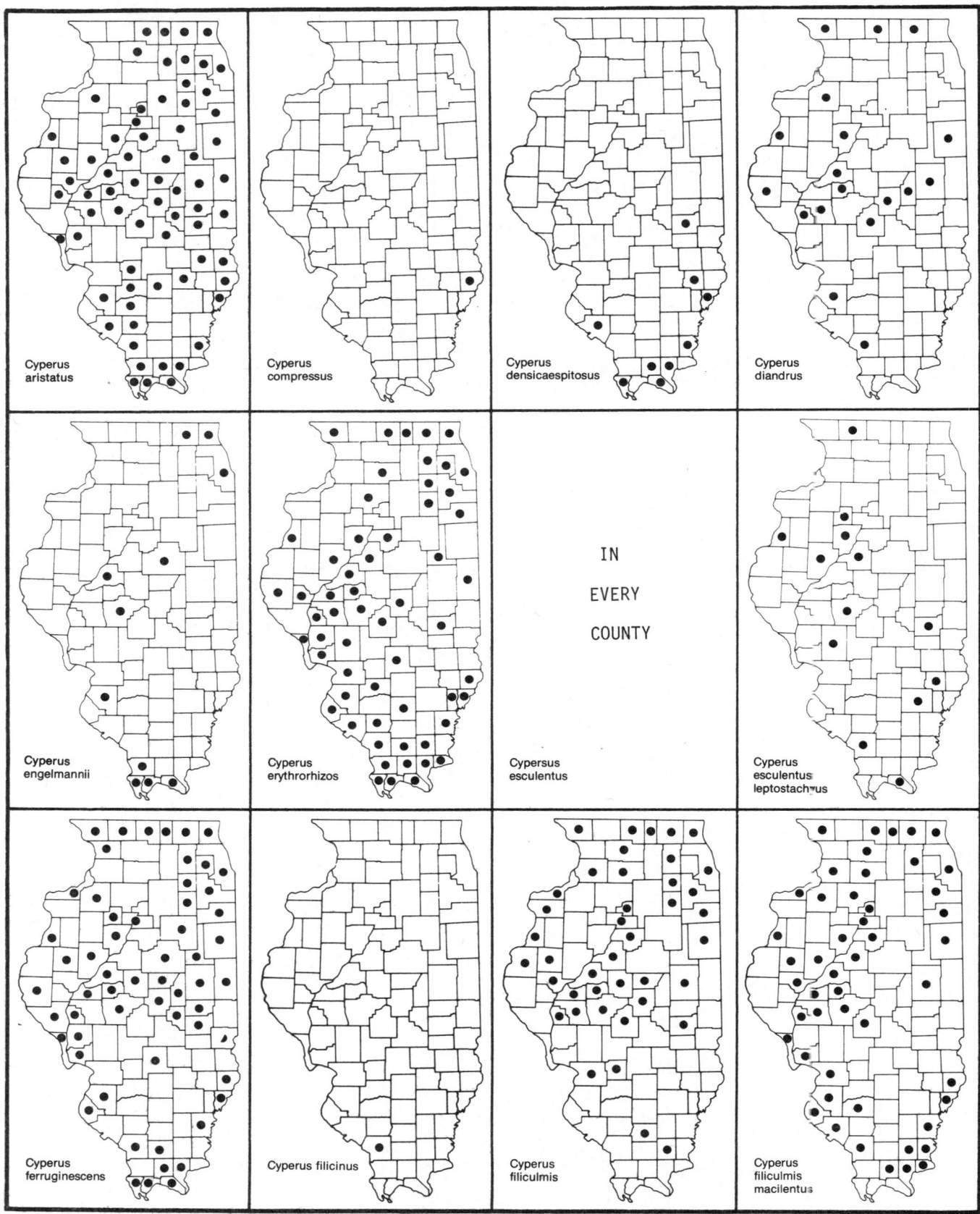

Cyperus
aristatus

Cyperus
compressus

Cyperus
densicaespitosus

Cyperus
diandrus

Cyperus
engelmannii

Cyperus
erythrorhizos

Cypersus
esculentus

IN

EVERY

COUNTY

Cyperus
esculentus
leptostachyus

Cyperus
ferruginescens

Cyperus
filicinus

Cyperus
filiculmis

Cyperus
filiculmis
macilentus

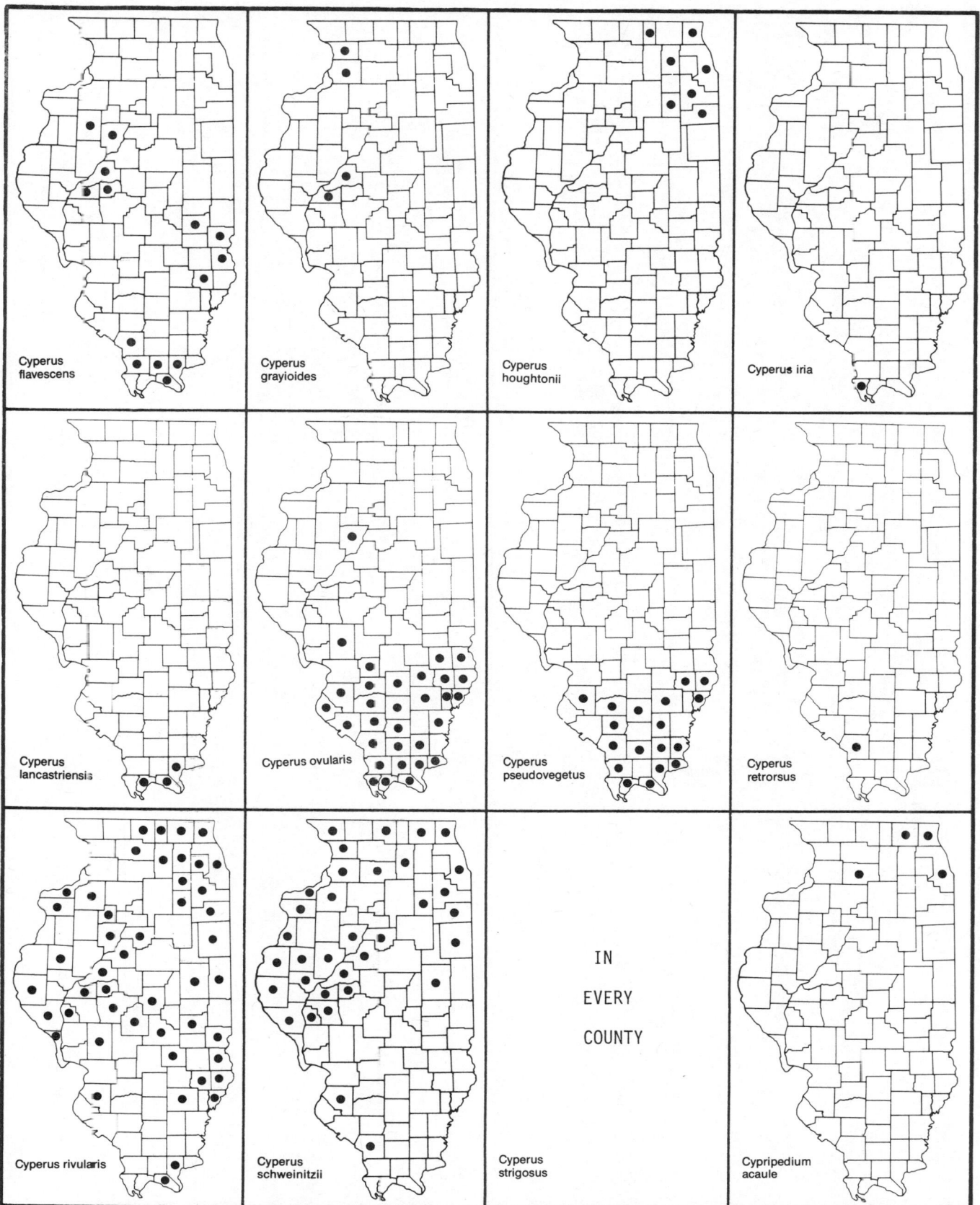

Cyperus
flavescens

Cyperus
grayioides

Cyperus
houghtonii

Cyperus iria

Cyperus
lancastriensis

Cyperus ovularis

Cyperus
pseudovegetus

Cyperus
retrorsus

Cyperus
rivularis

Cyperus
schweinitzii

Cyperus
strigosus

IN

EVERY

COUNTY

Cypripedium
acaule

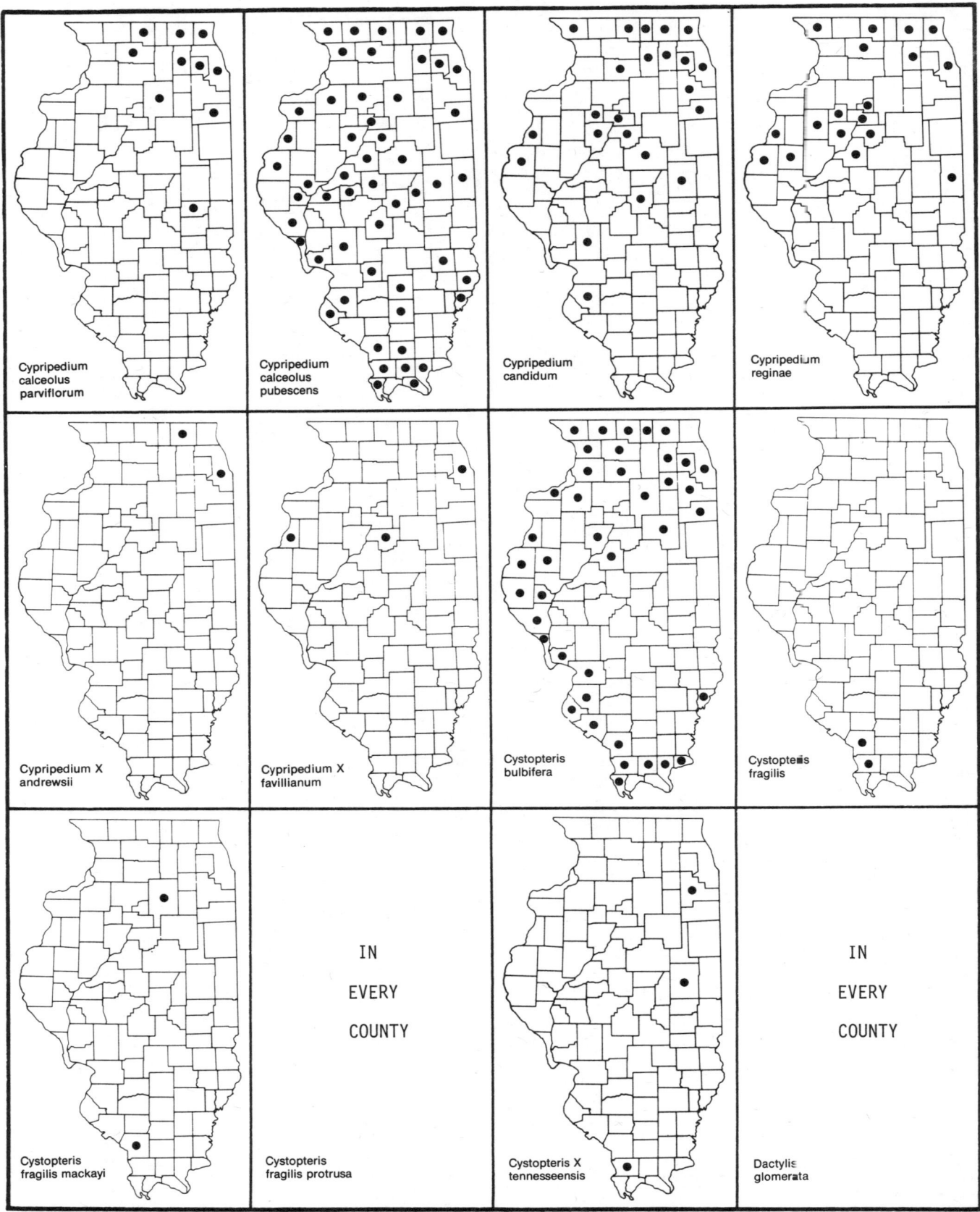

Cypripedium
calceolus
parviflorum

Cypripedium
calceolus
pubescens

Cypripedium
candidum

Cypripedium
reginae

Cypripedium X
andrewsii

Cypripedium X
favillianum

Cystopteris
bulbifera

Cystopteris
fragilis

Cystopteris
fragilis mackayi

Cystopteris
fragilis protrusa

IN

EVERY

COUNTY

Cystopteris X
tennesseensis

Dactylis
glomerata

IN

EVERY

COUNTY

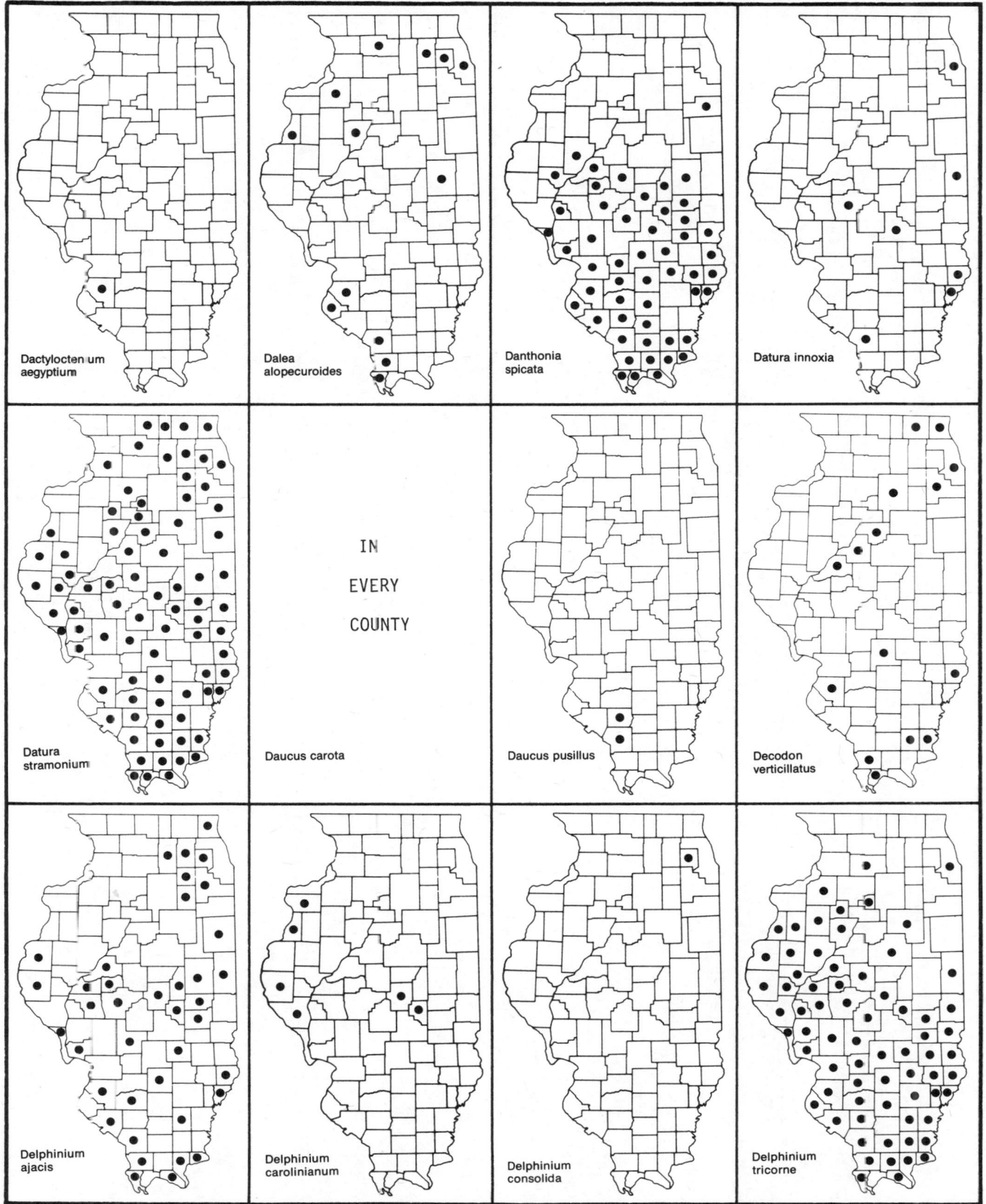

Dactyloctenium
aegyptium

Dalea
alopecuroides

Danthonia
spicata

Datura innoxia

Datura
stramonium

Daucus carota

IN

EVERY

COUNTY

Daucus pusillus

Decodon
verticillatus

Delphinium
ajacis

Delphinium
carolinianum

Delphinium
consolida

Delphinium
tricorne

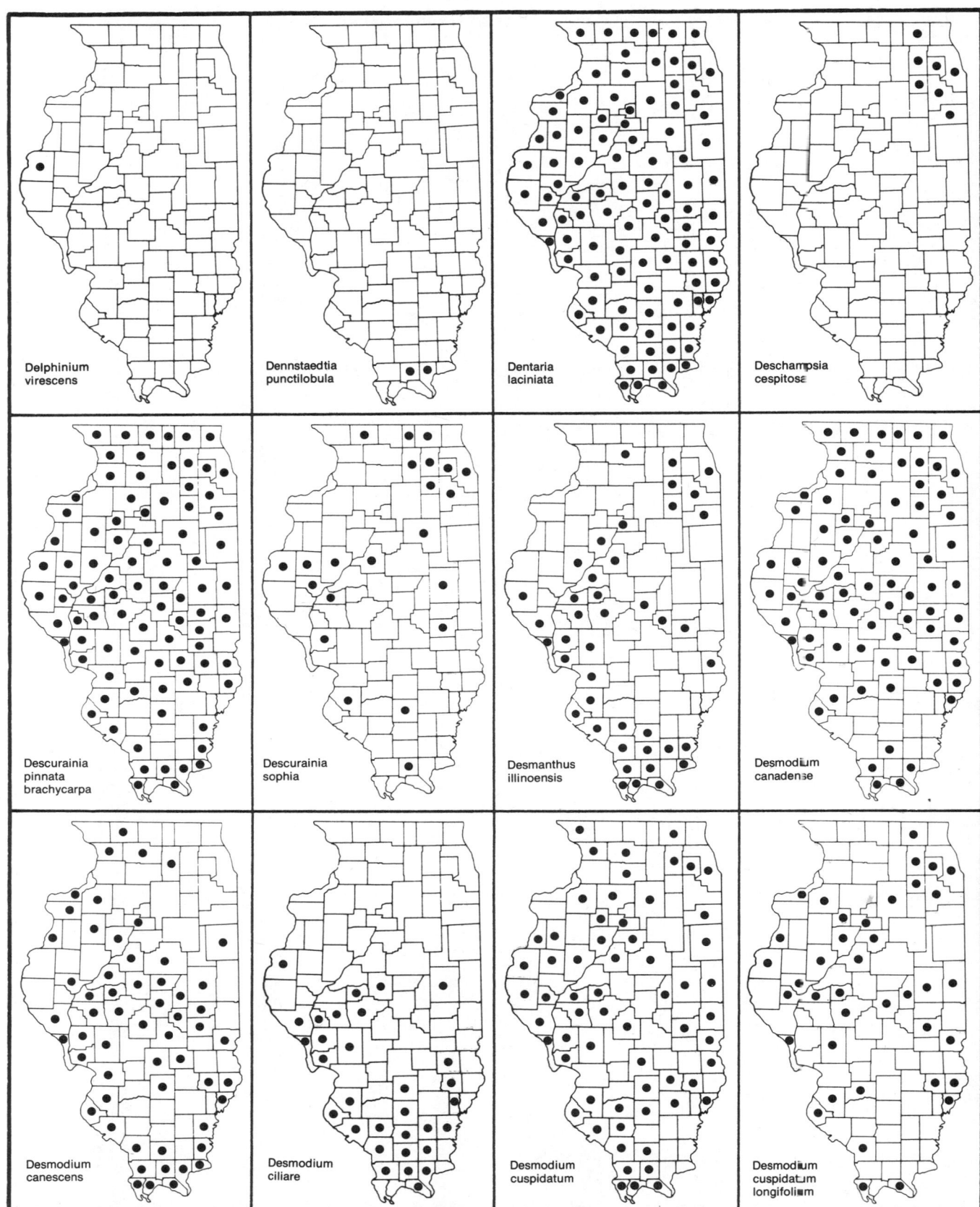

Delphinium
virescens

Dennstaedtia
punctilobula

Dentaria
laciniata

Deschampsia
cespitosa

Descurainia
pinnata
brachycarpa

Descurainia
sophia

Desmanthus
illinoensis

Desmodium
canadense

Desmodium
canescens

Desmodium
ciliare

Desmodium
cuspidatum

Desmodium
cuspidatum
longifolium

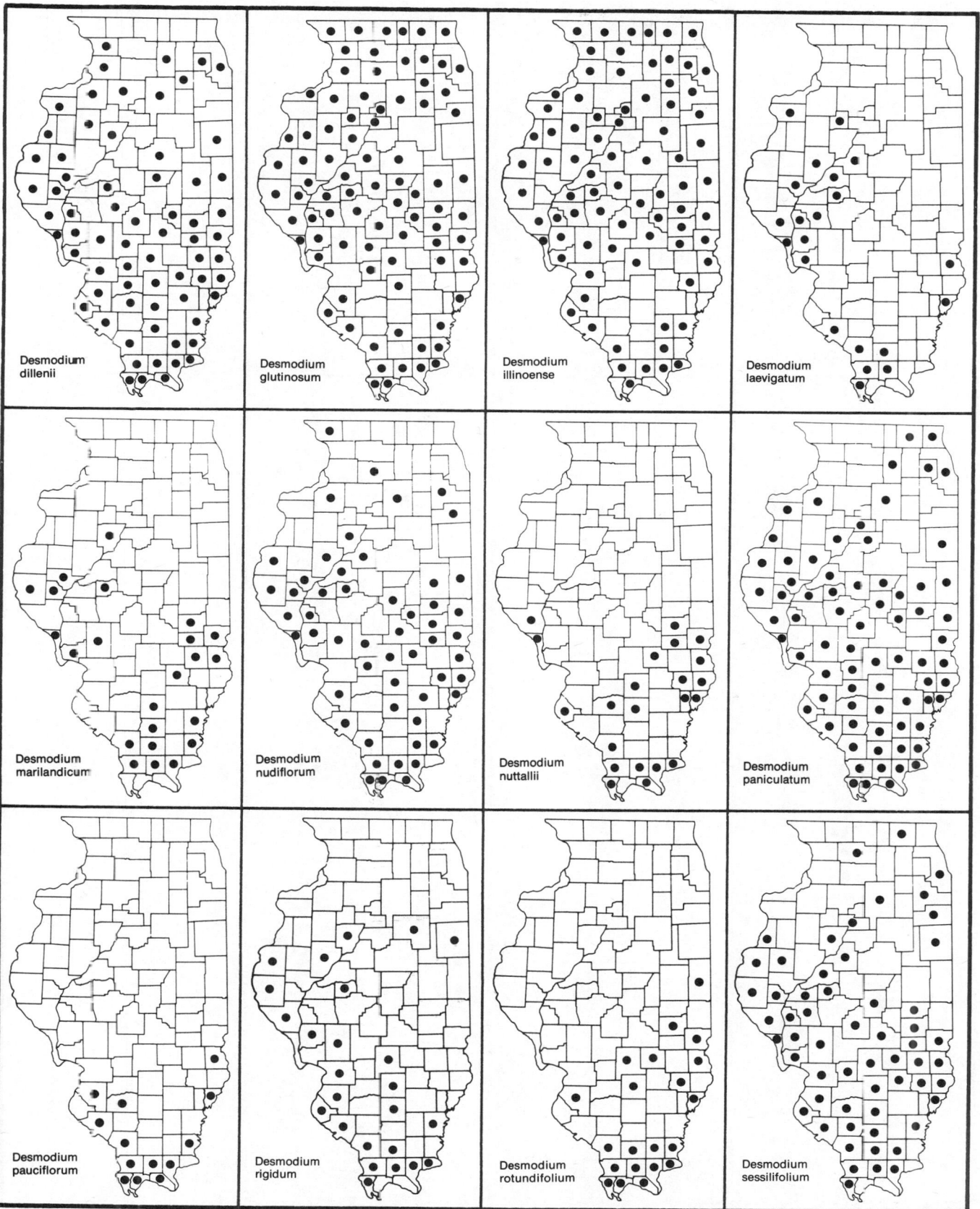

Desmodium
dillenii

Desmodium
glutinosum

Desmodium
illinoense

Desmodium
laevigatum

Desmodium
marilandicum

Desmodium
nudiflorum

Desmodium
nuttallii

Desmodium
paniculatum

Desmodium
pauciflorum

Desmodium
rigidum

Desmodium
rotundifolium

Desmodium
sessilifolium

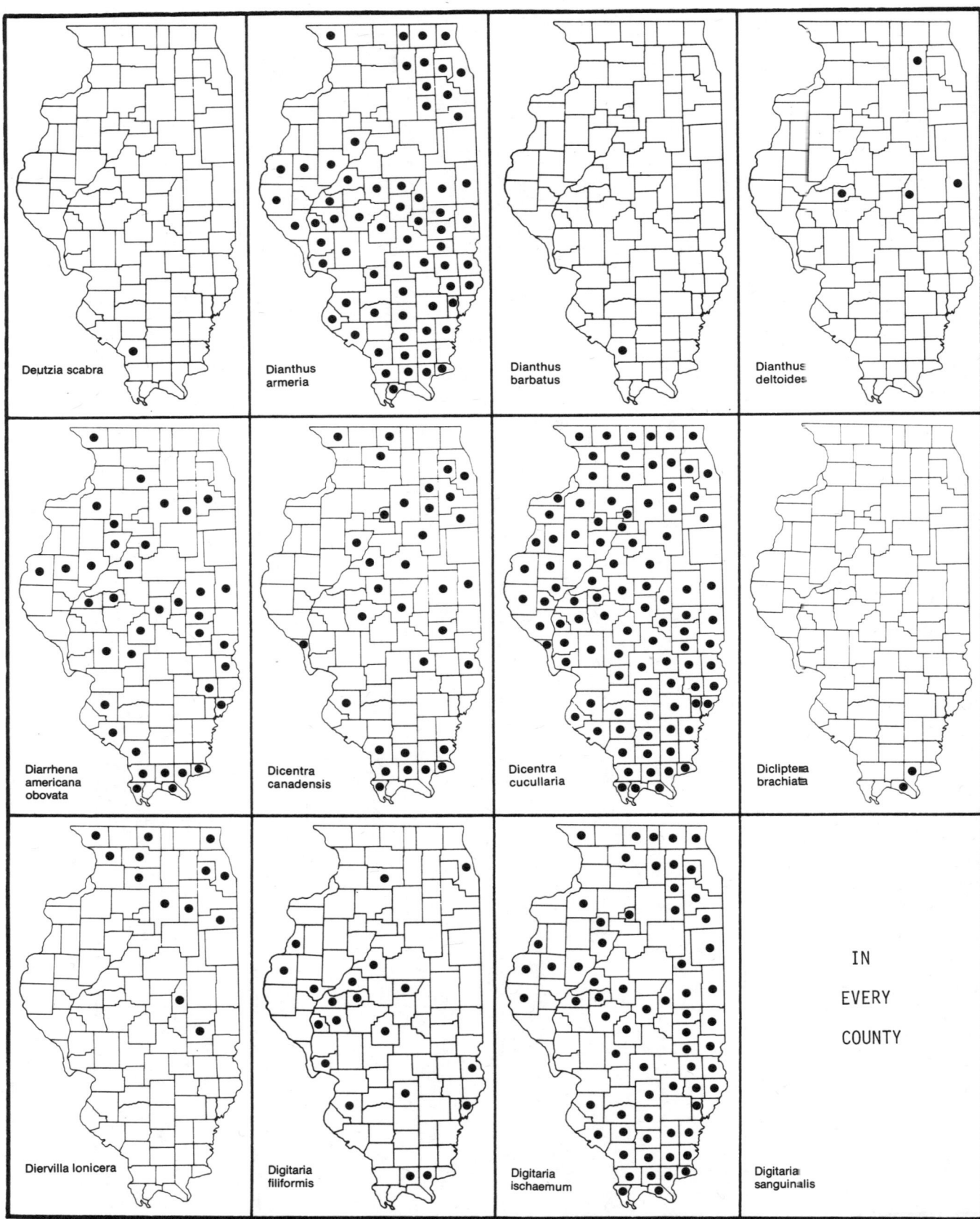

Deutzia scabra

Dianthus armeria

Dianthus barbatus

Dianthus deltoides

Diarrhena americana obovata

Dicentra canadensis

Dicentra cucullaria

Dicliptera brachiata

Diervilla lonicera

Digitaria filiformis

Digitaria ischaemum

Digitaria sanguinalis

IN EVERY COUNTY

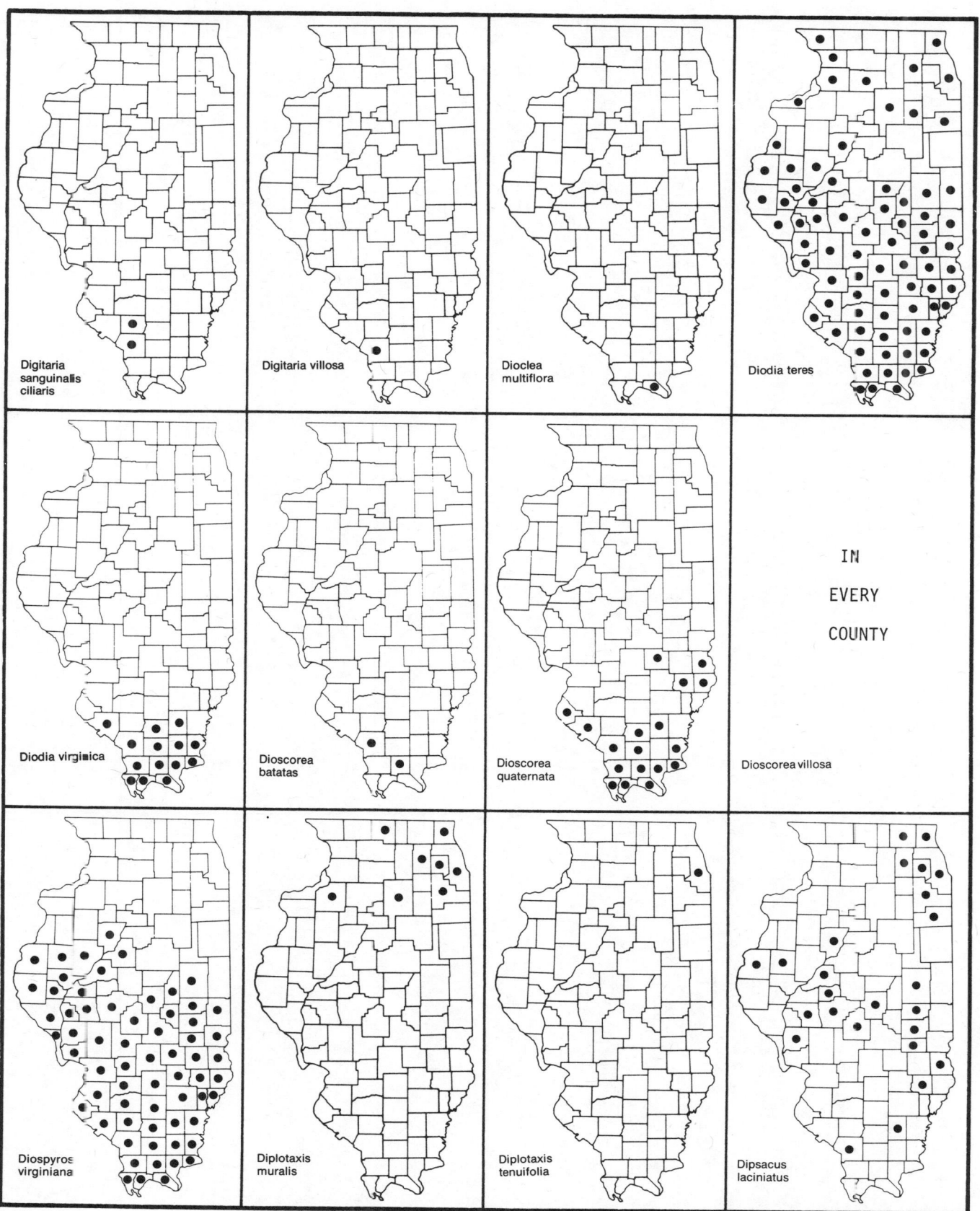

Digitaria
sanguinalis
ciliaris

Digitaria villosa

Dioclea
multiflora

Diodia teres

Diodia virginica

Dioscorea
batatas

Dioscorea
quaternata

Dioscorea villosa

IN

EVERY

COUNTY

Diospyros
virginiana

Diplotaxis
muralis

Diplotaxis
tenuifolia

Dipsacus
laciniatus

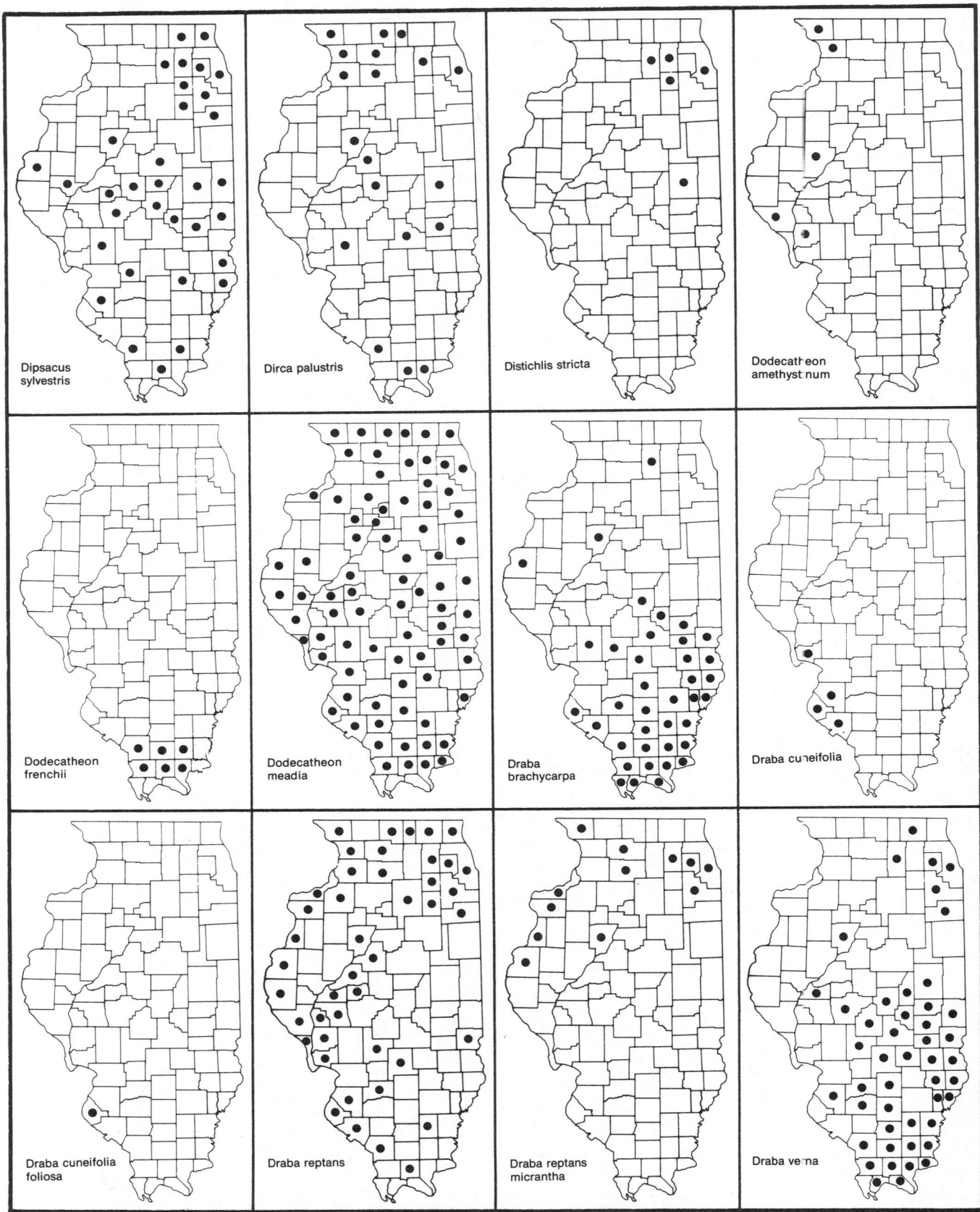

Dipsacus
sylvestris

Dirca palustris

Distichlis stricta

Dodecatheon
amethystinum

Dodecatheon
frenchii

Dodecatheon
meadia

Draba
brachycarpa

Draba cuneifolia

Draba cuneifolia
foliosa

Draba reptans

Draba reptans
micrantha

Draba verna

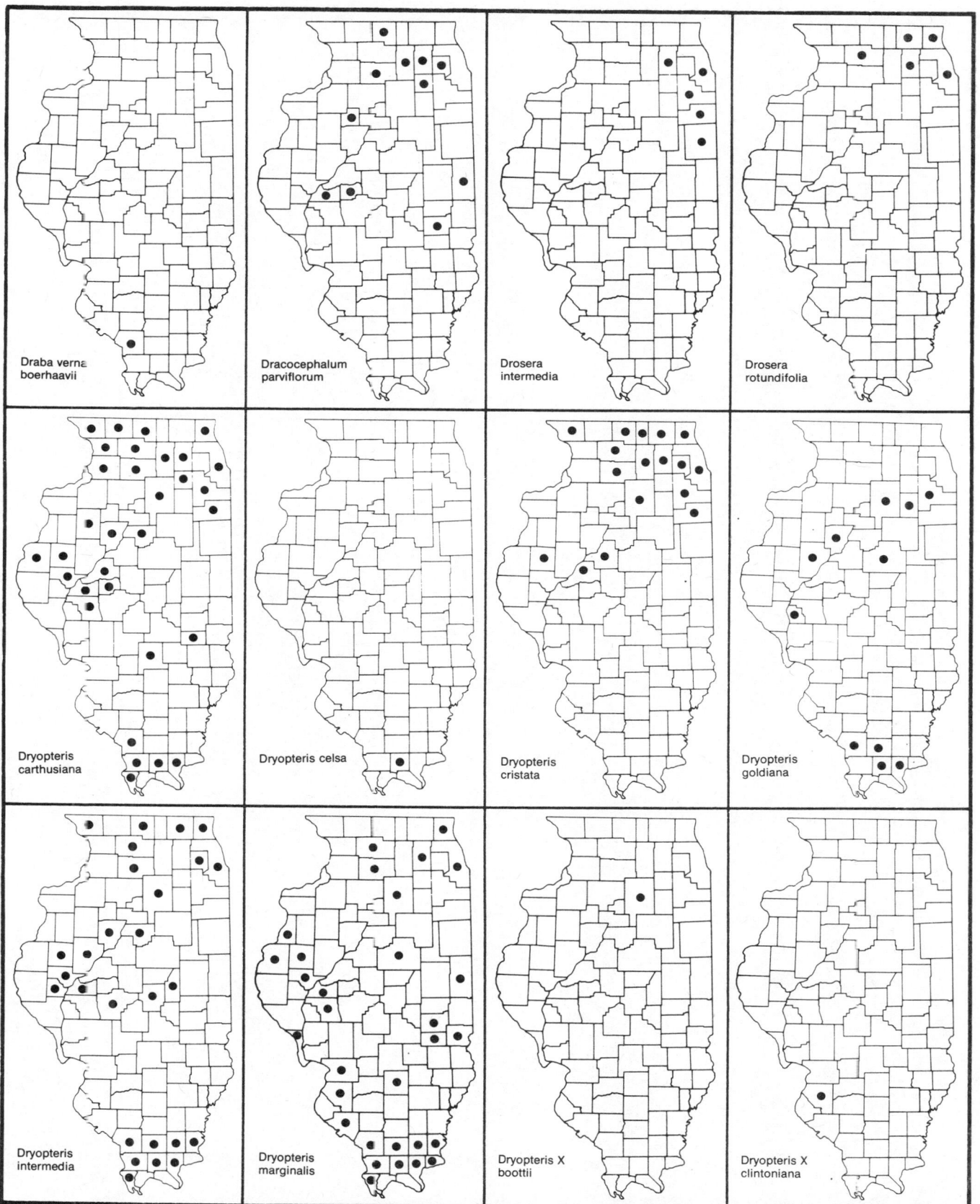

Draba verna
boerhaavii

Dracocephalum
parviflorum

Drosera
intermedia

Drosera
rotundifolia

Dryopteris
carthusiana

Dryopteris celsa

Dryopteris
cristata

Dryopteris
goldiana

Dryopteris
intermedia

Dryopteris
marginalis

Dryopteris X
boottii

Dryopteris X
clintoniana

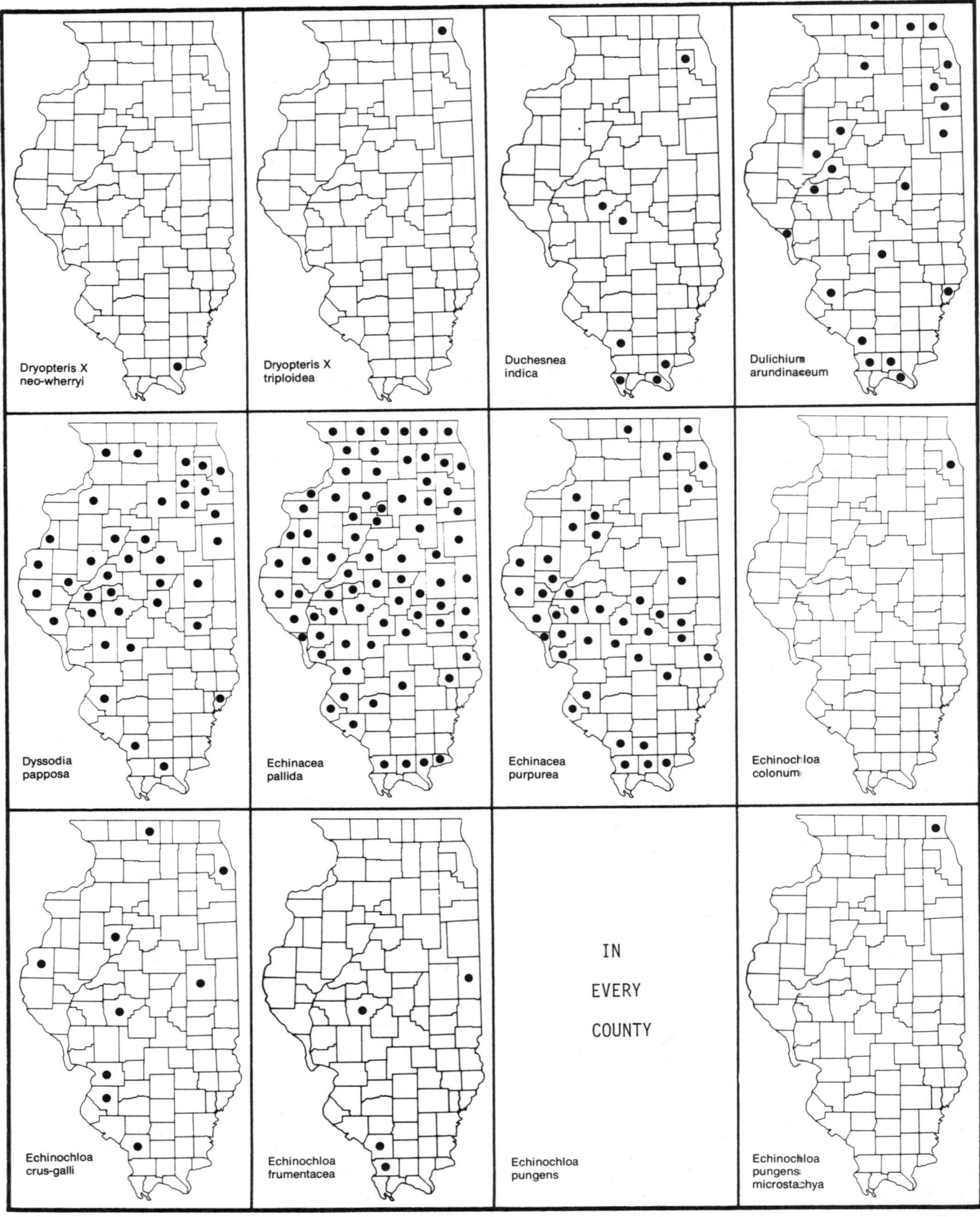

Dryopteris X
neo-wherryi

Dryopteris X
triploidea

Duchesnea
indica

Dulichium
arundinaceum

Dyssodia
papposa

Echinacea
pallida

Echinacea
purpurea

Echinochloa
colonum

Echinochloa
crus-galli

Echinochloa
frumentacea

Echinochloa
pungens

IN

EVERY

COUNTY

Echinochloa
pungens
microstachya

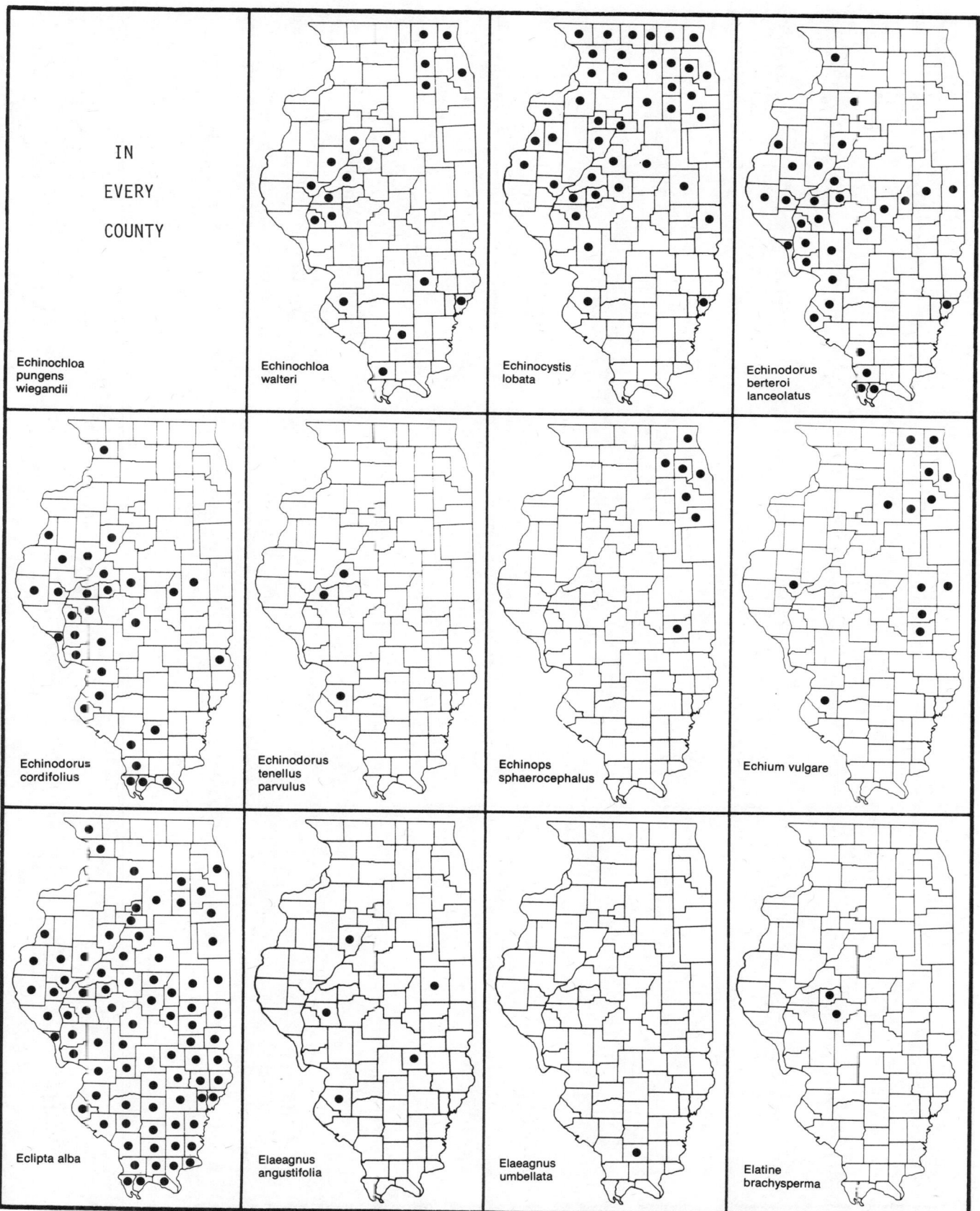

IN
EVERY
COUNTY

Echinochloa
pungens
wiegandii

Echinochloa
walteri

Echinocystis
lobata

Echinodorus
berteroi
lanceolatus

Echinodorus
cordifolius

Echinodorus
tenellus
parvulus

Echinops
sphaerocephalus

Echium vulgare

Eclipta alba

Elaeagnus
angustifolia

Elaeagnus
umbellata

Elatine
brachysperma

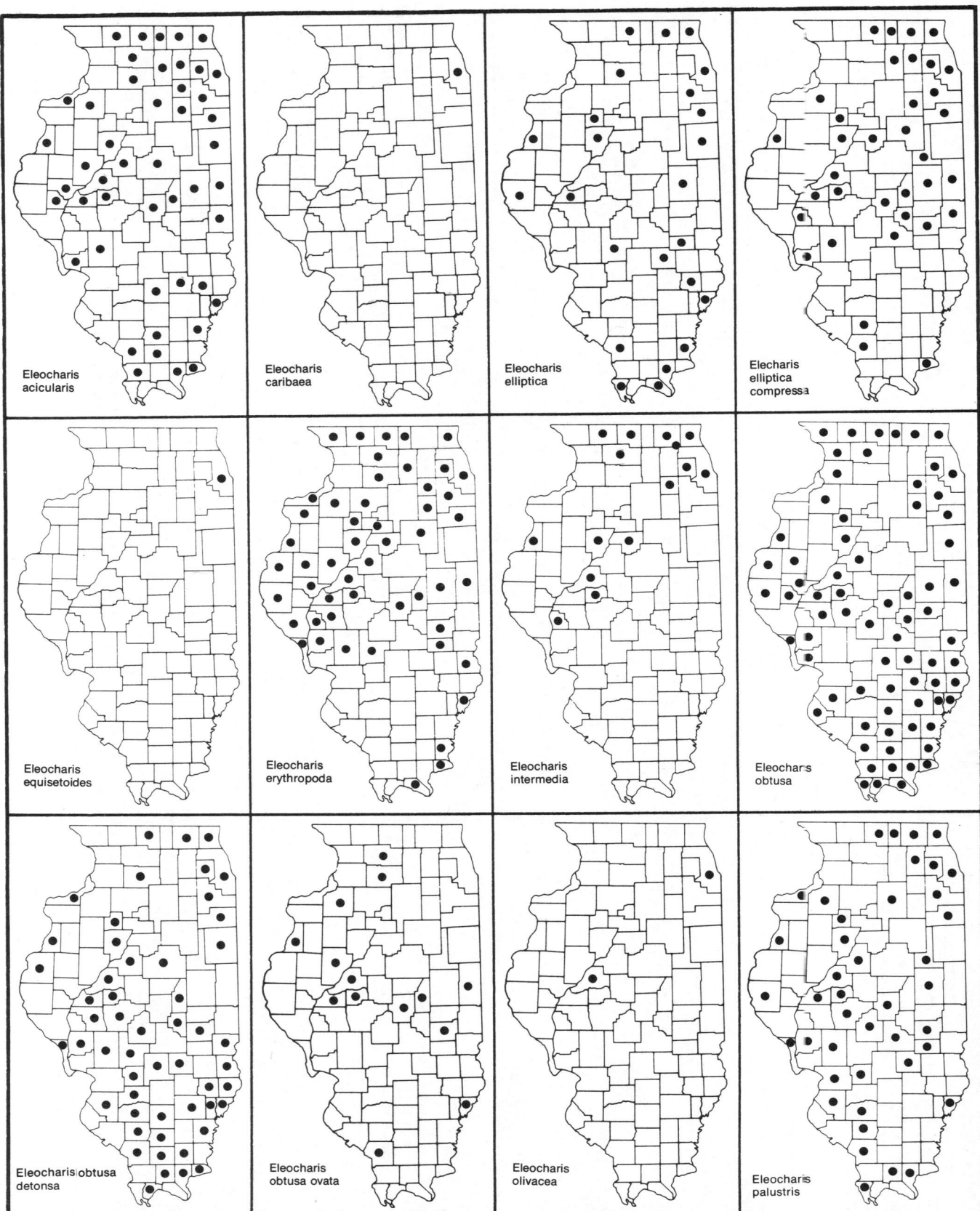

Eleocharis
acicularis

Eleocharis
caribaea

Eleocharis
elliptica

Elecharis
elliptica
compressa

Eleocharis
equisetoides

Eleocharis
erythropoda

Eleocharis
intermedia

Eleocharis
obtusa

Eleocharis obtusa
detonsa

Eleocharis
obtusa ovata

Eleocharis
olivacea

Eleocharis
palustris

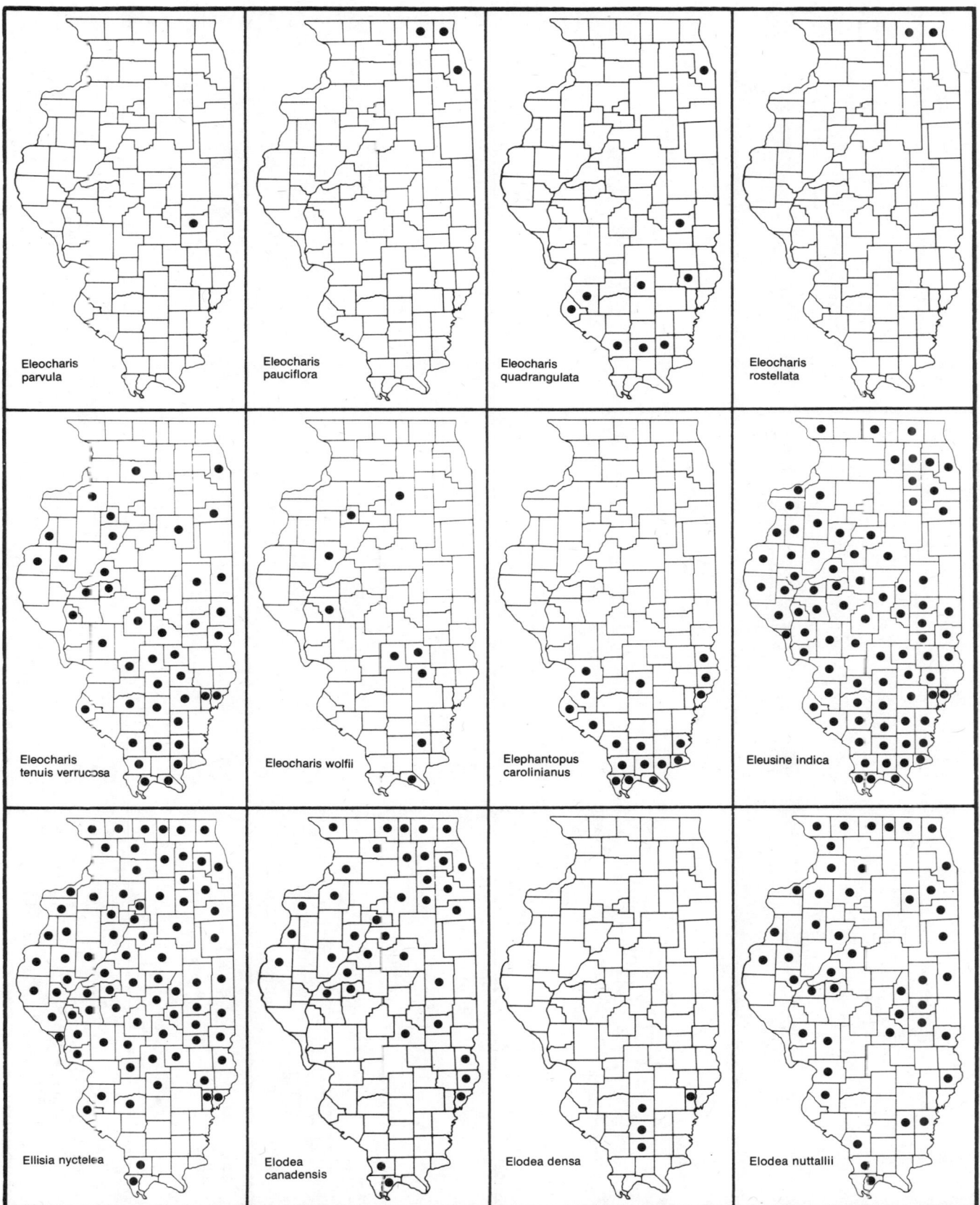

Eleocharis
parvula

Eleocharis
pauciflora

Eleocharis
quadrangulata

Eleocharis
rostellata

Eleocharis
tenuis verrucosa

Eleocharis wolfii

Elephantopus
carolinianus

Eleusine indica

Ellisia nyctelea

Elodea
canadensis

Elodea densa

Elodea nuttallii

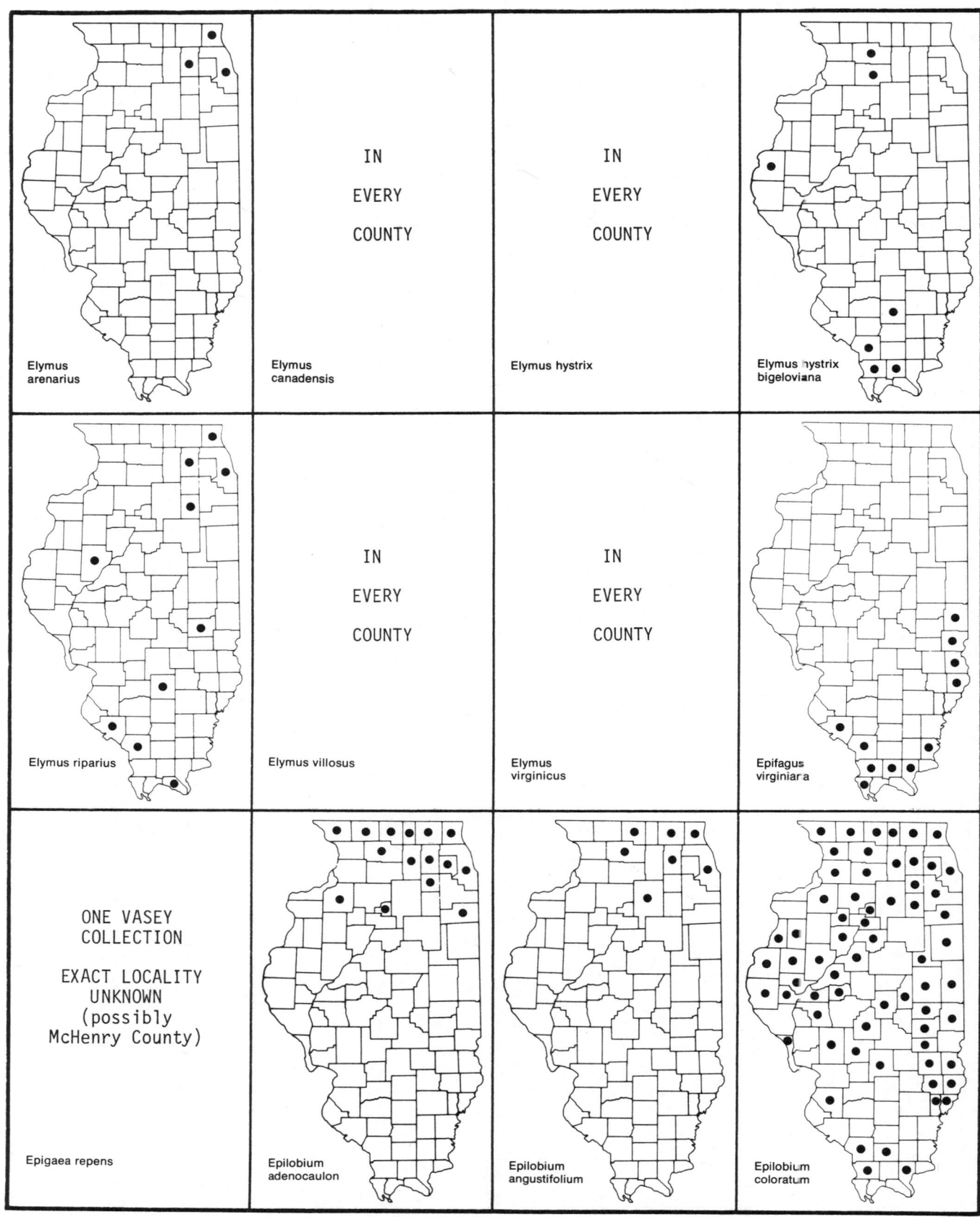

Elymus
arenarius

Elymus
canadensis

IN
EVERY
COUNTY

Elymus hystrix

IN
EVERY
COUNTY

Elymus hystrix
bigeloviana

Elymus riparius

Elymus villosus

IN
EVERY
COUNTY

Elymus
virginicus

IN
EVERY
COUNTY

Epifagus
virginiana

Epigaea repens

ONE VASEY
COLLECTION

EXACT LOCALITY
UNKNOWN
(possibly
McHenry County)

Epilobium
adenocaulon

Epilobium
angustifolium

Epilobium
coloratum

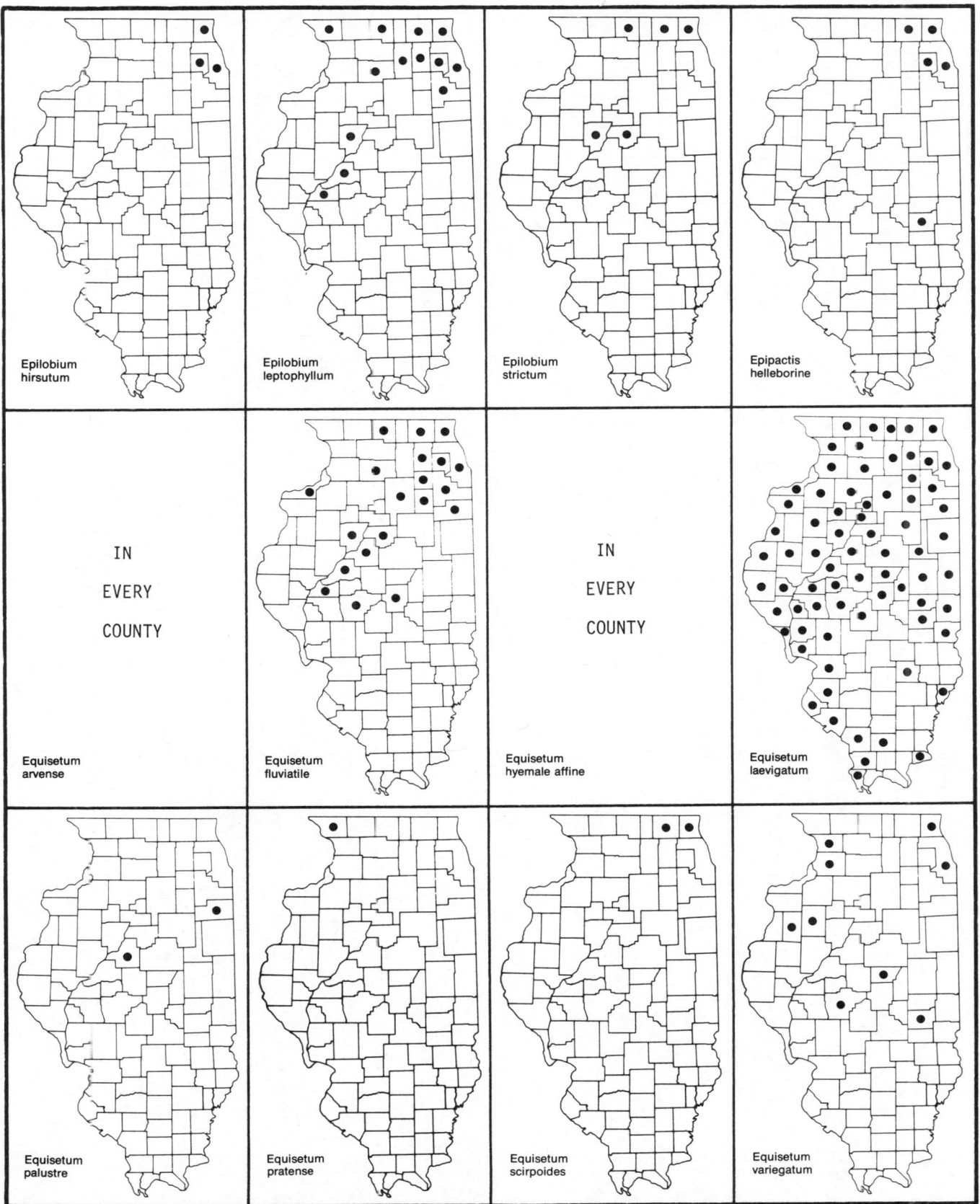

Epilobium
hirsutum

Epilobium
leptophyllum

Epilobium
strictum

Epipactis
helleborine

IN

EVERY

COUNTY

Equisetum
arvense

Equisetum
fluviatile

IN

EVERY

COUNTY

Equisetum
hyemale affine

Equisetum
laevigatum

Equisetum
palustre

Equisetum
pratense

Equisetum
scirpoides

Equisetum
variegatum

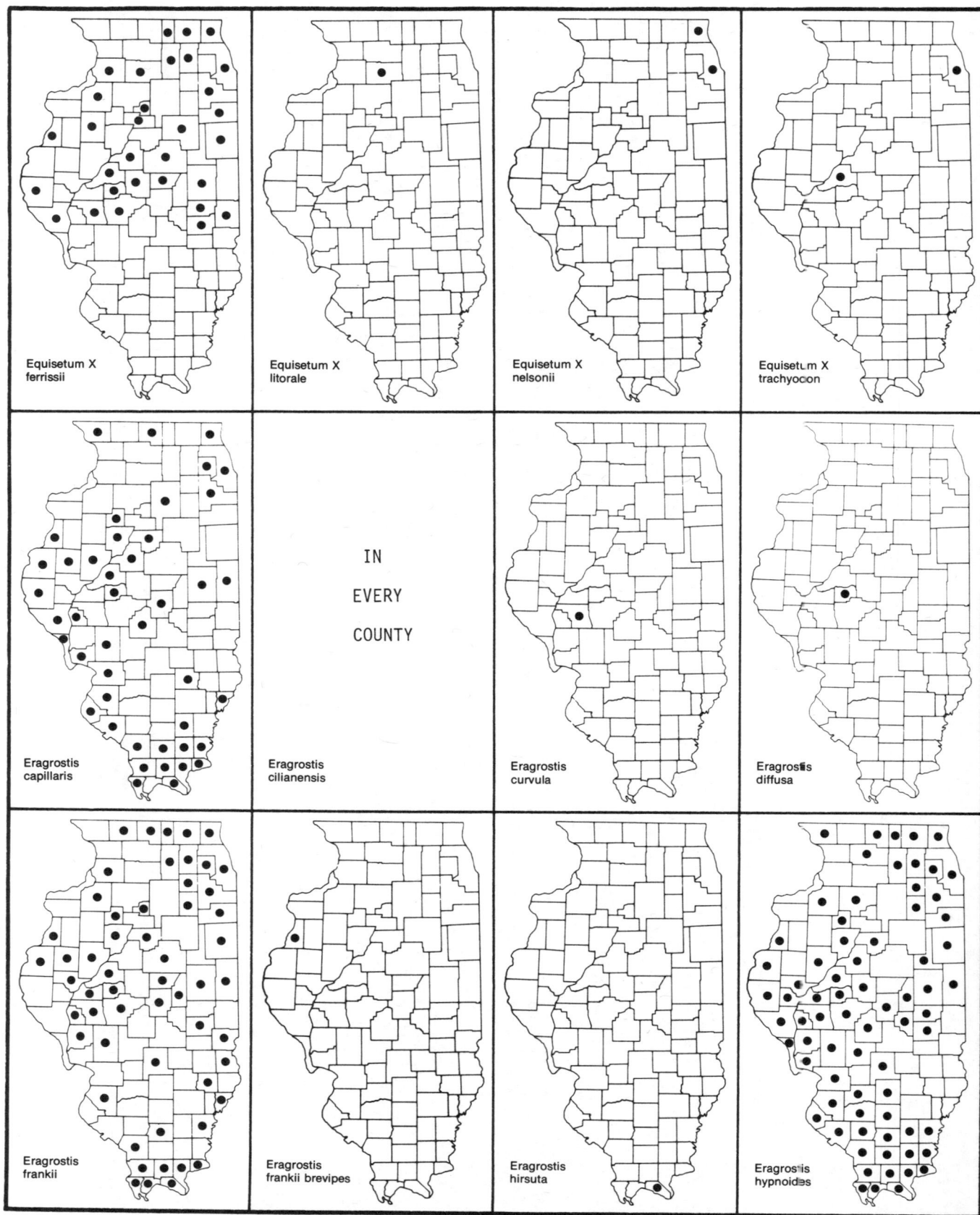

Equisetum X
ferrissii

Equisetum X
litorale

Equisetum X
nelsonii

Equisetum X
trachyocon

Eragrostis
capillaris

Eragrostis
cilianensis

IN

EVERY

COUNTY

Eragrostis
curvula

Eragrostis
diffusa

Eragrostis
frankii

Eragrostis
frankii brevipes

Eragrostis
hirsuta

Eragrostis
hypnoides

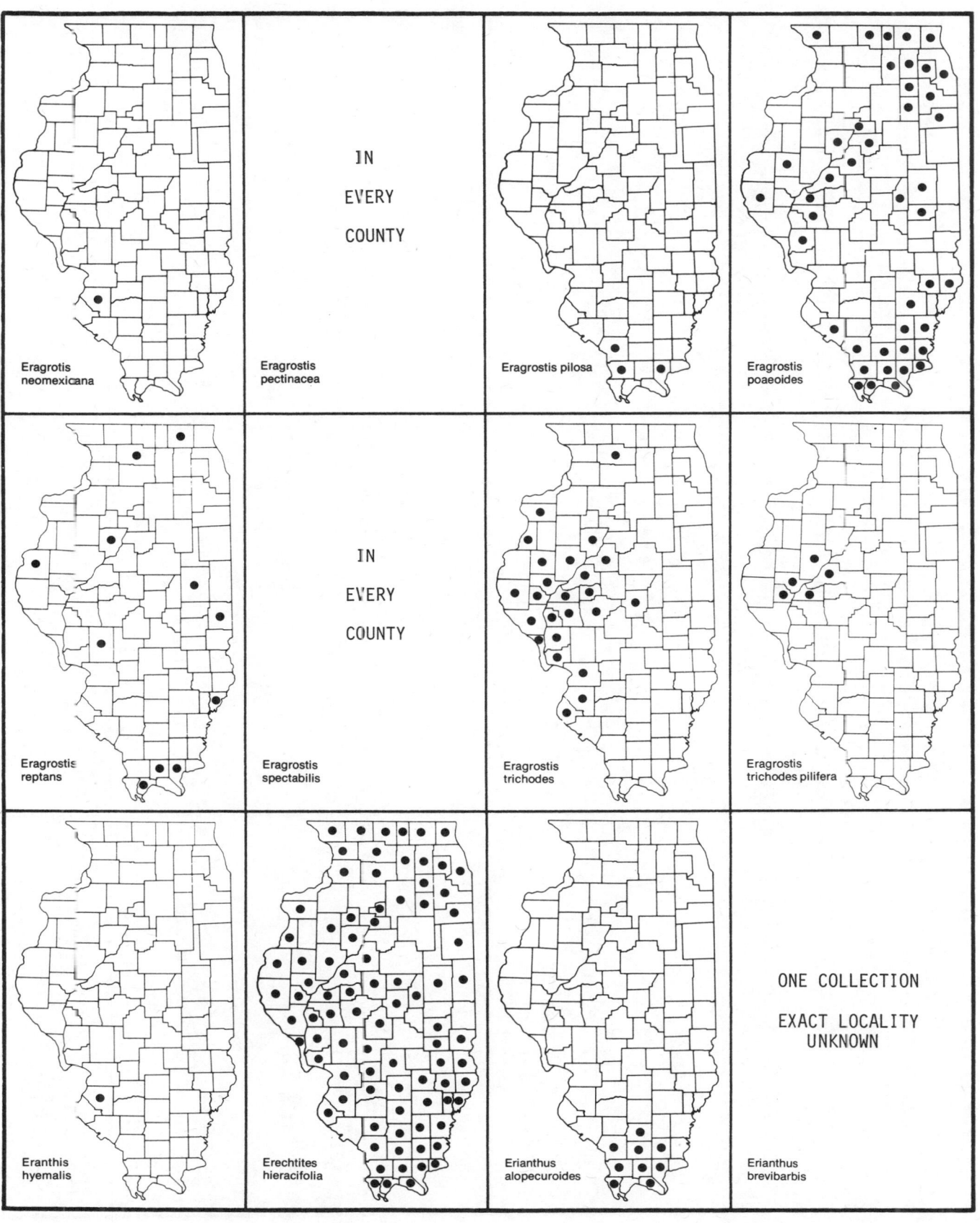

Eragrotis
neomexicana

Eragrostis
pectinacea

IN

EVERY

COUNTY

Eragrostis pilosa

Eragrostis
poaeoides

Eragrostis
reptans

Eragrostis
spectabilis

IN

EVERY

COUNTY

Eragrostis
trichodes

Eragrostis
trichodes pilifera

Eranthis
hyemalis

Erechtites
hieracifolia

Erianthus
alopecuroides

Erianthus
brevibarbis

ONE COLLECTION

EXACT LOCALITY
UNKNOWN

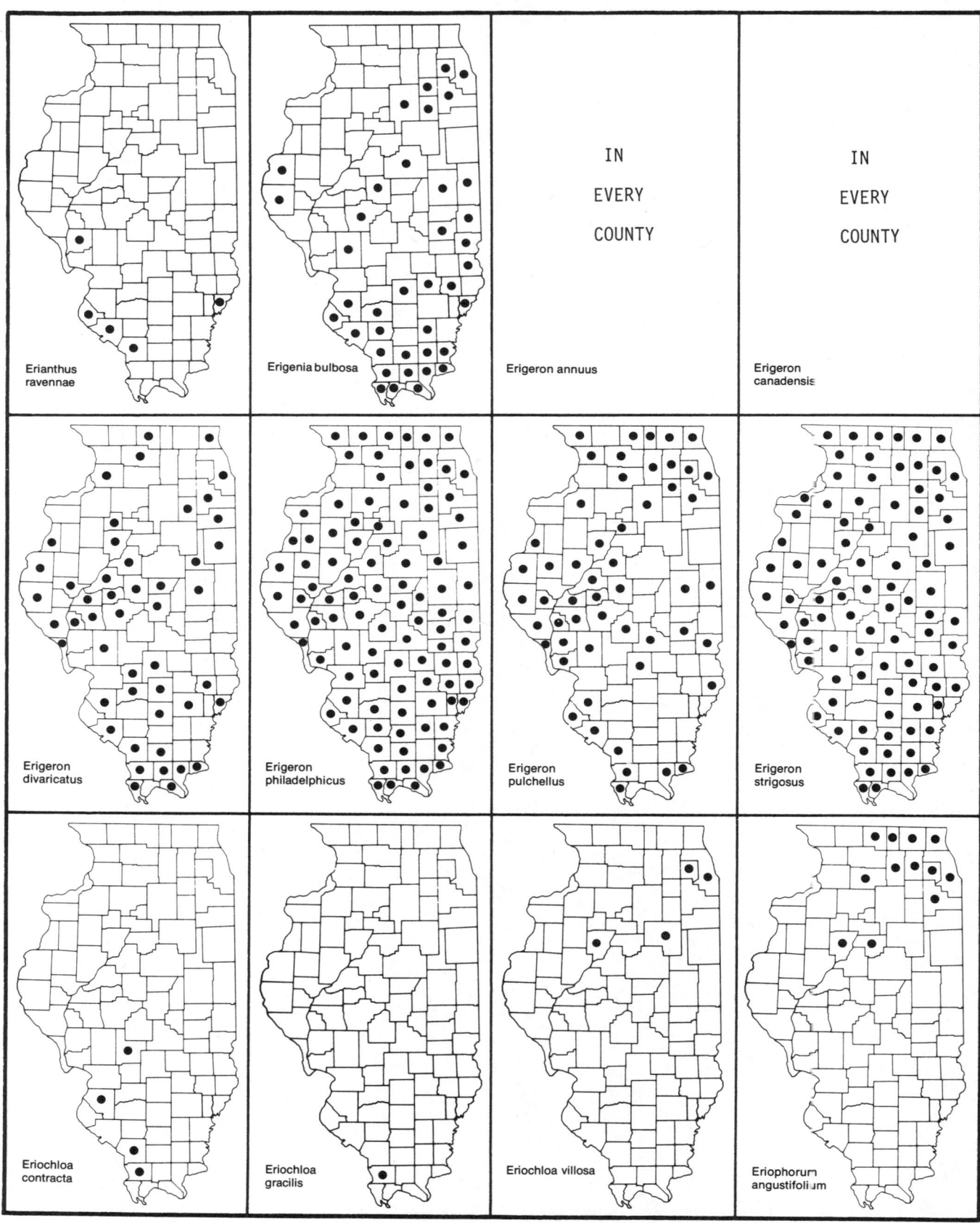

Erianthus
ravennae

Erigenia bulbosa

IN

EVERY

COUNTY

Erigeron annuus

IN

EVERY

COUNTY

Erigeron
canadensis

Erigeron
divaricatus

Erigeron
philadelphicus

Erigeron
pulchellus

Erigeron
strigosus

Eriochloa
contracta

Eriochloa
gracilis

Eriochloa villosa

Eriophorum
angustifolium

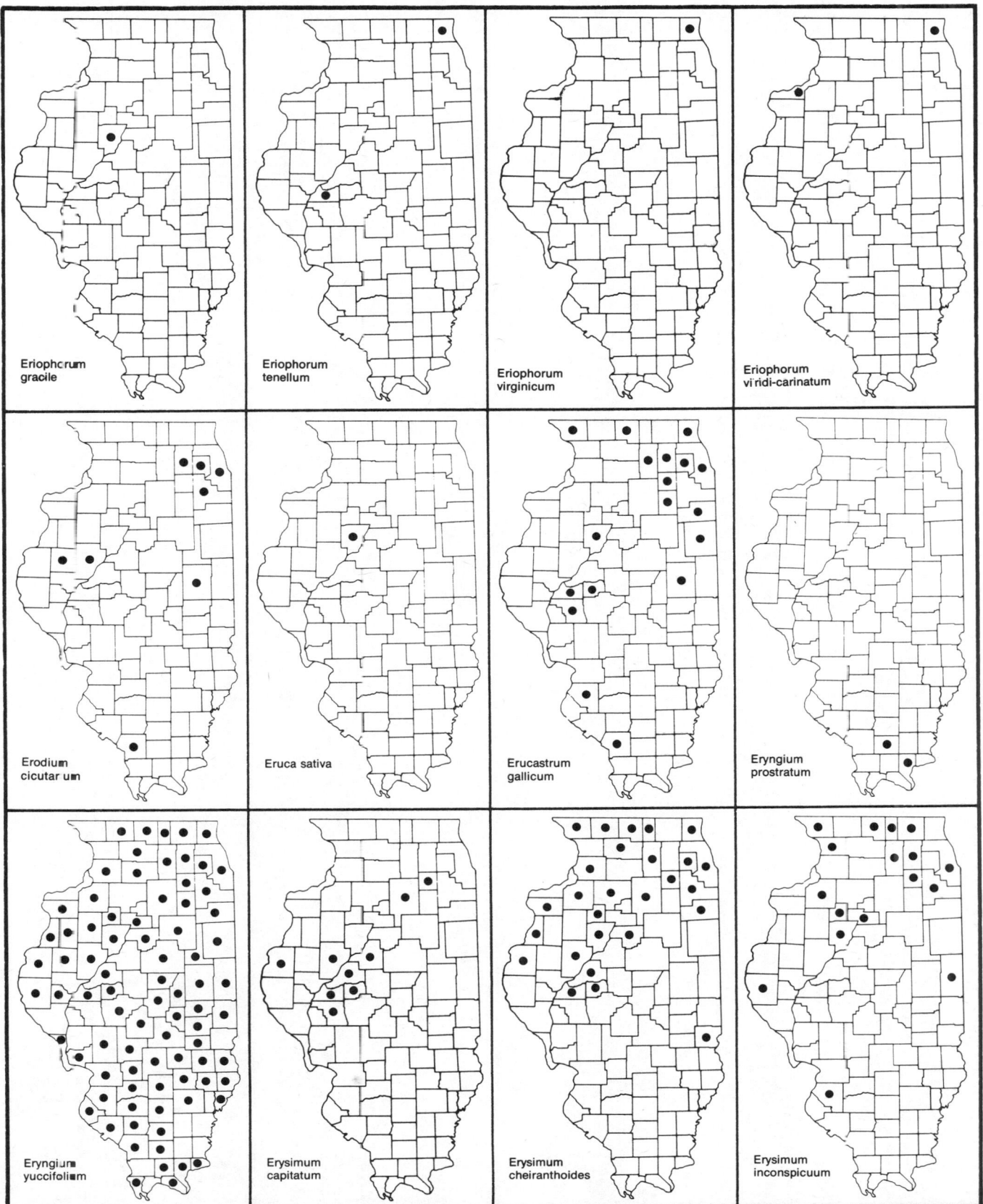

Eriophorum
gracile

Eriophorum
tenellum

Eriophorum
virginicum

Eriophorum
viridi-carinatum

Erodium
cicutarum

Eruca sativa

Erucastrum
gallicum

Eryngium
prostratum

Eryngium
yuccifolium

Erysimum
capitatum

Erysimum
cheiranthoides

Erysimum
inconspicuum

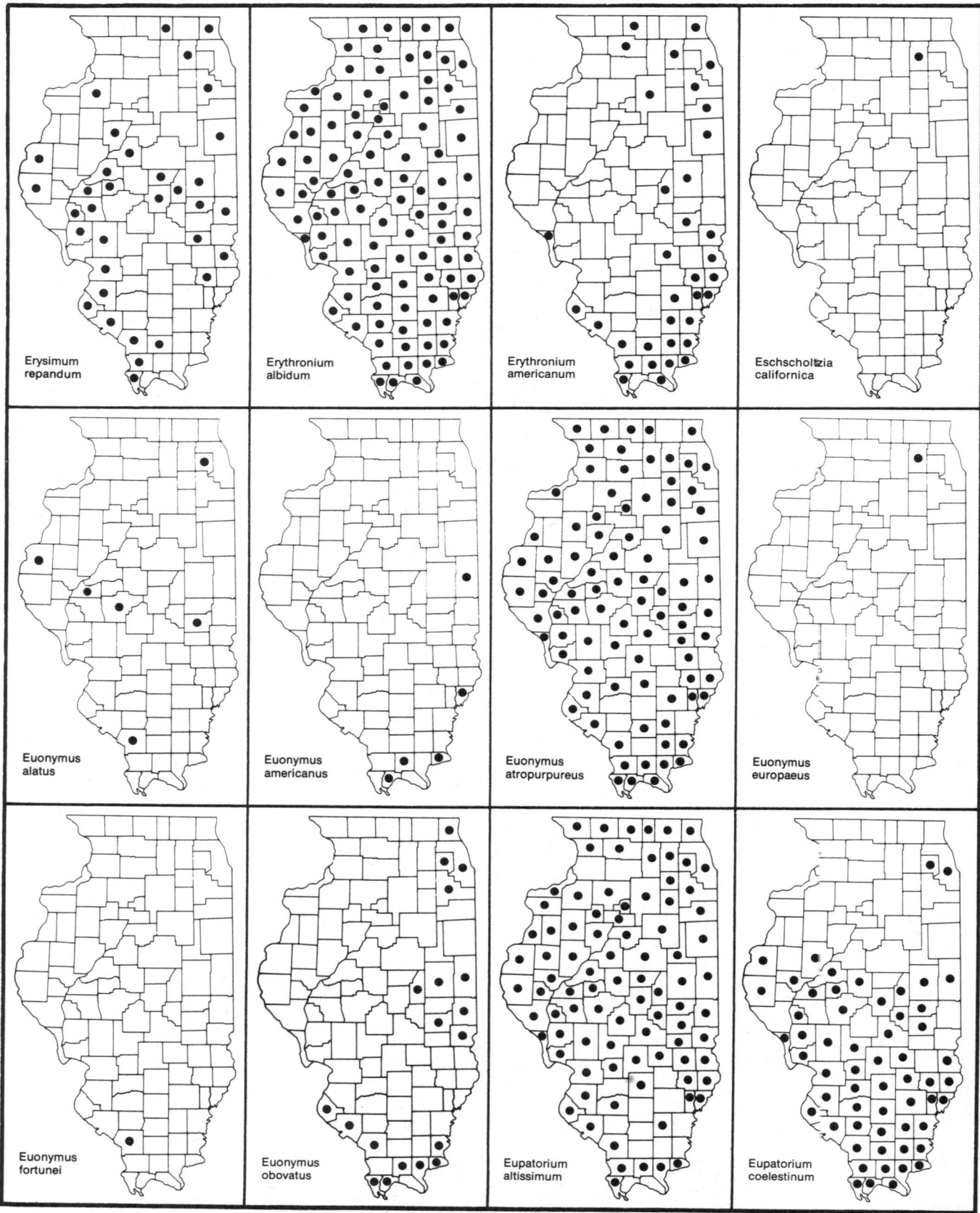

Erysimum
repandum

Erythronium
albidum

Erythronium
americanum

Eschscholtzia
californica

Euonymus
alatus

Euonymus
americanus

Euonymus
atropurpureus

Euonymus
europaeus

Euonymus
fortunei

Euonymus
obovatus

Eupatorium
altissimum

Eupatorium
coelestinum

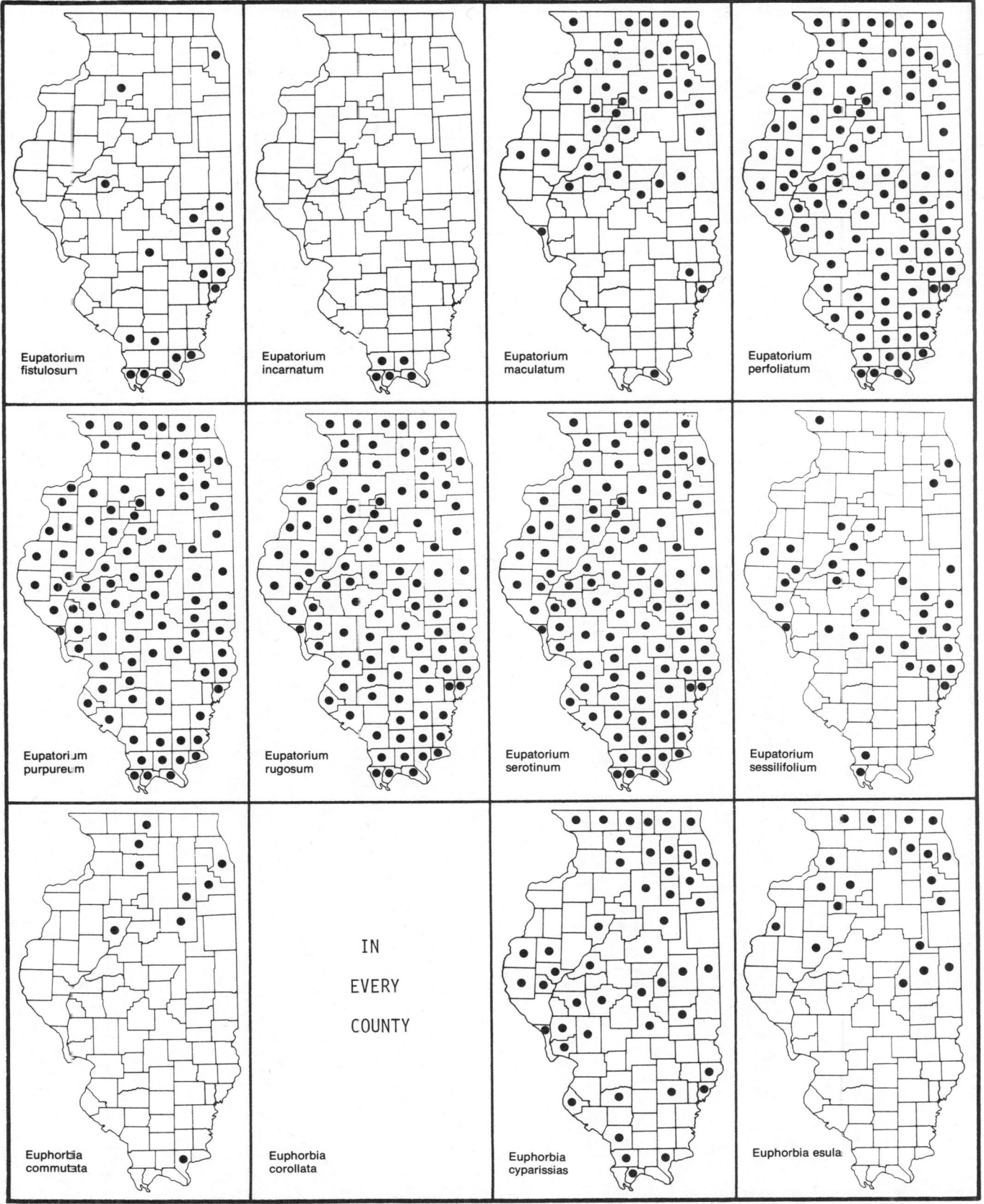

Eupatorium
fistulosum

Eupatorium
incarnatum

Eupatorium
maculatum

Eupatorium
perfoliatum

Eupatorium
purpureum

Eupatorium
rugosum

Eupatorium
serotinum

Eupatorium
sessilifolium

Euphorbia
commutata

Euphorbia
corollata

IN

EVERY

COUNTY

Euphorbia
cyparissias

Euphorbia esula

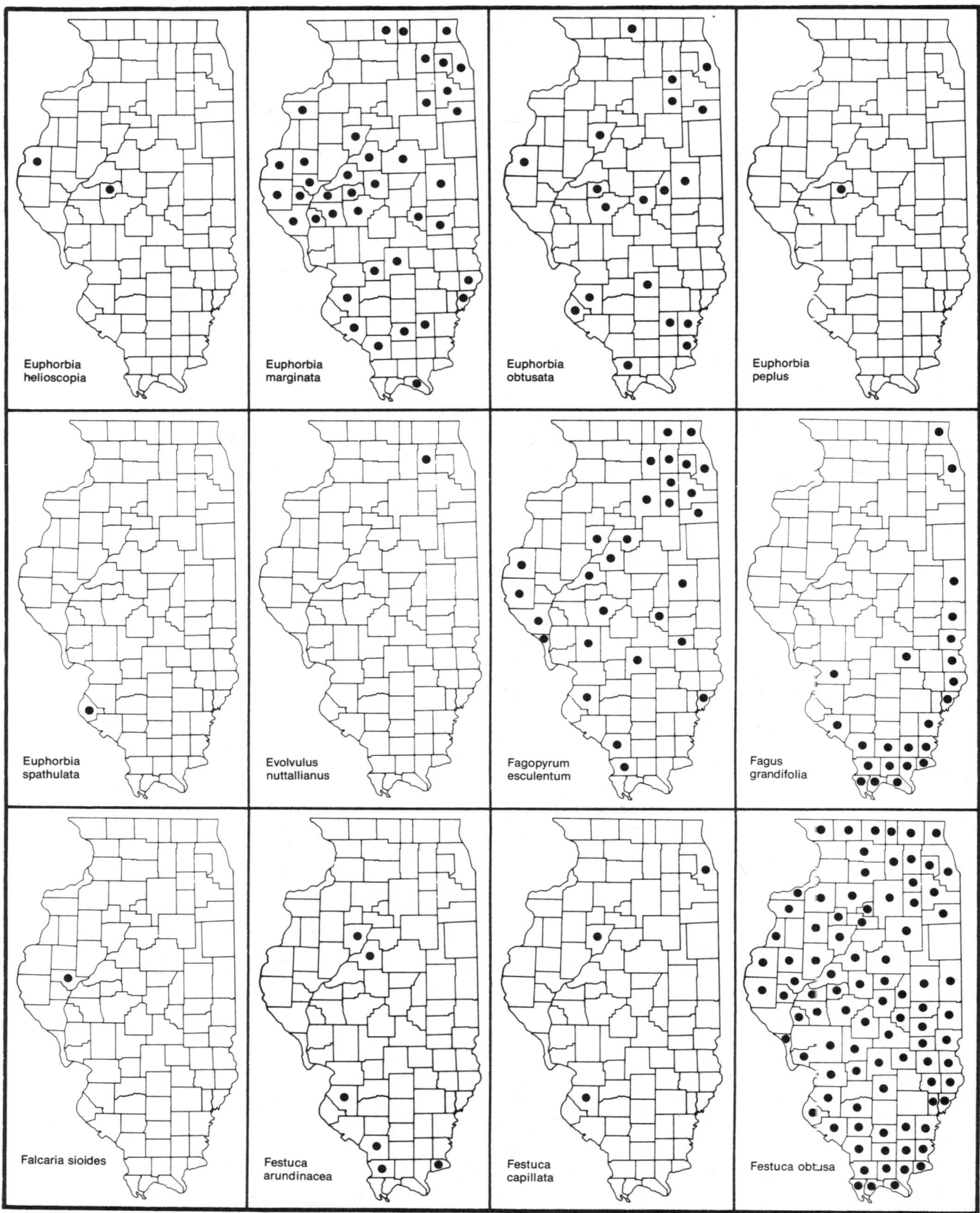

Euphorbia
helioscopia

Euphorbia
marginata

Euphorbia
obtusata

Euphorbia
peplus

Euphorbia
spathulata

Evolvulus
nuttallianus

Fagopyrum
esculentum

Fagus
grandifolia

Falcaria sioides

Festuca
arundinacea

Festuca
capillata

Festuca obtusa

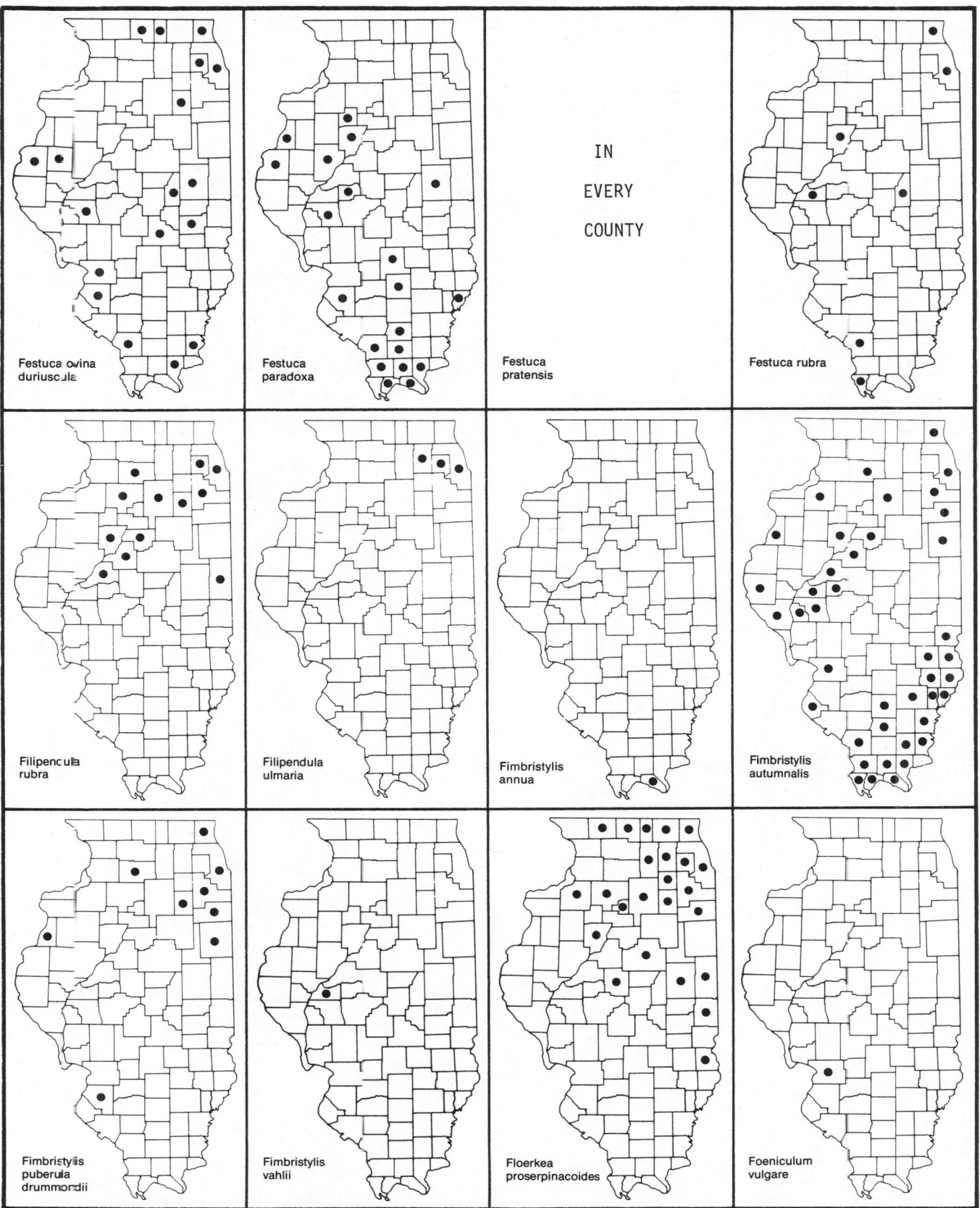

Festuca ovina
duriuscula

Festuca
paradoxa

IN

EVERY

COUNTY

Festuca
pratensis

Festuca rubra

Filipencula
rubra

Filipendula
ulmaria

Fimbristylis
annua

Fimbristylis
autumnalis

Fimbristylis
puberula
drummondii

Fimbristylis
vahlii

Floerkea
proserpinacoides

Foeniculum
vulgare

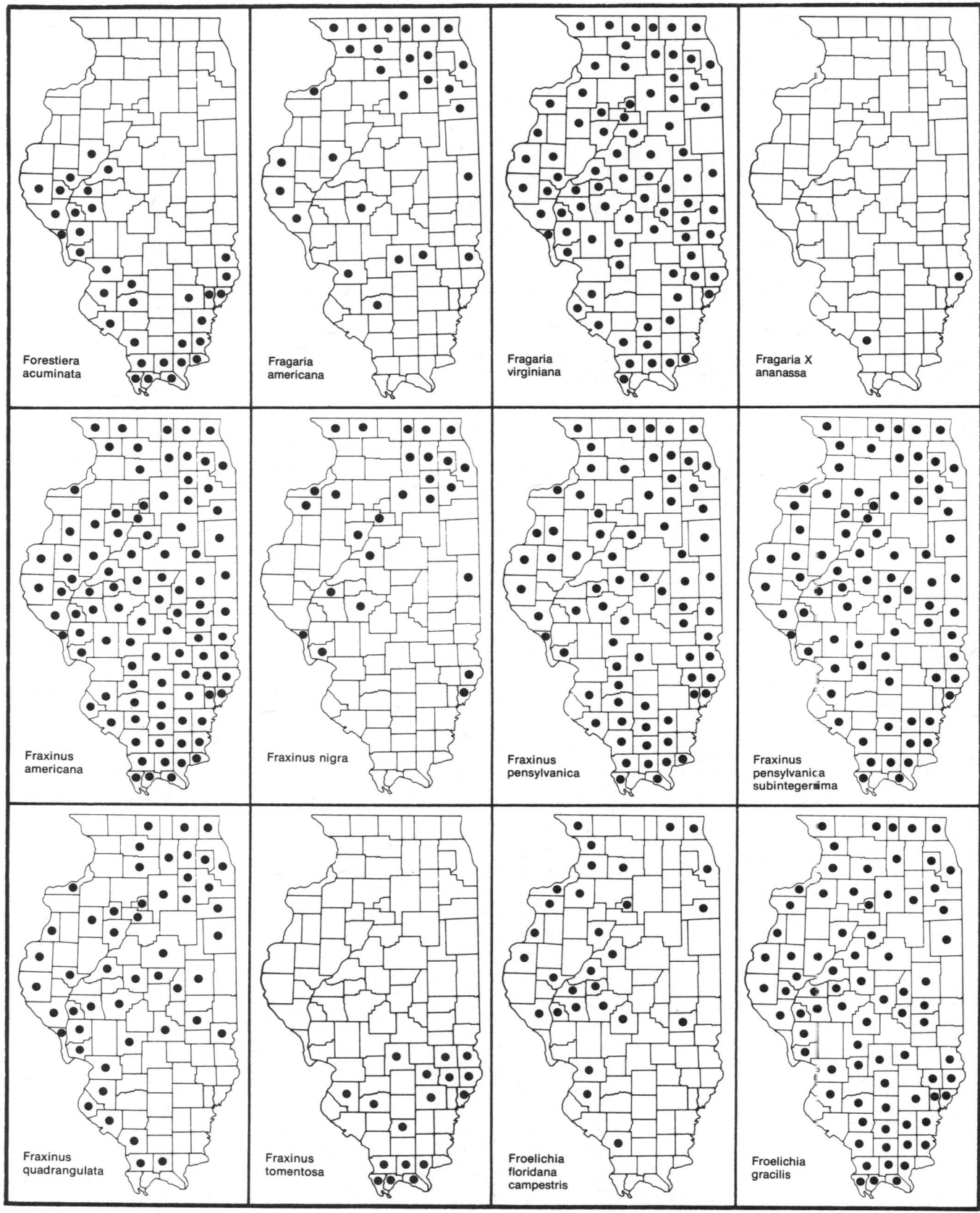

Forestiera
acuminata

Fragaria
americana

Fragaria
virginiana

Fragaria X
ananassa

Fraxinus
americana

Fraxinus nigra

Fraxinus
pensylvanica

Fraxinus
pensylvanica
subintegerrima

Fraxinus
quadrangulata

Fraxinus
tomentosa

Froelichia
floridana
campestris

Froelichia
gracilis

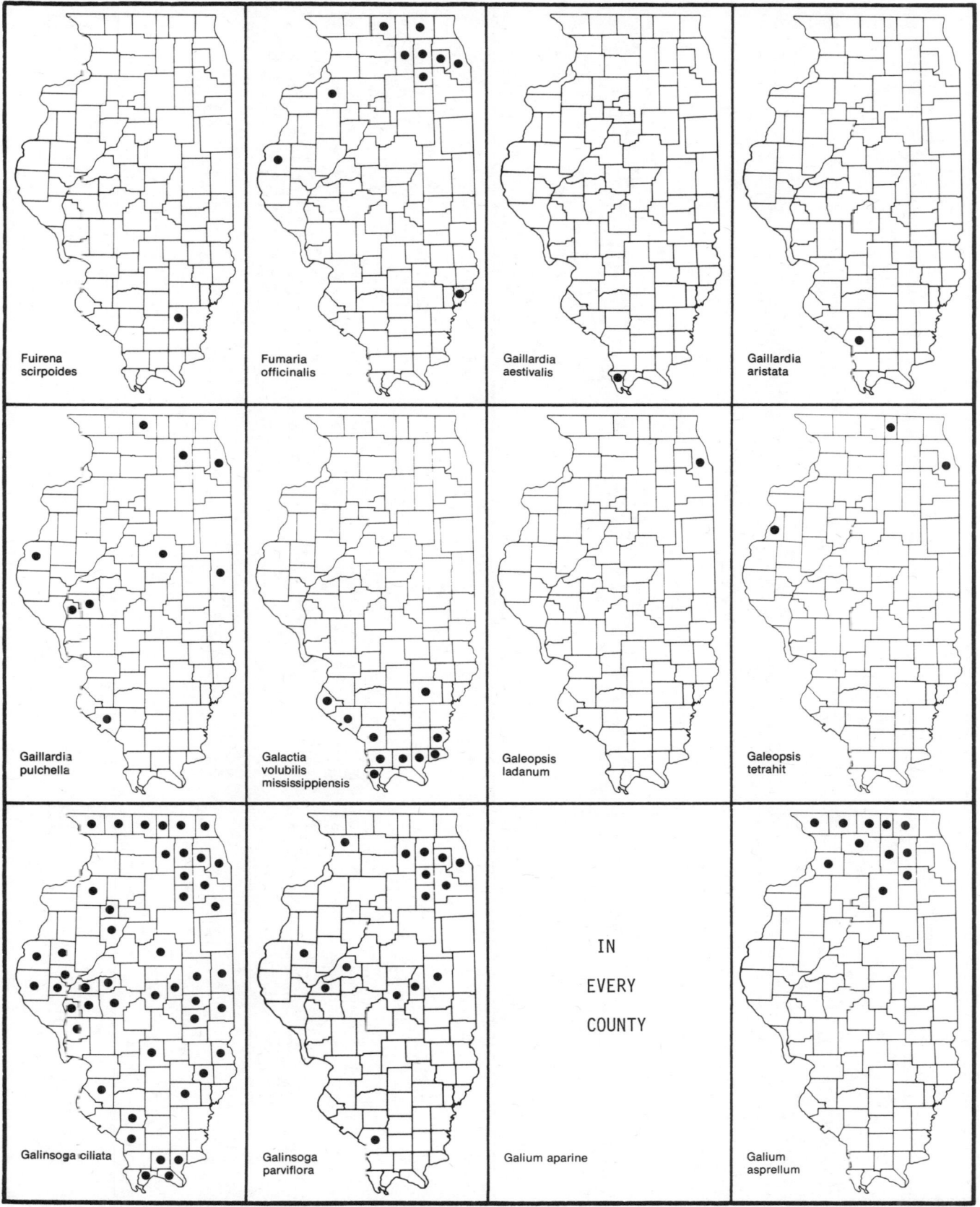

Fuirena
scirpoides

Fumaria
officinalis

Gaillardia
aestivalis

Gaillardia
aristata

Gaillardia
pulchella

Galactia
volubilis
mississippiensis

Galeopsis
ladanum

Galeopsis
tetrahit

Galinsoga ciliata

Galinsoga
parviflora

Galium aparine

IN

EVERY

COUNTY

Galium
asprellum

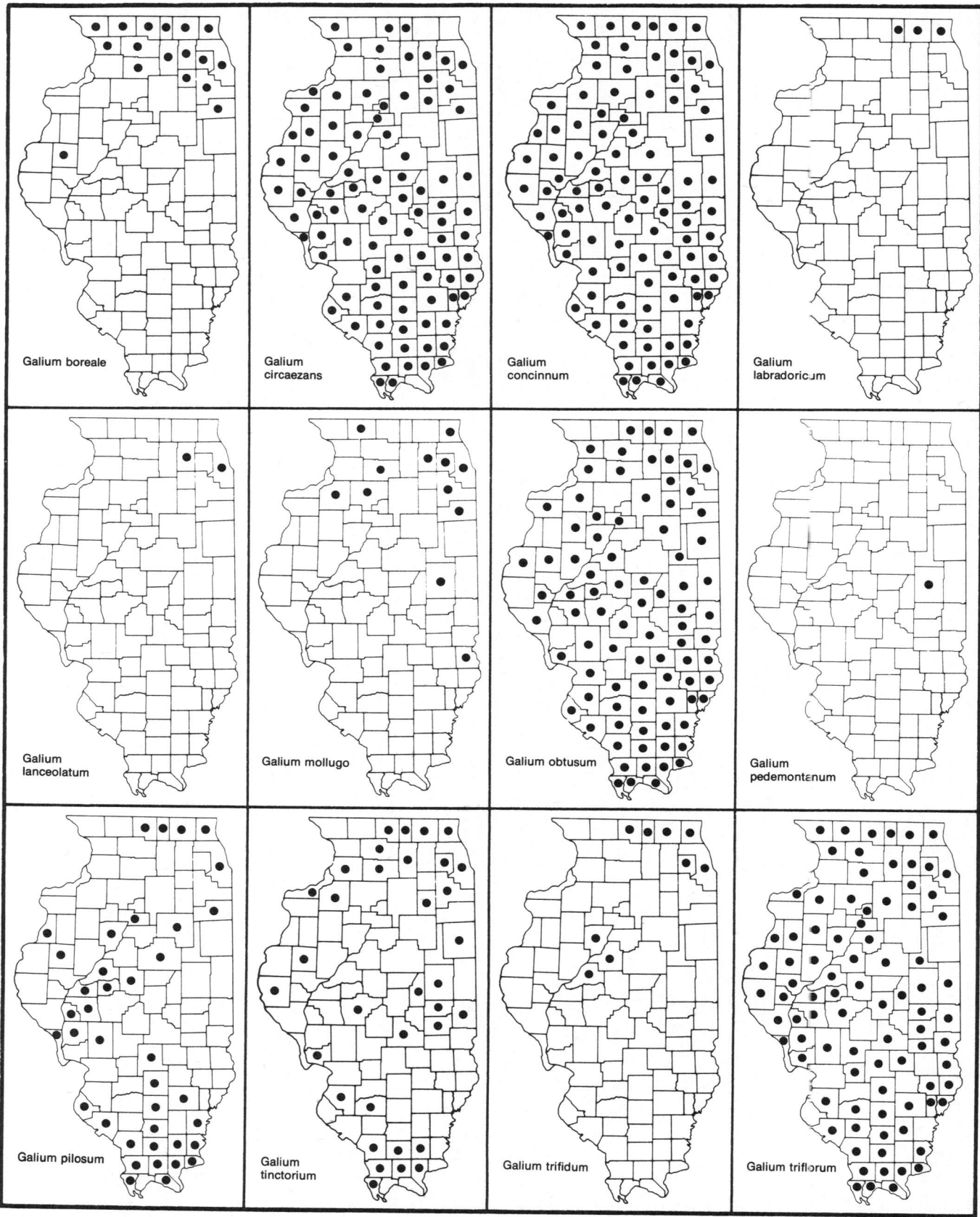

Galium boreale

Galium circaezans

Galium concinnum

Galium labradoricum

Galium lanceolatum

Galium mollugo

Galium obtusum

Galium pedemontanum

Galium pilosum

Galium tinctorium

Galium trifidum

Galium triflorum

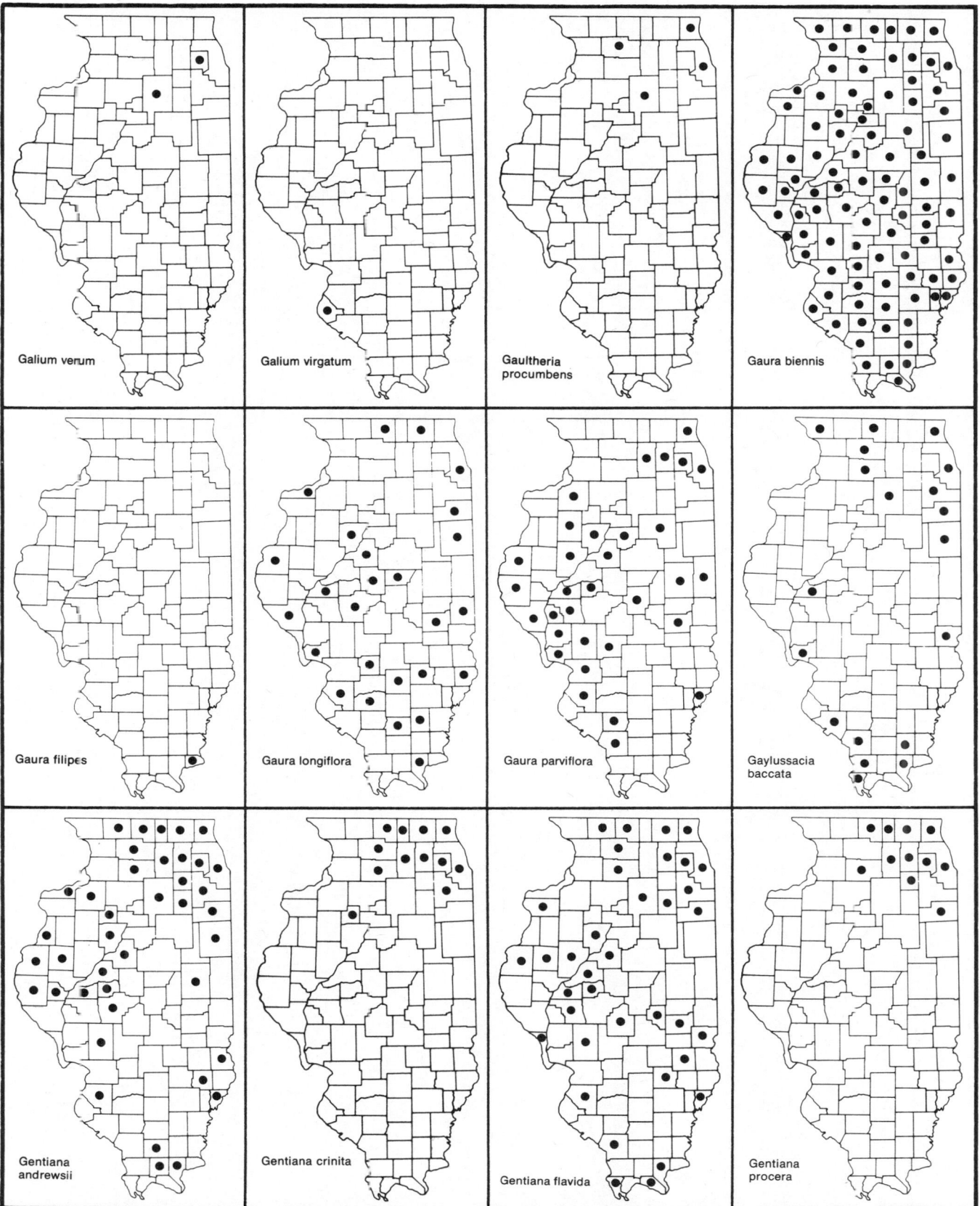

Galium verum

Galium virgatum

Gaultheria
procumbens

Gaura biennis

Gaura filipes

Gaura longiflora

Gaura parviflora

Gaylussacia
baccata

Gentiana
andrewsii

Gentiana crinita

Gentiana flavida

Gentiana
procera

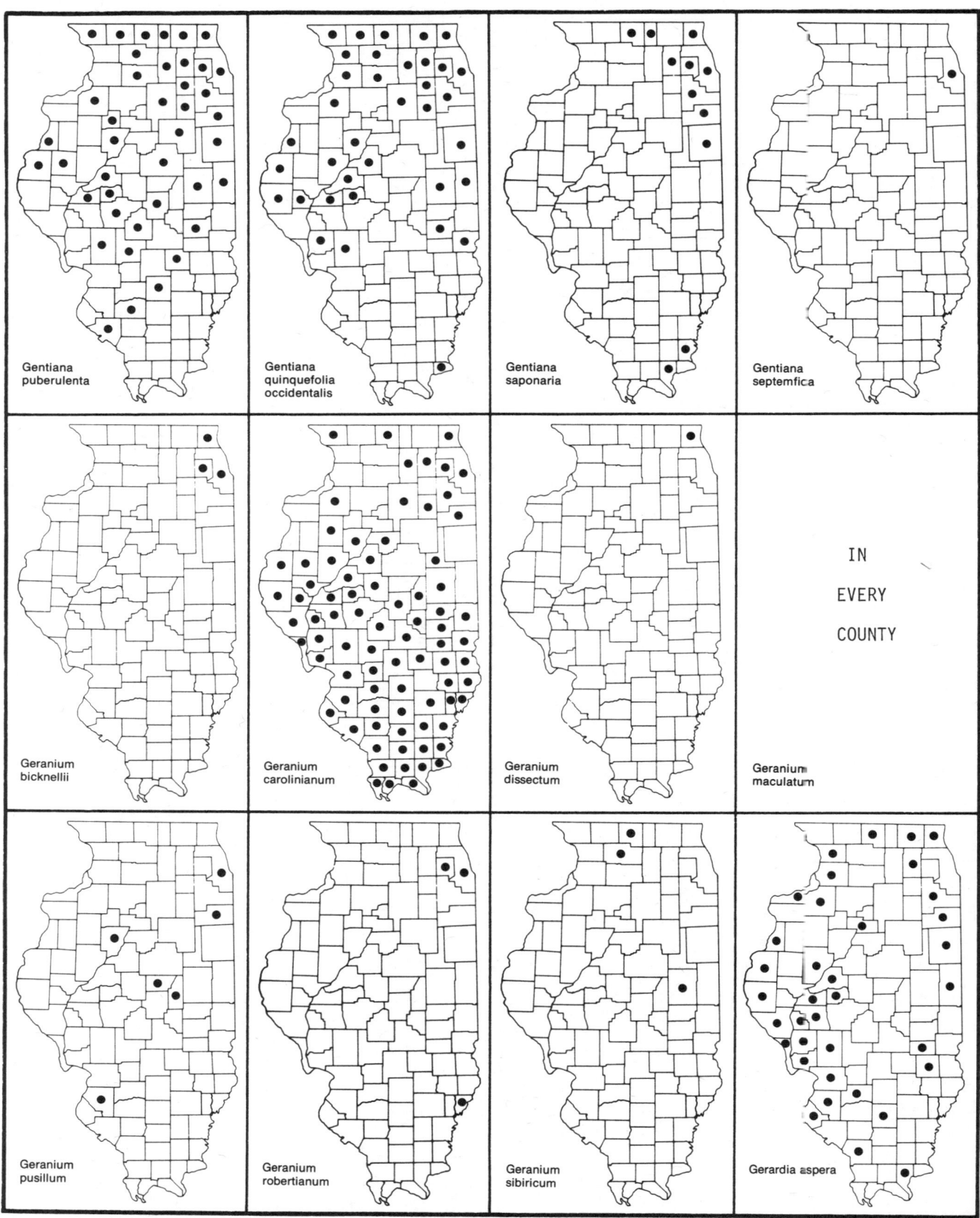

Gentiana
puberulenta

Gentiana
quinquefolia
occidentalis

Gentiana
saponaria

Gentiana
septemfica

Geranium
bicknellii

Geranium
carolinianum

Geranium
dissectum

Geranium
maculatum

IN

EVERY

COUNTY

Geranium
pusillum

Geranium
robertianum

Geranium
sibiricum

Gerardia aspera

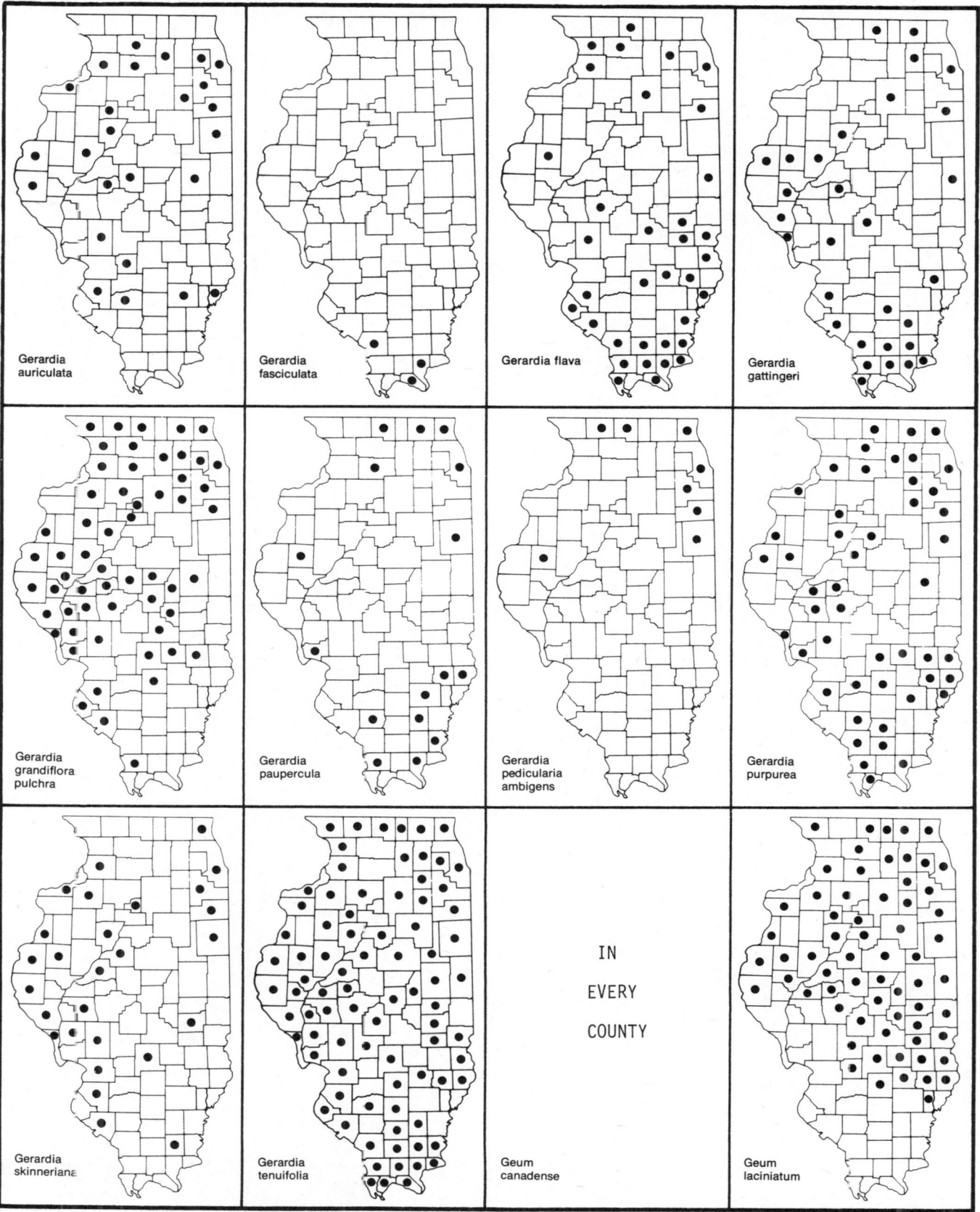

Gerardia
auriculata

Gerardia
fasciculata

Gerardia flava

Gerardia
gattingeri

Gerardia
grandiflora
pulchra

Gerardia
paupercula

Gerardia
pedicularia
ambigens

Gerardia
purpurea

Gerardia
skinneriana

Gerardia
tenuifolia

Geum
canadense

IN

EVERY

COUNTY

Geum
laciniatum

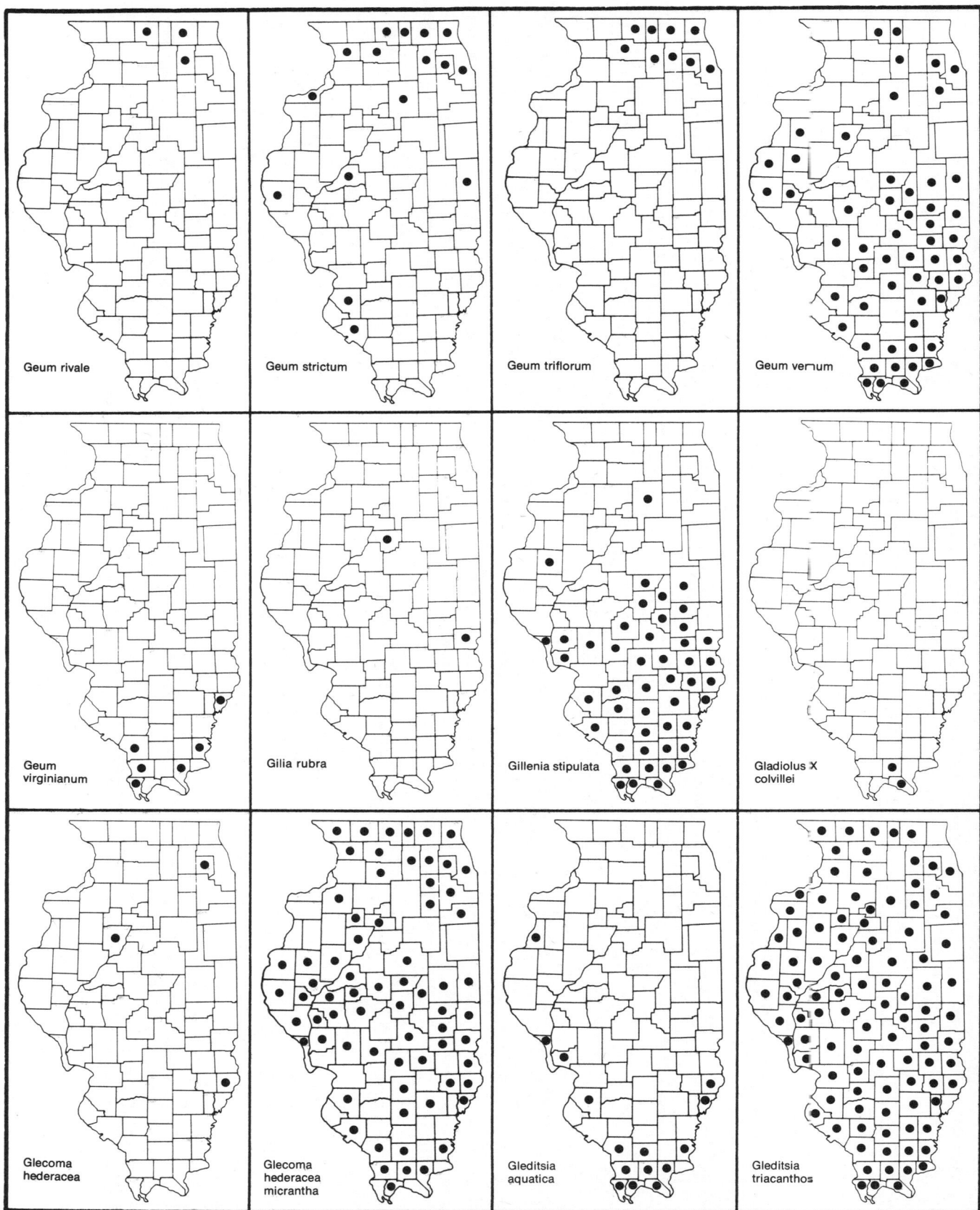

Geum rivale

Geum strictum

Geum triflorum

Geum venum

Geum virginianum

Gilia rubra

Gillenia stipulata

Gladiolus X colvillei

Glecoma hederacea

Glecoma hederacea micrantha

Gleditsia aquatica

Gleditsia triacanthos

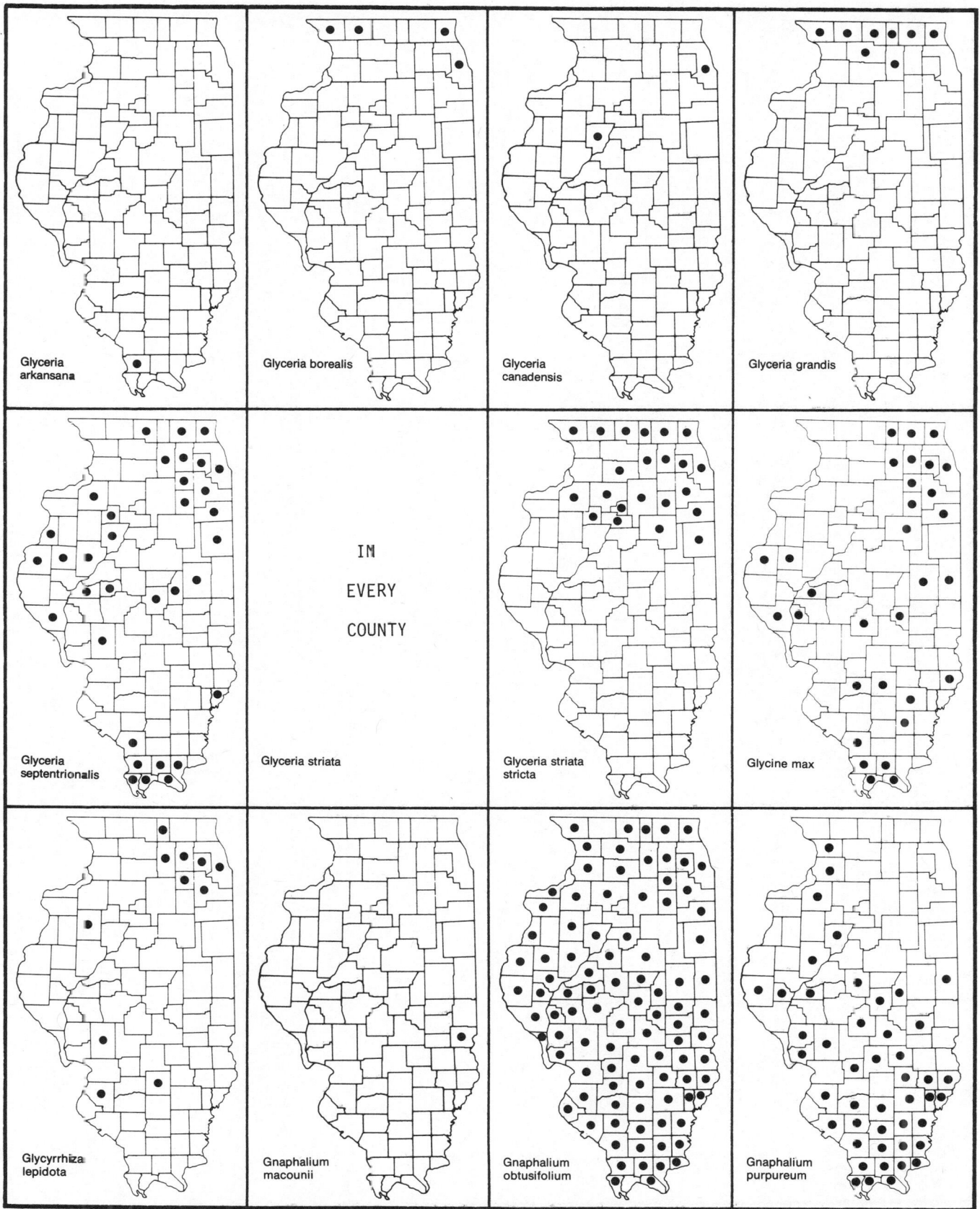

Glyceria
arkansana

Glyceria borealis

Glyceria
canadensis

Glyceria grandis

Glyceria
septentrionalis

IM

EVERY

COUNTY

Glyceria striata

Glyceria striata
stricta

Glycine max

Glycyrrhiza
lepidota

Gnaphalium
macounii

Gnaphalium
obtusifolium

Gnaphalium
purpureum

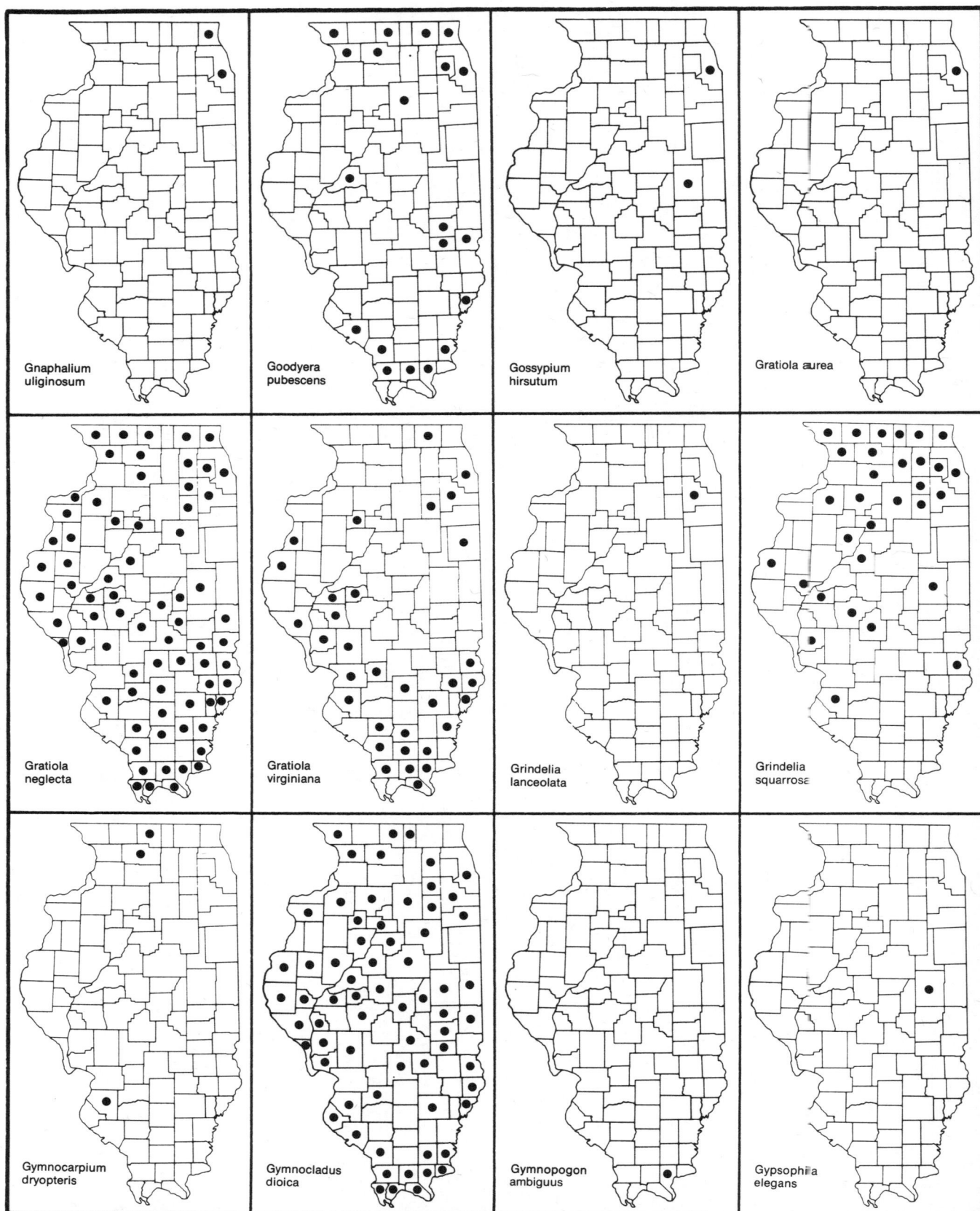

Gnaphalium
uliginosum

Goodyera
pubescens

Gossypium
hirsutum

Gratiola aurea

Gratiola
neglecta

Gratiola
virginiana

Grindelia
lanceolata

Grindelia
squarrosa

Gymnocarpium
dryopteris

Gymnocladus
dioica

Gymnopogon
ambiguus

Gypsophila
elegans

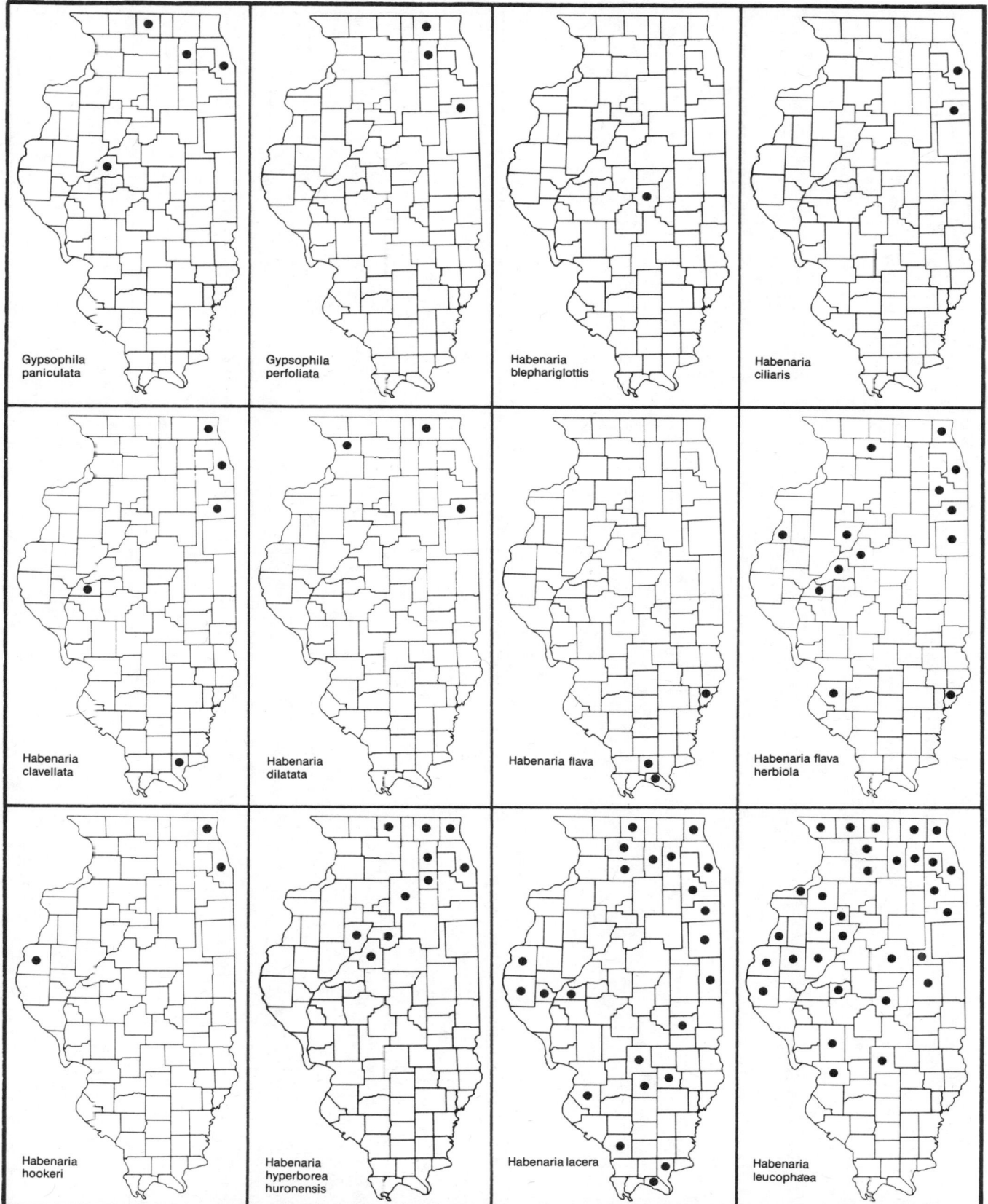

Gypsophila
paniculata

Gypsophila
perfoliata

Habenaria
blephariglottis

Habenaria
ciliaris

Habenaria
clavellata

Habenaria
dilatata

Habenaria
flava

Habenaria flava
herbiola

Habenaria
hookeri

Habenaria
hyperborea
huronensis

Habenaria lacera

Habenaria
leucophæa

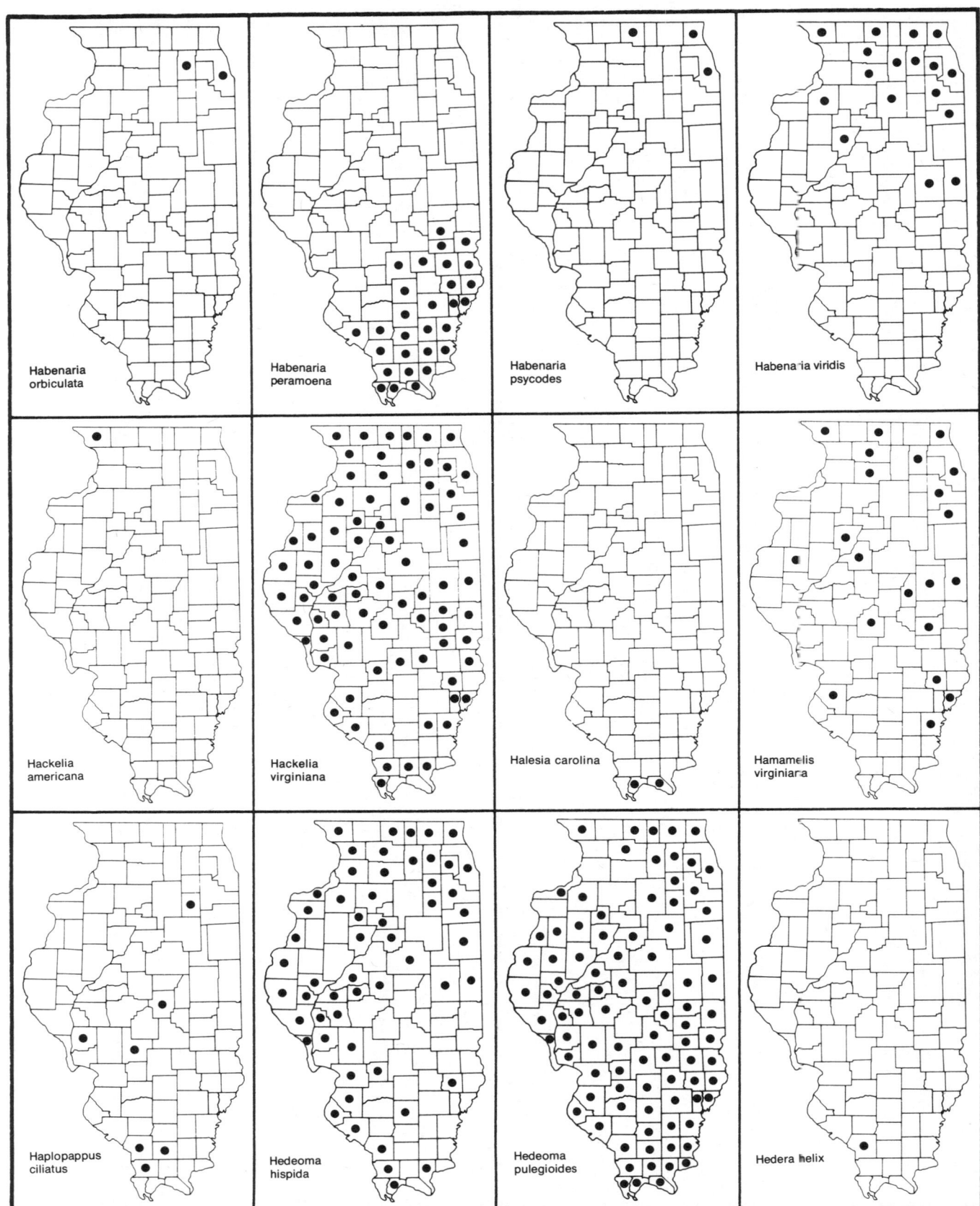

Habenaria
orbiculata

Habenaria
peramoena

Habenaria
psycodes

Habenaria viridis

Hackelia
americana

Hackelia
virginiana

Halesia carolina

Hamamelis
virginiana

Haplopappus
ciliatus

Hedeoma
hispida

Hedeoma
pulegioides

Hedera helix

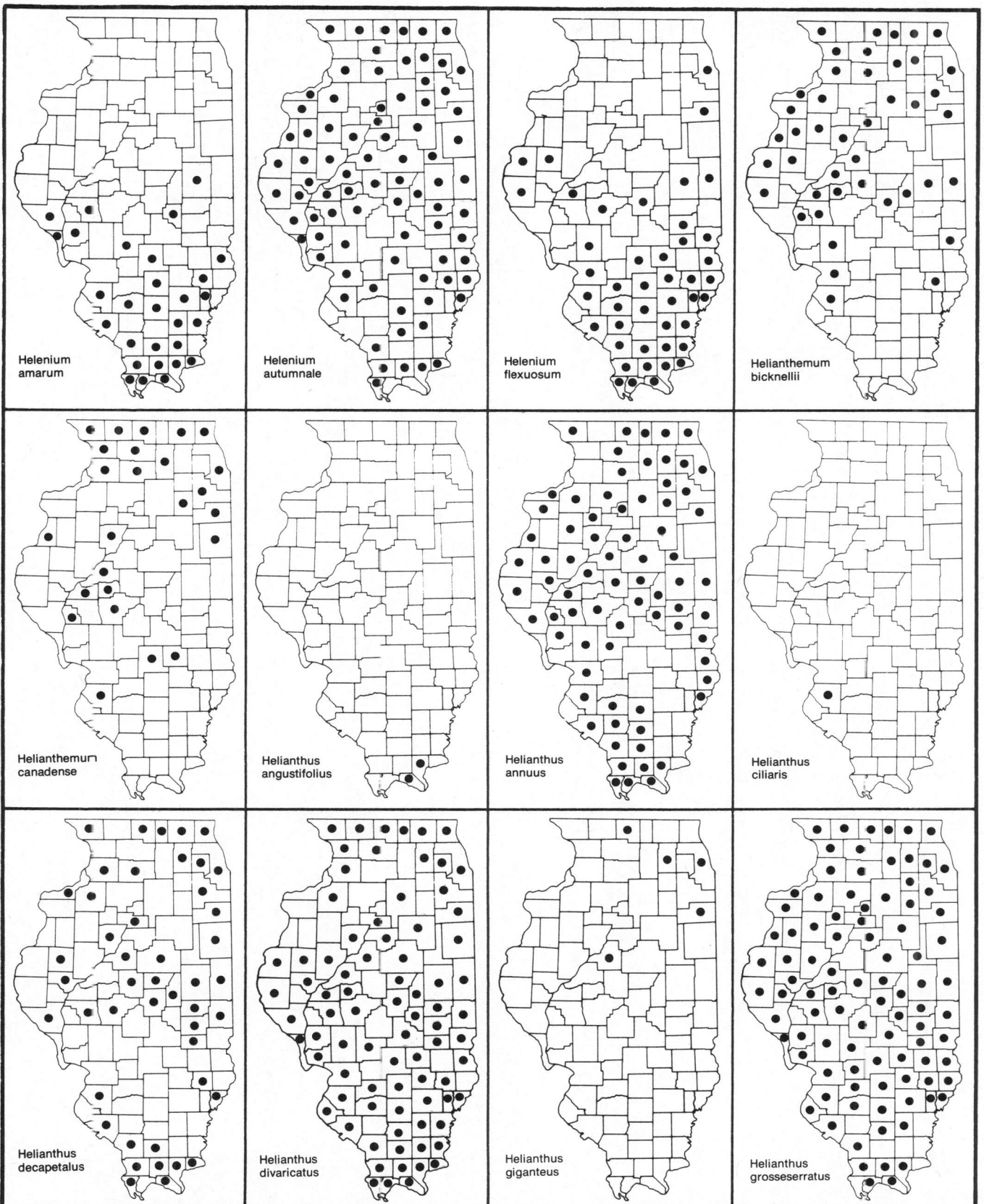

Helenium
amarum

Helenium
autumnale

Helenium
flexuosum

Helianthemum
bicknellii

Helianthemum
canadense

Helianthus
angustifolius

Helianthus
annuus

Helianthus
ciliaris

Helianthus
decapetalus

Helianthus
divaricatus

Helianthus
giganteus

Helianthus
grosseserratus

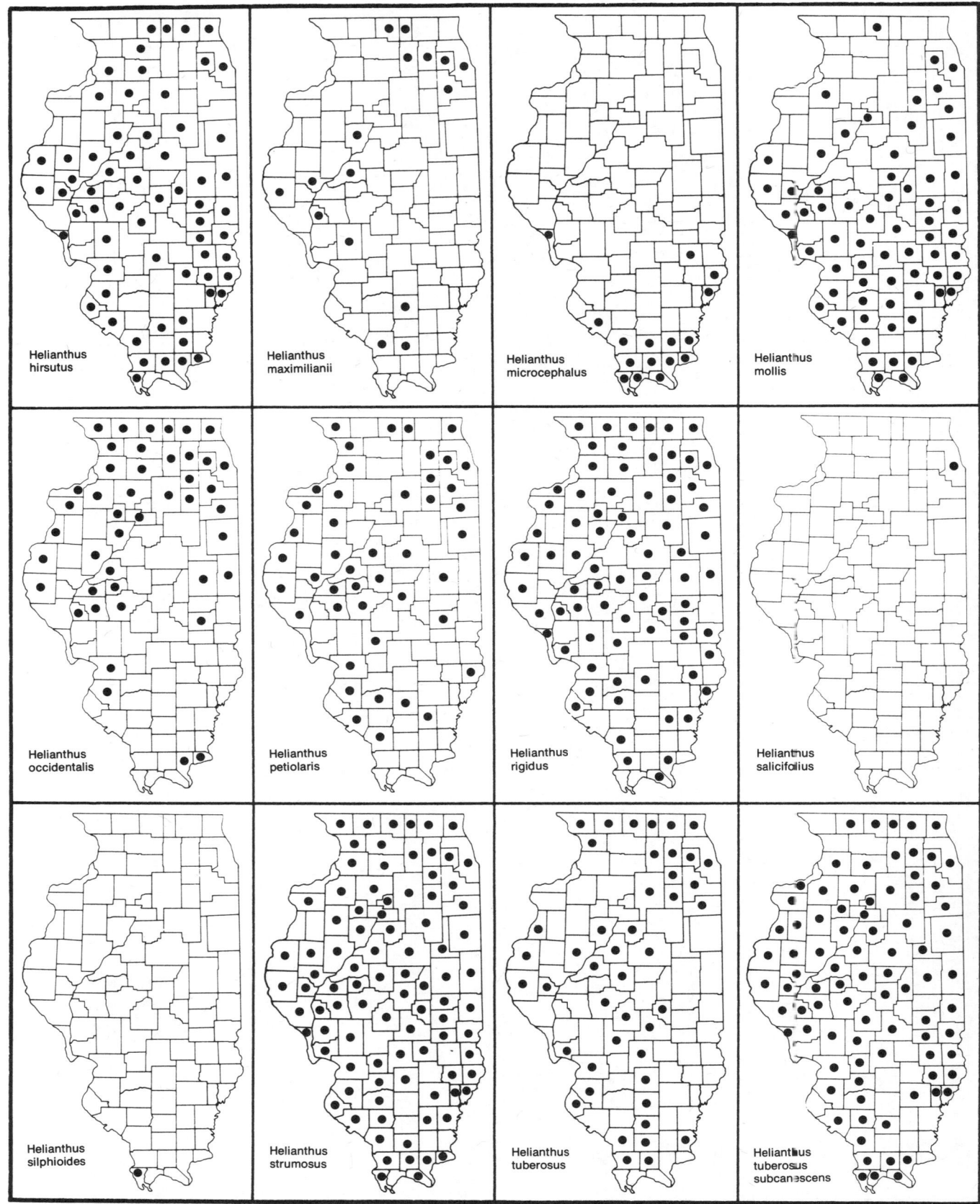

Helianthus
hirsutus

Helianthus
maximilianii

Helianthus
microcephalus

Helianthus
mollis

Helianthus
occidentalis

Helianthus
petiolaris

Helianthus
rigidus

Helianthus
salicifolius

Helianthus
silphioides

Helianthus
strumosus

Helianthus
tuberosus

Helianthus
tuberosus
subcanescens

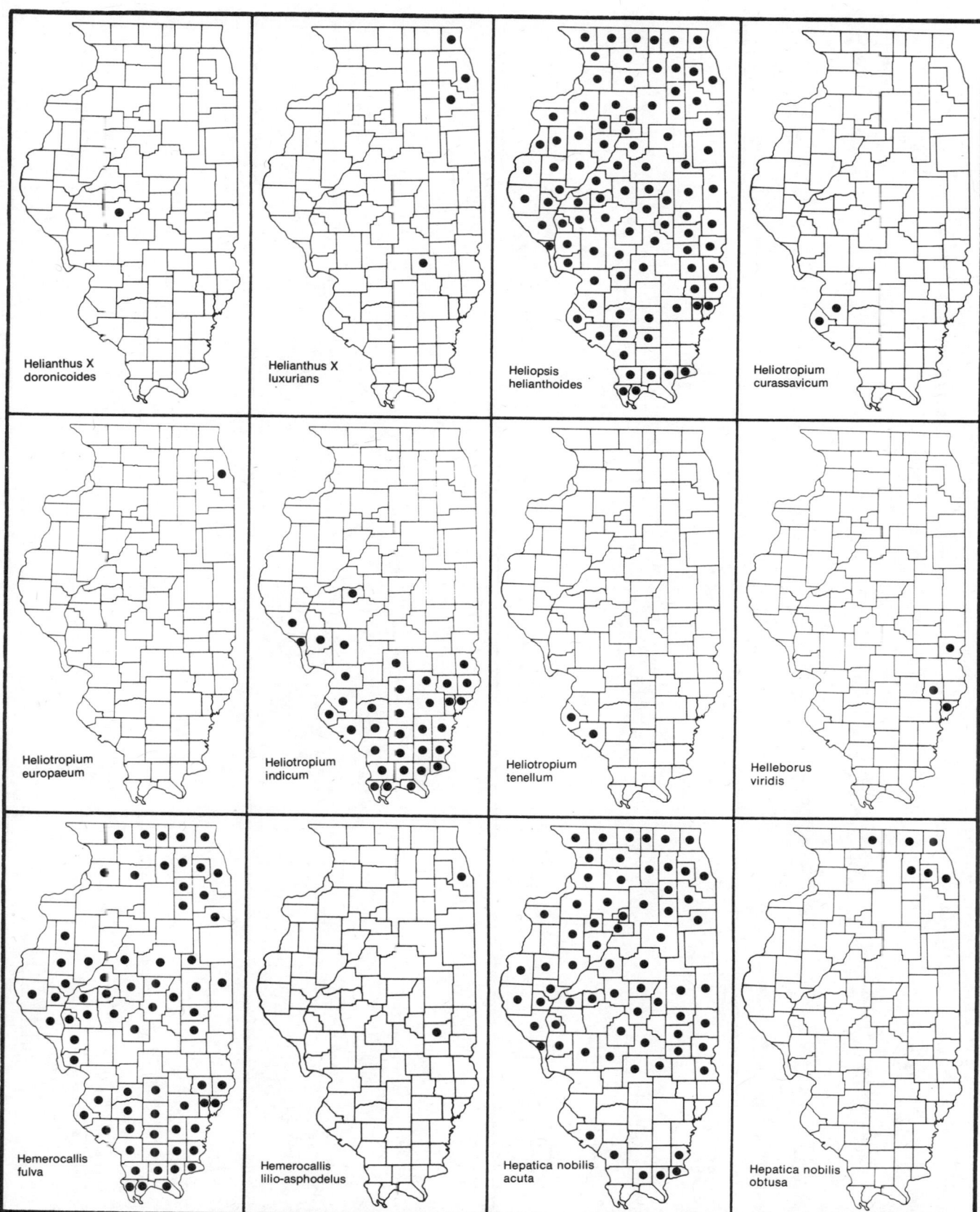

Helianthus X
doronicoides

Helianthus X
luxurians

Heliopsis
helianthoides

Heliotropium
curassavicum

Heliotropium
europaeum

Heliotropium
indicum

Heliotropium
tenellum

Helleborus
viridis

Hemerocallis
fulva

Hemerocallis
lilio-asphodelus

Hepatica nobilis
acuta

Hepatica nobilis
obtusa

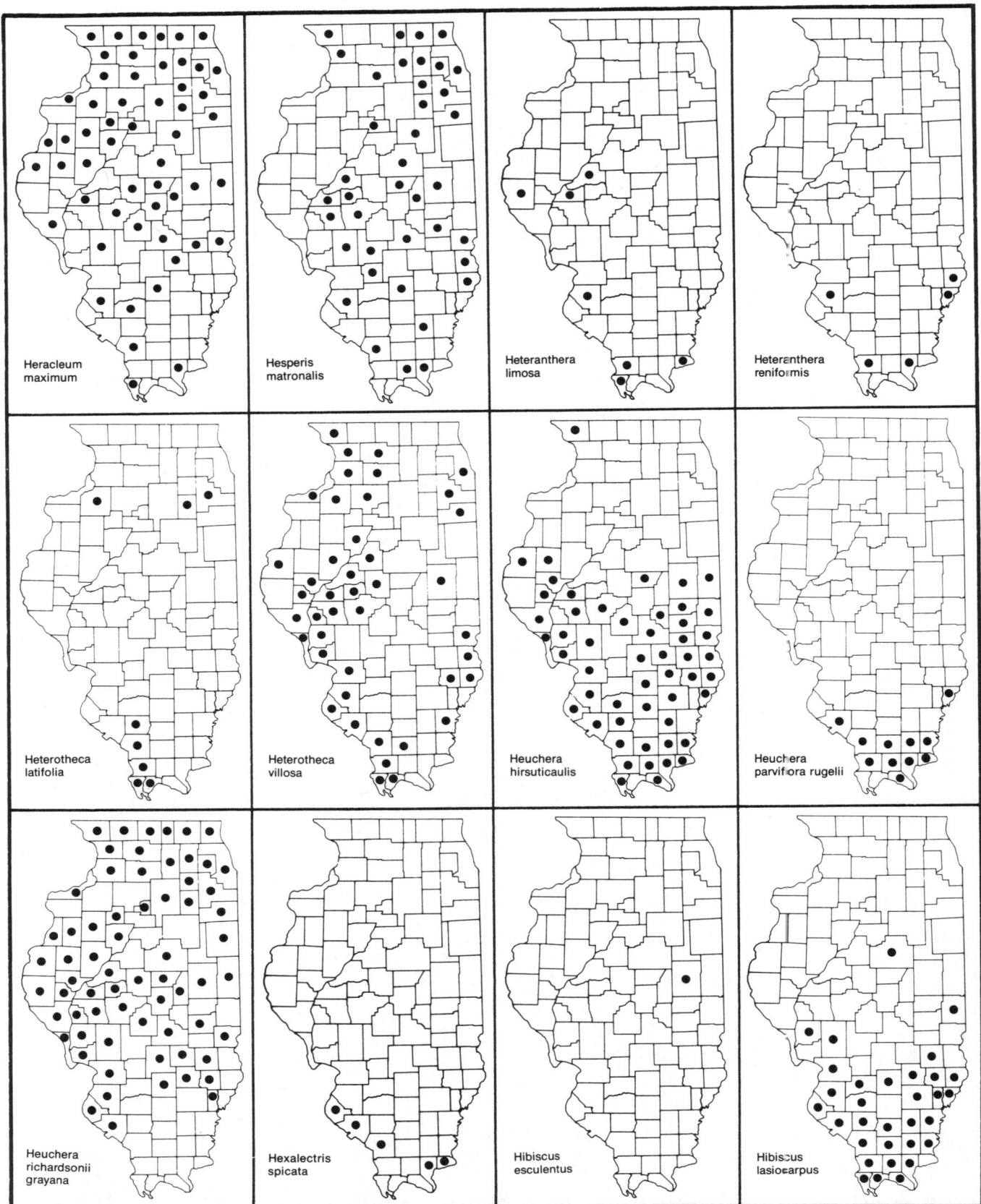

Heracleum
maximum

Hesperis
matronalis

Heteranthera
limosa

Heteranthera
reniformis

Heterotheca
latifolia

Heterotheca
villosa

Heuchera
hirsuticaulis

Heuchera
parviflora rugelii

Heuchera
richardsonii
grayana

Hexalectris
spicata

Hibiscus
esculentus

Hibiscus
lasiocarpus

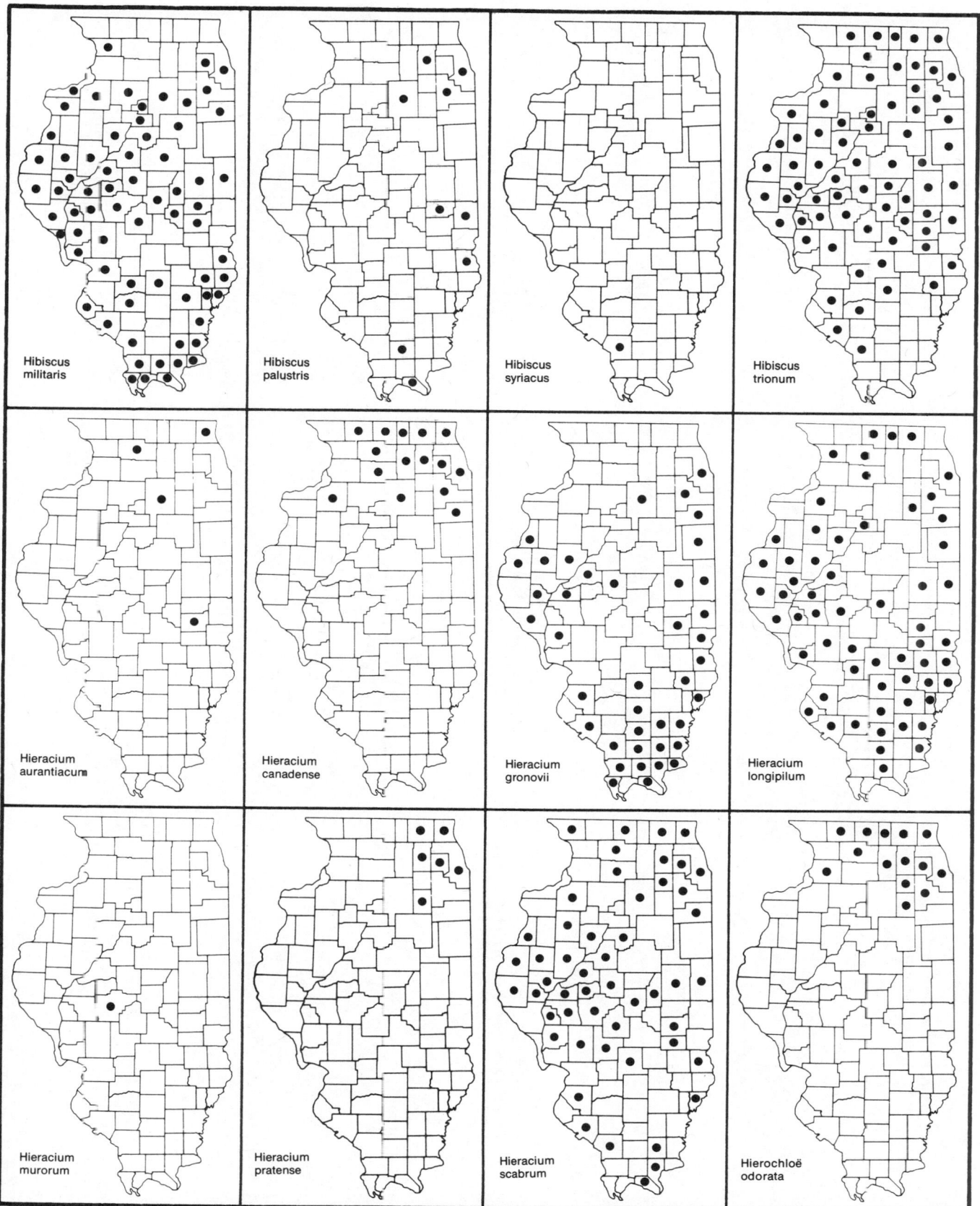

Hibiscus
militaris

Hibiscus
palustris

Hibiscus
syriacus

Hibiscus
trionum

Hieracium
aurantiacum

Hieracium
canadense

Hieracium
gronovii

Hieracium
longipilum

Hieracium
murorum

Hieracium
pratense

Hieracium
scabrum

Hierochloë
odorata

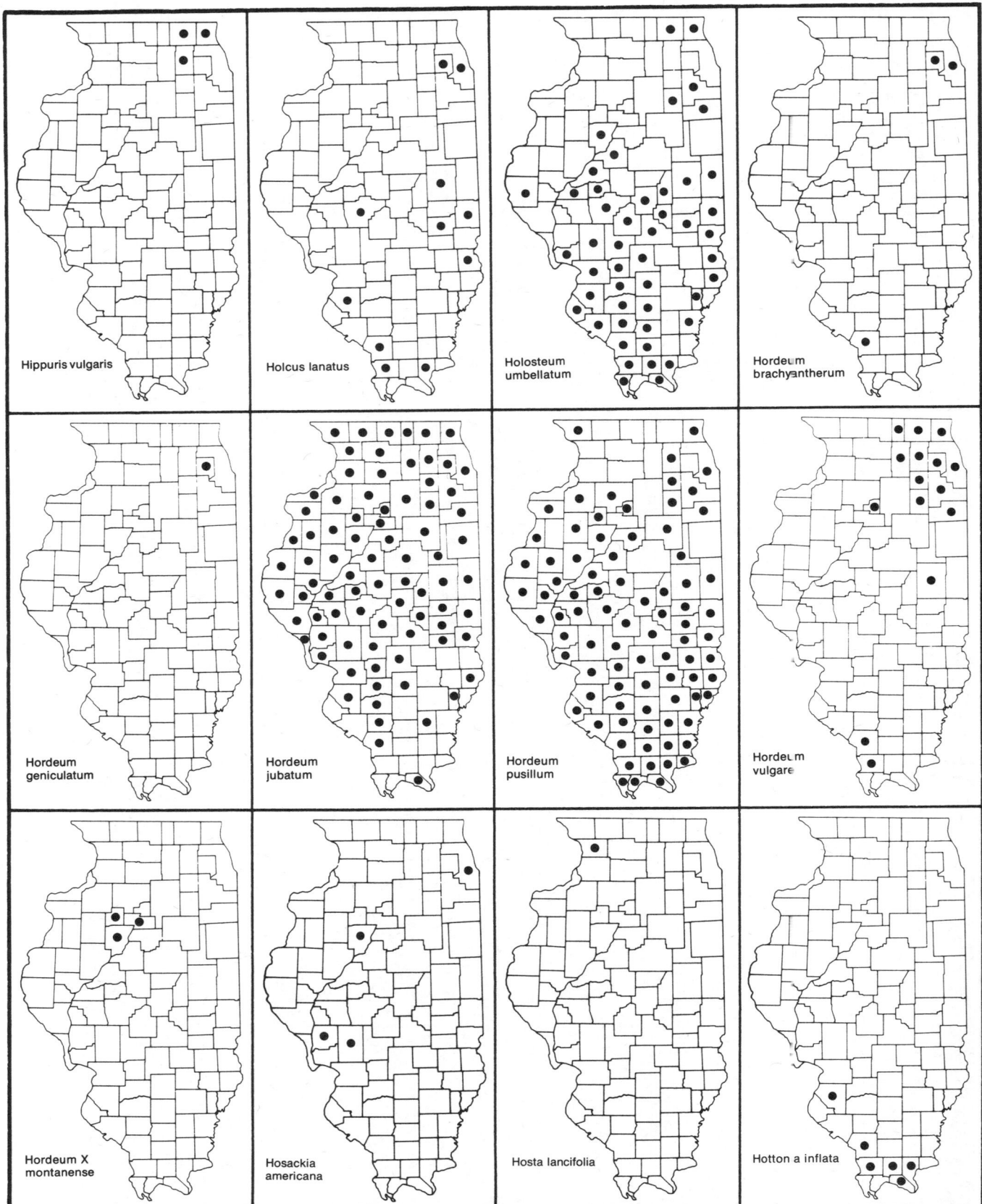

Hippuris vulgaris

Holcus lanatus

Holosteum
umbellatum

Hordeum
brachyantherum

Hordeum
geniculatum

Hordeum
jubatum

Hordeum
pusillum

Hordeum
vulgare

Hordeum X
montanense

Hosackia
americana

Hosta lancifolia

Hottonia inflata

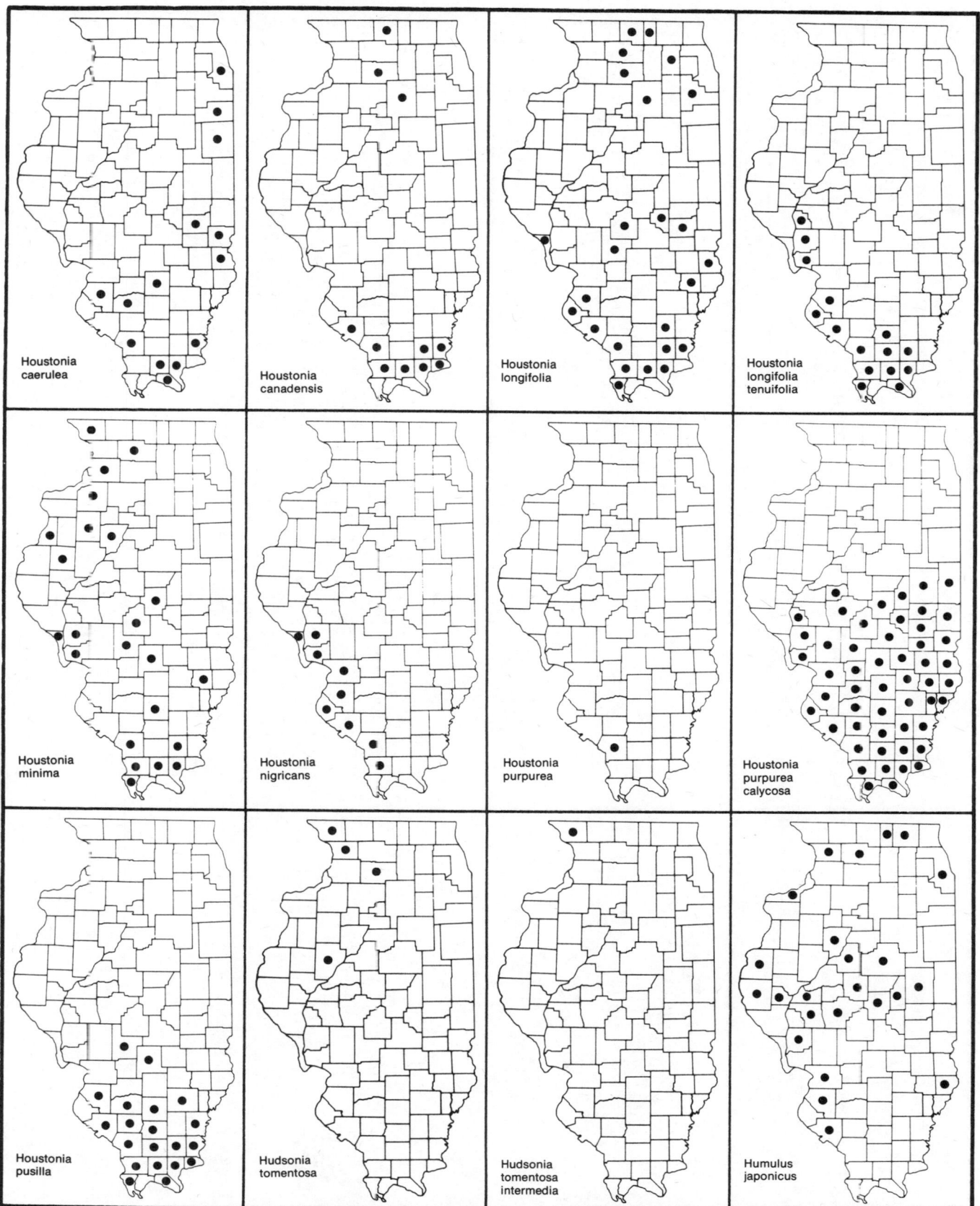

Houstonia
caerulea

Houstonia
canadensis

Houstonia
longifolia

Houstonia
longifolia
tenuifolia

Houstonia
minima

Houstonia
nigricans

Houstonia
purpurea

Houstonia
purpurea
calycosa

Houstonia
pusilla

Hudsonia
tomentosa

Hudsonia
tomentosa
intermedia

Humulus
japonicus

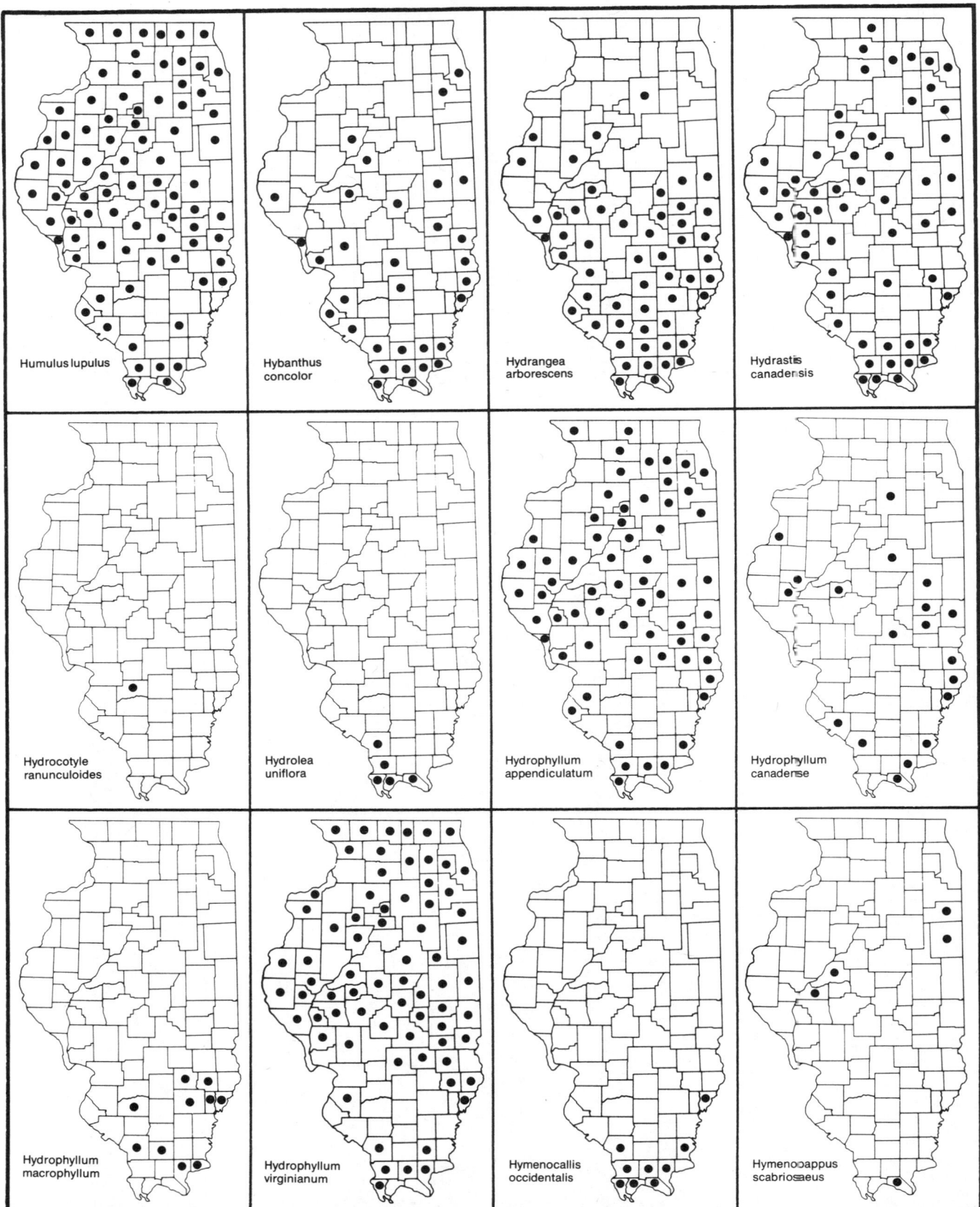

Humulus lupulus

Hybanthus concolor

Hydrangea arborescens

Hydrastis canadensis

Hydrocotyle ranunculoides

Hydrolea uniflora

Hydrophyllum appendiculatum

Hydrophyllum canadense

Hydrophyllum macrophyllum

Hydrophyllum virginianum

Hymenocallis occidentalis

Hymenopappus scabriosaeus

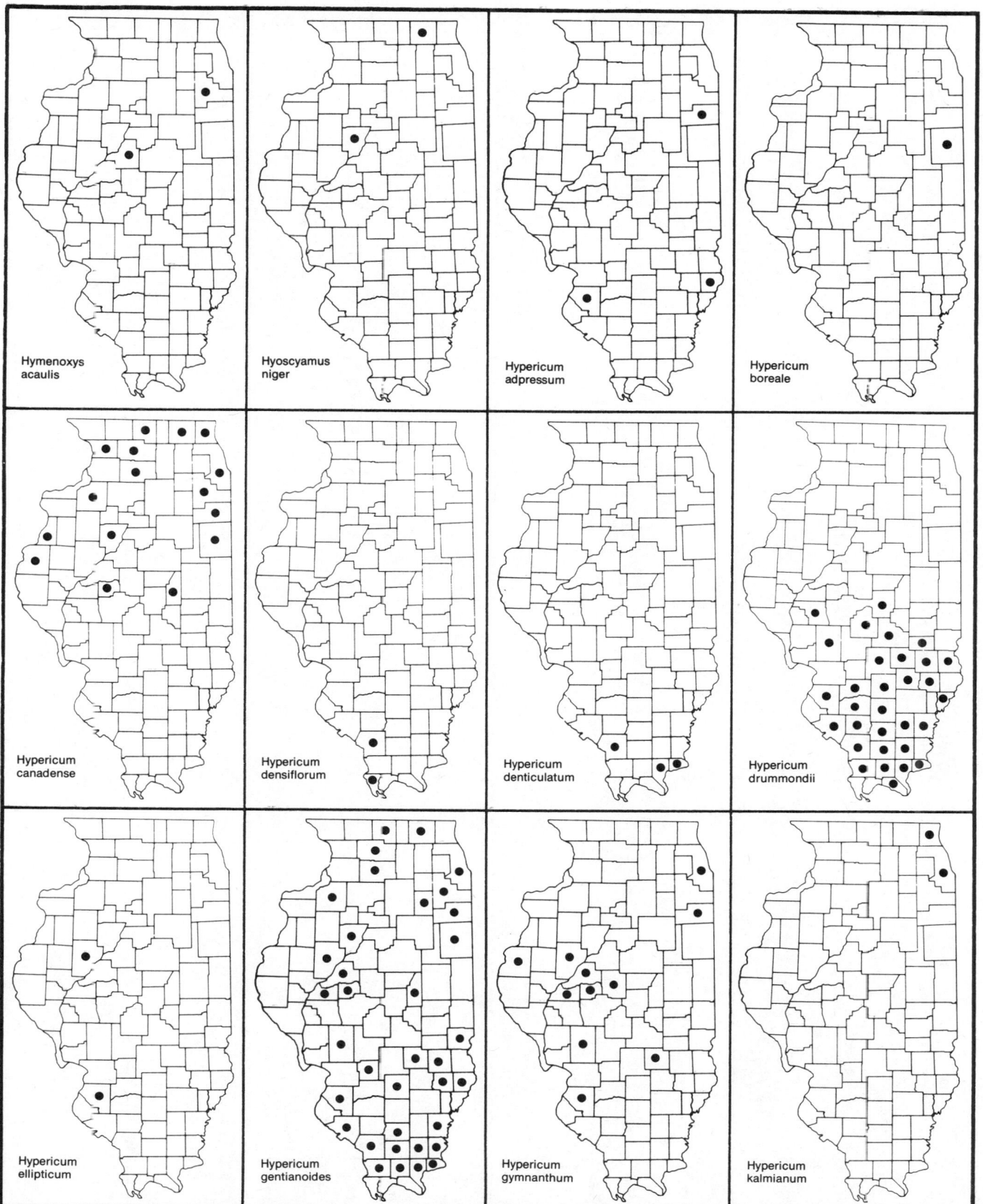

Hymenoxys
acaulis

Hyoscyamus
niger

Hypericum
adpressum

Hypericum
boreale

Hypericum
canadense

Hypericum
densiflorum

Hypericum
denticulatum

Hypericum
drummondii

Hypericum
ellipticum

Hypericum
gentianoides

Hypericum
gymnanthum

Hypericum
kalmianum

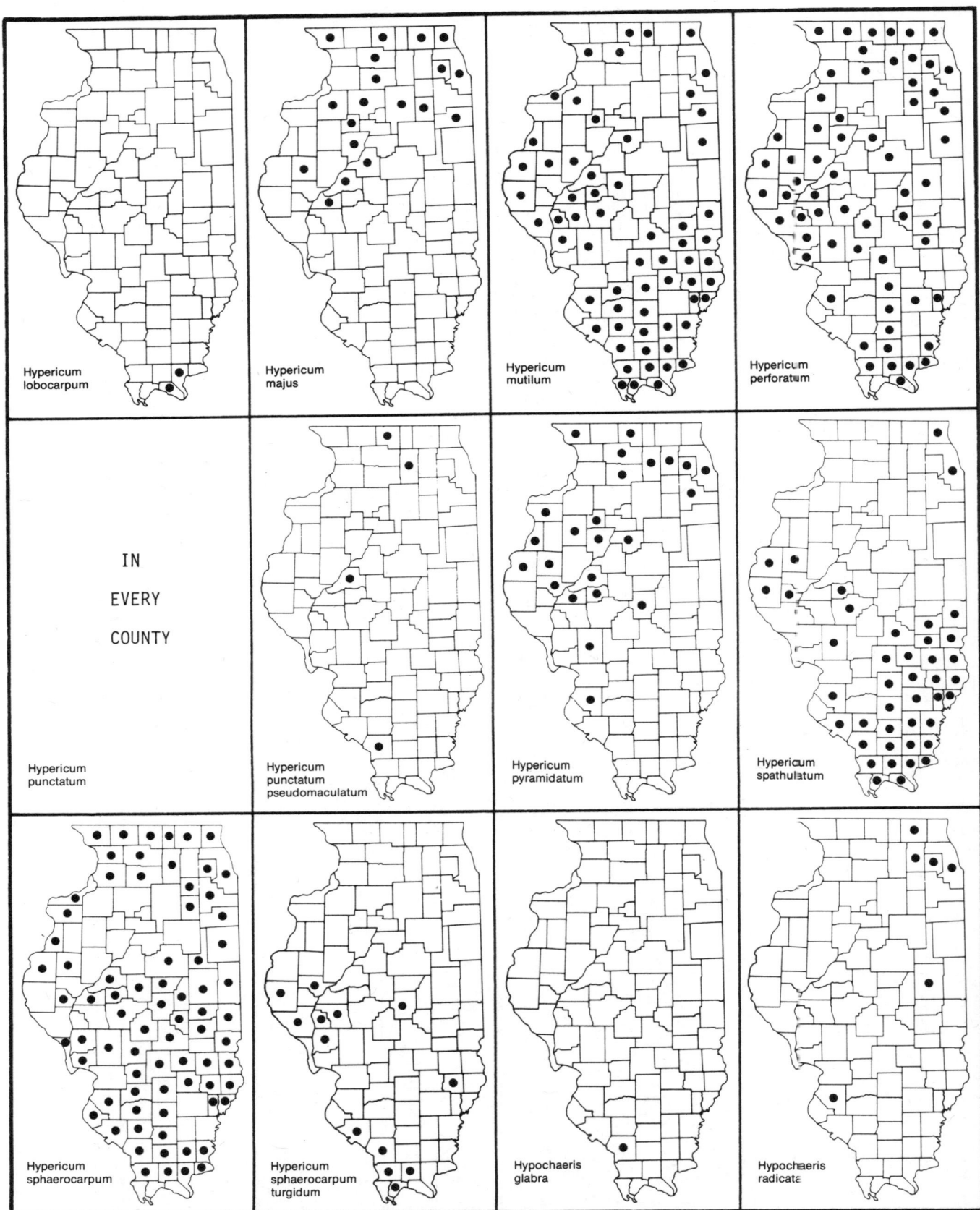

Hypericum
lobocarpum

Hypericum
majus

Hypericum
mutilum

Hypericum
perforatum

IN

EVERY

COUNTY

Hypericum
punctatum

Hypericum
punctatum
pseudomaculatum

Hypericum
pyramidatum

Hypericum
spathulatum

Hypericum
sphaerocarpum

Hypericum
sphaerocarpum
turgidum

Hypochaeris
glabra

Hypochaeris
radicata

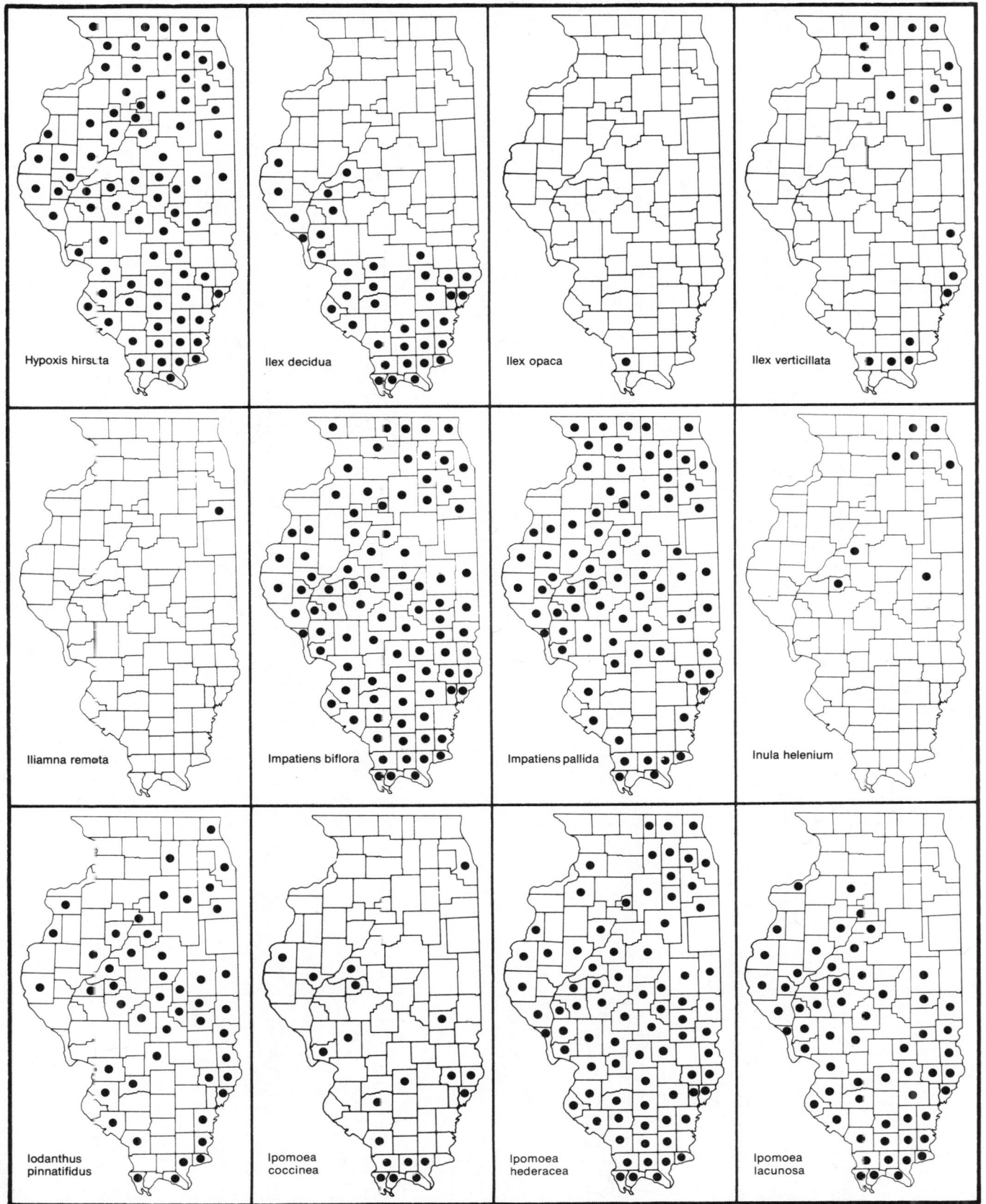

Hypoxis hirsuta

Ilex decidua

Ilex opaca

Ilex verticillata

Iliamna remota

Impatiens biflora

Impatiens pallida

Inula helenium

Iodanthus
pinnatifidus

Ipomoea
coccinea

Ipomoea
hederacea

Ipomoea
lacunosa

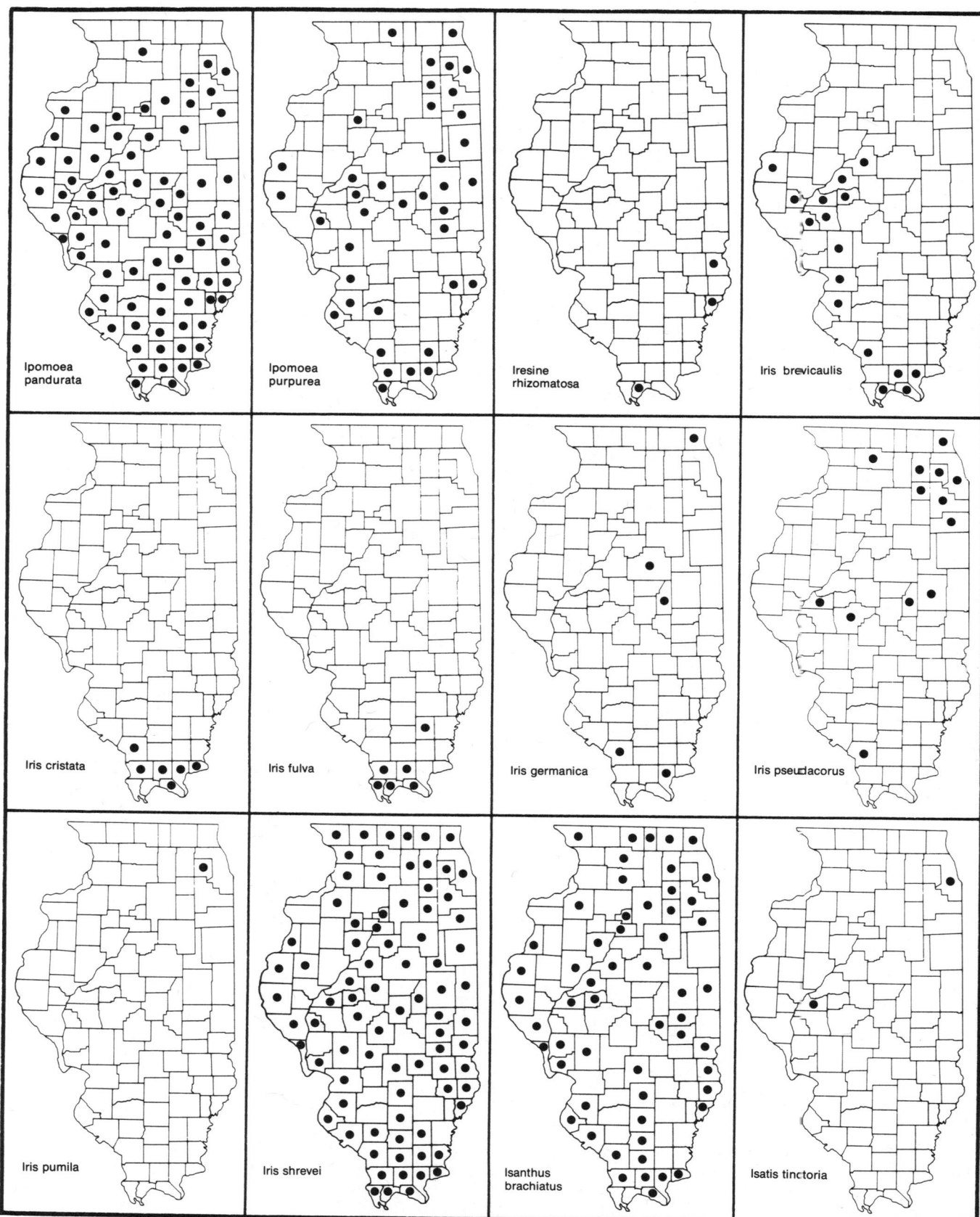

Ipomoea
pandurata

Ipomoea
purpurea

Iresine
rhizomatosa

Iris brevicaulis

Iris cristata

Iris fulva

Iris germanica

Iris pseudacorus

Iris pumila

Iris shrevei

Isanthus
brachiatus

Isatis tinctoria

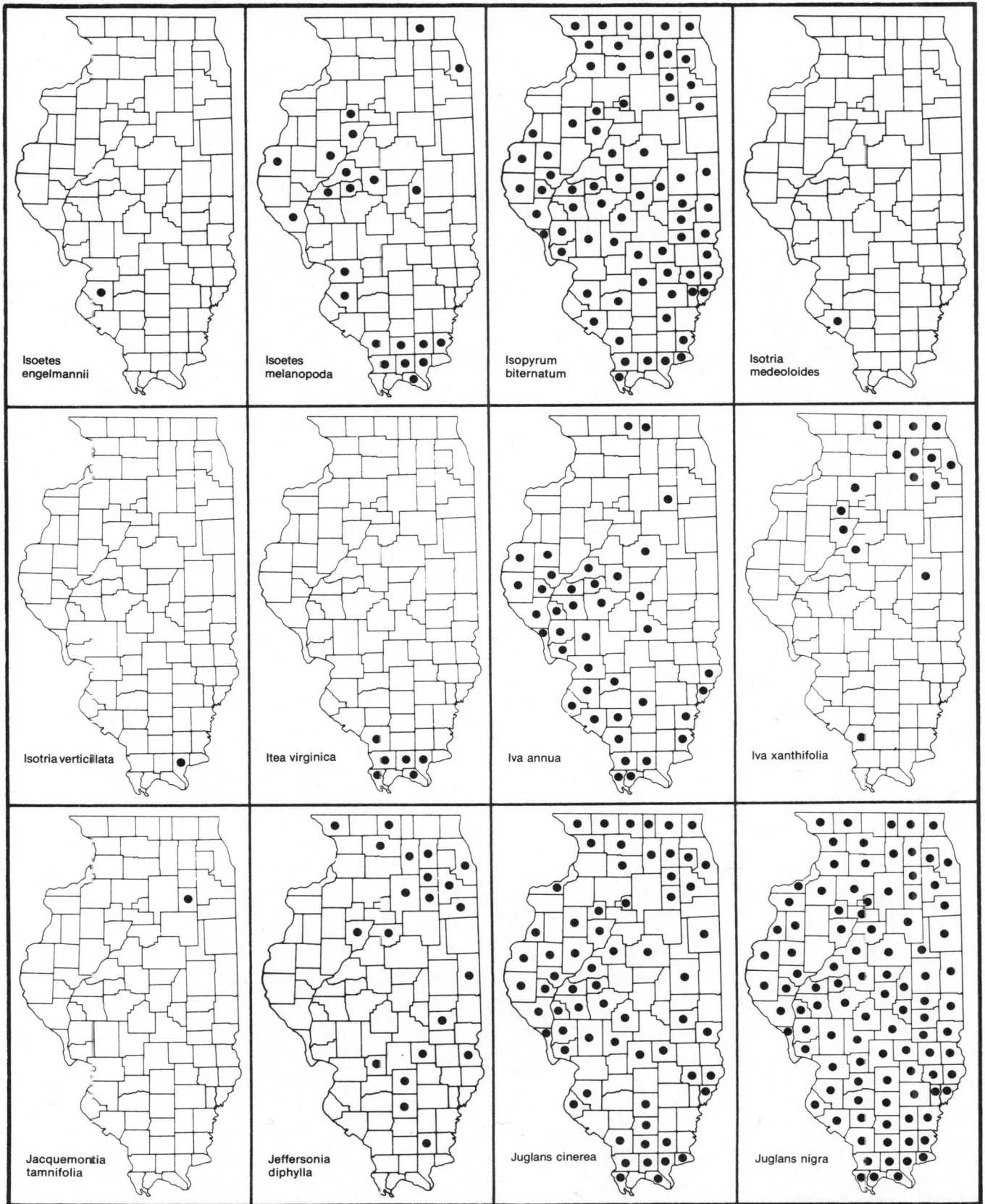

Isoetes
engelmannii

Isoetes
melanopoda

Isopyrum
biternatum

Isotria
medeoloides

Isotria verticillata

Itea virginica

Iva annua

Iva xanthifolia

Jacquemontia
tamnifolia

Jeffersonia
diphylla

Juglans cinerea

Juglans nigra

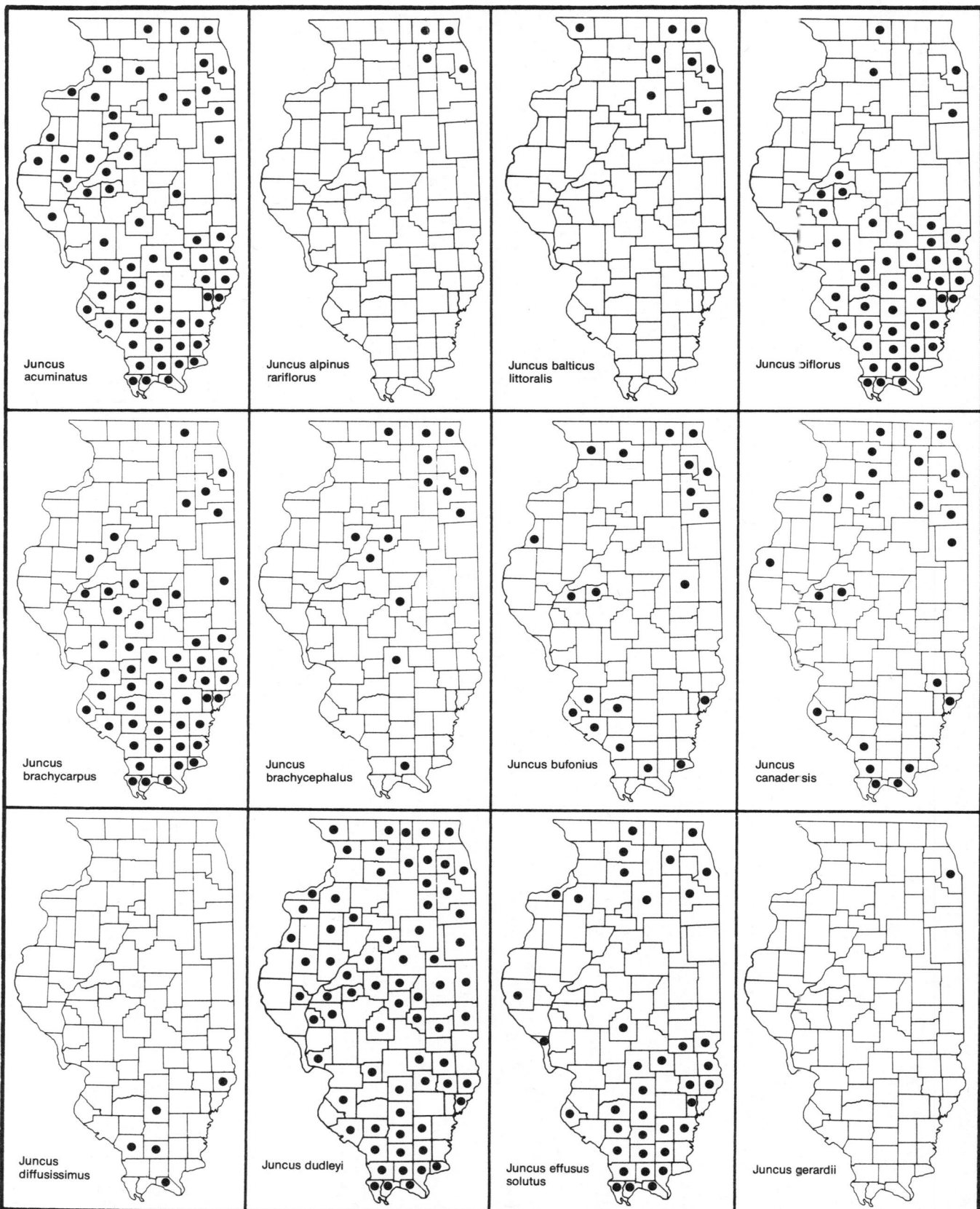

Juncus
acuminatus

Juncus alpinus
rariflorus

Juncus balticus
littoralis

Juncus biflorus

Juncus
brachycarpus

Juncus
brachycephalus

Juncus bufonius

Juncus
canadensis

Juncus
diffusissimus

Juncus dudleyi

Juncus effusus
solutus

Juncus gerardii

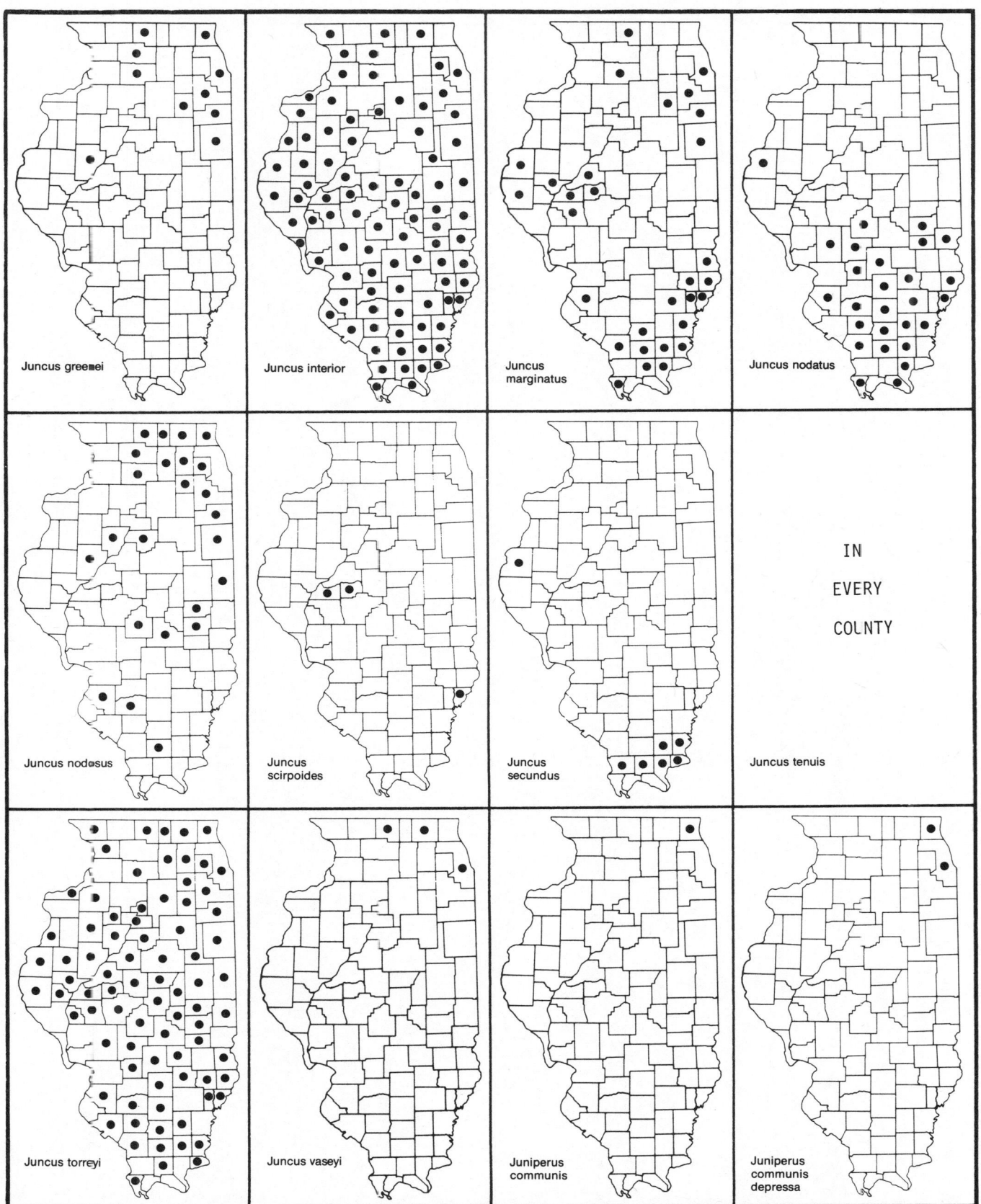

Juncus greenei

Juncus interior

Juncus
marginatus

Juncus nodatus

Juncus nodosus

Juncus
scirpoides

Juncus
secundus

Juncus tenuis

IN

EVERY

COUNTY

Juncus torreyi

Juncus vaseyi

Juniperus
communis

Juniperus
communis
depressa

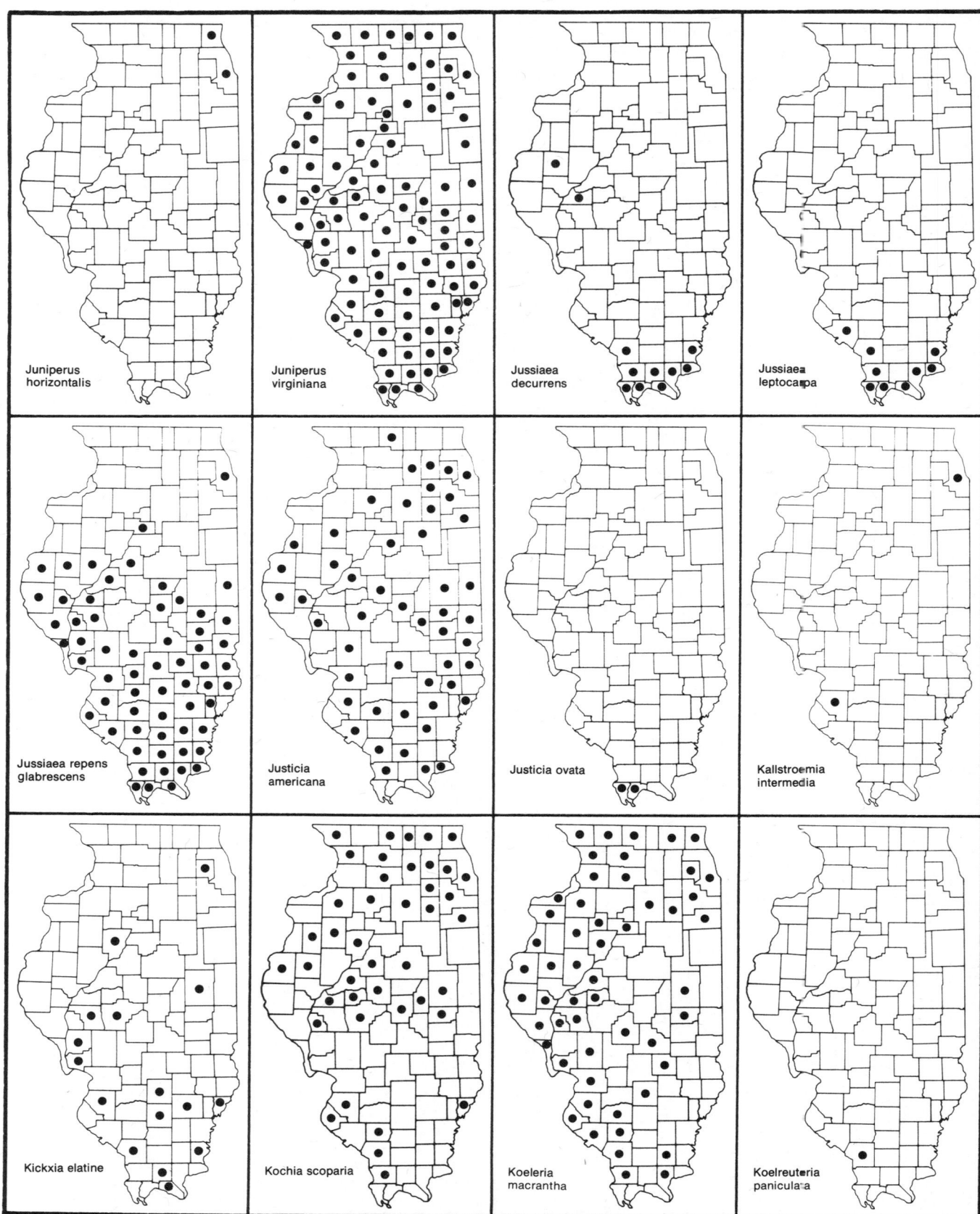

Juniperus
horizontalis

Juniperus
virginiana

Jussiaea
decurrens

Jussiaea
leptocarpa

Jussiaea repens
glabrescens

Justicia
americana

Justicia ovata

Kallstroemia
intermedia

Kickxia elatine

Kochia scoparia

Koeleria
macrantha

Koelreuteria
paniculata

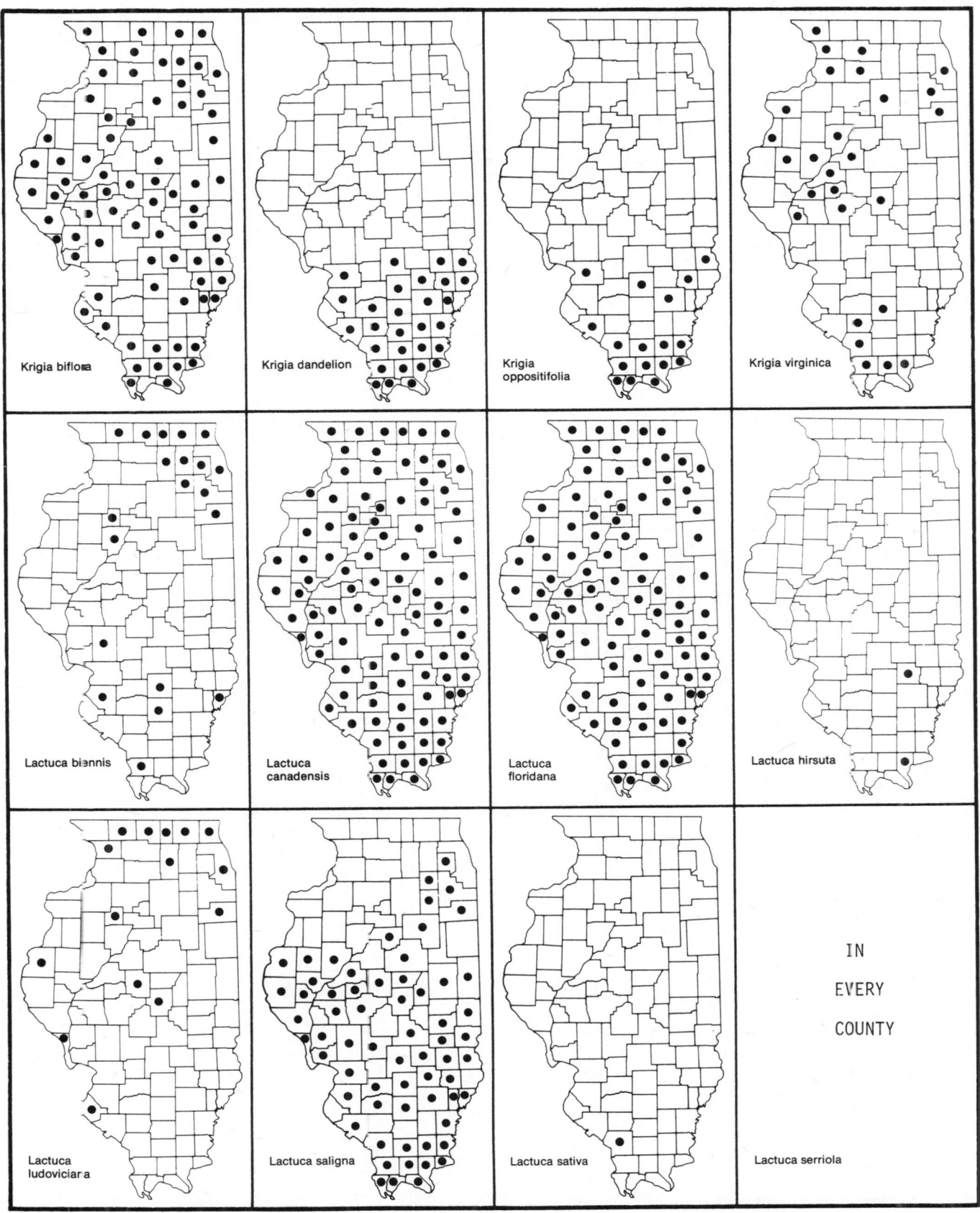

Krigia biflora

Krigia dandelion

Krigia oppositifolia

Krigia virginica

Lactuca biennis

Lactuca canadensis

Lactuca floridana

Lactuca hirsuta

Lactuca ludoviciana

Lactuca saligna

Lactuca sativa

Lactuca serriola

IN

EVERY

COUNTY

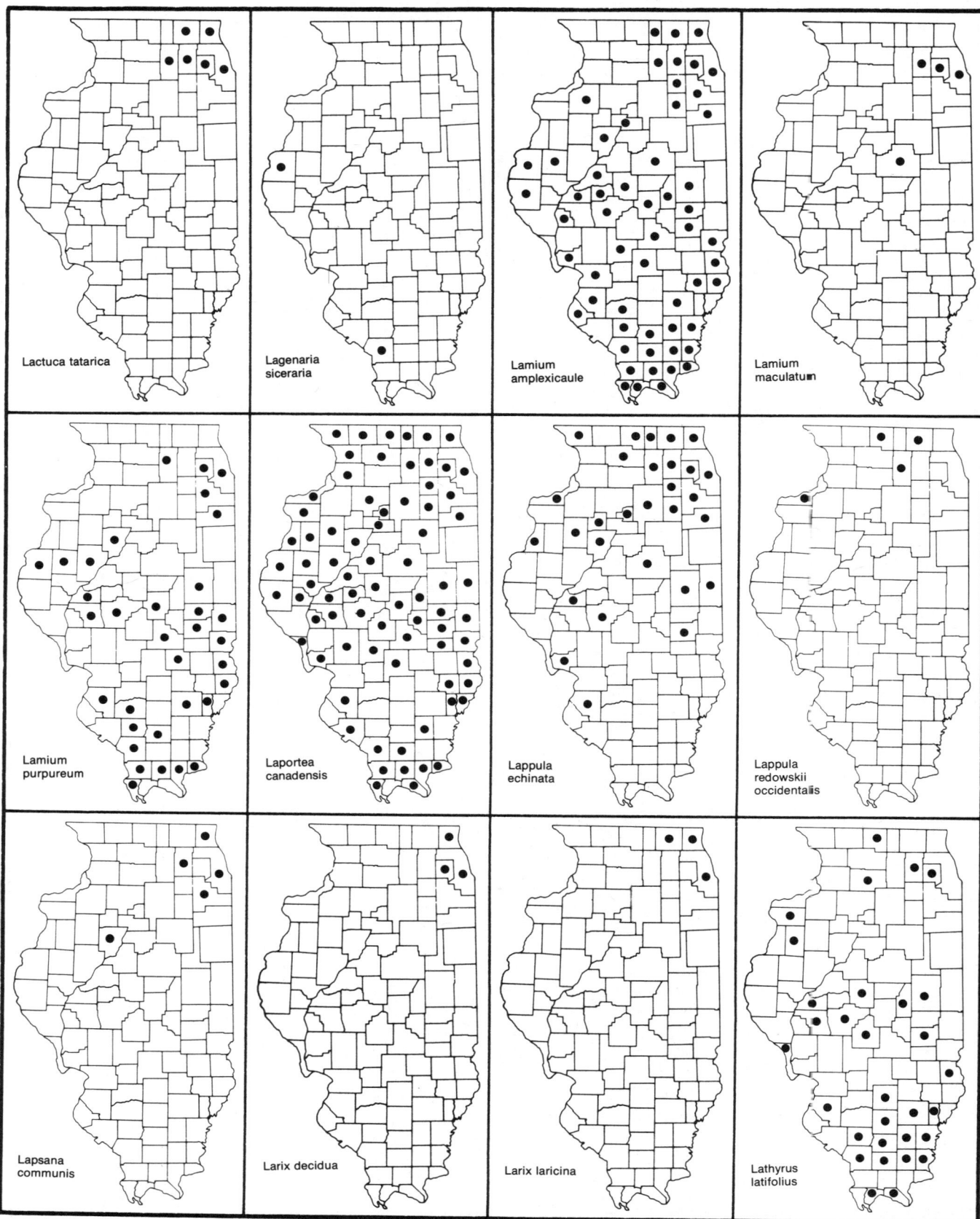

Lactuca tatarica

Lagenaria
siceraria

Lamium
amplexicaule

Lamium
maculatum

Lamium
purpureum

Laportea
canadensis

Lappula
echinata

Lappula
redowskii
occidentalis

Lapsana
communis

Larix decidua

Larix laricina

Lathyrus
latifolius

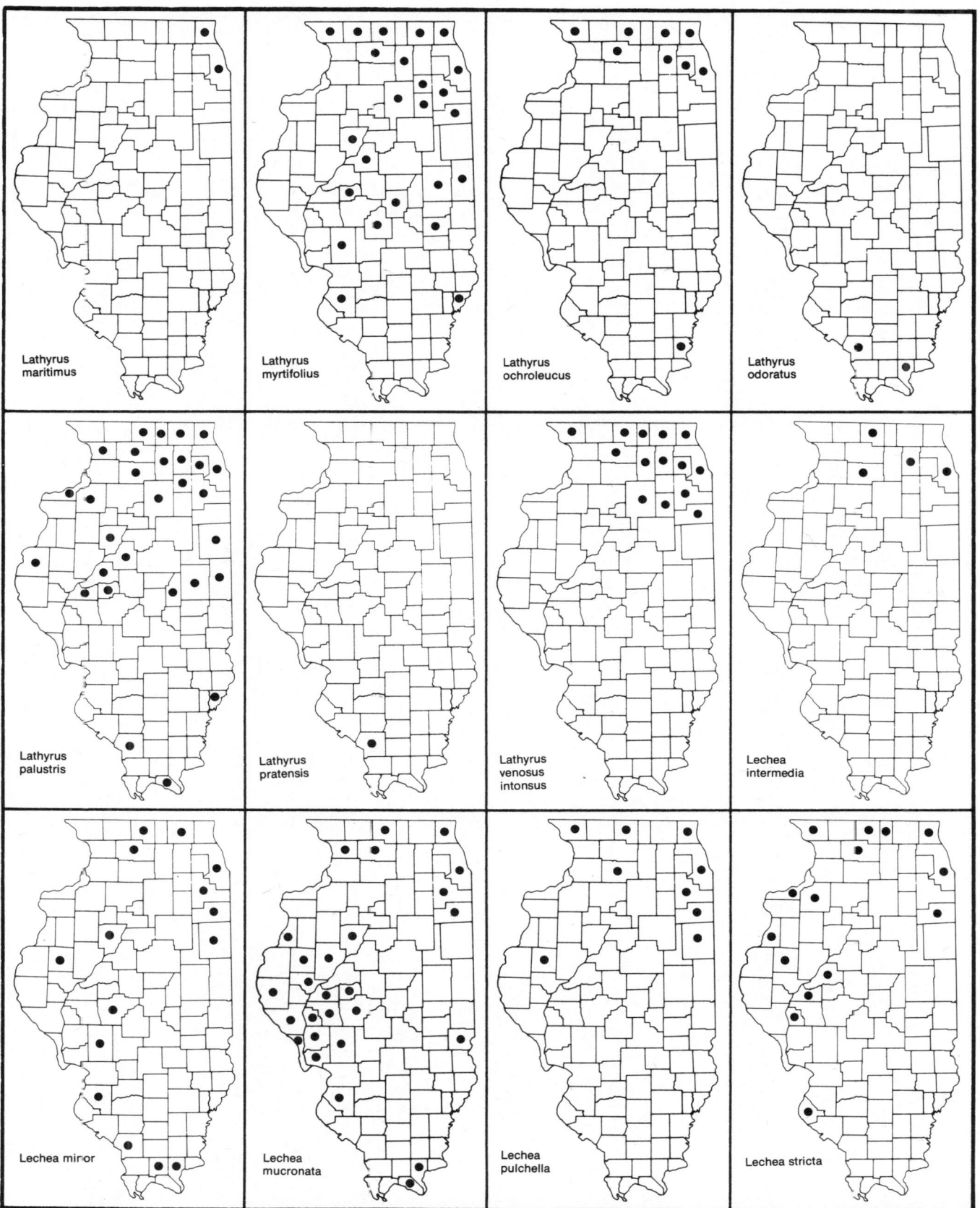

Lathyrus
maritimus

Lathyrus
myrtifolius

Lathyrus
ochroleucus

Lathyrus
odoratus

Lathyrus
palustris

Lathyrus
pratensis

Lathyrus
venosus
intonsus

Lechea
intermedia

Lechea
minor

Lechea
mucronata

Lechea
pulchella

Lechea
stricta

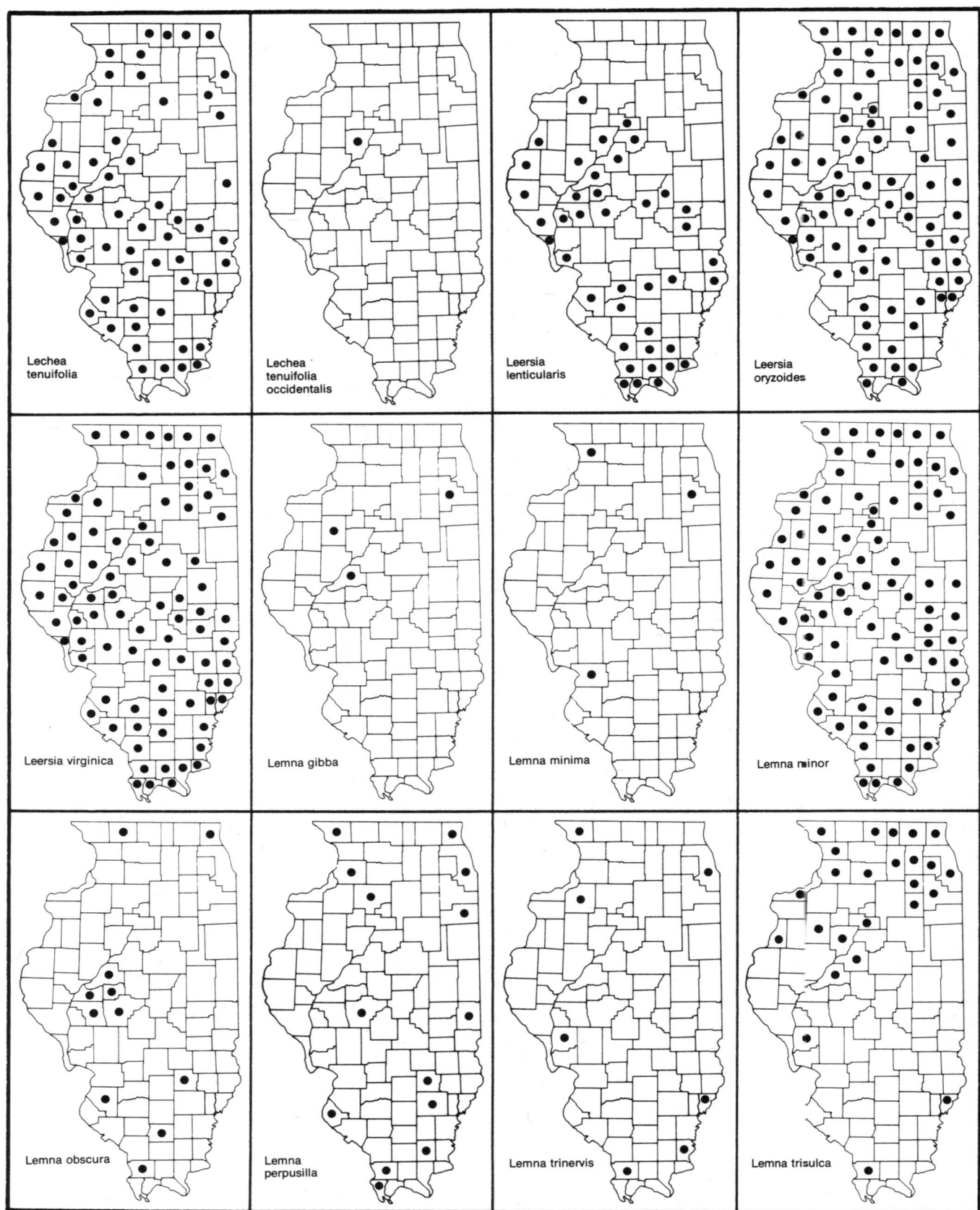

Lechea
tenuifolia

Lechea
tenuifolia
occidentalis

Leersia
lenticularis

Leersia
oryzoides

Leersia virginica

Lemna gibba

Lemna minima

Lemna minor

Lemna obscura

Lemna
perpusilla

Lemna trinervis

Lemna trisulca

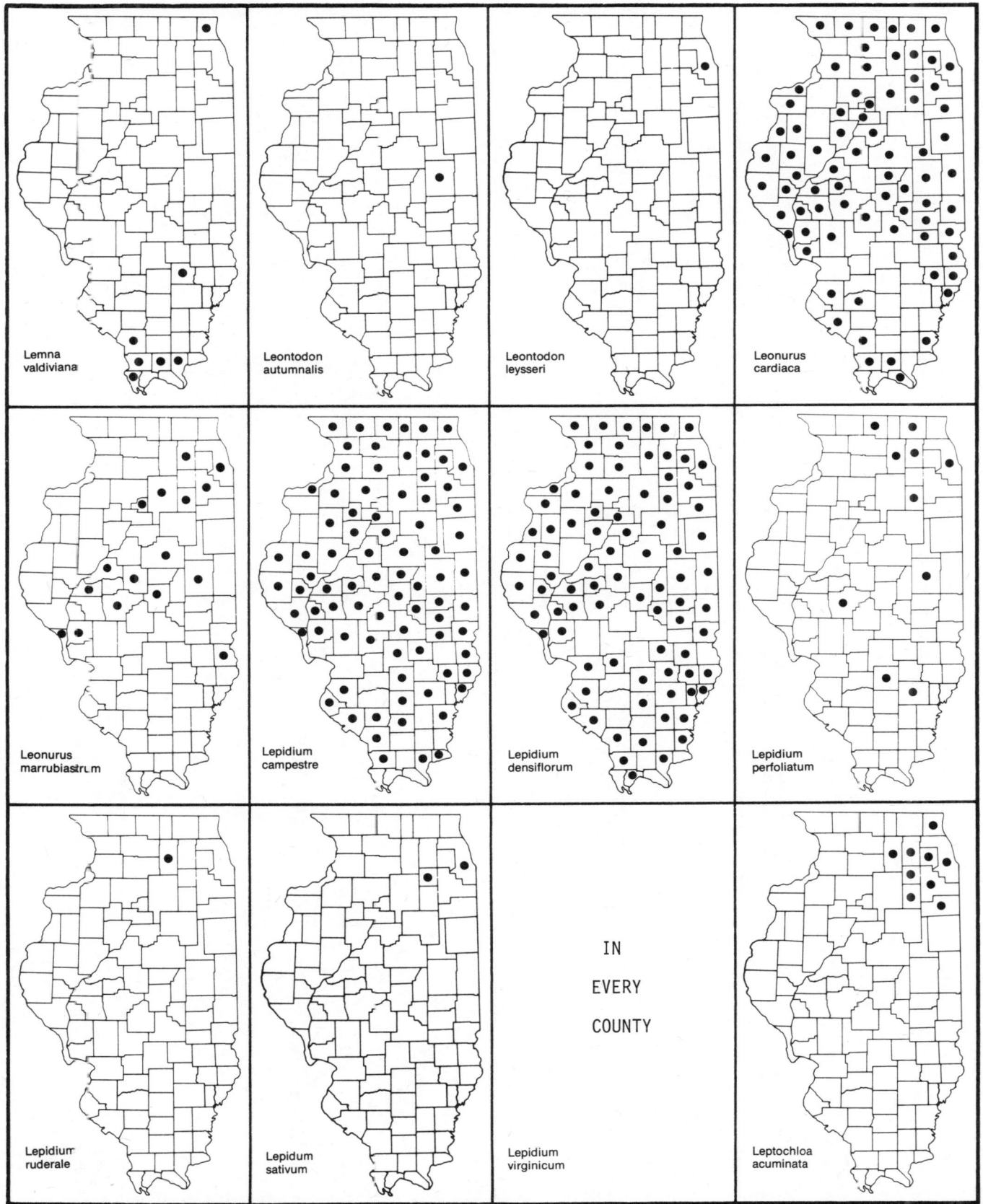

Lemna
valdiviana

Leontodon
autumnalis

Leontodon
leysseri

Leonurus
cardiaca

Leonurus
marrubiastrum

Lepidium
campestre

Lepidium
densiflorum

Lepidium
perfoliatum

Lepidium
ruderale

Lepidum
sativum

Lepidium
virginicum

IN

EVERY

COUNTY

Leptochloa
acuminata

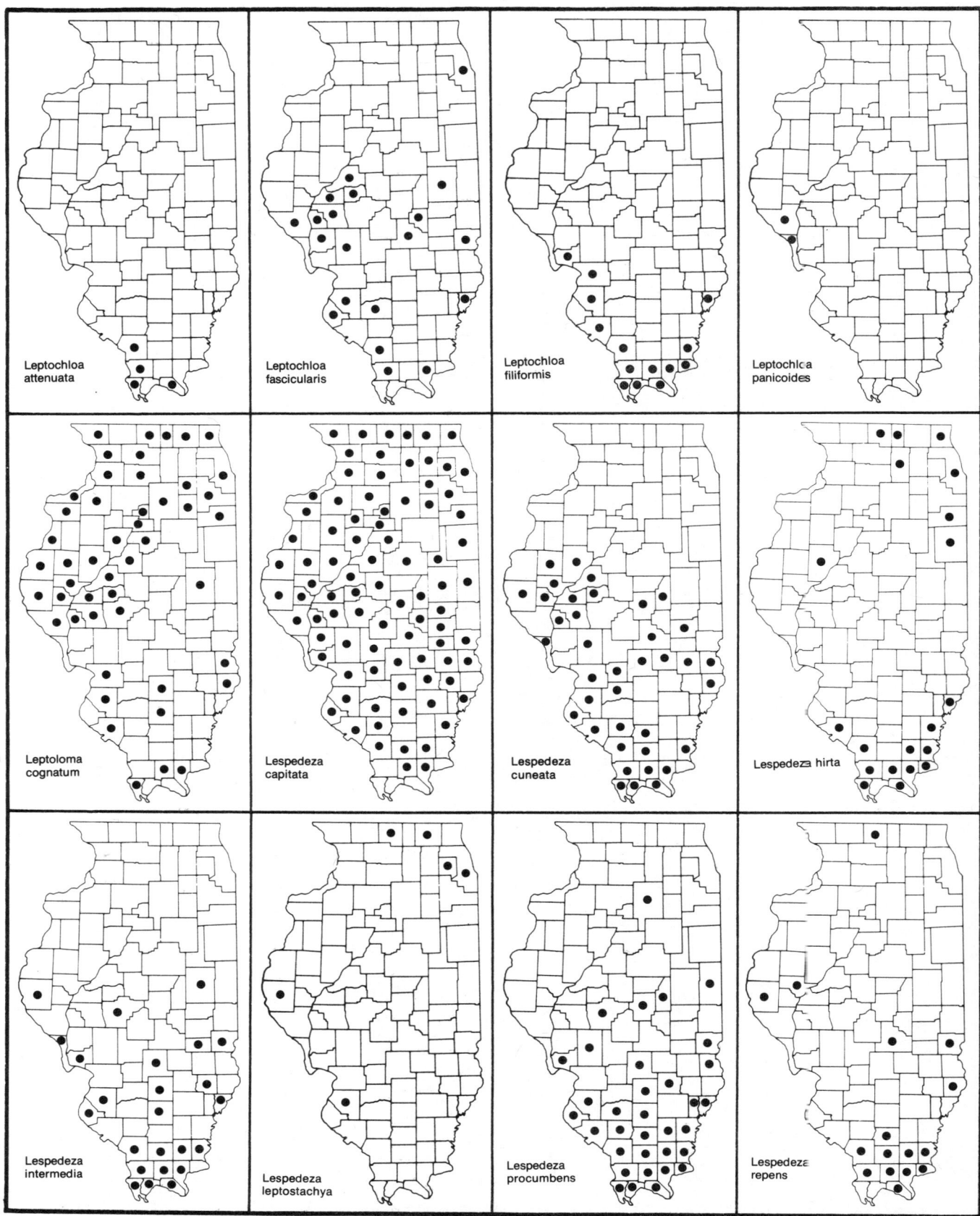

Leptochloa
attenuata

Leptochloa
fascicularis

Leptochloa
filiformis

Leptochloa
panicoides

Leptoloma
cognatum

Lespedeza
capitata

Lespedeza
cuneata

Lespedeza hirta

Lespedeza
intermedia

Lespedeza
leptostachya

Lespedeza
procumbens

Lespedeza
repens

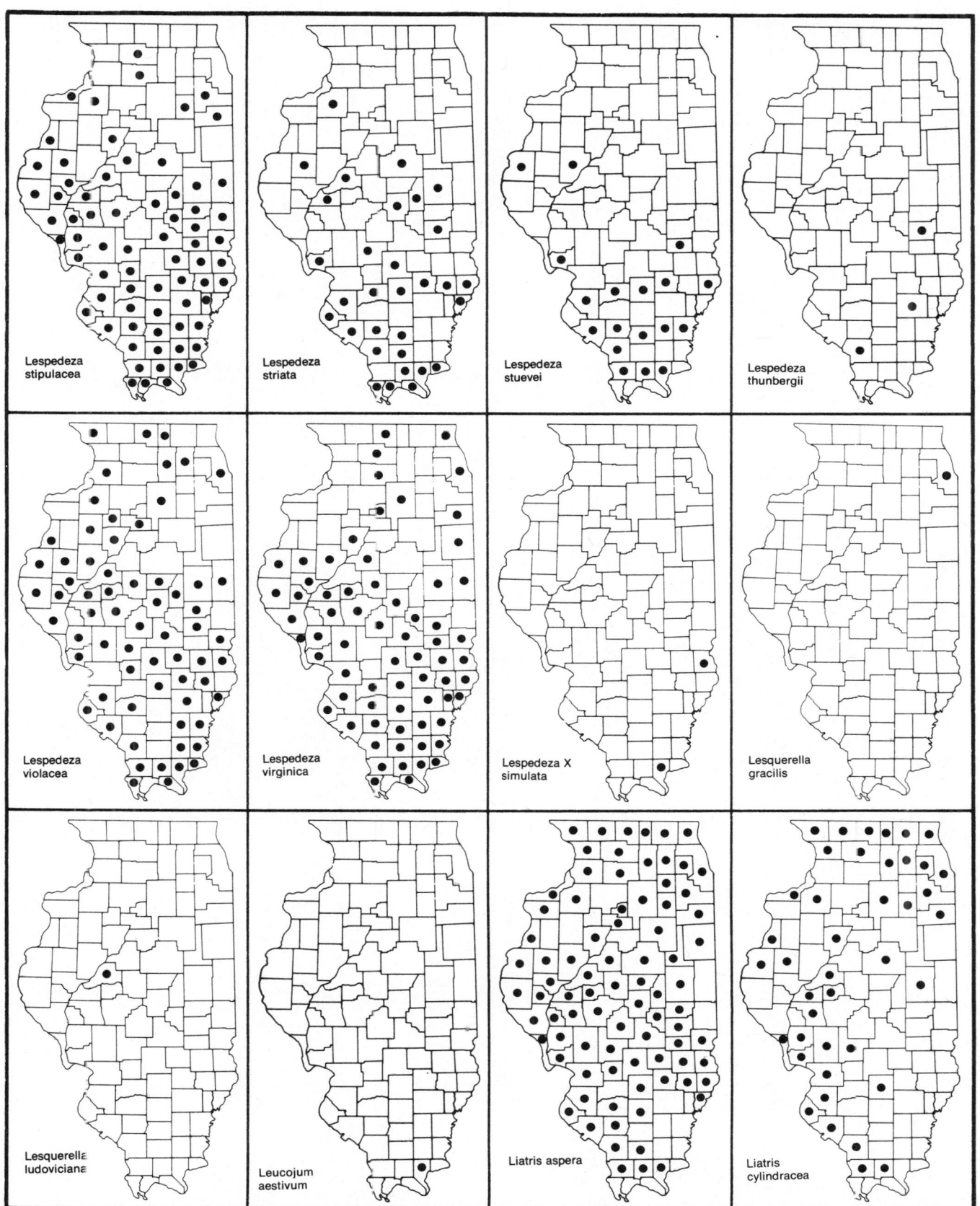

Lespedeza
stipulacea

Lespedeza
striata

Lespedeza
stuevei

Lespedeza
thunbergii

Lespedeza
violacea

Lespedeza
virginica

Lespedeza X
simulata

Lesquerella
gracilis

Lesquerella
ludoviciana

Leucojum
aestivum

Liatris
aspera

Liatris
cylindracea

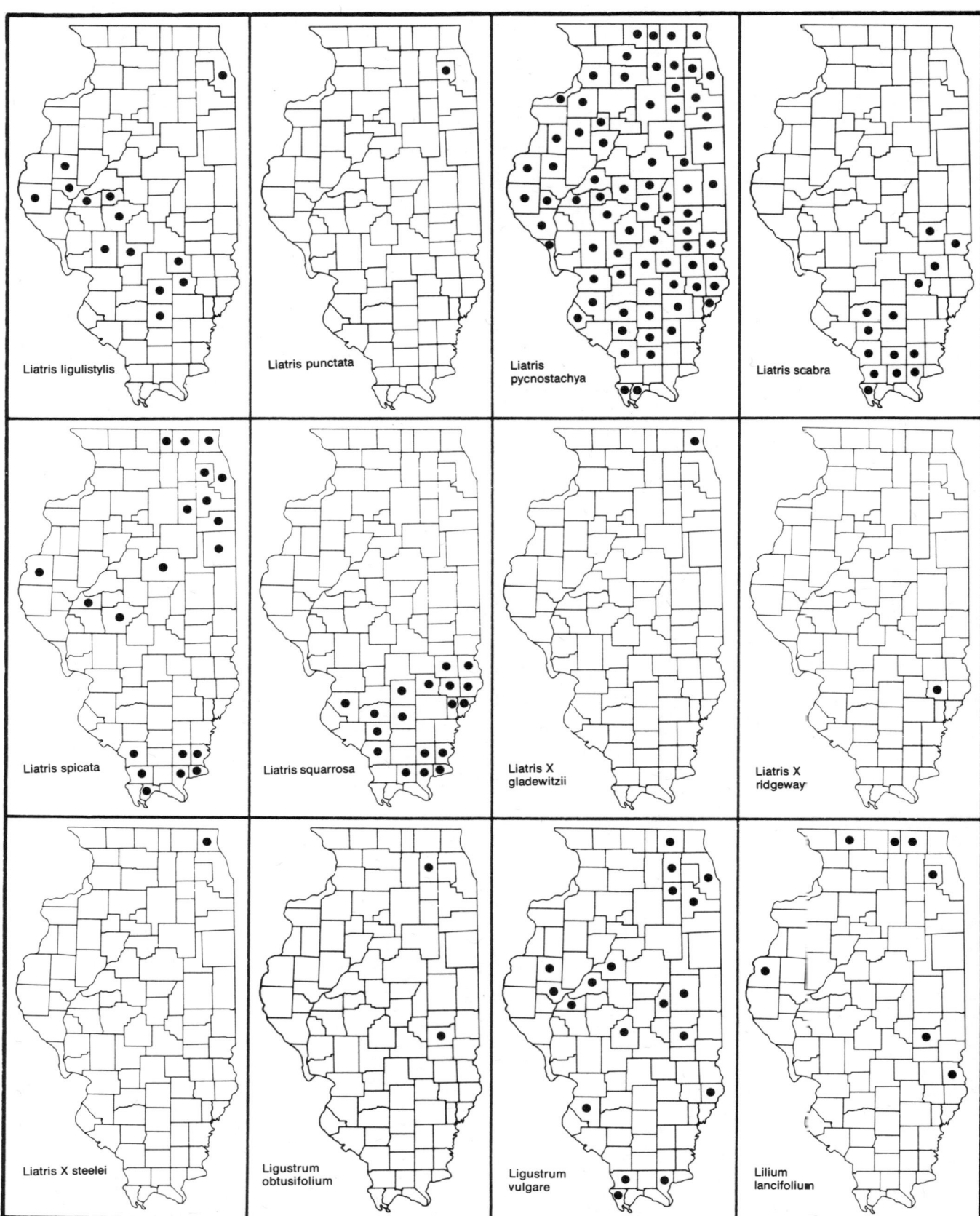

Liatris ligulistylis

Liatris punctata

Liatris
pycnostachya

Liatris scabra

Liatris spicata

Liatris squarrosa

Liatris X
gladewitzii

Liatris X
ridgeway

Liatris X steelei

Ligustrum
obtusifolium

Ligustrum
vulgare

Lilium
lancifolium

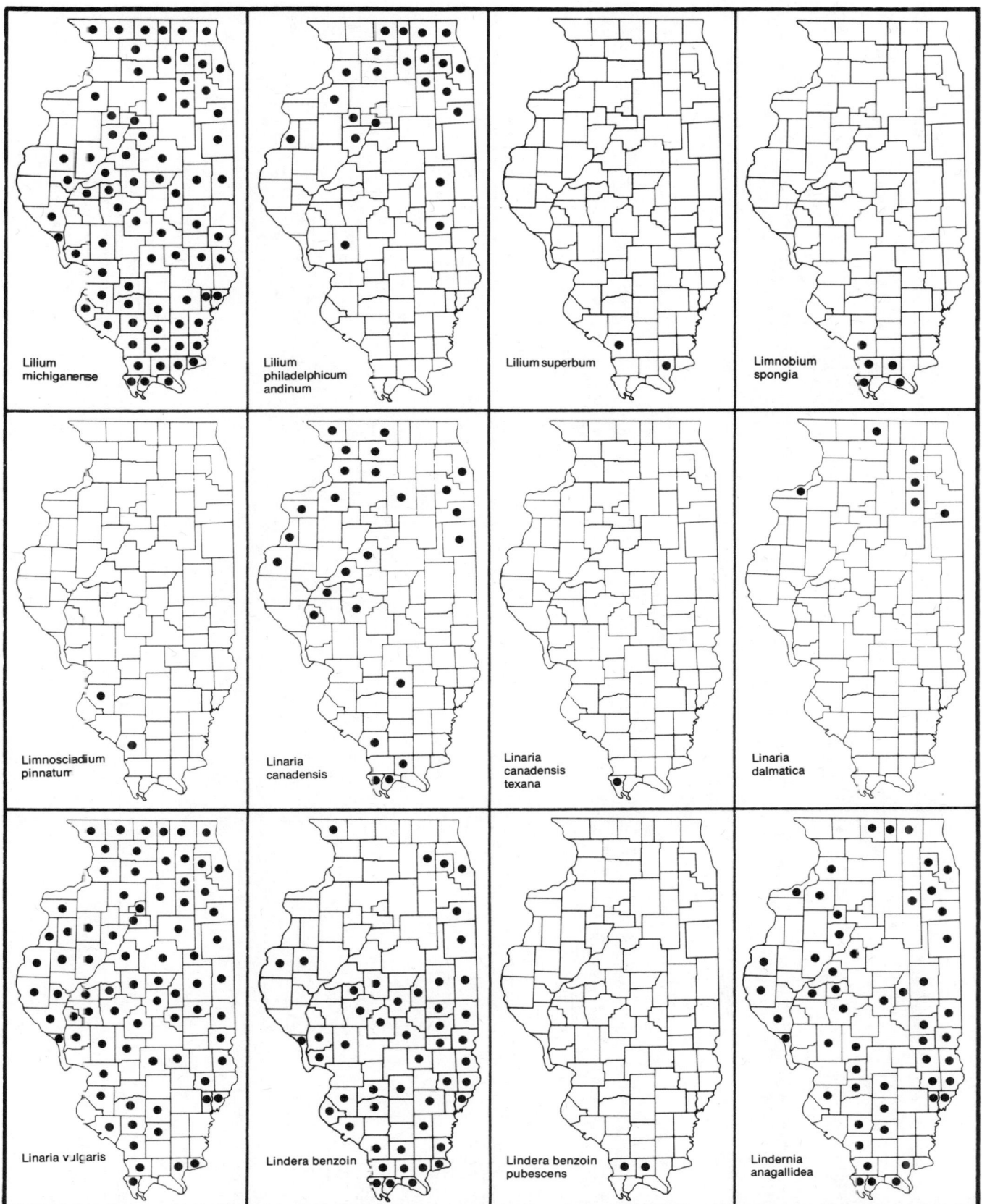

Lilium
michiganense

Lilium
philadelphicum
andinum

Lilium superbum

Limnobium
spongia

Limnosciadium
pinnatum

Linaria
canadensis

Linaria
canadensis
texana

Linaria
dalmatica

Linaria vulgaris

Lindera benzoin

Lindera benzoin
pubescens

Lindernia
anagallidea

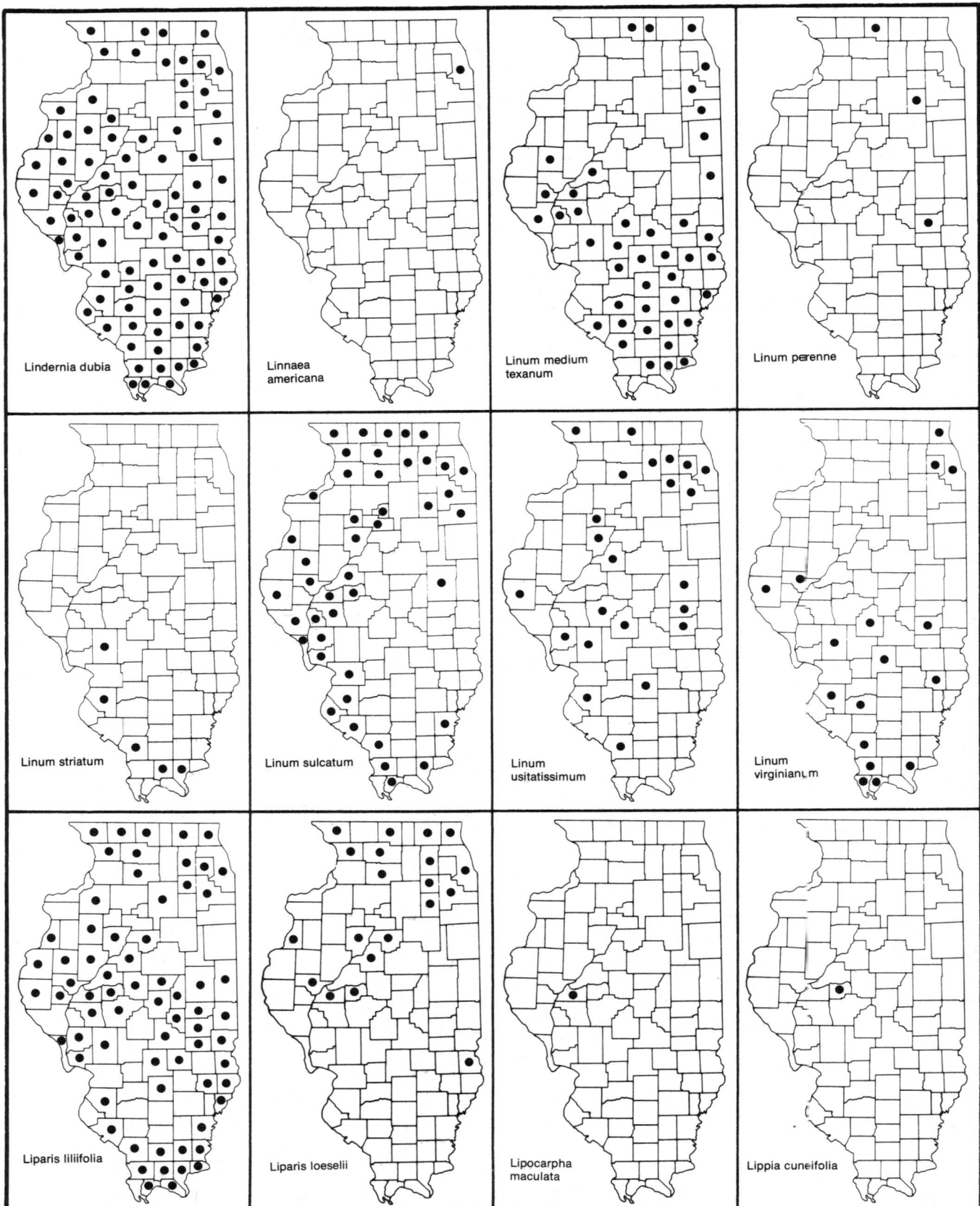

Lindernia dubia

Linnaea americana

Linum medium texanum

Linum perenne

Linum striatum

Linum sulcatum

Linum usitatissimum

Linum virginianum

Liparis liliifolia

Liparis loeselii

Lipocarpha maculata

Lippia cuneifolia

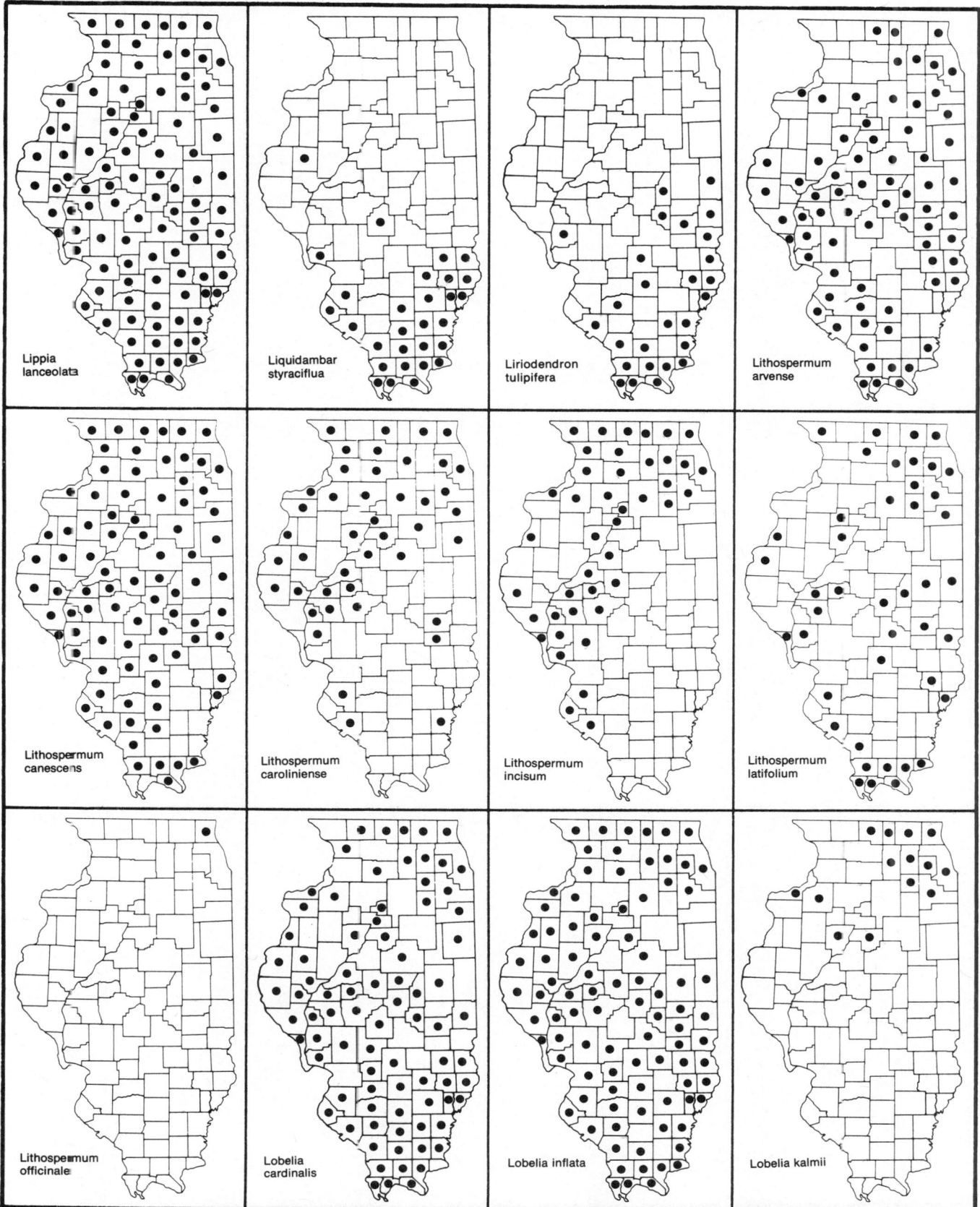

Lippia
lanceolata

Liquidambar
styraciflua

Liriodendron
tulipifera

Lithospermum
arvense

Lithospermum
canescens

Lithospermum
caroliniense

Lithospermum
incisum

Lithospermum
latifolium

Lithospermum
officinale

Lobelia
cardinalis

Lobelia inflata

Lobelia kalmii

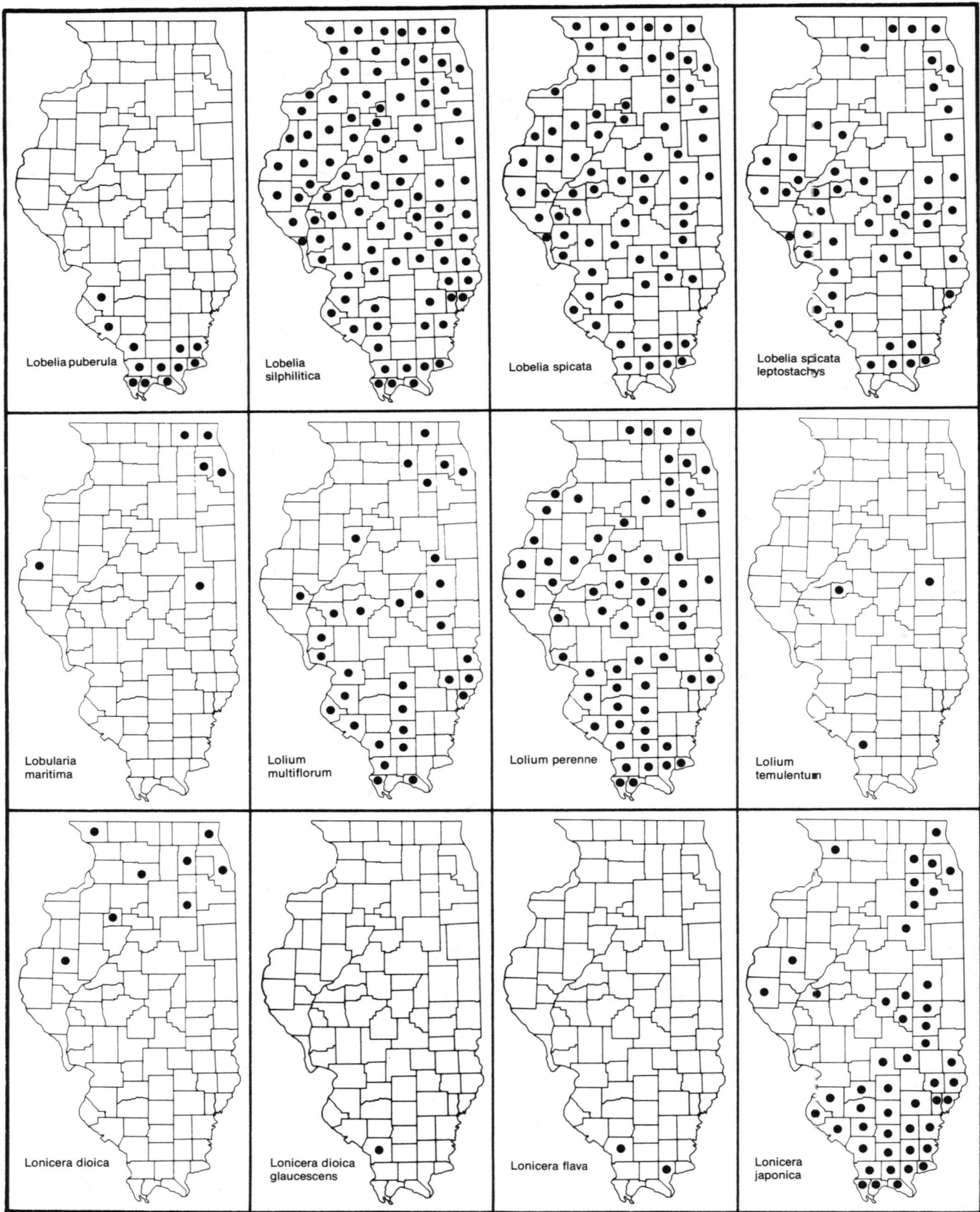

Lobelia puberula

Lobelia silphilitica

Lobelia spicata

Lobelia spicata leptostachys

Lobularia maritima

Lolium multiflorum

Lolium perenne

Lolium temulentum

Lonicera dioica

Lonicera dioica glaucescens

Lonicera flava

Lonicera japonica

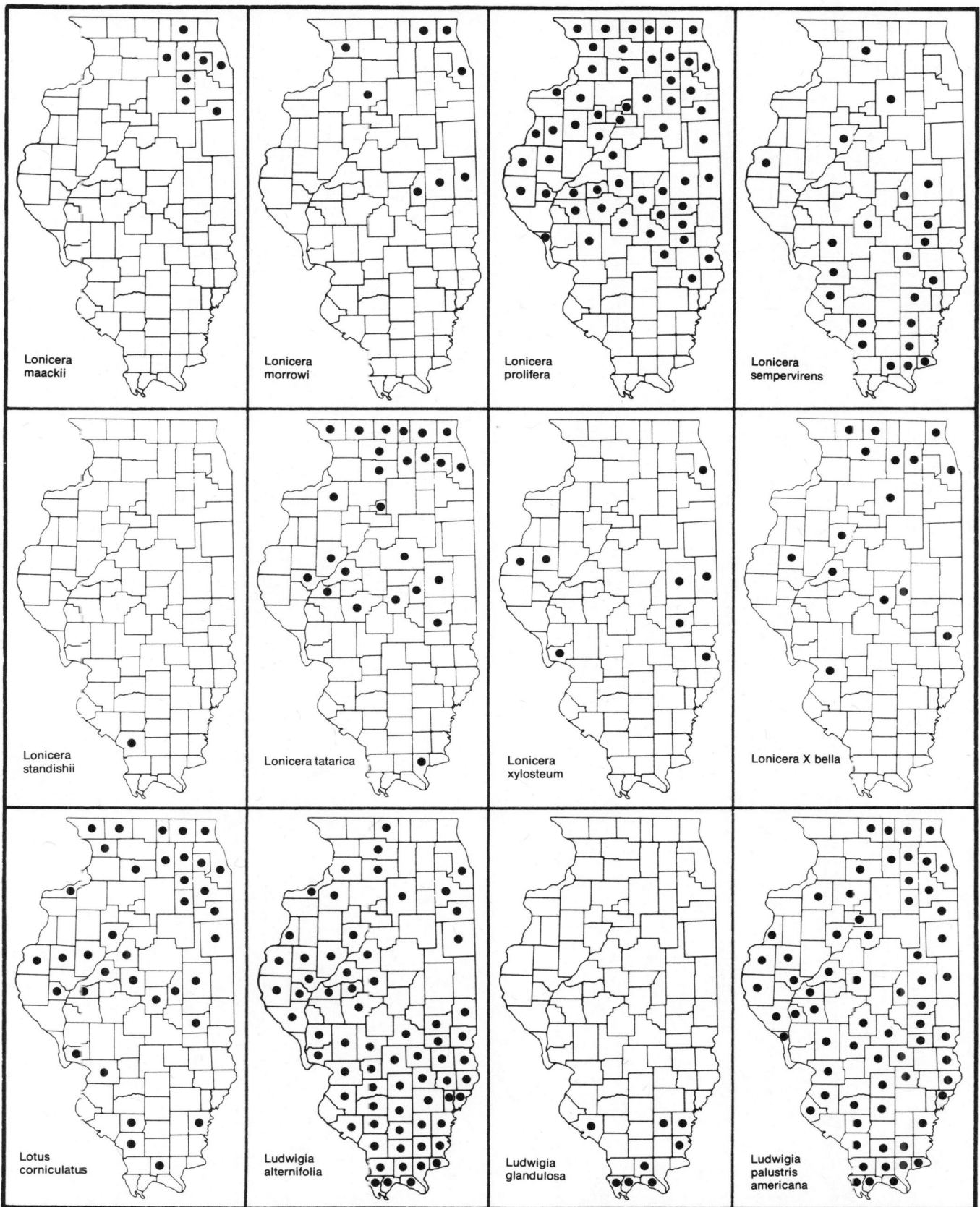

Lonicera maackii

Lonicera morrowi

Lonicera prolifera

Lonicera sempervirens

Lonicera standishii

Lonicera tatarica

Lonicera xylosteum

Lonicera X bella

Lotus corniculatus

Ludwigia alternifolia

Ludwigia glandulosa

Ludwigia palustris americana

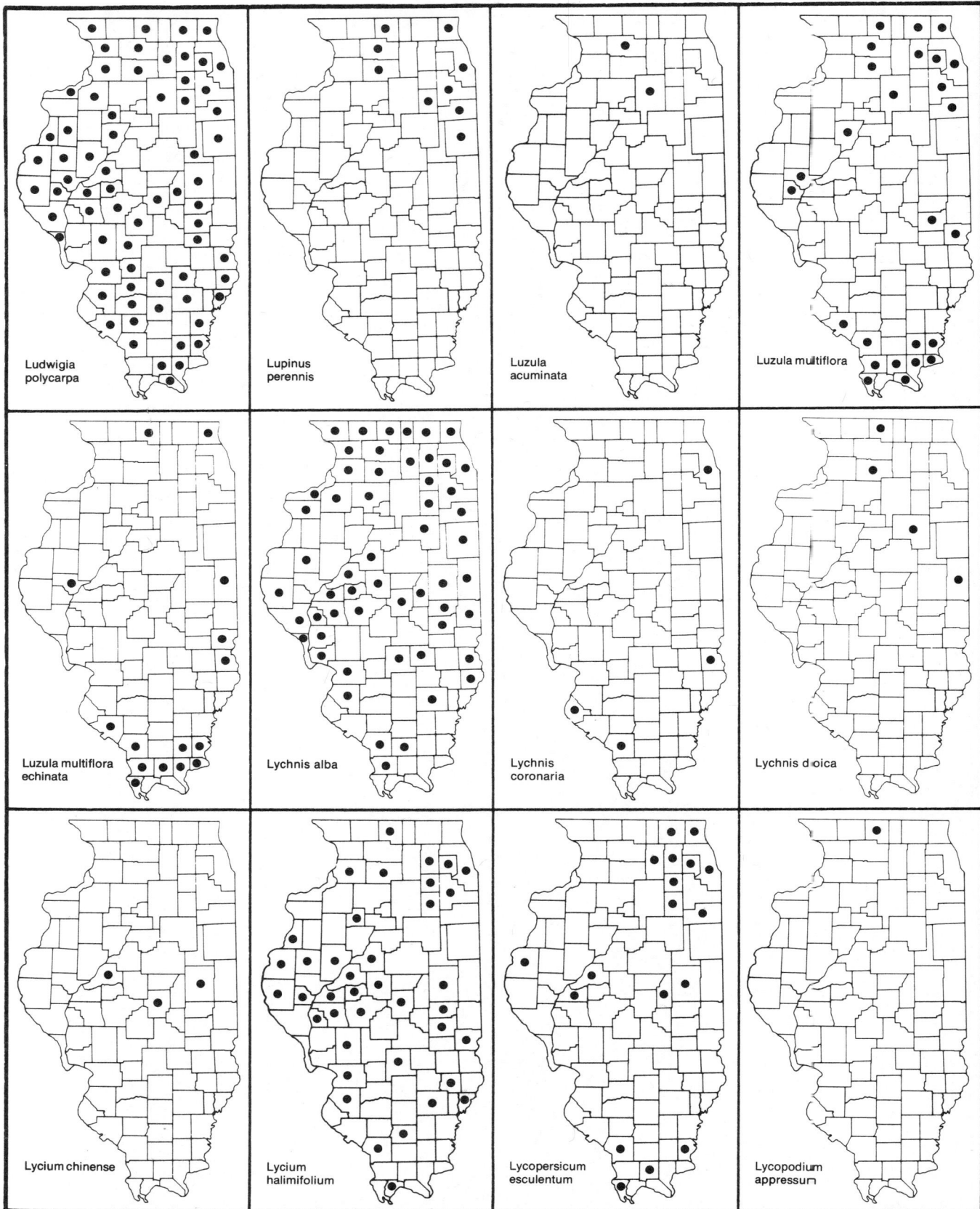

Ludwigia
polycarpa

Lupinus
perennis

Luzula
acuminata

Luzula multiflora

Luzula multiflora
echinata

Lychnis alba

Lychnis
coronaria

Lychnis dioica

Lycium chinense

Lycium
halimifolium

Lycopersicum
esculentum

Lycopodium
appressum

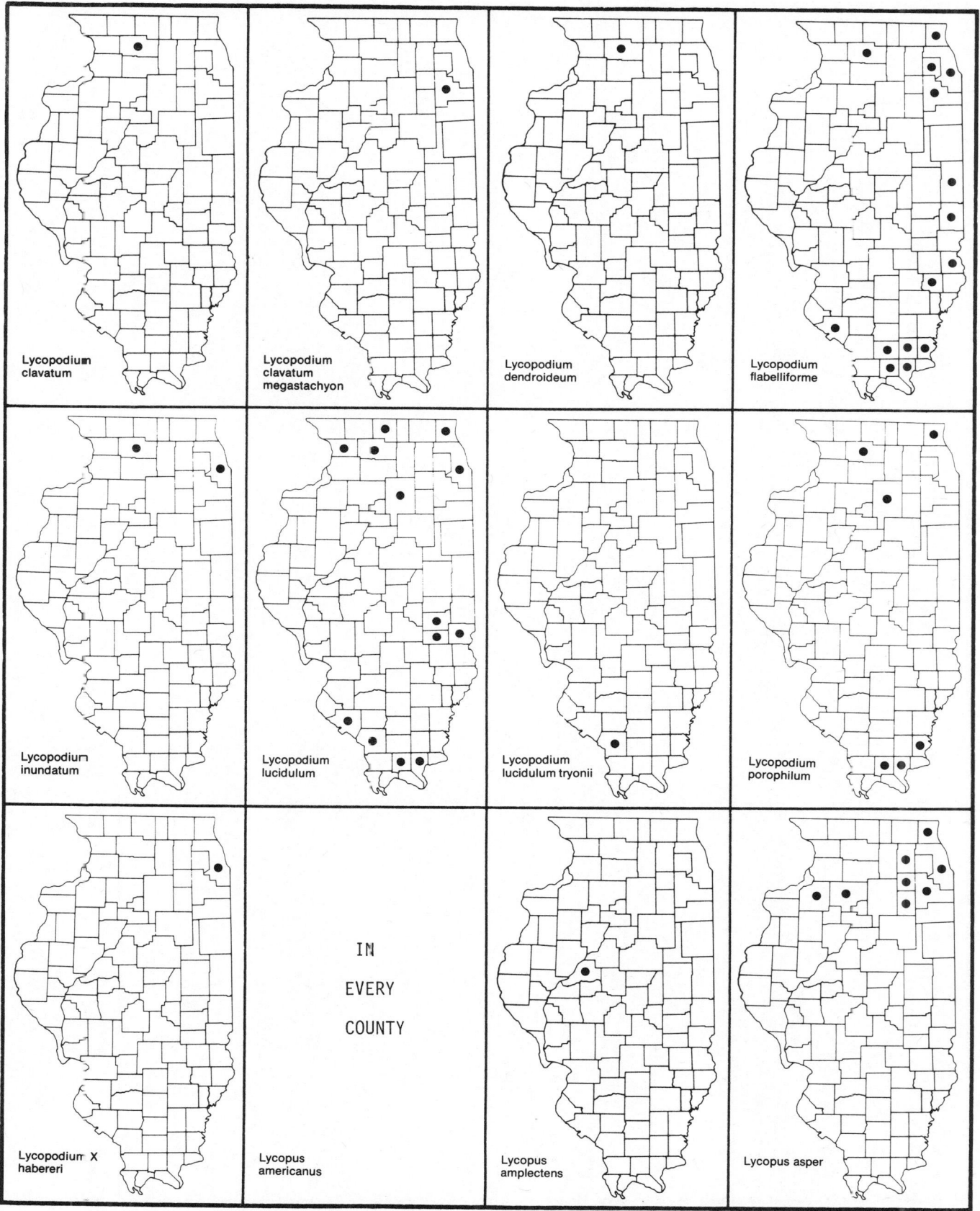

Lycopodium
clavatum

Lycopodium
clavatum
megastachyon

Lycopodium
dendroideum

Lycopodium
flabelliforme

Lycopodium
inundatum

Lycopodium
lucidulum

Lycopodium
lucidulum tryonii

Lycopodium
porophilum

Lycopodium X
habereri

Lycopus
americanus

IN

EVERY

COUNTY

Lycopus
amplectens

Lycopus asper

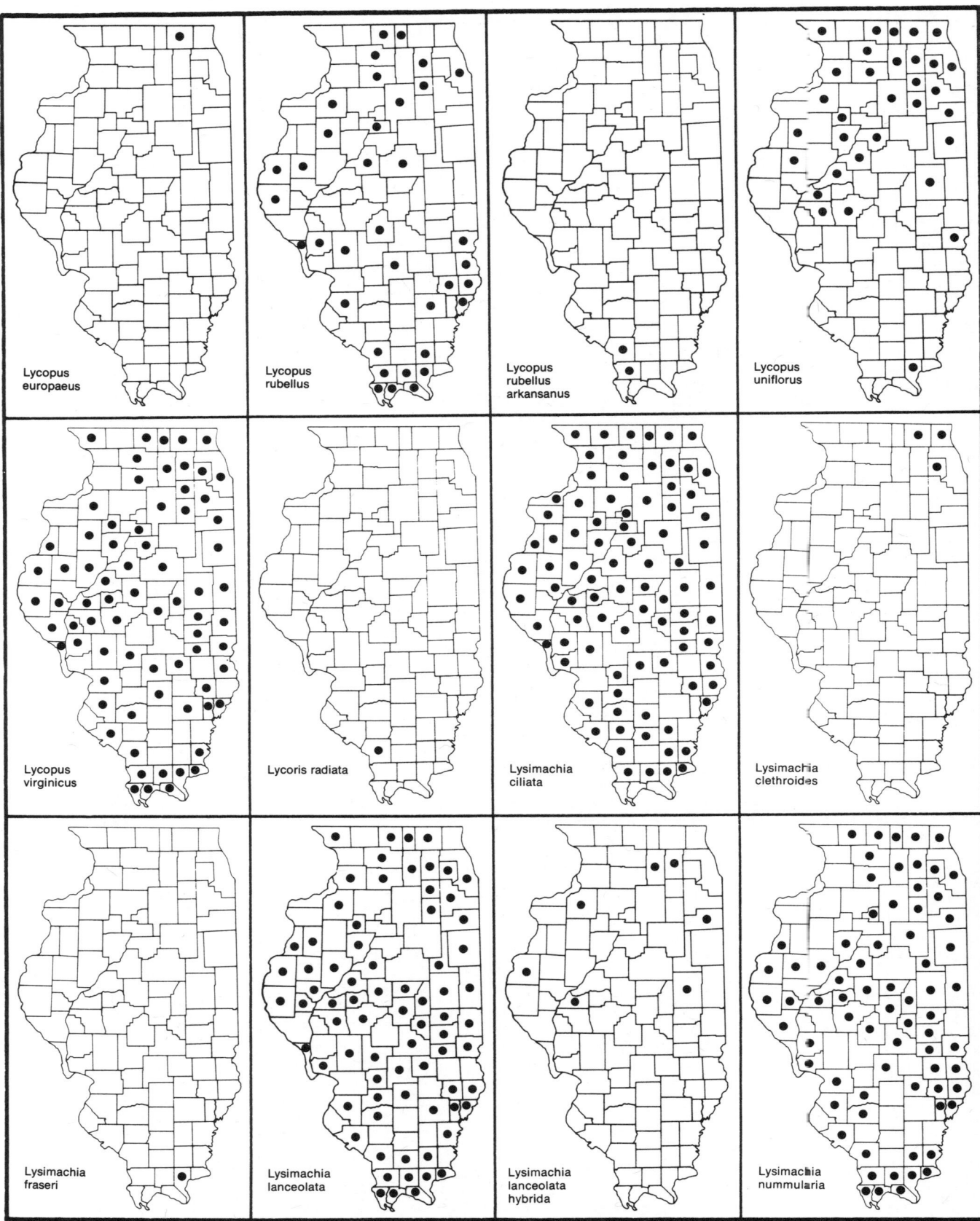

Lycopus
europaeus

Lycopus
rubellus

Lycopus
rubellus
arkansanus

Lycopus
uniflorus

Lycopus
virginicus

Lycoris radiata

Lysimachia
ciliata

Lysimachia
clethroides

Lysimachia
fraseri

Lysimachia
lanceolata

Lysimachia
lanceolata
hybrida

Lysimachia
nummularia

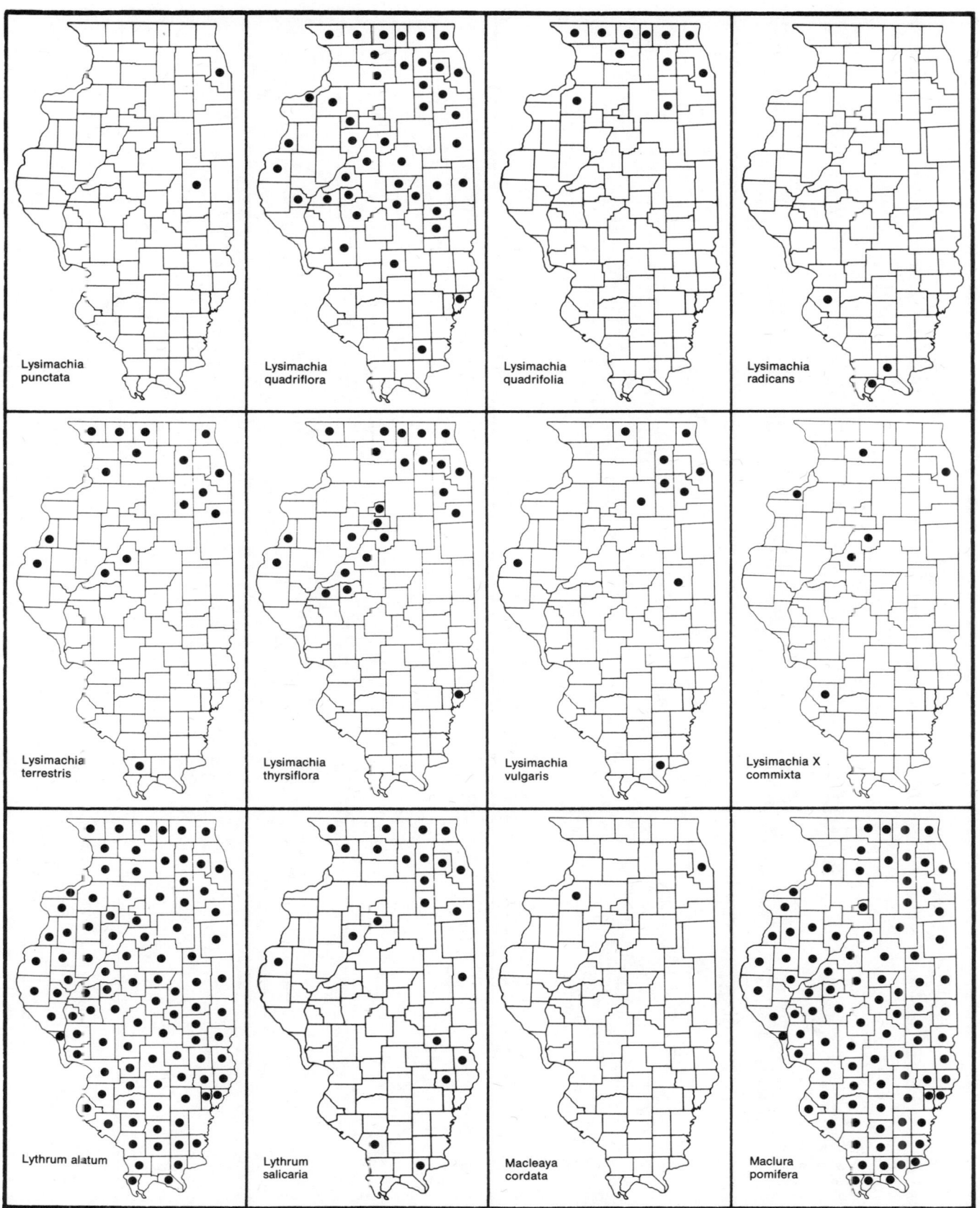

Lysimachia
punctata

Lysimachia
quadriflora

Lysimachia
quadrifolia

Lysimachia
radicans

Lysimachia
terrestris

Lysimachia
thyrsiflora

Lysimachia
vulgaris

Lysimachia X
commixta

Lythrum alatum

Lythrum
salicaria

Macleaya
cordata

Maclura
pomifera

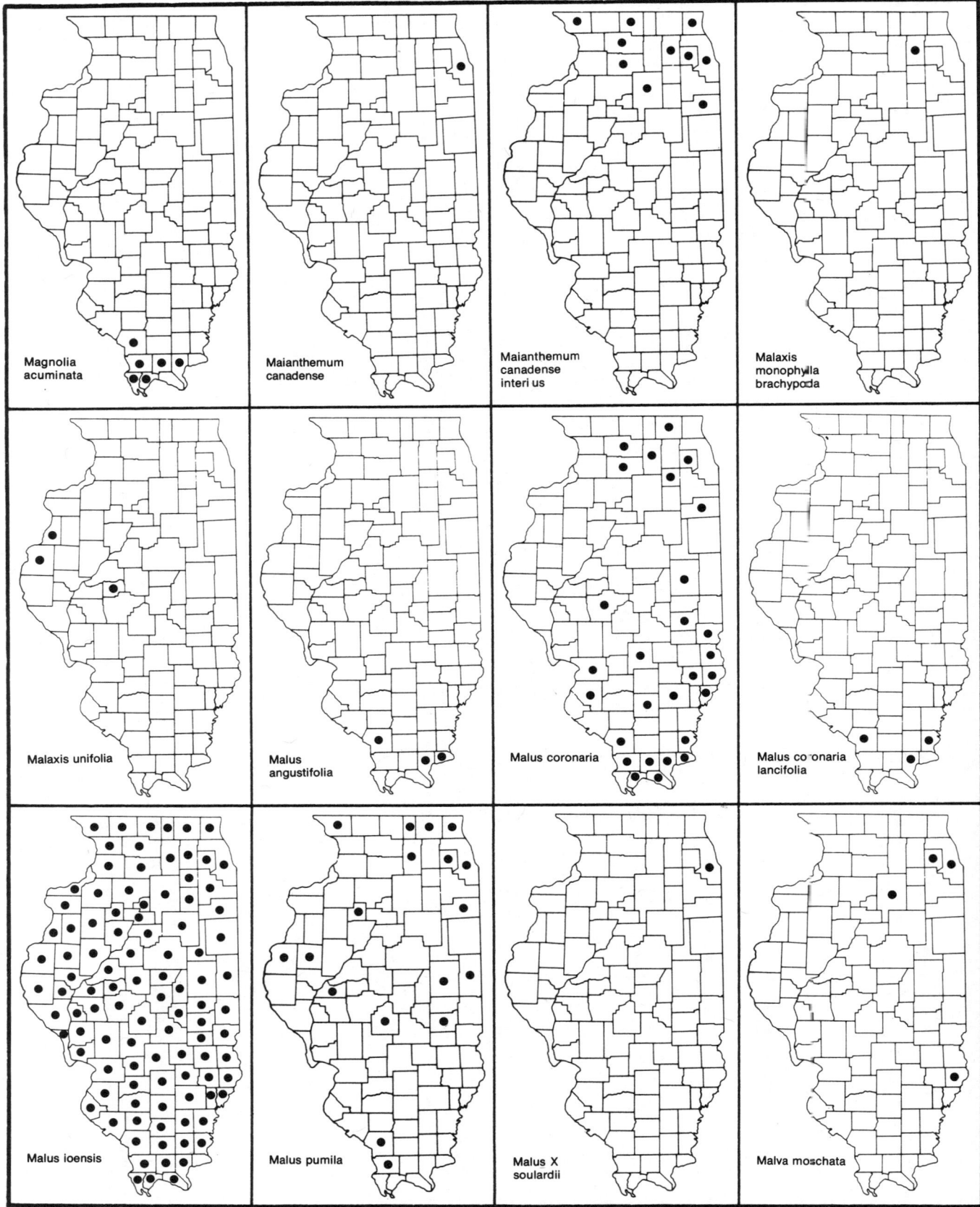

Magnolia
acuminata

Maianthemum
canadense

Maianthemum
canadense
interi us

Malaxis
monophylla
brachypoda

Malaxis unifolia

Malus
angustifolia

Malus coronaria

Malus coronaria
lancifolia

Malus ioensis

Malus pumila

Malus X
soulardii

Malva moschata

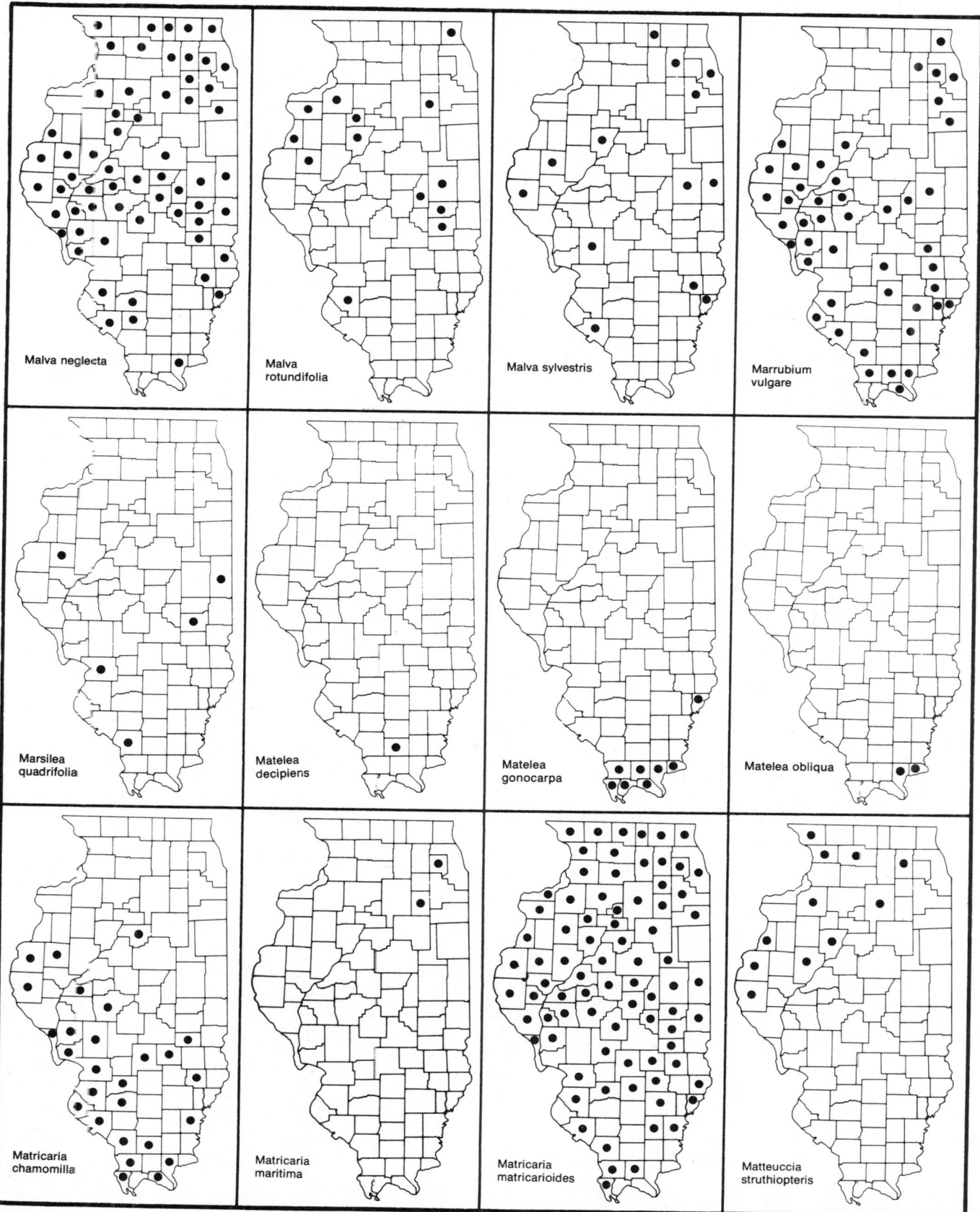

Malva neglecta

Malva
rotundifolia

Malva sylvestris

Marrubium
vulgare

Marsilea
quadrifolia

Matelea
decipiens

Matelea
gonocarpa

Matelea obliqua

Matricaria
chamomilla

Matricaria
maritima

Matricaria
matricarioides

Matteuccia
struthiopteris

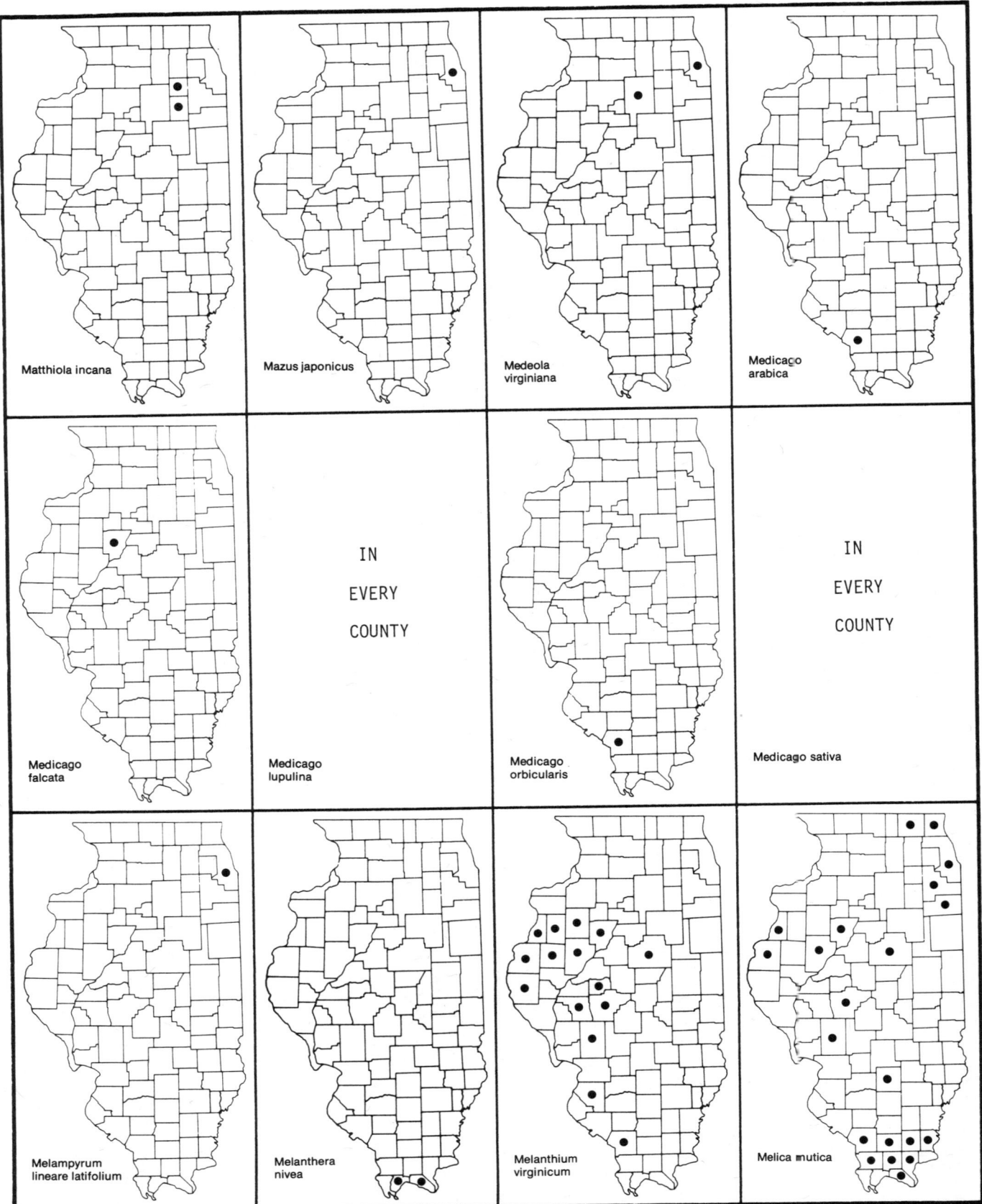

Matthiola incana

Mazus japonicus

Medeola
virginiana

Medicago
arabica

Medicago
falcata

Medicago
lupulina

IN

EVERY

COUNTY

Medicago
orbicularis

Medicago sativa

IN

EVERY

COUNTY

Melampyrum
lineare latifolium

Melanthera
nivea

Melanthium
virginicum

Melica mutica

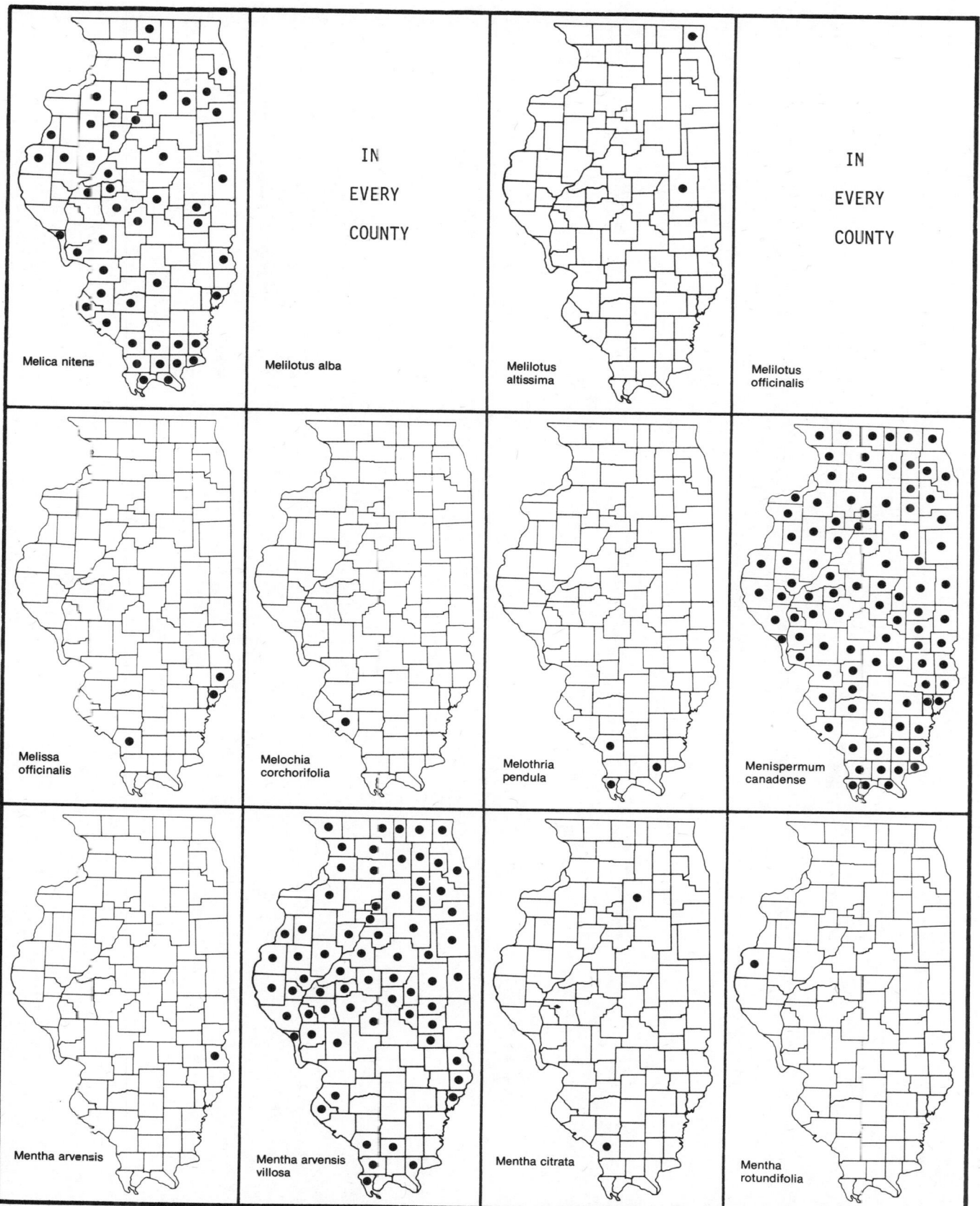

Melica nitens

Melilotus alba

IN

EVERY

COUNTY

Melilotus
altissima

Melilotus
officinalis

IN

EVERY

COUNTY

Melissa
officinalis

Melochia
corchorifolia

Melothria
pendula

Menispermum
canadense

Mentha arvensis

Mentha arvensis
villosa

Mentha citrata

Mentha
rotundifolia

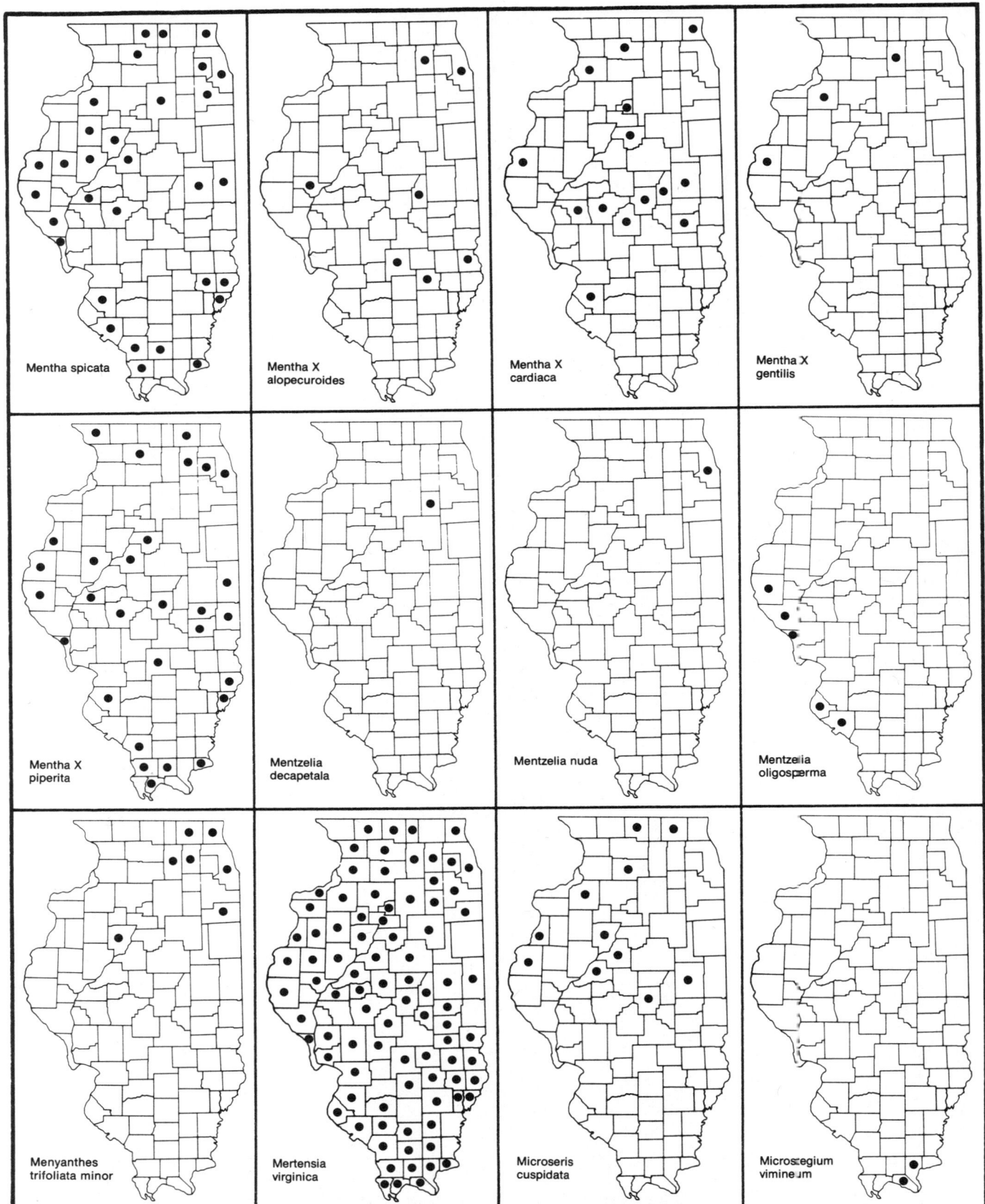

Mentha spicata

Mentha X
alopecuroides

Mentha X
cardiaca

Mentha X
gentilis

Mentha X
piperita

Mentzelia
decapetala

Mentzelia nuda

Mentzelia
oligosperma

Menyanthes
trifoliata minor

Mertensia
virginica

Microseris
cuspidata

Microstegium
vimineum

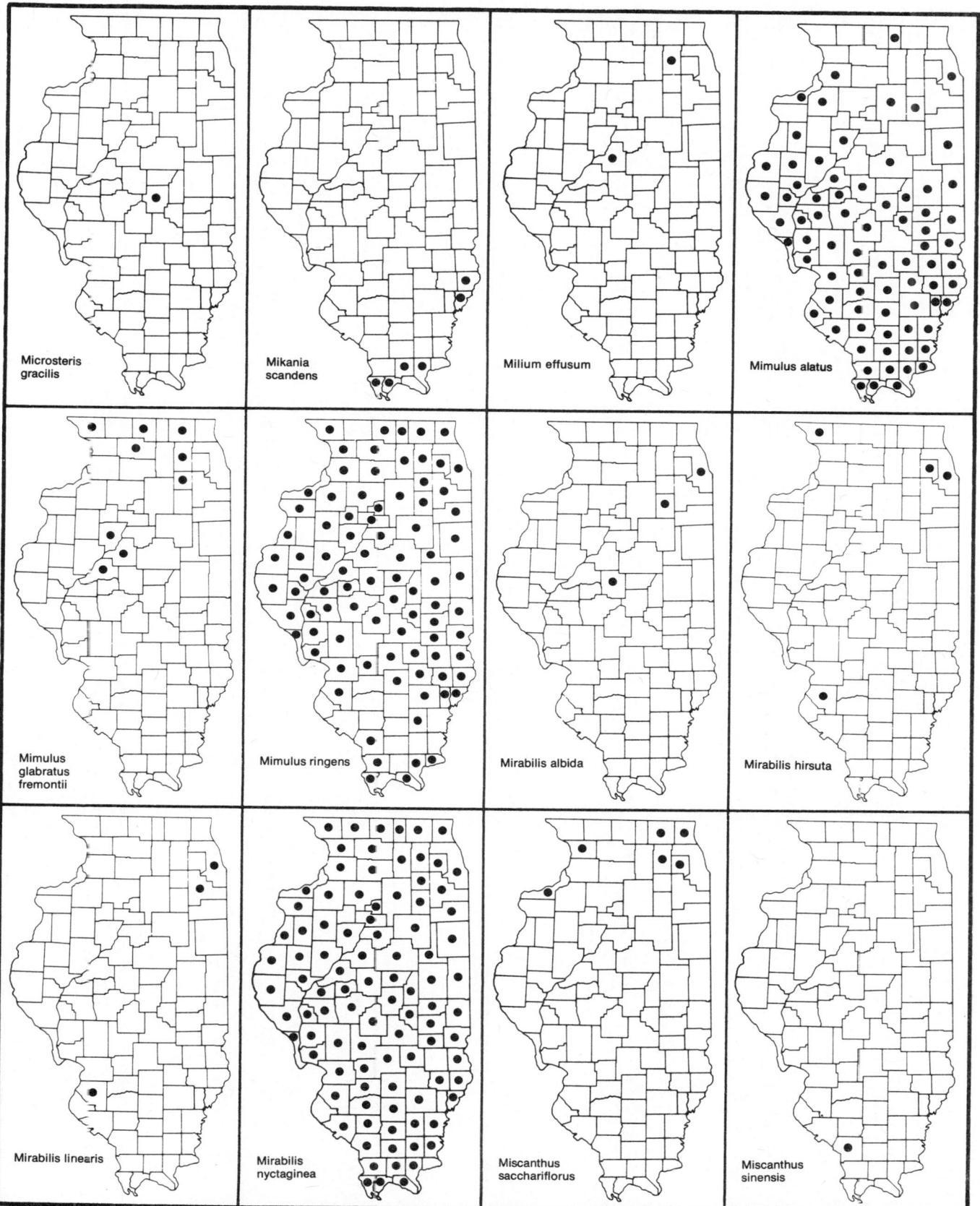

Microsteris
gracilis

Mikania
scandens

Milium effusum

Mimulus alatus

Mimulus
glabratus
fremontii

Mimulus ringens

Mirabilis albida

Mirabilis hirsuta

Mirabilis linearis

Mirabilis
nyctaginea

Miscanthus
sacchariflorus

Miscanthus
sinensis

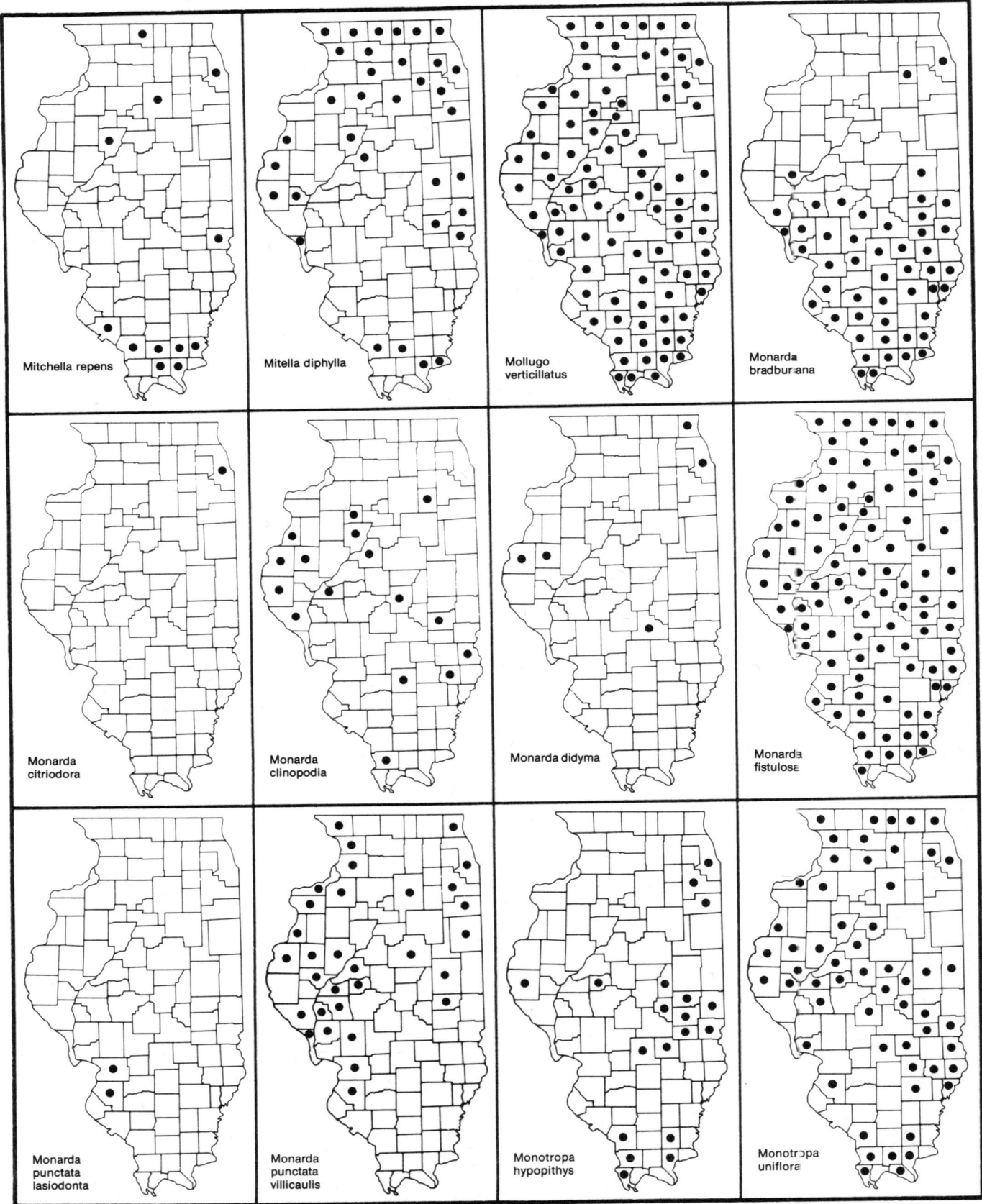

Mitchella repens

Mitella diphylla

Mollugo
verticillatus

Monarda
bradburiana

Monarda
citriodora

Monarda
clinopodia

Monarda didyma

Monarda
fistulosa

Monarda
punctata
lasiodonta

Monarda
punctata
villicaulis

Monotropa
hypopithys

Monotropa
uniflora

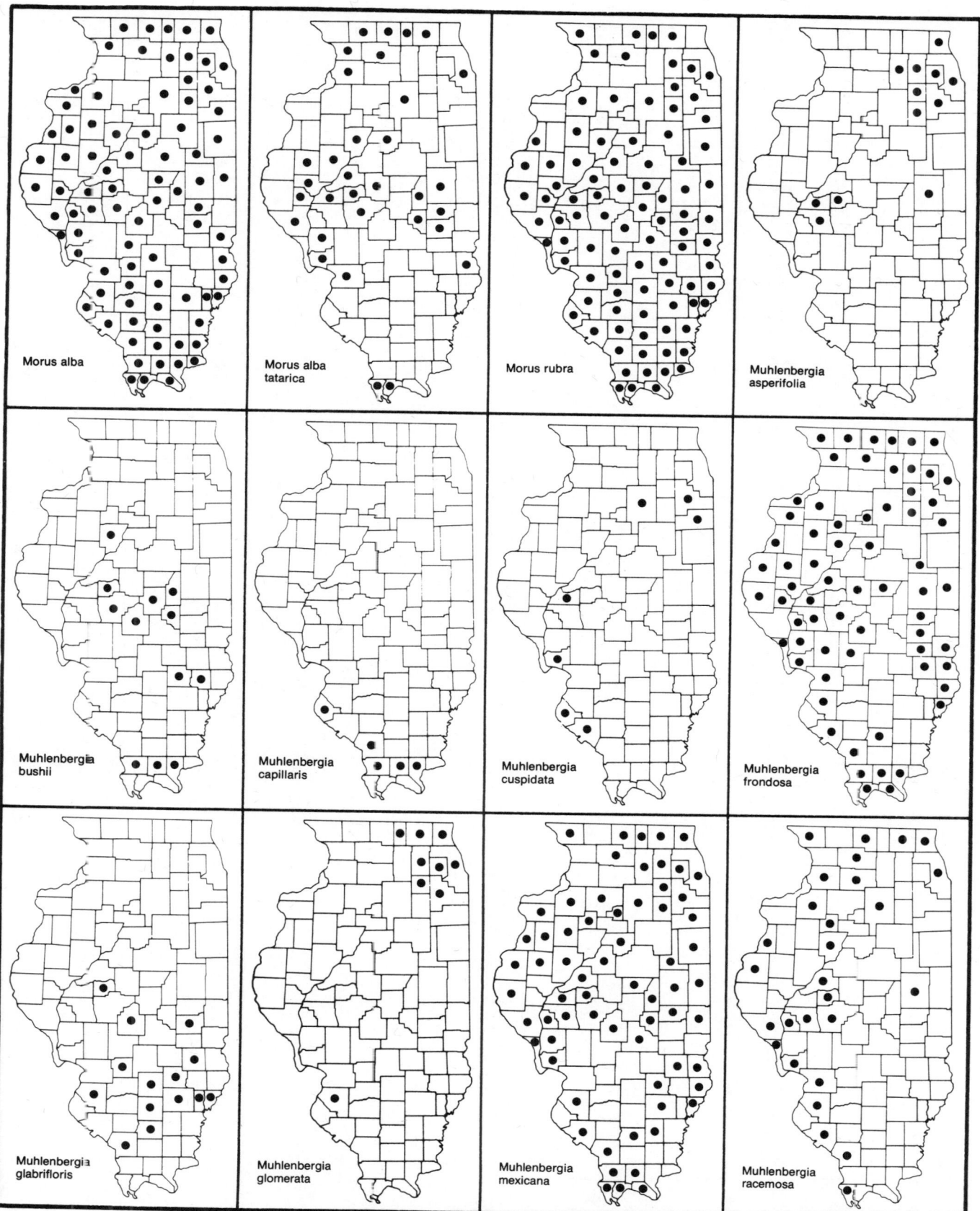

Morus alba

Morus alba
tatarica

Morus rubra

Muhlenbergia
asperifolia

Muhlenbergia
bushii

Muhlenbergia
capillaris

Muhlenbergia
cuspidata

Muhlenbergia
frondosa

Muhlenbergia
glabrifloris

Muhlenbergia
glomerata

Muhlenbergia
mexicana

Muhlenbergia
racemosa

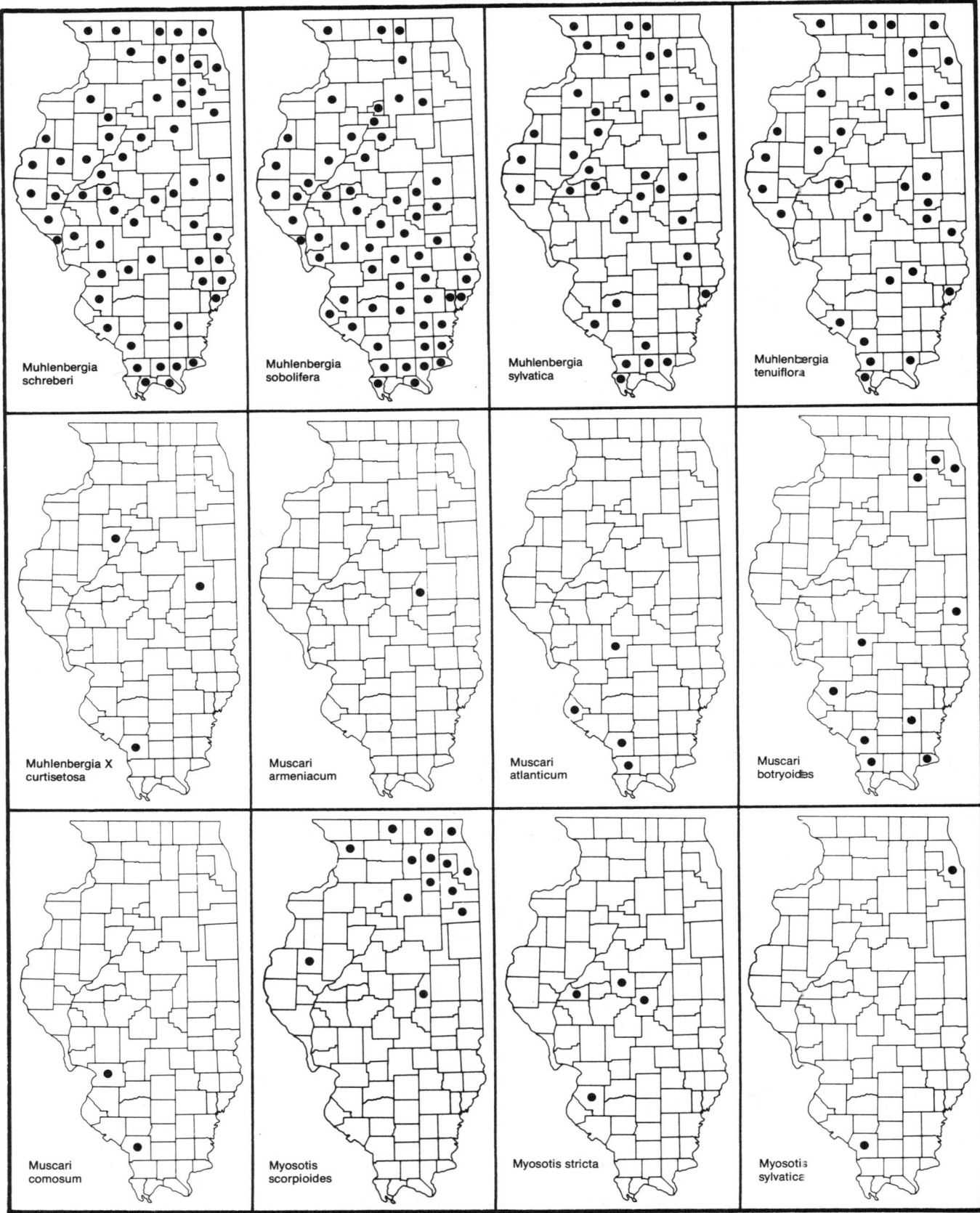

Muhlenbergia
schreberi

Muhlenbergia
sobolifera

Muhlenbergia
sylvatica

Muhlenbergia
tenuiflora

Muhlenbergia X
curtisetosa

Muscari
armeniacum

Muscari
atlanticum

Muscari
botryoides

Muscari
comosum

Myosotis
scorpioides

Myosotis stricta

Myosotis
sylvatica

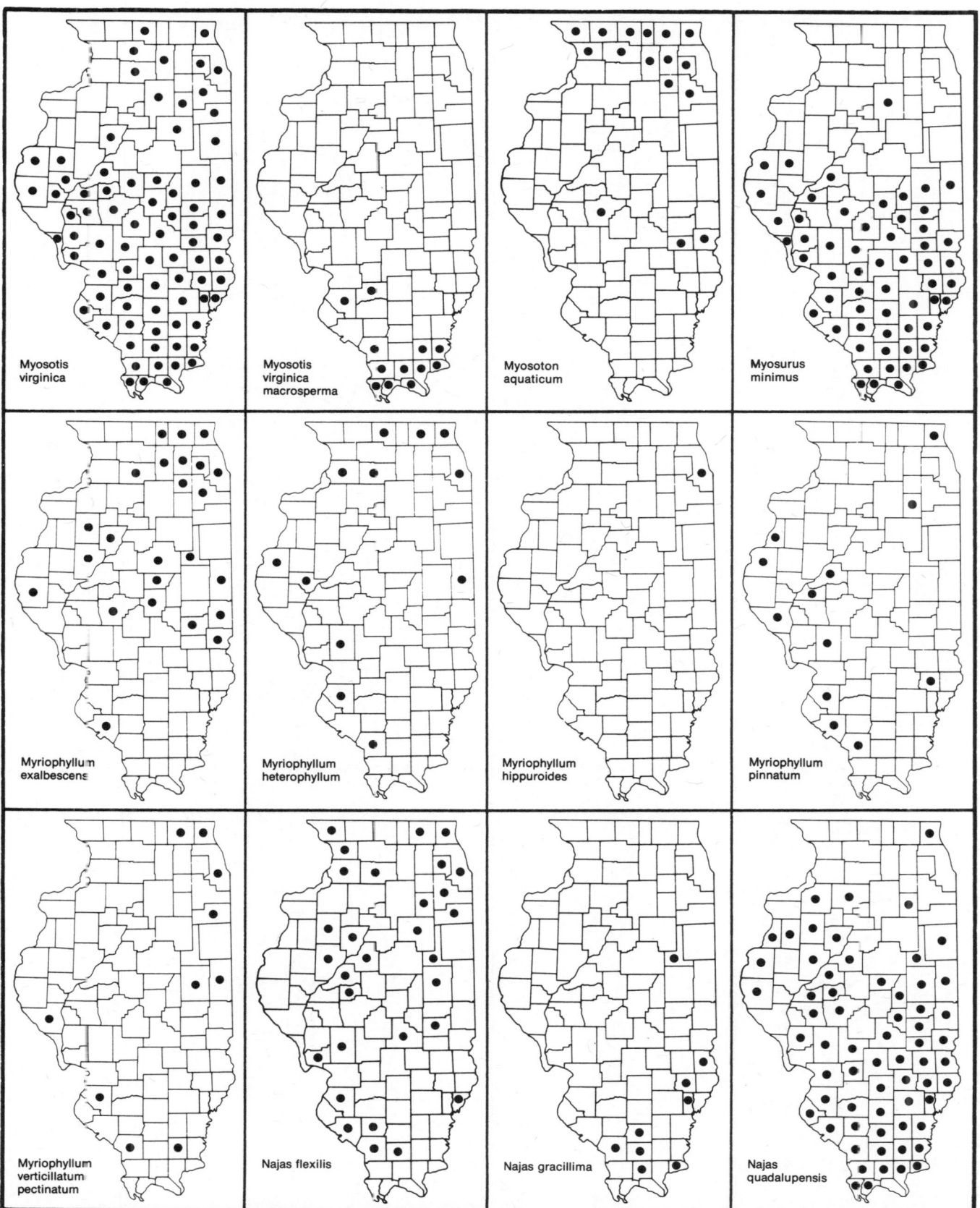

Myosotis
virginica

Myosotis
virginica
macrosperma

Myosoton
aquaticum

Myosurus
minimus

Myriophyllum
exalbescens

Myriophyllum
heterophyllum

Myriophyllum
hippuroides

Myriophyllum
pinnatum

Myriophyllum
verticillatum
pectinatum

Najas flexilis

Najas gracillima

Najas
quadalupensis

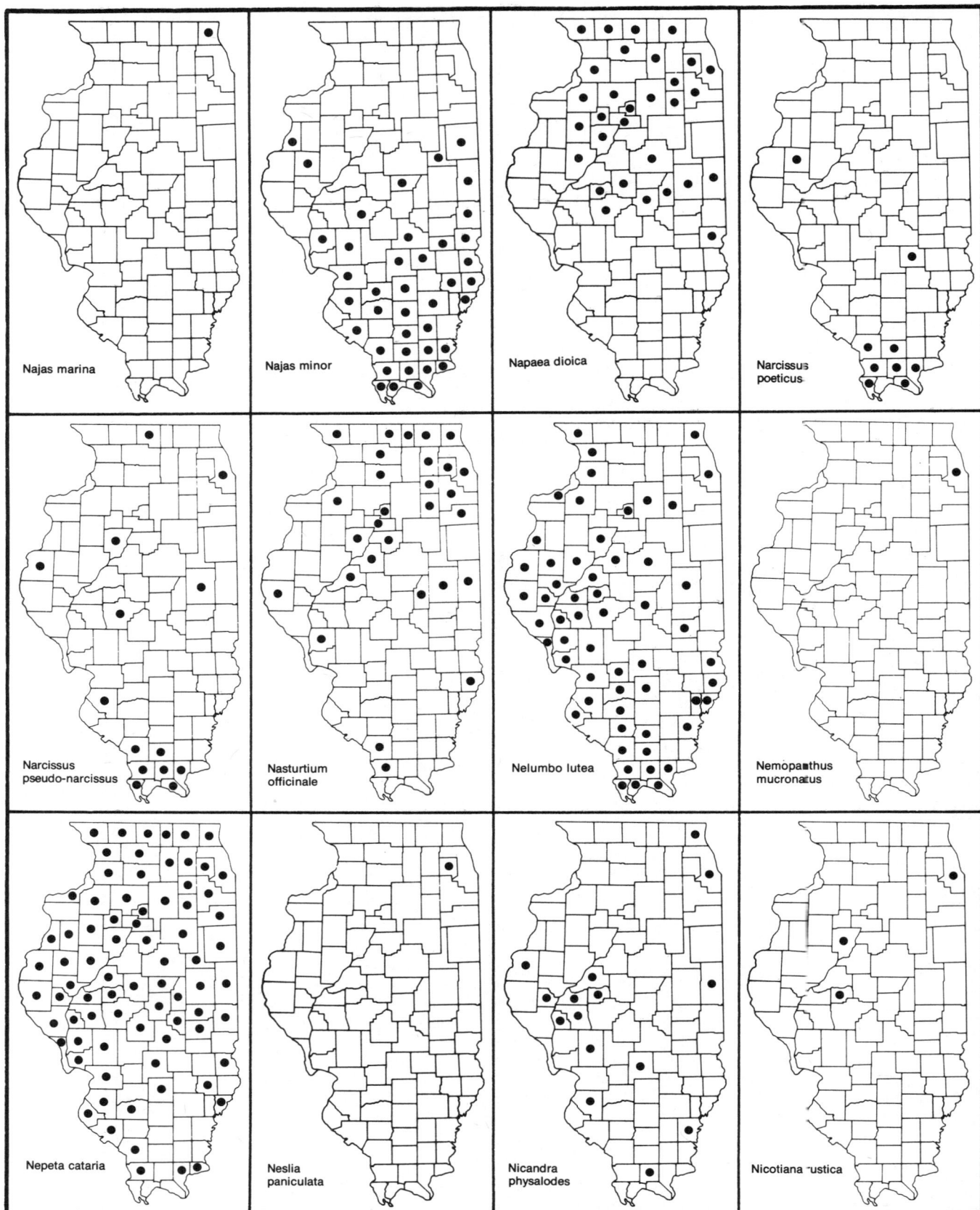

Najas marina

Najas minor

Napaea dioica

Narcissus poeticus

Narcissus pseudo-narcissus

Nasturtium officinale

Nelumbo lutea

Nemopanthus mucronatus

Nepeta cataria

Neslia paniculata

Nicandra physalodes

Nicotiana rustica

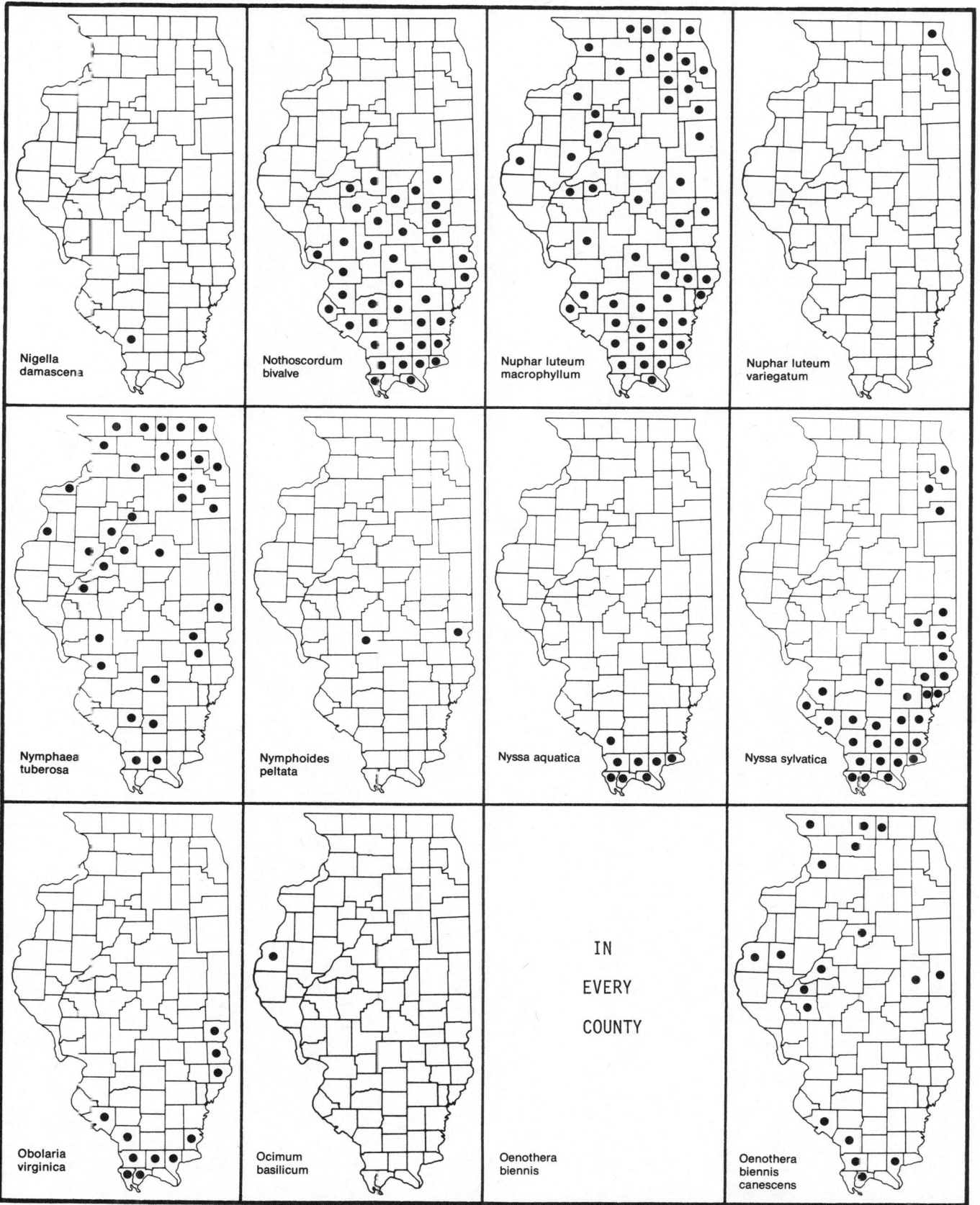

Nigella
damascena

Nothoscordum
bivalve

Nuphar luteum
macrophyllum

Nuphar luteum
variegatum

Nymphaea
tuberosa

Nymphoides
peltata

Nyssa aquatica

Nyssa sylvatica

Obolaria
virginica

Ocimum
basilicum

Oenothera
biennis

IN

EVERY

COUNTY

Oenothera
biennis
canescens

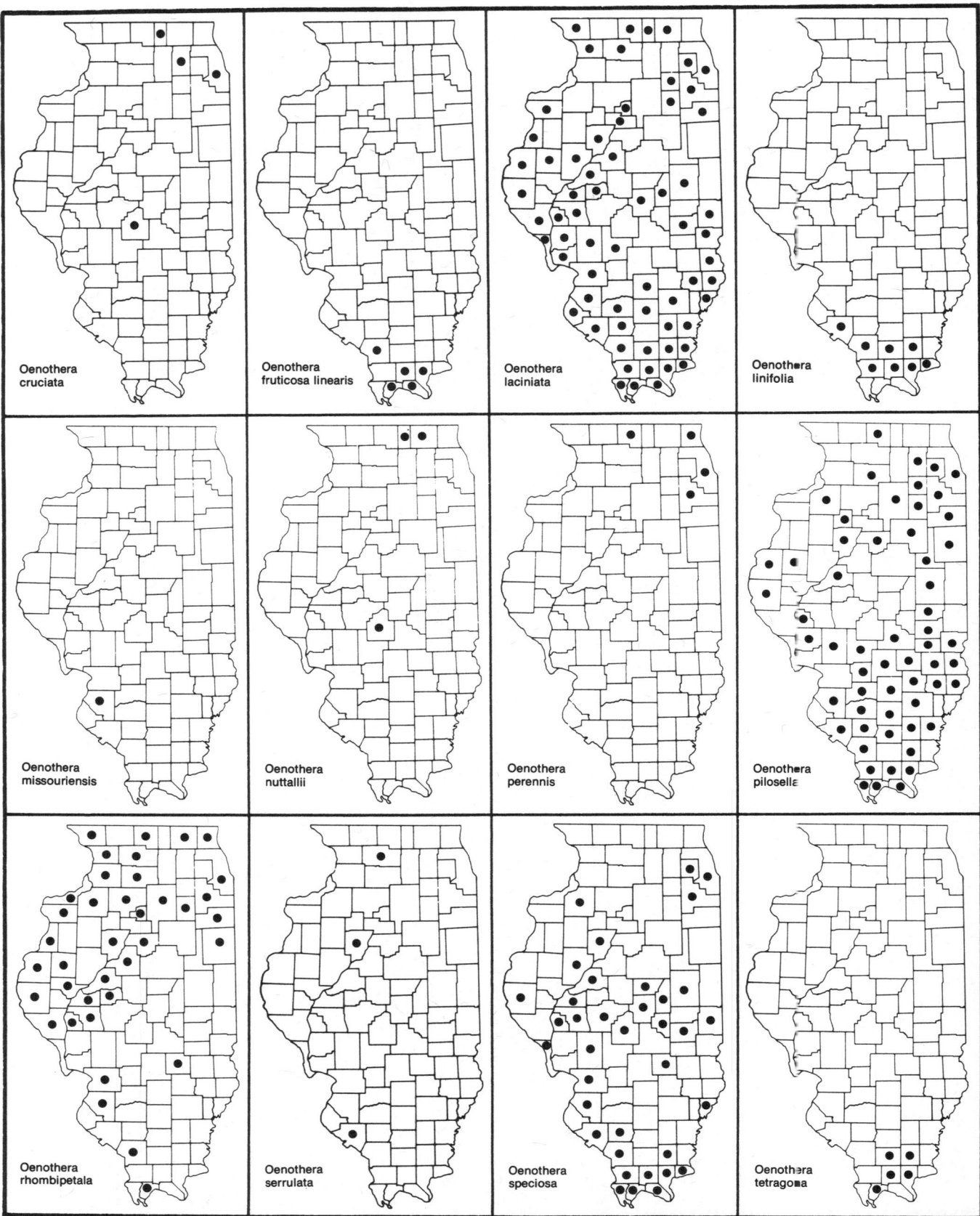

Oenothera
cruciata

Oenothera
fruticosa linearis

Oenothera
laciniata

Oenothera
linifolia

Oenothera
missouriensis

Oenothera
nuttallii

Oenothera
perennis

Oenothera
pilosella

Oenothera
rhombipetala

Oenothera
serrulata

Oenothera
speciosa

Oenothera
tetragona

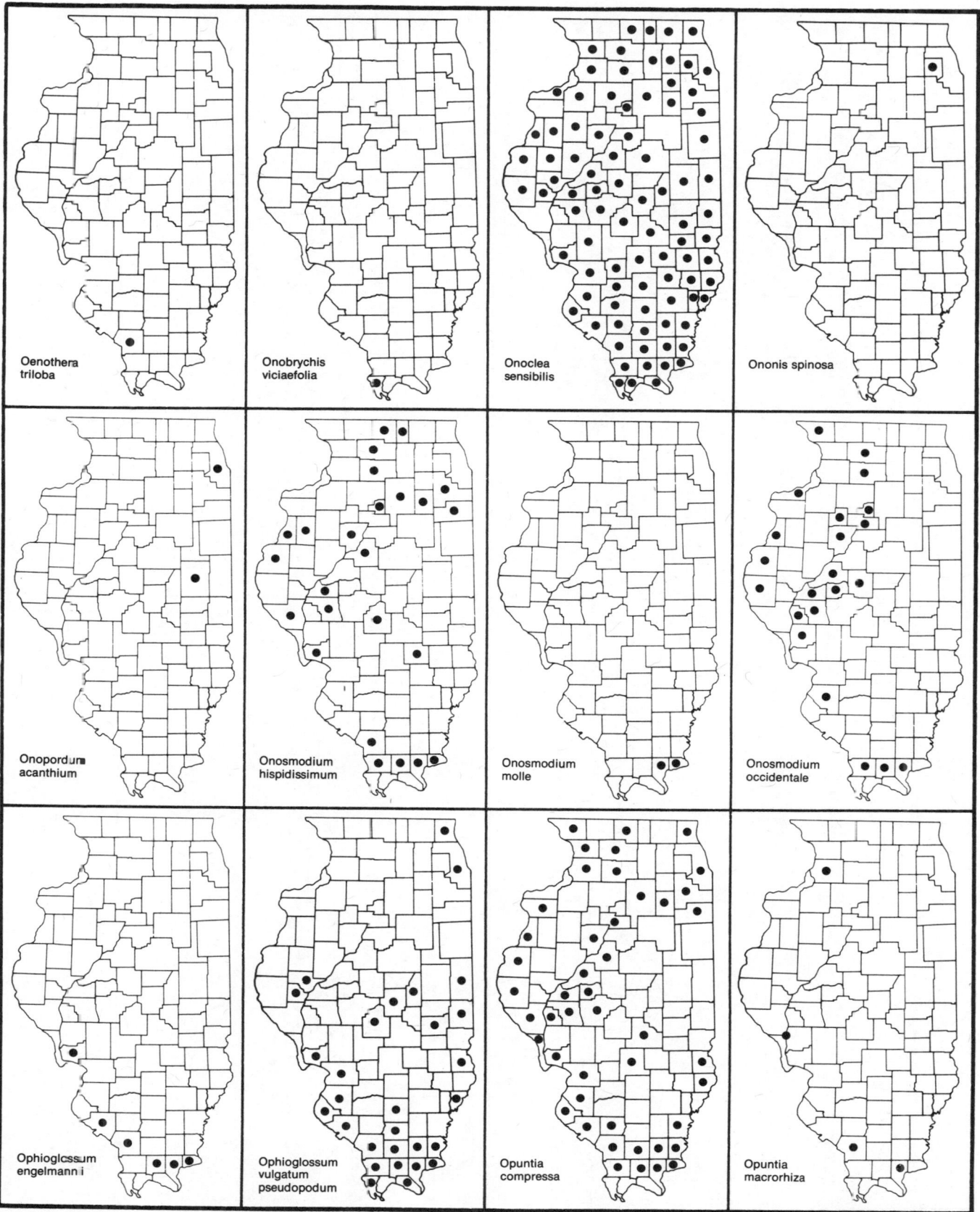

Oenothera
triloba

Onobrychis
viciaefolia

Onoclea
sensibilis

Ononis spinosa

Onopordum
acanthium

Onosmodium
hispidissimum

Onosmodium
molle

Onosmodium
occidentale

Ophioglossum
engelmanni

Ophioglossum
vulgatum
pseudopodum

Opuntia
compressa

Opuntia
macrorhiza

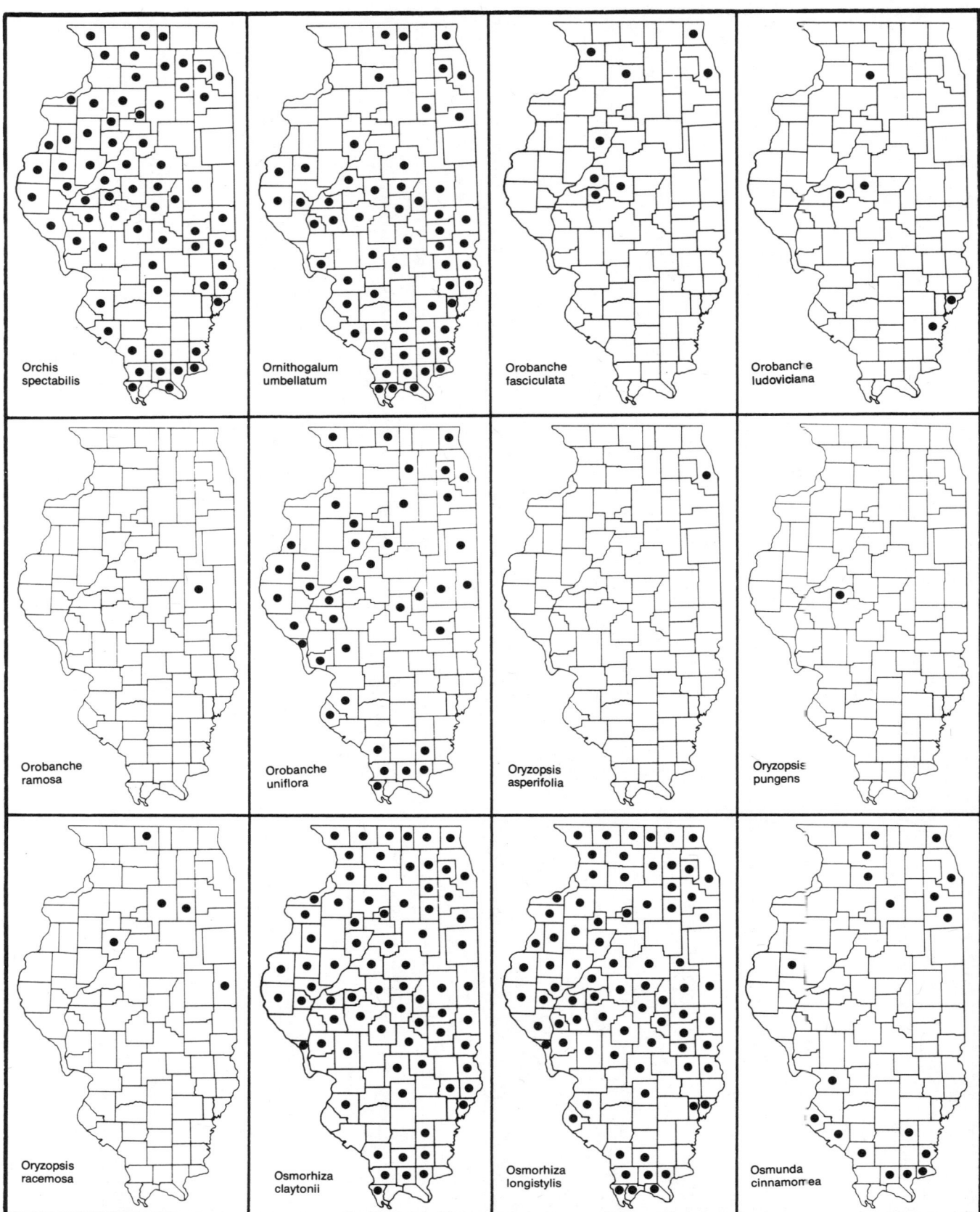

Orchis
spectabilis

Ornithogalum
umbellatum

Orobanche
fasciculata

Orobanche
ludoviciana

Orobanche
ramosa

Orobanche
uniflora

Oryzopsis
asperifolia

Oryzopsis
pungens

Oryzopsis
racemosa

Osmorhiza
claytonii

Osmorhiza
longistylis

Osmunda
cinnamomea

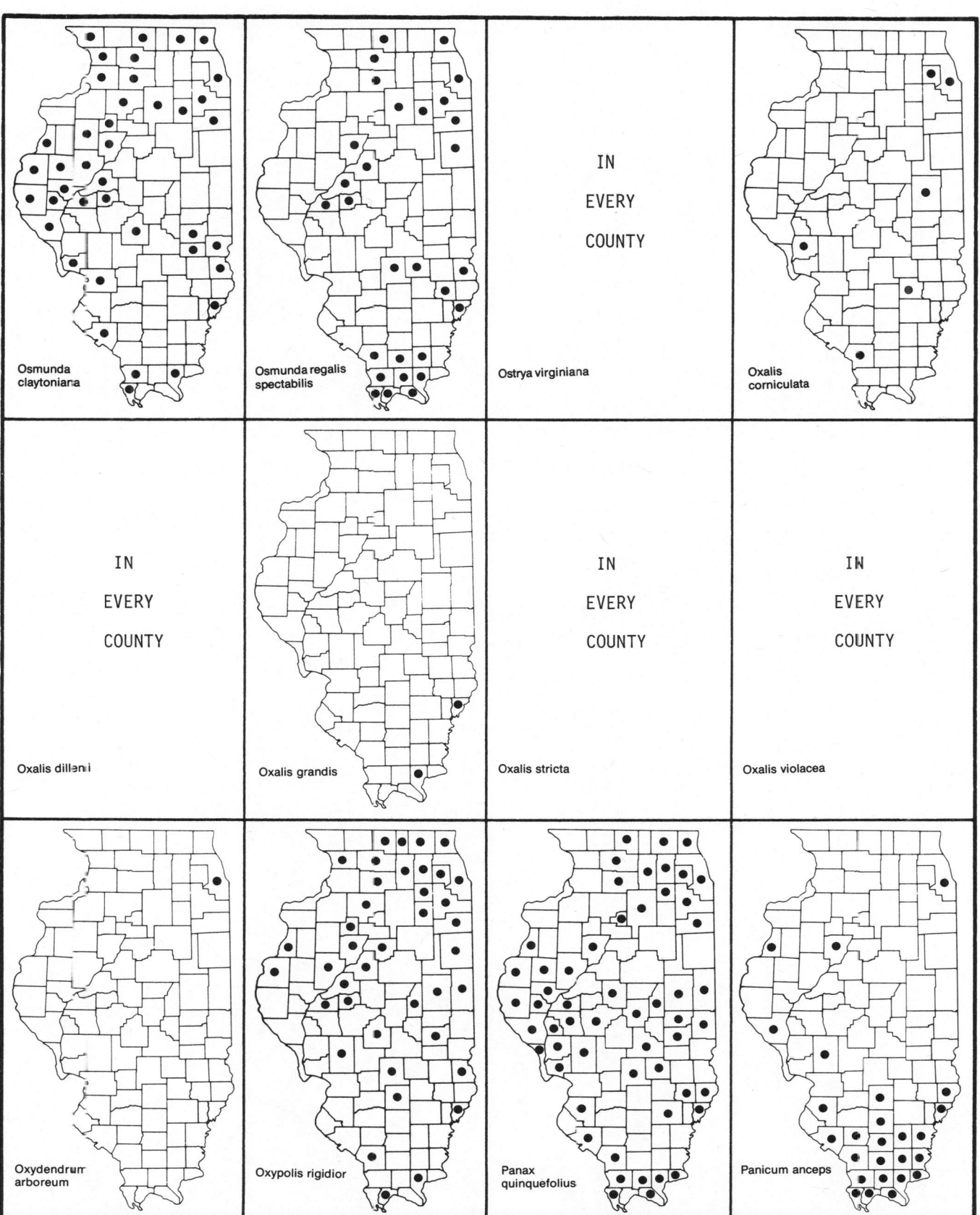

Osmunda
claytoniana

Osmunda regalis
spectabilis

Ostrya virginiana

IN

EVERY

COUNTY

Oxalis
corniculata

Oxalis dilleni

IN

EVERY

COUNTY

Oxalis grandis

Oxalis stricta

IN

EVERY

COUNTY

Oxalis violacea

IN

EVERY

COUNTY

Oxydendrum
arboreum

Oxypolis rigidior

Panax
quinquefolius

Panicum anceps

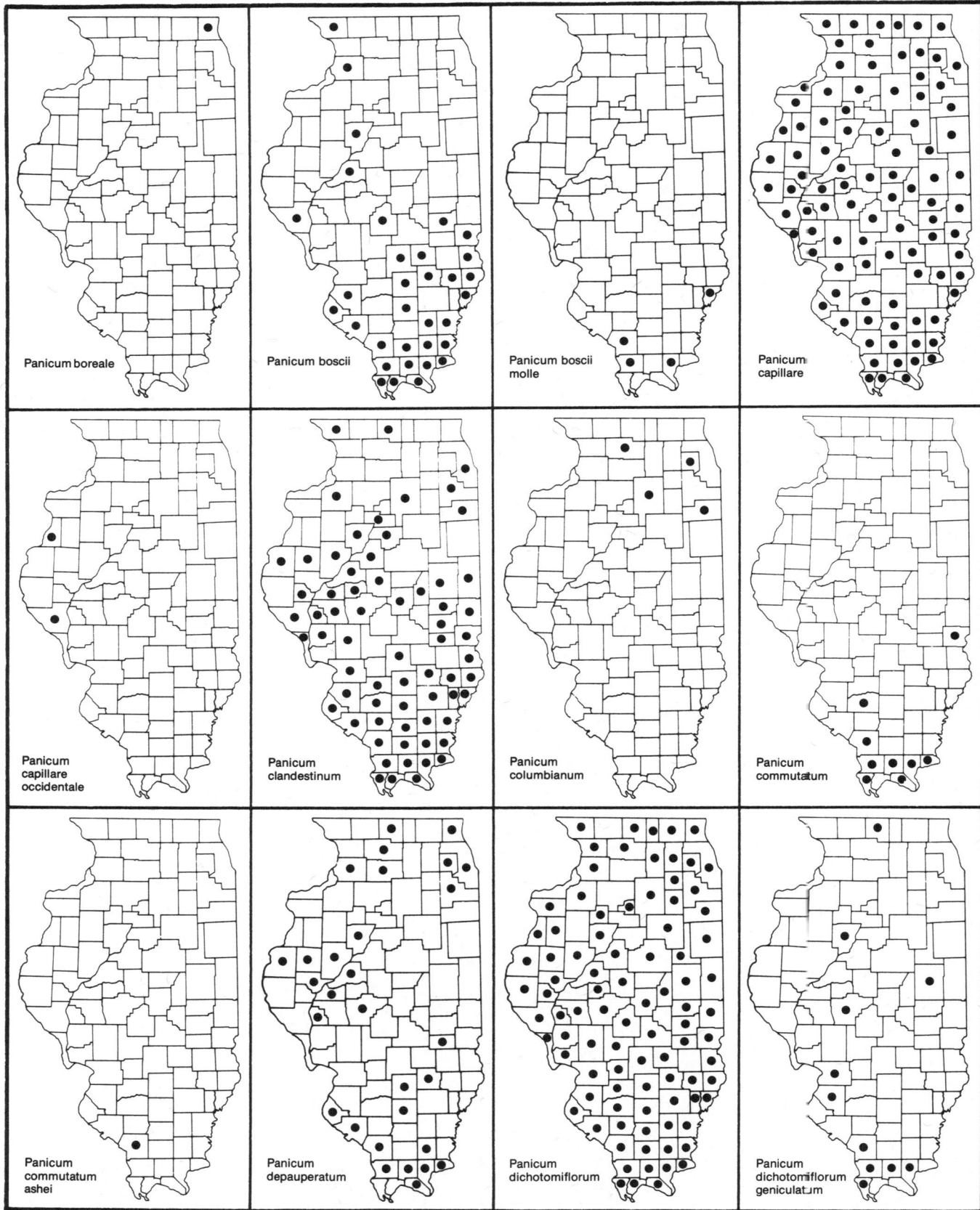

Panicum boreale

Panicum boscii

Panicum boscii molle

Panicum capillare

Panicum capillare occidentale

Panicum clandestinum

Panicum columbianum

Panicum commutatum

Panicum commutatum ashei

Panicum depauperatum

Panicum dichotomiflorum

Panicum dichotomiflorum geniculatum

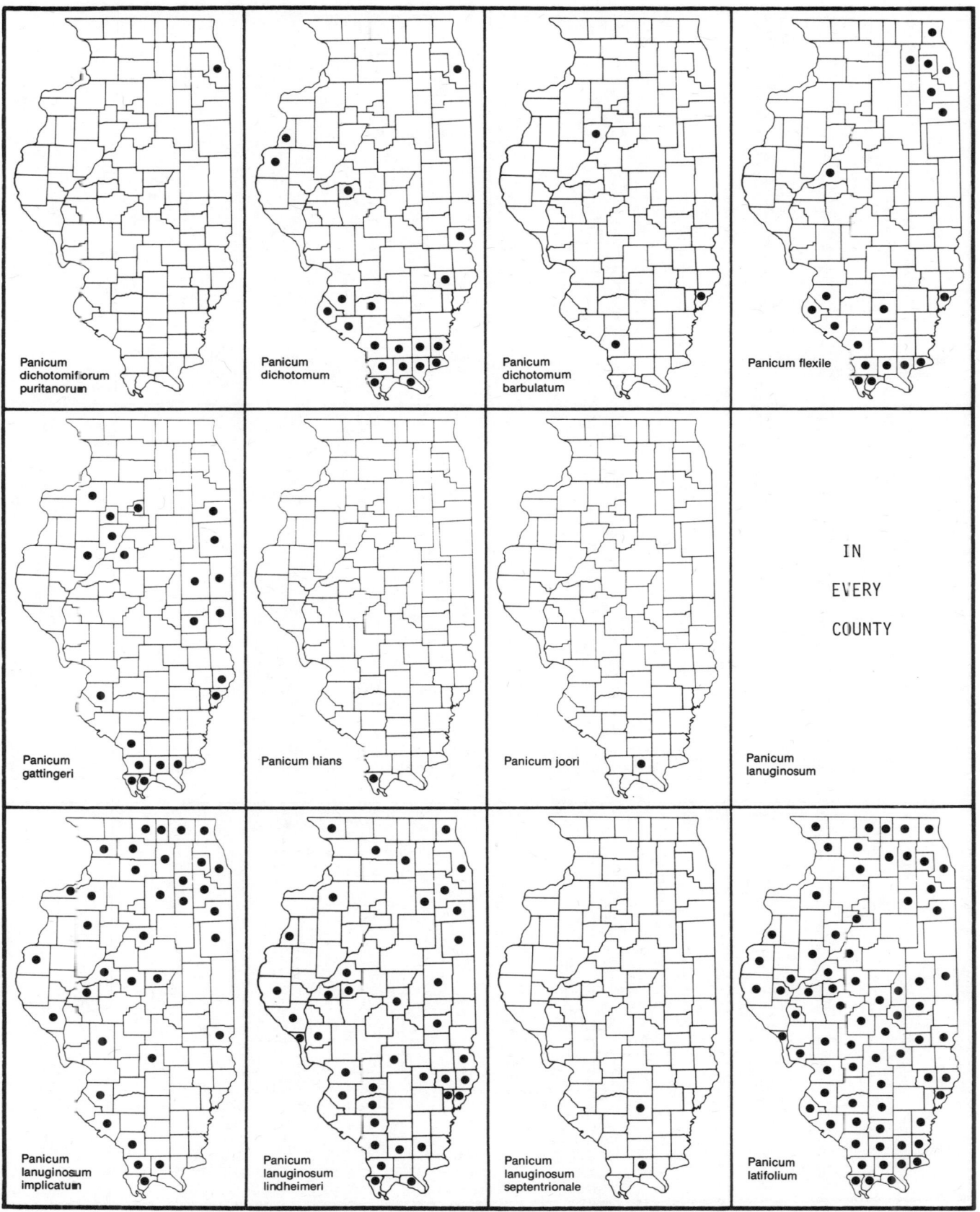

Panicum
dichotomiflorum
puritanorum

Panicum
dichotomum

Panicum
dichotomum
barbulatum

Panicum flexile

Panicum
gattingeri

Panicum hians

Panicum joori

IN

EVERY

COUNTY

Panicum
lanuginosum

Panicum
lanuginosum
implicatum

Panicum
lanuginosum
lindheimeri

Panicum
lanuginosum
septentrionale

Panicum
latifolium

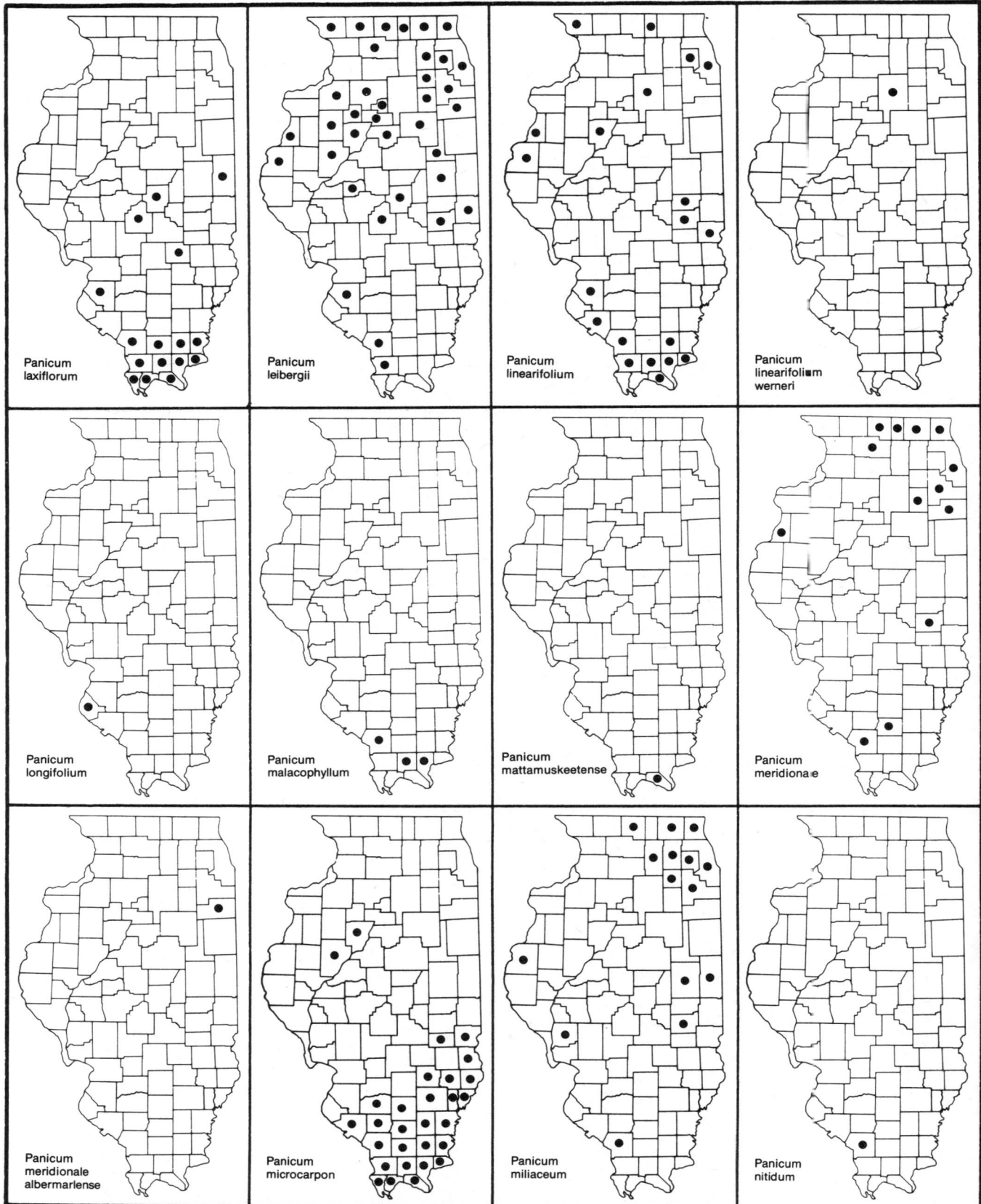

Panicum
laxiflorum

Panicum
leibergii

Panicum
linearifolium

Panicum
linearifolium
werneri

Panicum
longifolium

Panicum
malacophyllum

Panicum
mattamuskeetense

Panicum
meridionale

Panicum
meridionale
albermarlense

Panicum
microcarpon

Panicum
miliaceum

Panicum
nitidum

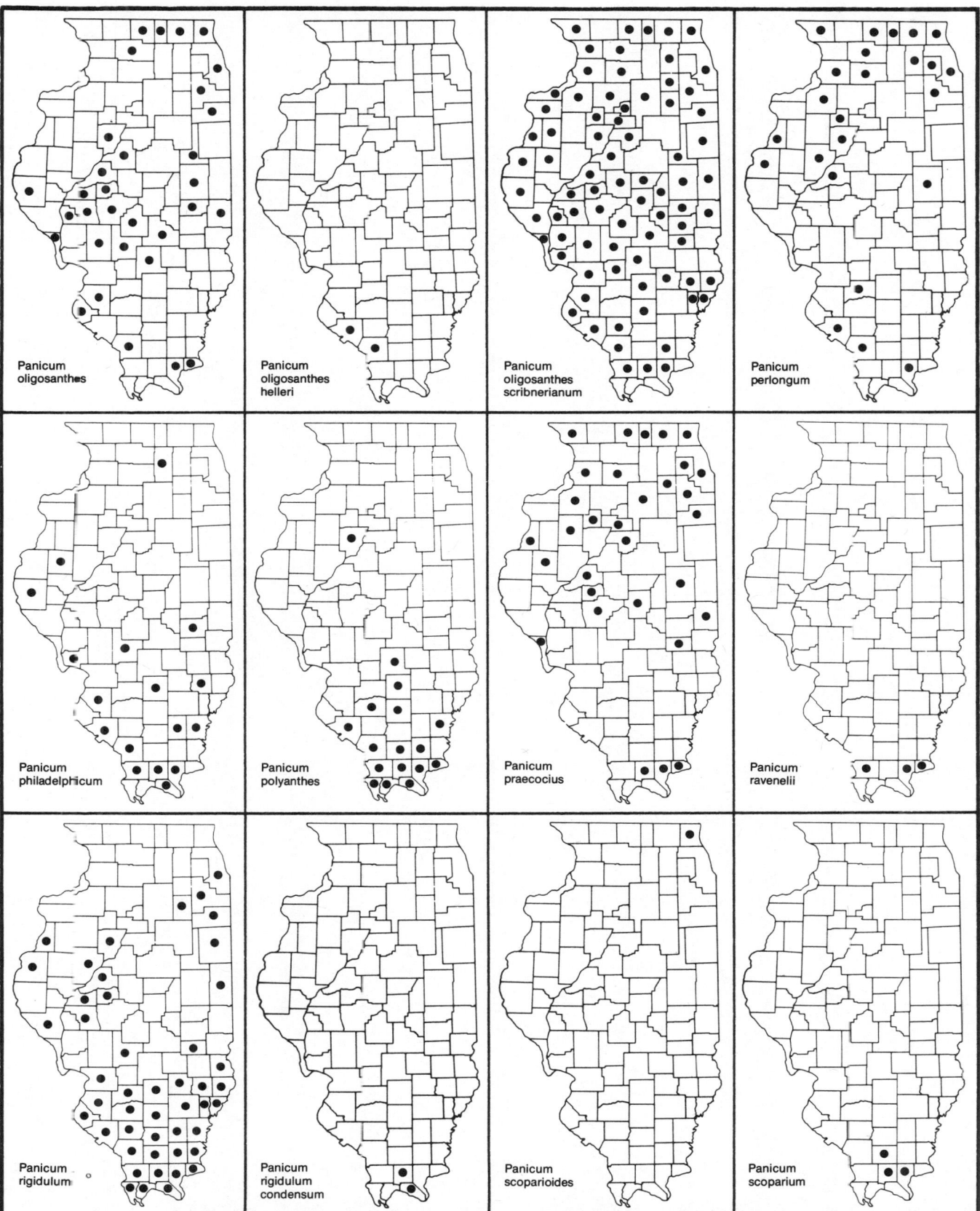

Panicum
oligosanthes

Panicum
oligosanthes
helleri

Panicum
oligosanthes
scribnerianum

Panicum
perlongum

Panicum
philadelphicum

Panicum
polyanthes

Panicum
praecocius

Panicum
ravenelii

Panicum
rigidulum

Panicum
rigidulum
condensum

Panicum
scoparioides

Panicum
scoparium

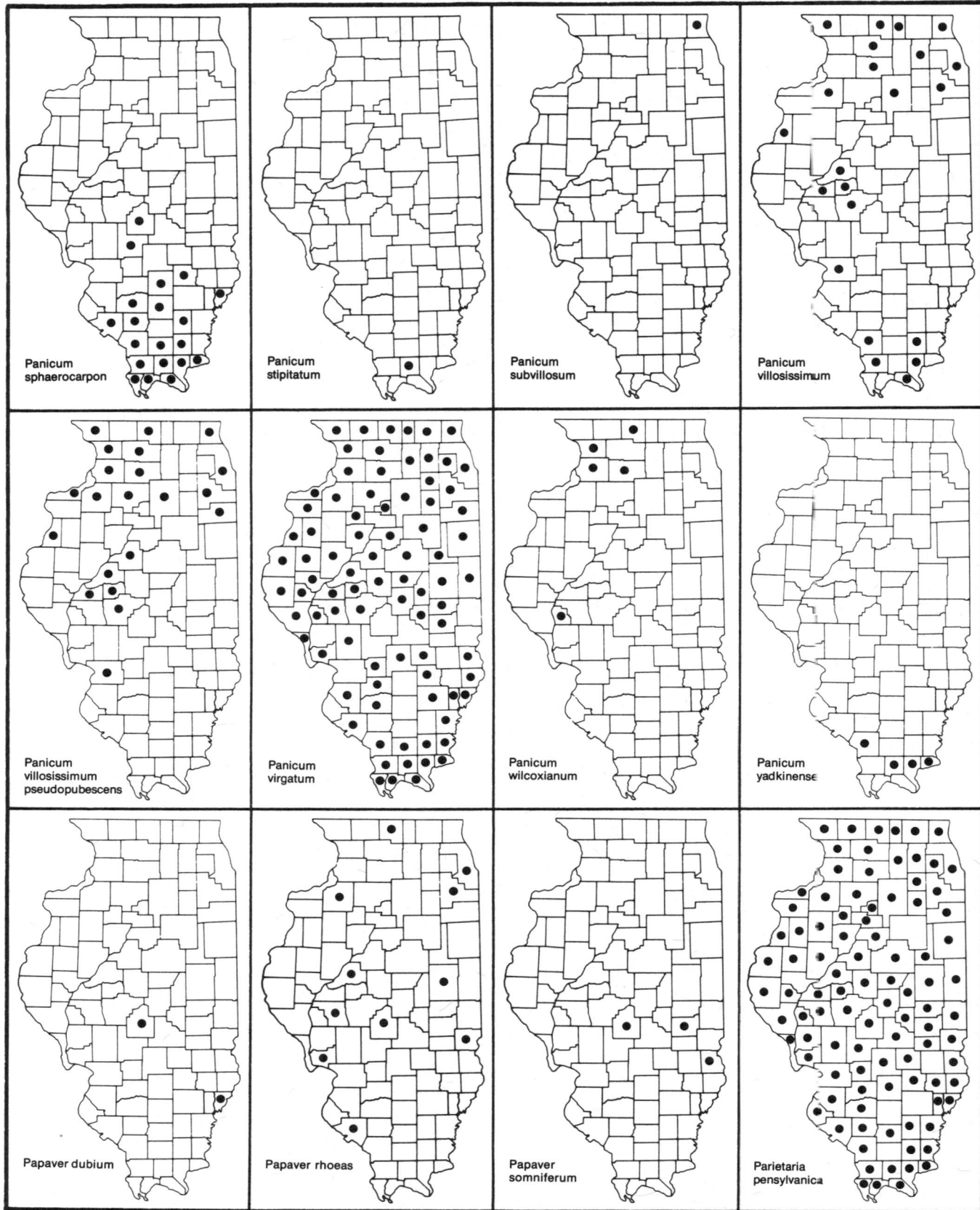

Panicum
sphaerocarpon

Panicum
stipitatum

Panicum
subvillosum

Panicum
villosissimum

Panicum
villosissimum
pseudopubescens

Panicum
virgatum

Panicum
wilcoxianum

Panicum
yadkinense

Papaver dubium

Papaver rhoeas

Papaver
somniferum

Parietaria
pensylvanica

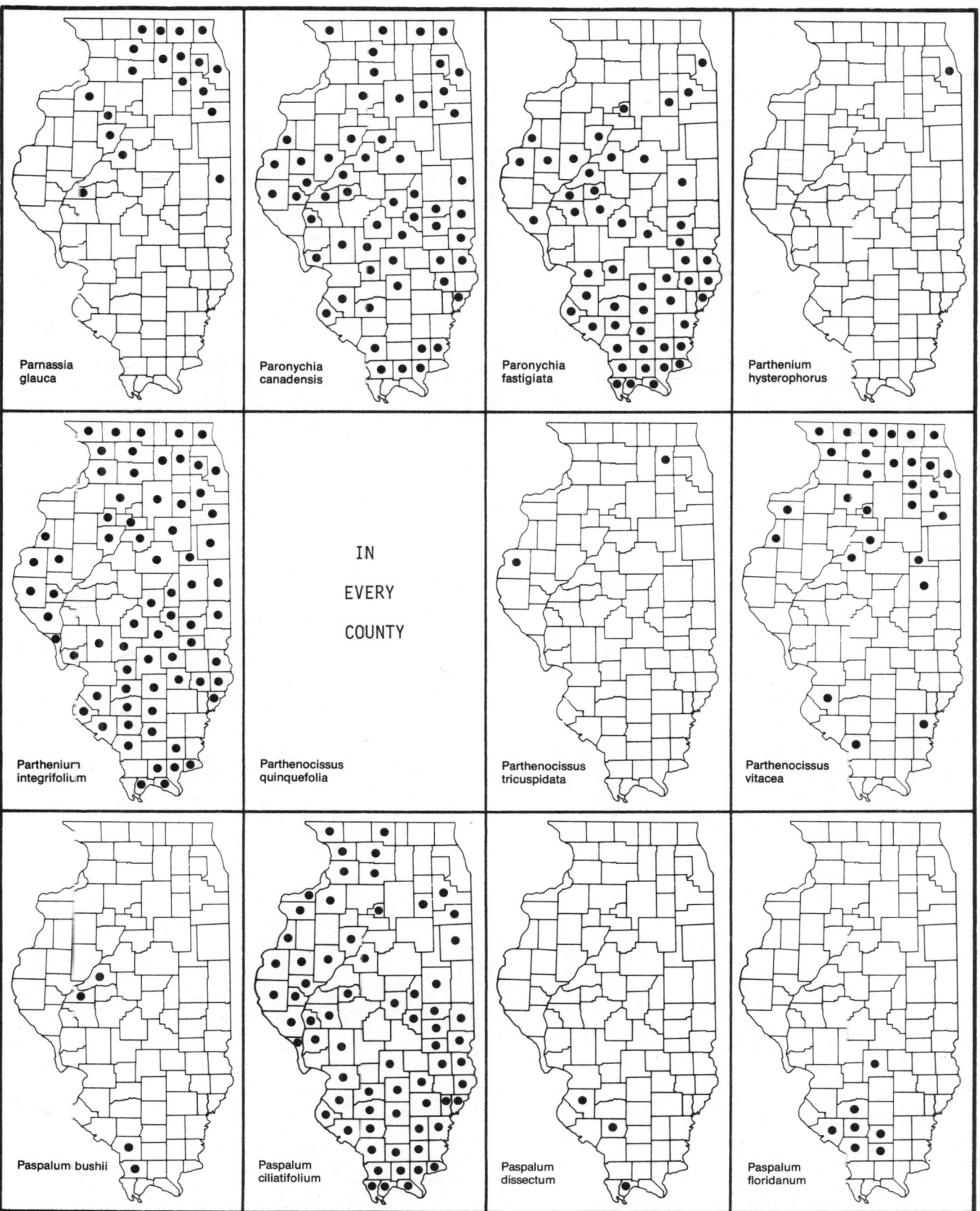

Parnassia
glauca

Paronychia
canadensis

Paronychia
fastigiata

Parthenium
hysterophorus

Parthenium
integrifolium

IN

EVERY

COUNTY

Parthenocissus
quinquefolia

Parthenocissus
tricuspidata

Parthenocissus
vitacea

Paspalum bushii

Paspalum
ciliatifolium

Paspalum
dissectum

Paspalum
floridanum

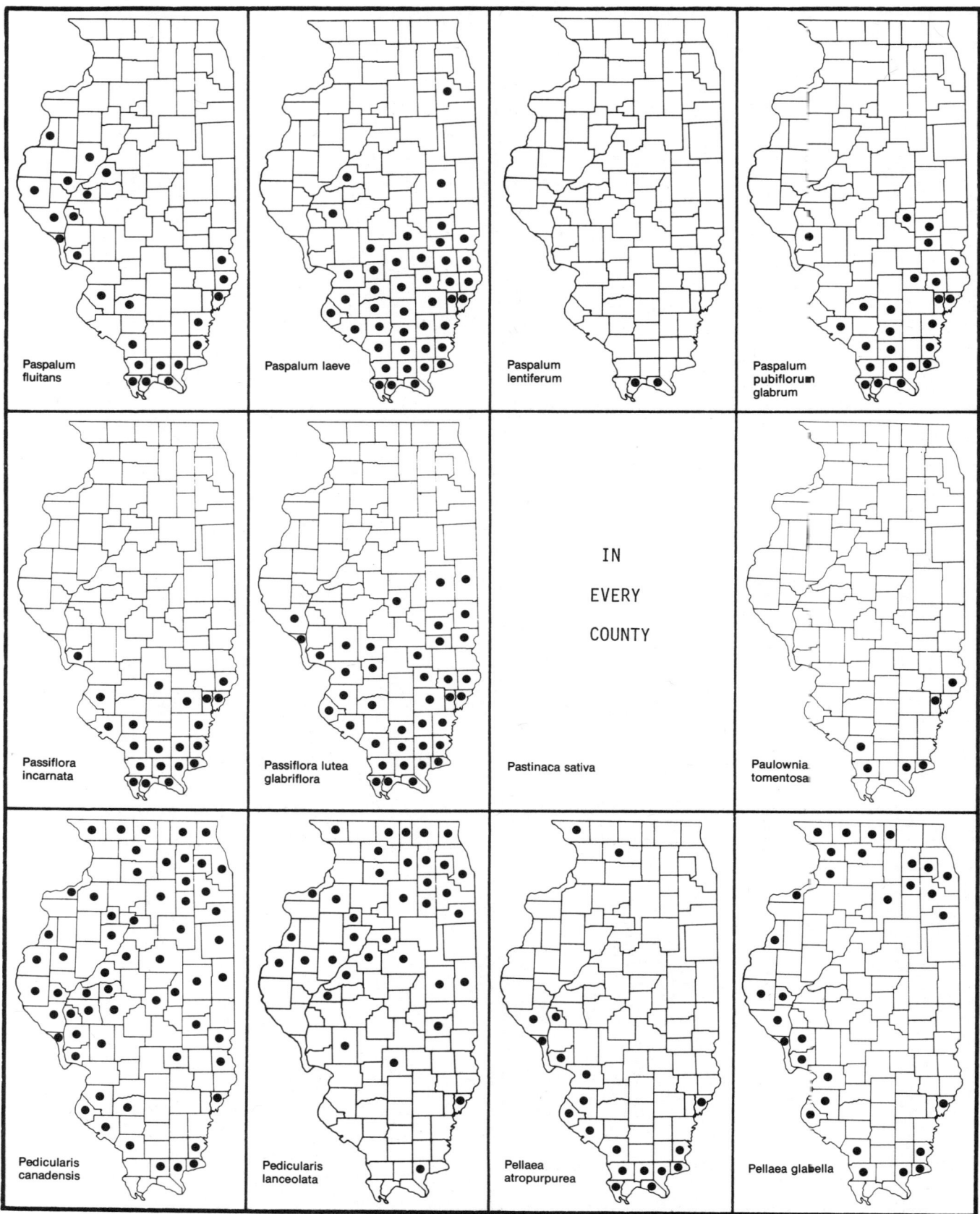

Paspalum
fluitans

Paspalum laeve

Paspalum
lentiferum

Paspalum
pubiflorum
glabrum

Passiflora
incarnata

Passiflora lutea
glabriflora

IN

EVERY

COUNTY

Pastinaca sativa

Paulownia
tomentosa

Pedicularis
canadensis

Pedicularis
lanceolata

Pellaea
atropurpurea

Pellaea glabella

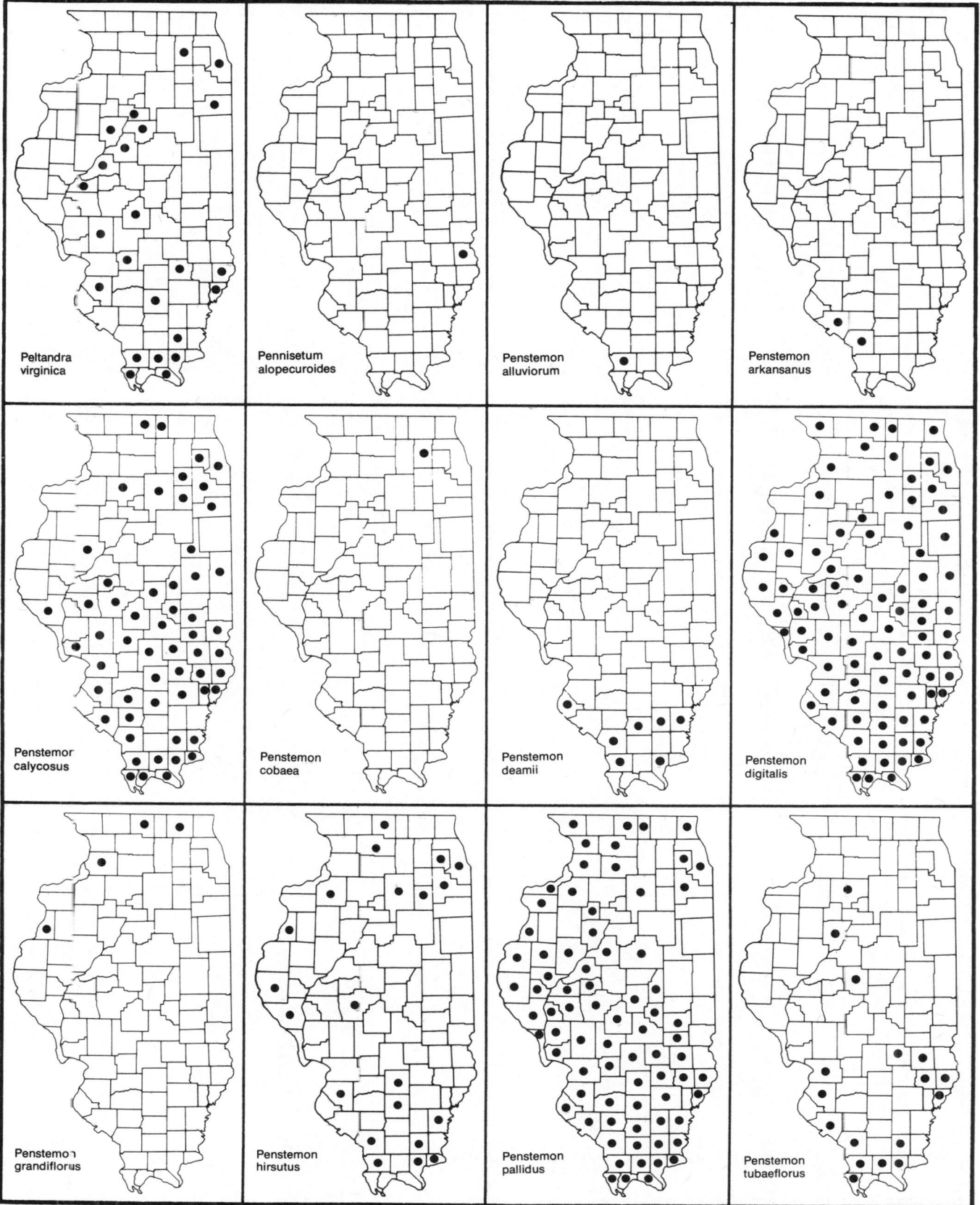

Peltandra
virginica

Pennisetum
alopecuroides

Penstemon
alluviorum

Penstemon
arkansanus

Penstemon
calycosus

Penstemon
cobaea

Penstemon
deamii

Penstemon
digitalis

Penstemon
grandiflorus

Penstemon
hirsutus

Penstemon
pallidus

Penstemon
tubaeflorus

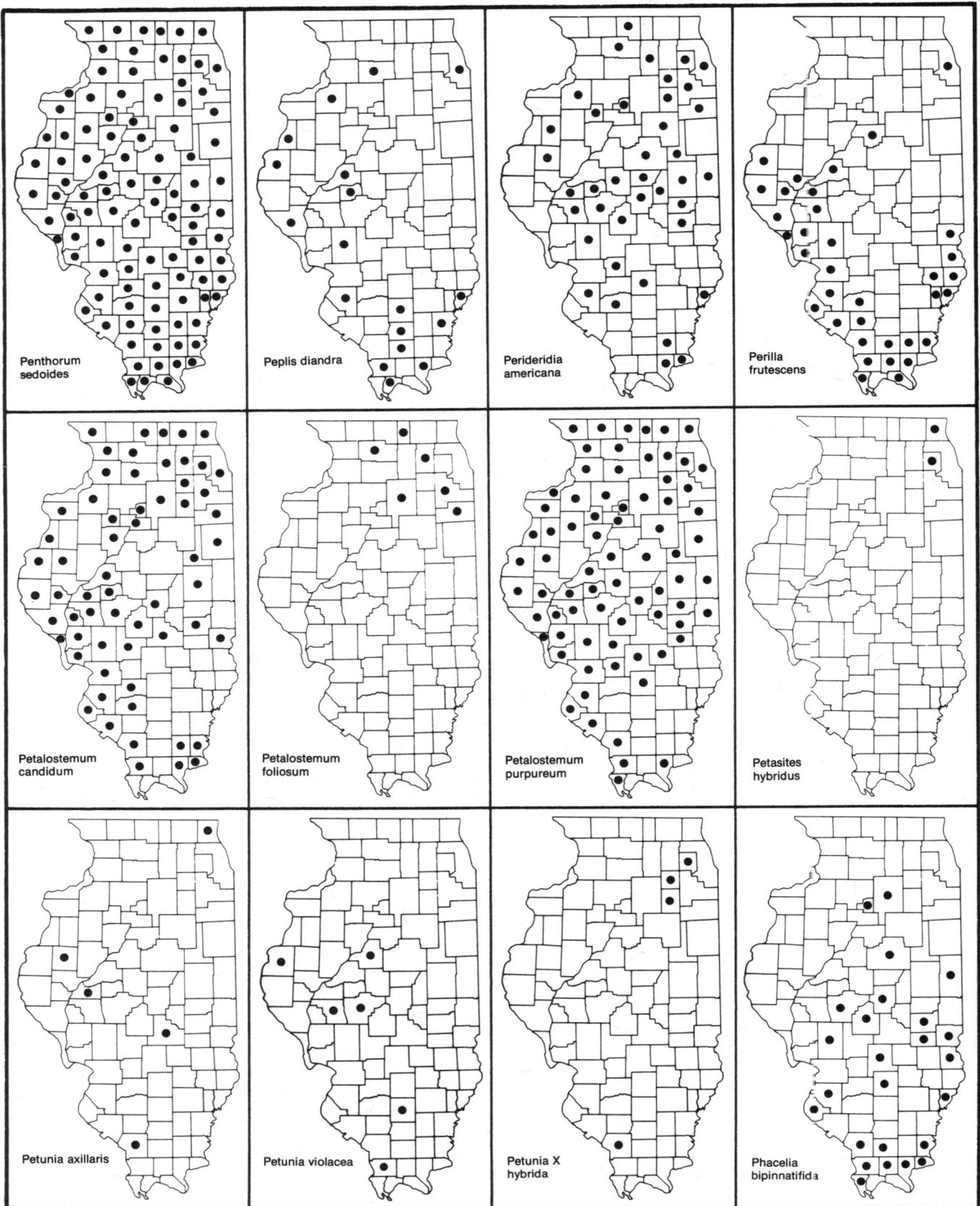

Penthorum
sedoides

Peplis diandra

Perideridia
americana

Perilla
frutescens

Petalostemum
candidum

Petalostemum
foliosum

Petalostemum
purpureum

Petasites
hybridus

Petunia axillaris

Petunia violacea

Petunia X
hybrida

Phacelia
bipinnatifida

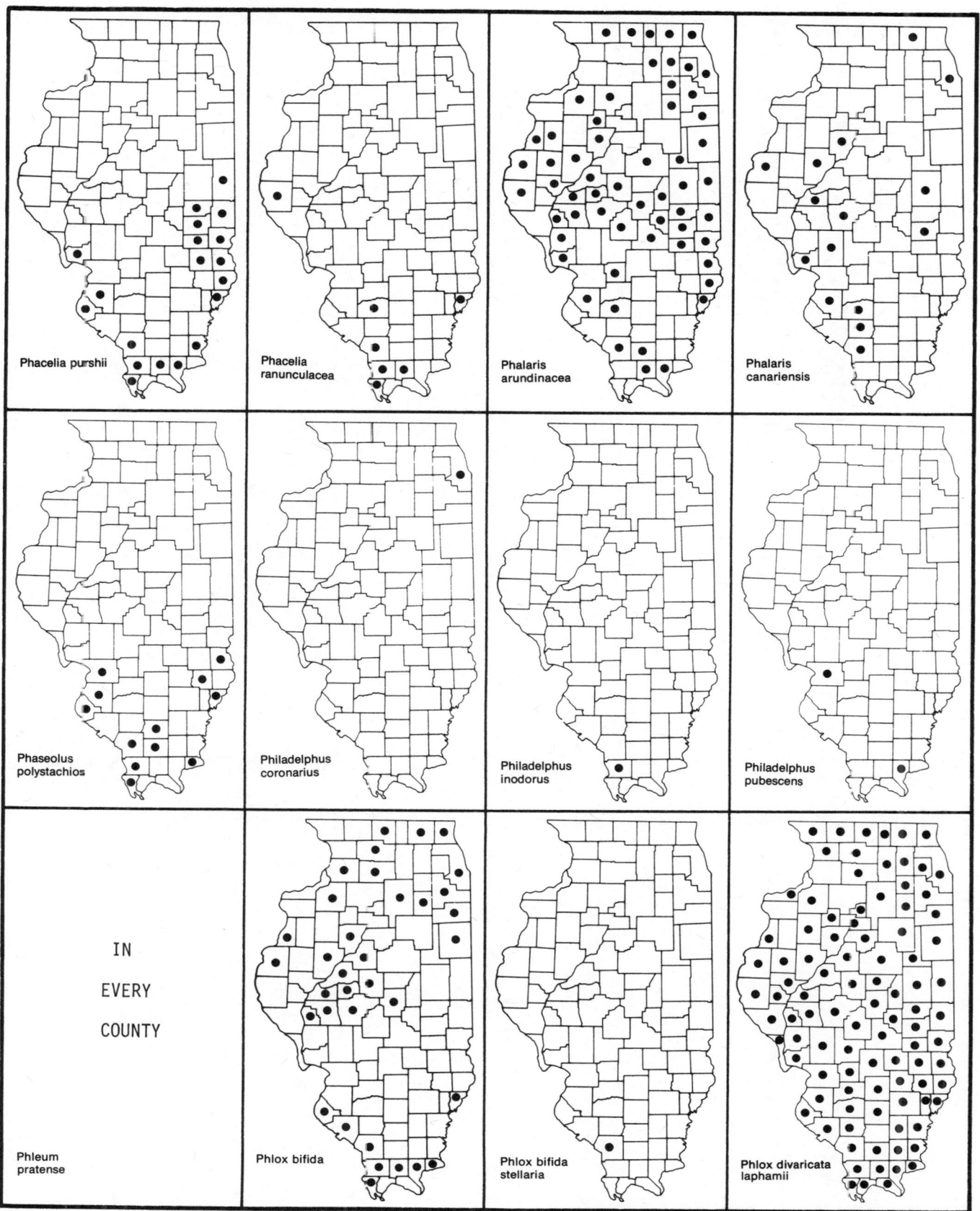

Phacelia purshii

Phacelia ranunculacea

Phalaris arundinacea

Phalaris canariensis

Phaseolus polystachios

Philadelphus coronarius

Philadelphus inodorus

Philadelphus pubescens

IN

EVERY

COUNTY

Phleum pratense

Phlox bifida

Phlox bifida stellaria

Phlox divaricata laphamii

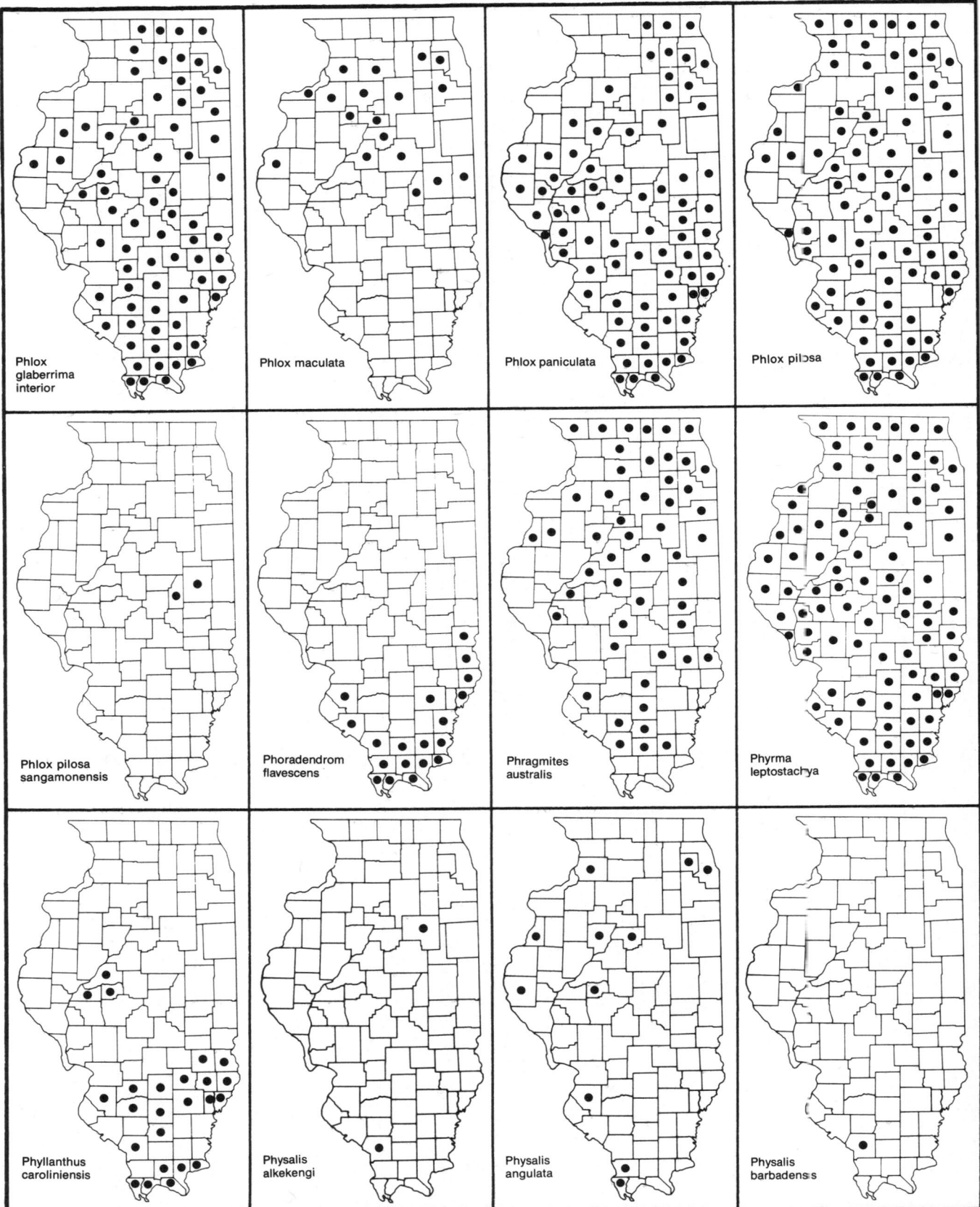

Phlox
glaberrima
interior

Phlox maculata

Phlox paniculata

Phlox pilosa

Phlox pilosa
sangamonensis

Phoradendrom
flavescens

Phragmites
australis

Phyrma
leptostachya

Phyllanthus
caroliniensis

Physalis
alkekengi

Physalis
angulata

Physalis
barbadensis

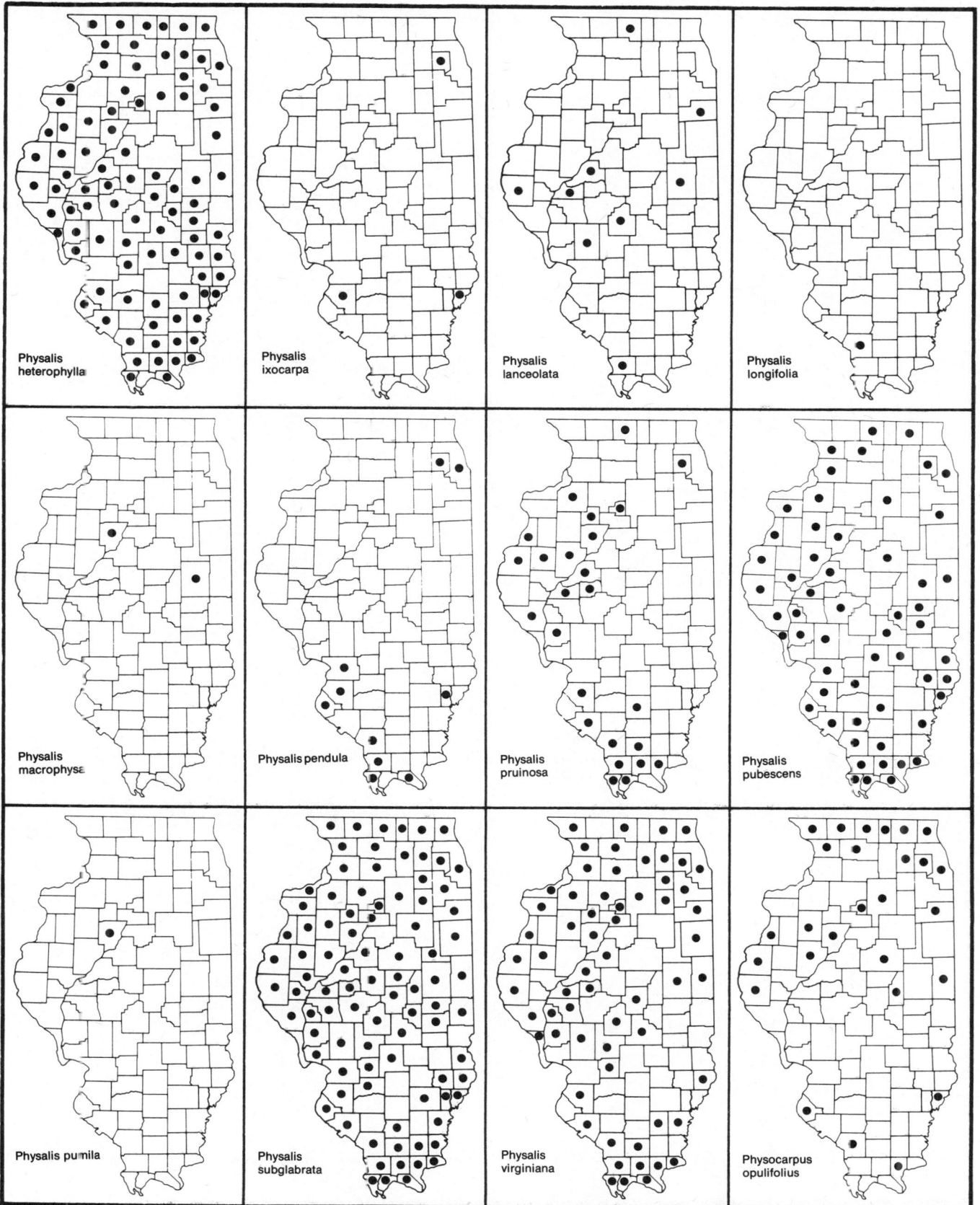

Physalis
heterophylla

Physalis
ixocarpa

Physalis
lanceolata

Physalis
longifolia

Physalis
macrophysa

Physalis
pendula

Physalis
pruinosa

Physalis
pubescens

Physalis pumila

Physalis
subglabrata

Physalis
virginiana

Physocarpus
opulifolius

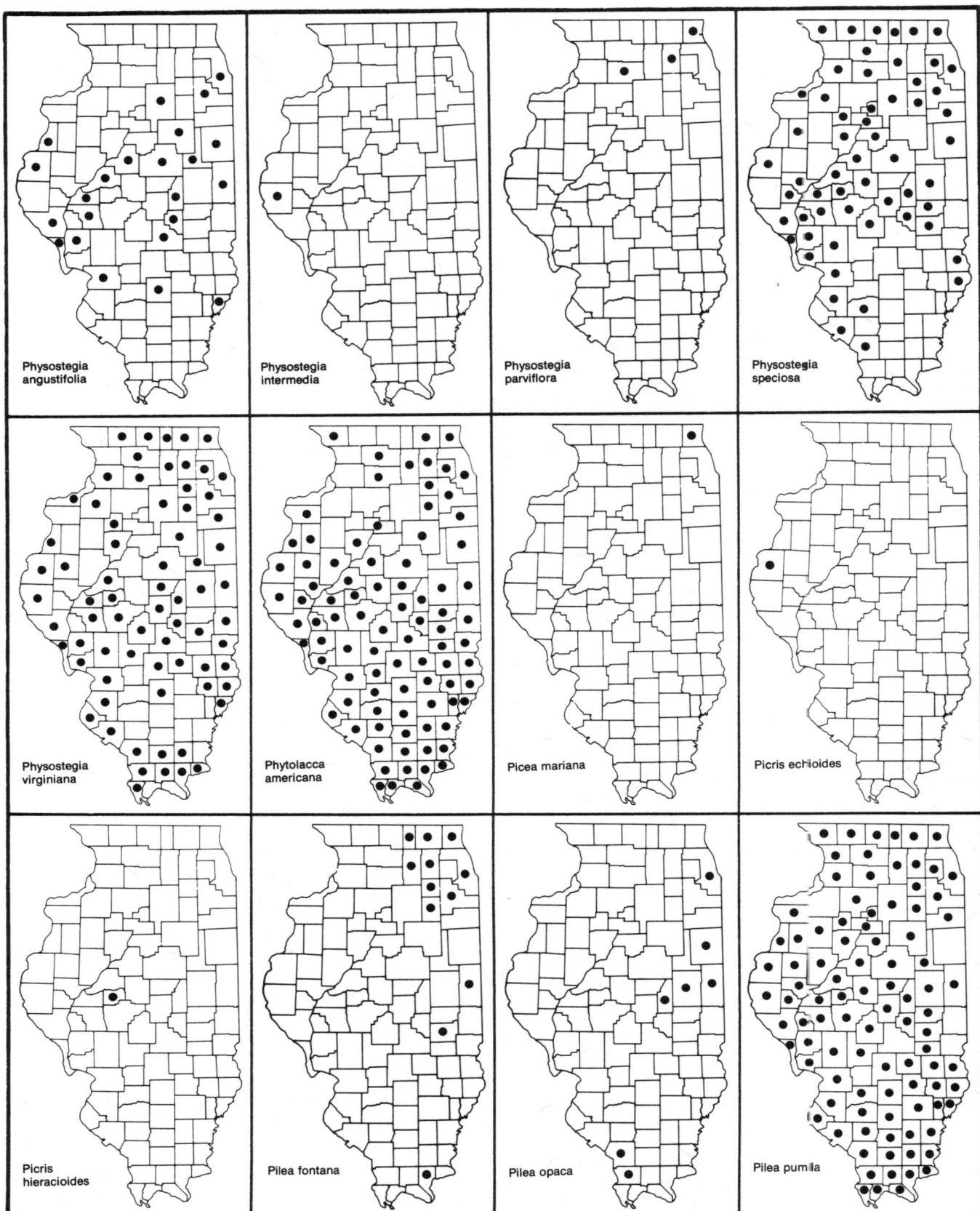

Physostegia
angustifolia

Physostegia
intermedia

Physostegia
parviflora

Physostegia
speciosa

Physostegia
virginiana

Phytolacca
americana

Picea mariana

Picris echioides

Picris
hieracioides

Pilea fontana

Pilea opaca

Pilea pumila

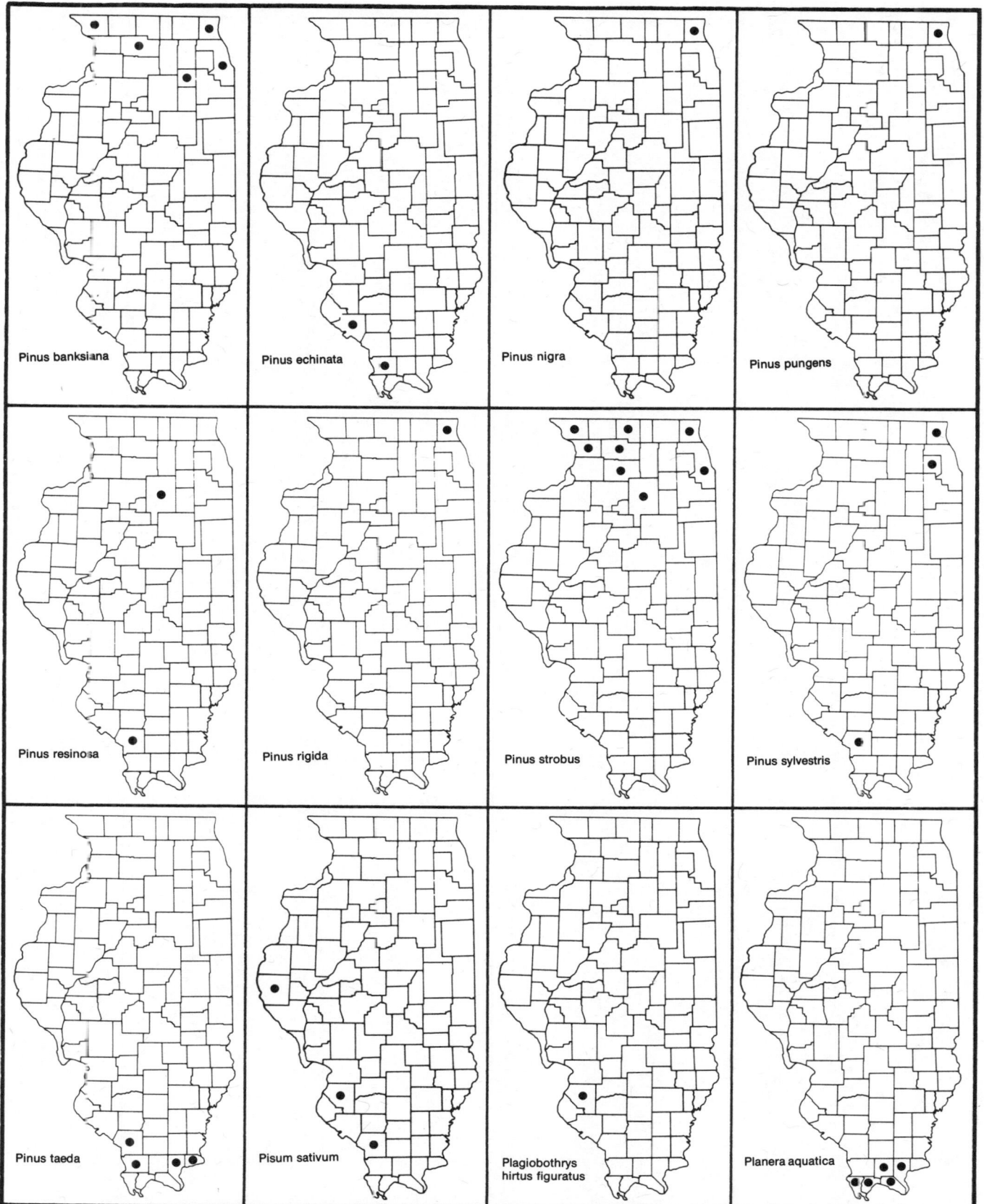

Pinus banksiana

Pinus echinata

Pinus nigra

Pinus pungens

Pinus resinosa

Pinus rigida

Pinus strobus

Pinus sylvestris

Pinus taeda

Pisum sativum

Plagiobothrys
hirtus figuratus

Planera aquatica

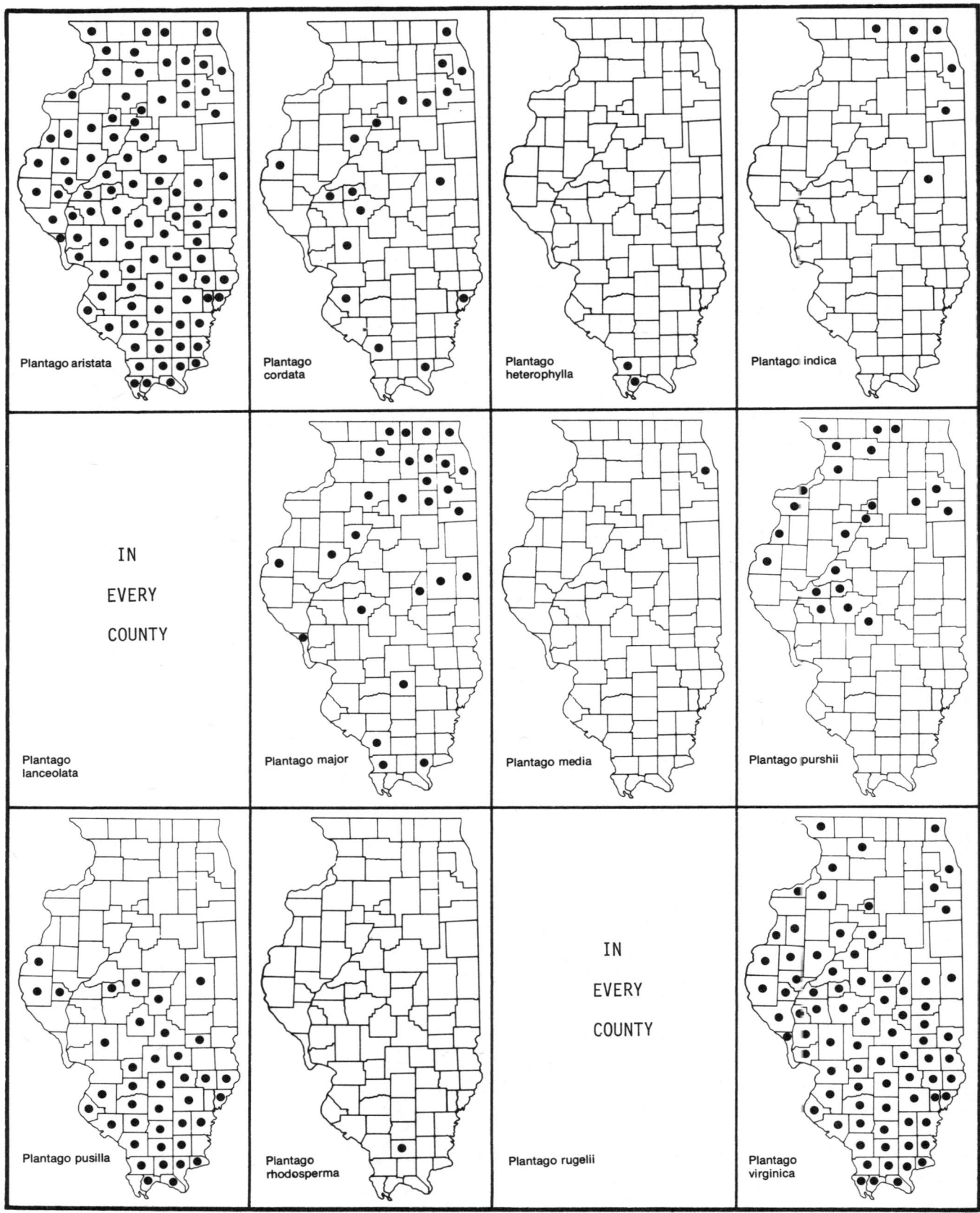

Plantago aristata

Plantago cordata

Plantago heterophylla

Plantago indica

Plantago lanceolata

IN EVERY COUNTY

Plantago major

Plantago media

Plantago purshii

Plantago pusilla

Plantago rhodosperma

Plantago rugelii

IN EVERY COUNTY

Plantago virginica

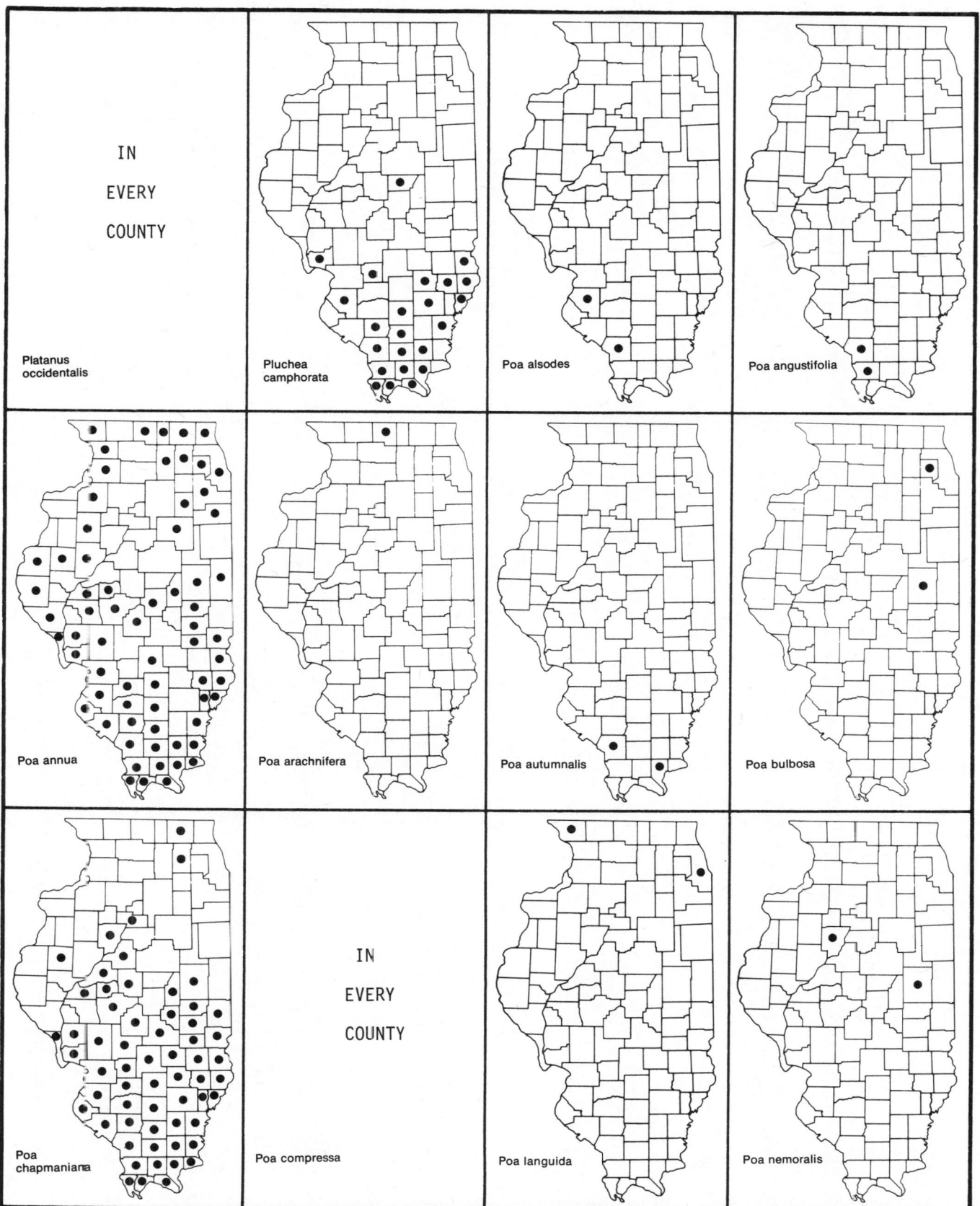

IN

EVERY

COUNTY

Platanus
occidentalis

Pluchea
camphorata

Poa alsodes

Poa angustifolia

Poa annua

Poa arachnifera

Poa autumnalis

Poa bulbosa

Poa
chapmaniana

Poa compressa

IN

EVERY

COUNTY

Poa languida

Poa nemoralis

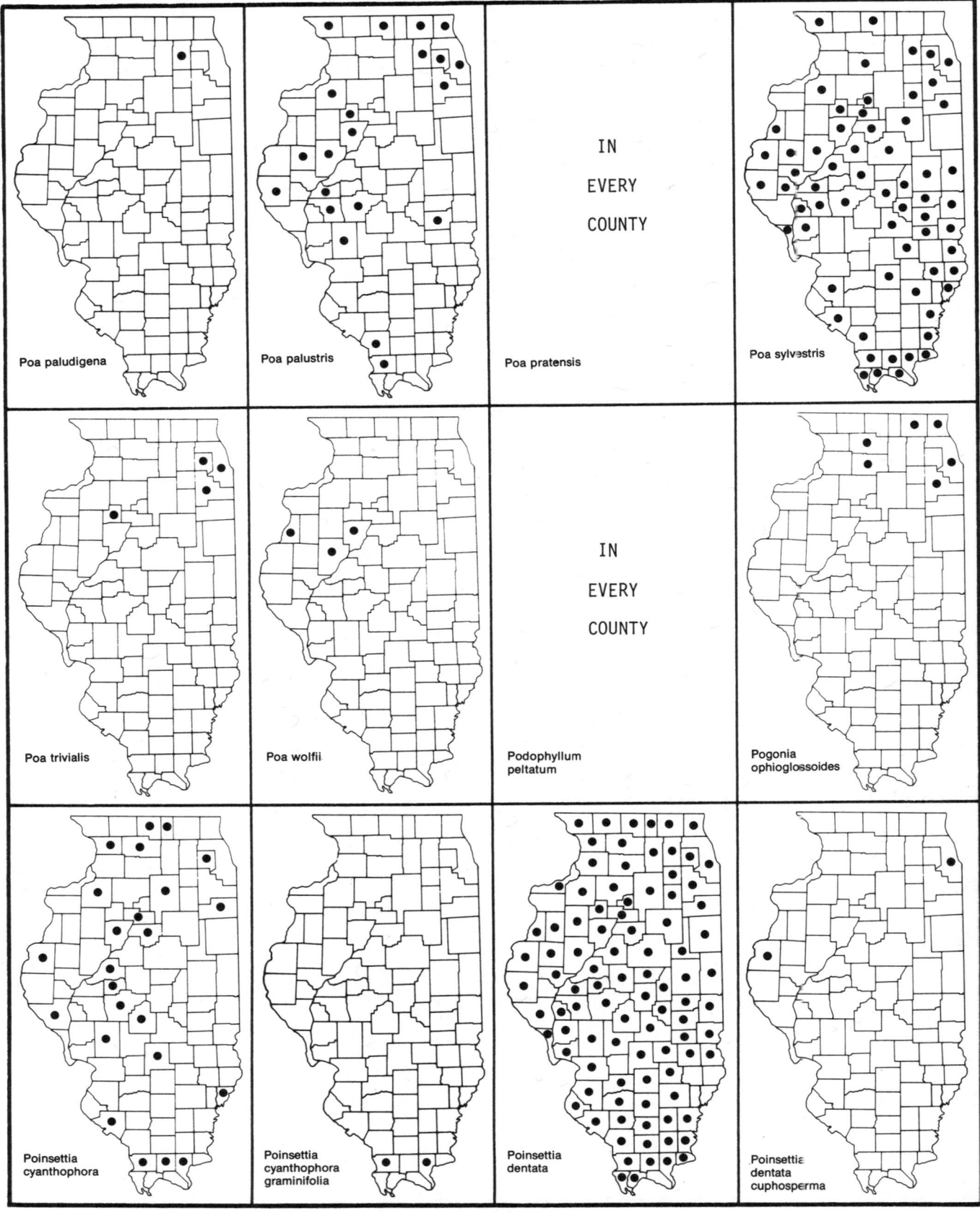

Poa paludigena

Poa palustris

Poa pratensis

IN

EVERY

COUNTY

Poa sylvestris

Poa trivialis

Poa wolfii

Podophyllum
peltatum

IN

EVERY

COUNTY

Pogonia
ophioglossoides

Poinsettia
cyanthophora

Poinsettia
cyanthophora
graminifolia

Poinsettia
dentata

Poinsettia
dentata
cuphosperma

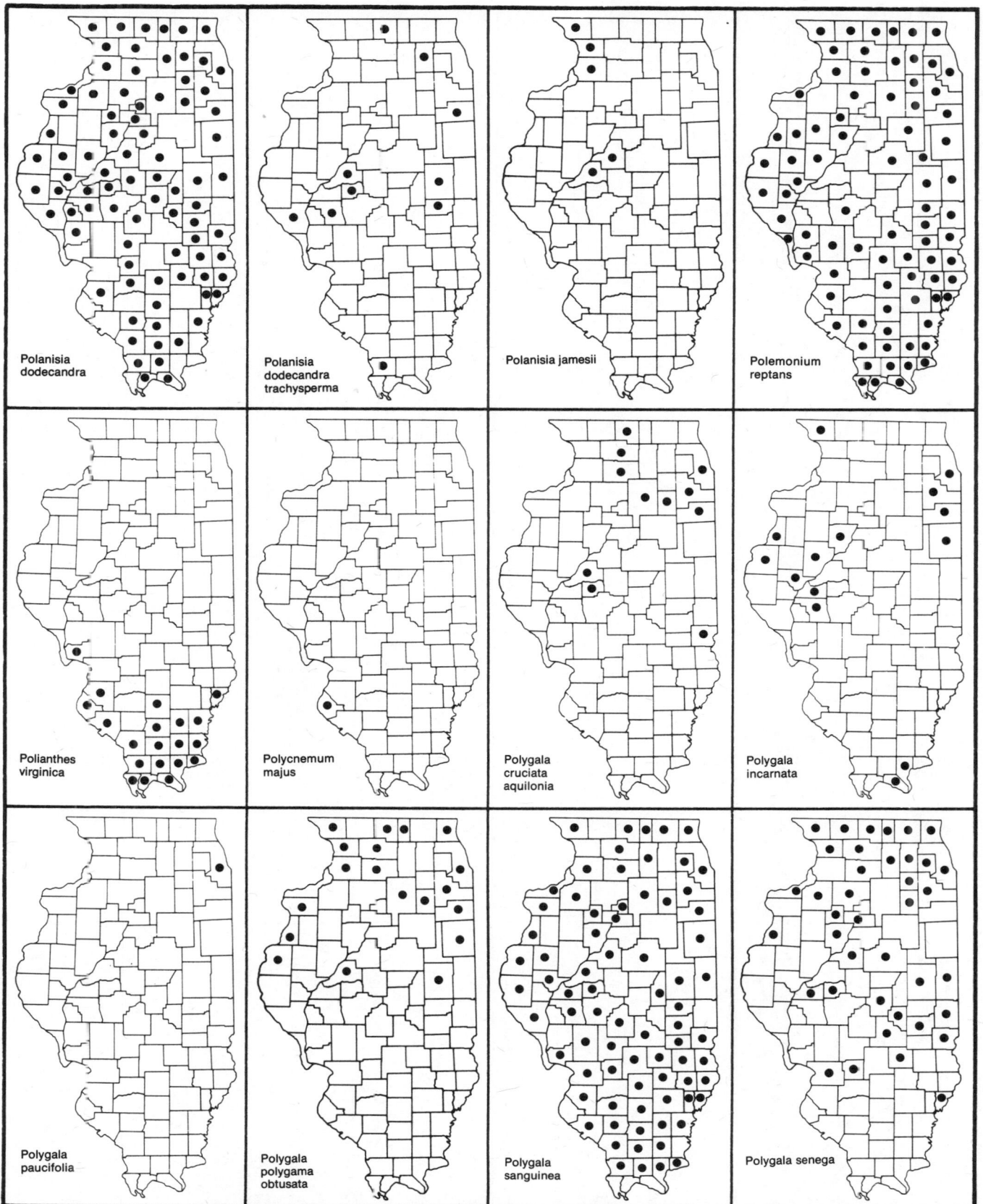

Polanisia
dodecandra

Polanisia
dodecandra
trachysperma

Polanisia jamesii

Polemonium
reptans

Polianthes
virginica

Polycnemum
majus

Polygala
cruciata
aquilonia

Polygala
incarnata

Polygala
paucifolia

Polygala
polygama
obtusata

Polygala
sanguinea

Polygala senega

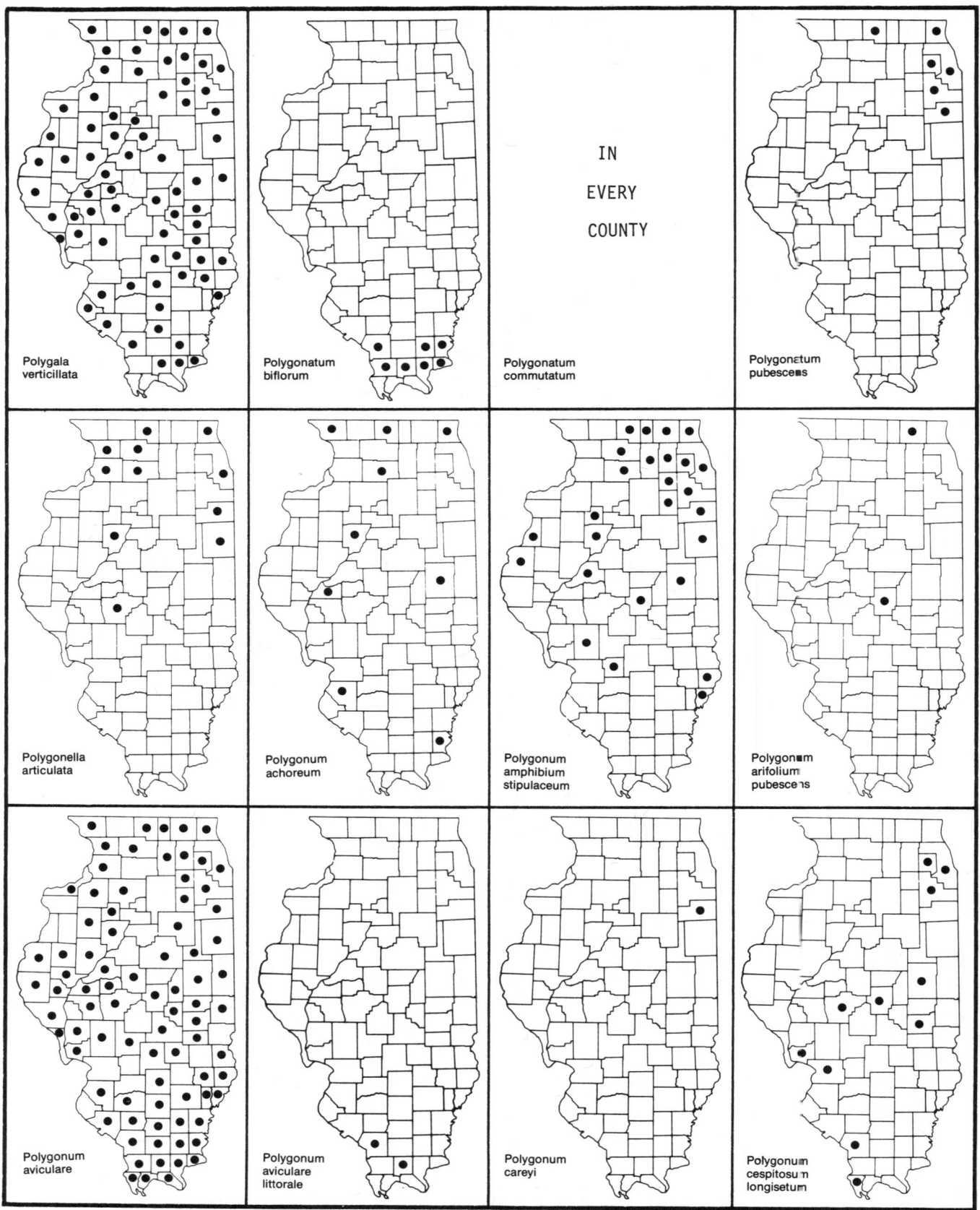

Polygala
verticillata

Polygonatum
biflorum

IN

EVERY

COUNTY

Polygonatum
commutatum

Polygonatum
pubescens

Polygonella
articulata

Polygonum
achoreum

Polygonum
amphibium
stipulaceum

Polygonum
arifolium
pubescens

Polygonum
aviculare

Polygonum
aviculare
littorale

Polygonum
careyi

Polygonum
cespitosum
longisetum

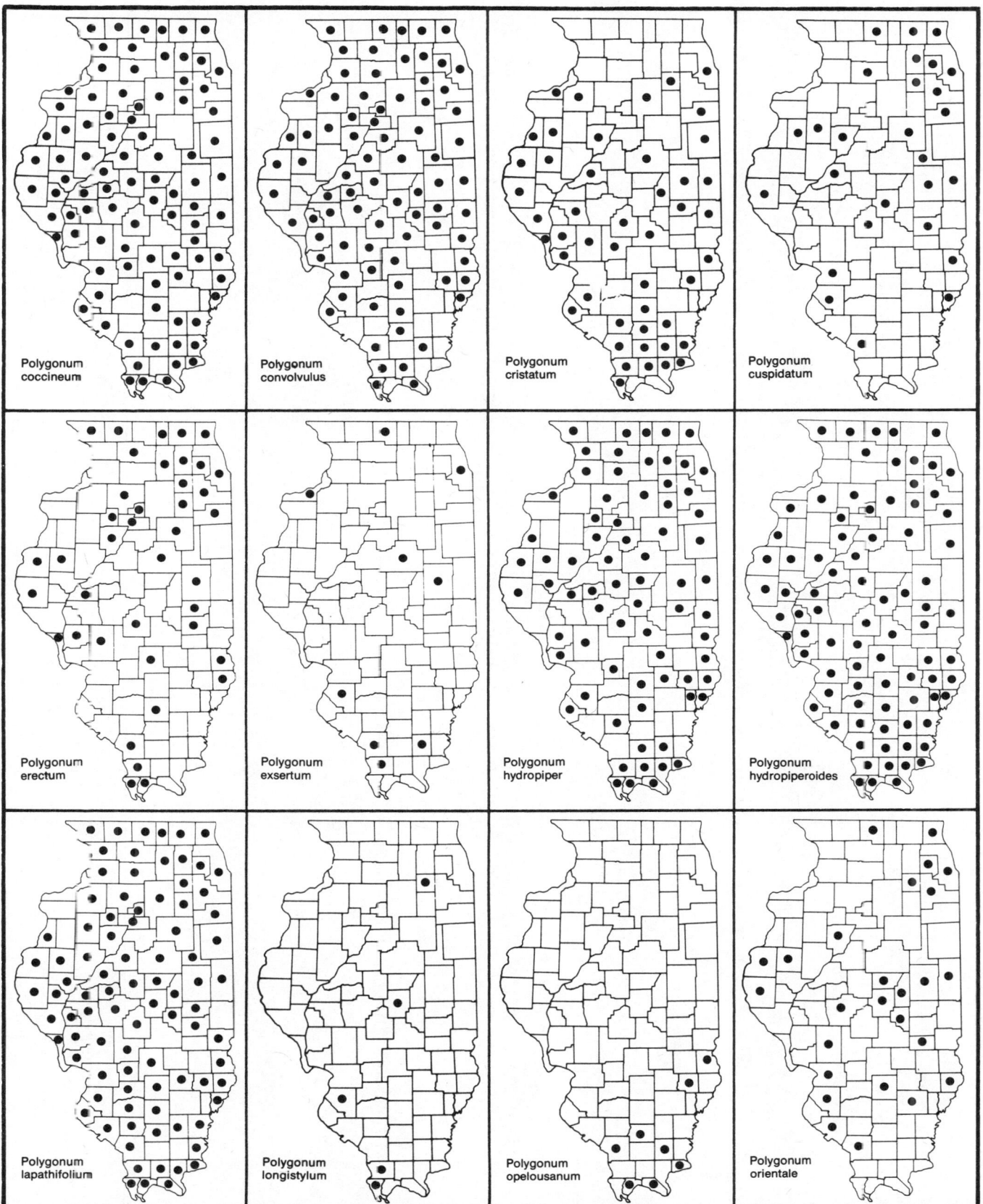

Polygonum
coccineum

Polygonum
convolvulus

Polygonum
cristatum

Polygonum
cuspidatum

Polygonum
erectum

Polygonum
exsertum

Polygonum
hydropiper

Polygonum
hydropiperoides

Polygonum
lapathifolium

Polygonum
longistylum

Polygonum
opelousanum

Polygonum
orientale

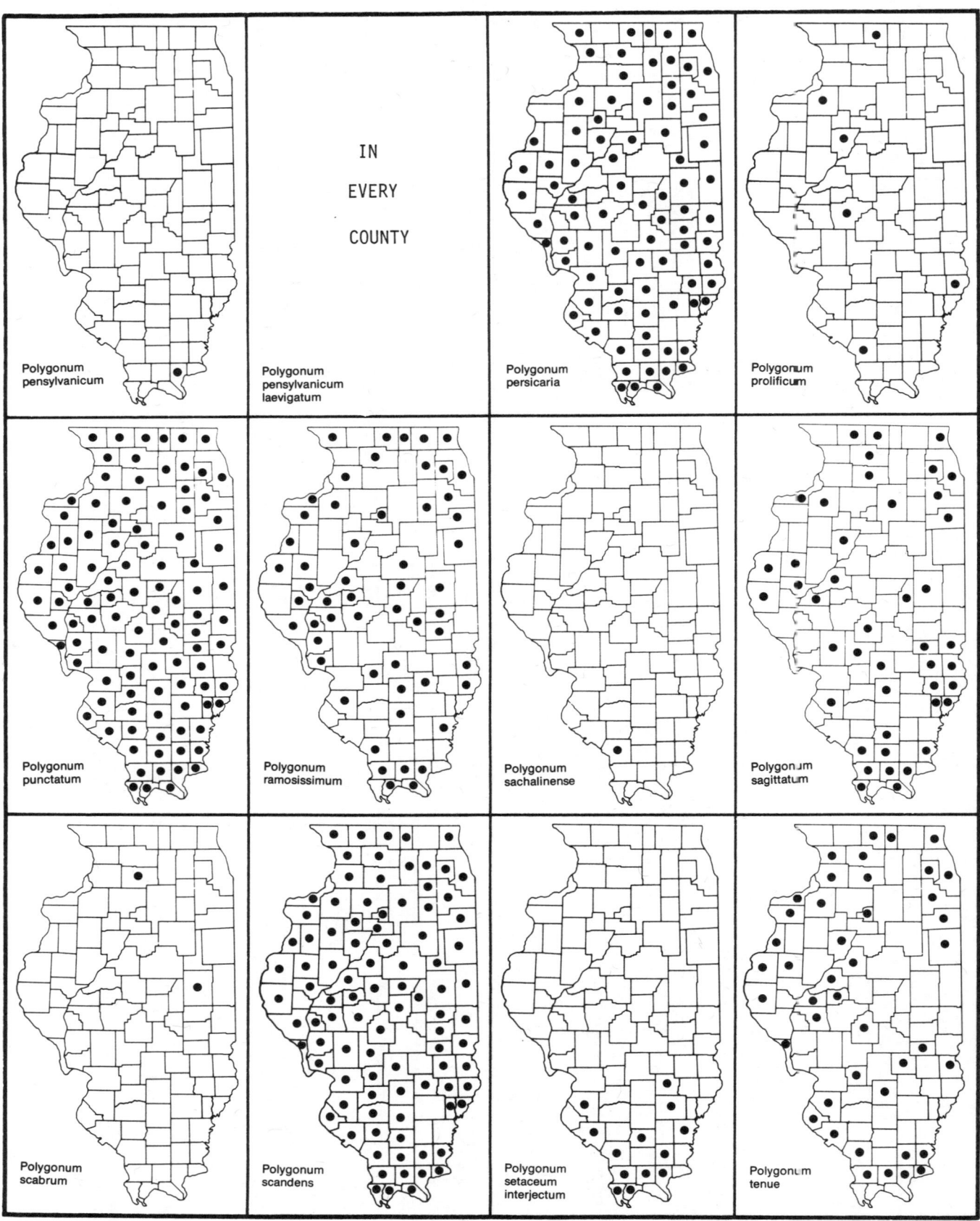

IN

EVERY

COUNTY

Polygonum
pensylvanicum

Polygonum
pensylvanicum
laevigatum

Polygonum
persicaria

Polygonum
prolificum

Polygonum
punctatum

Polygonum
ramosissimum

Polygonum
sachalinense

Polygonum
sagittatum

Polygonum
scabrum

Polygonum
scandens

Polygonum
setaceum
interjectum

Polygonum
tenue

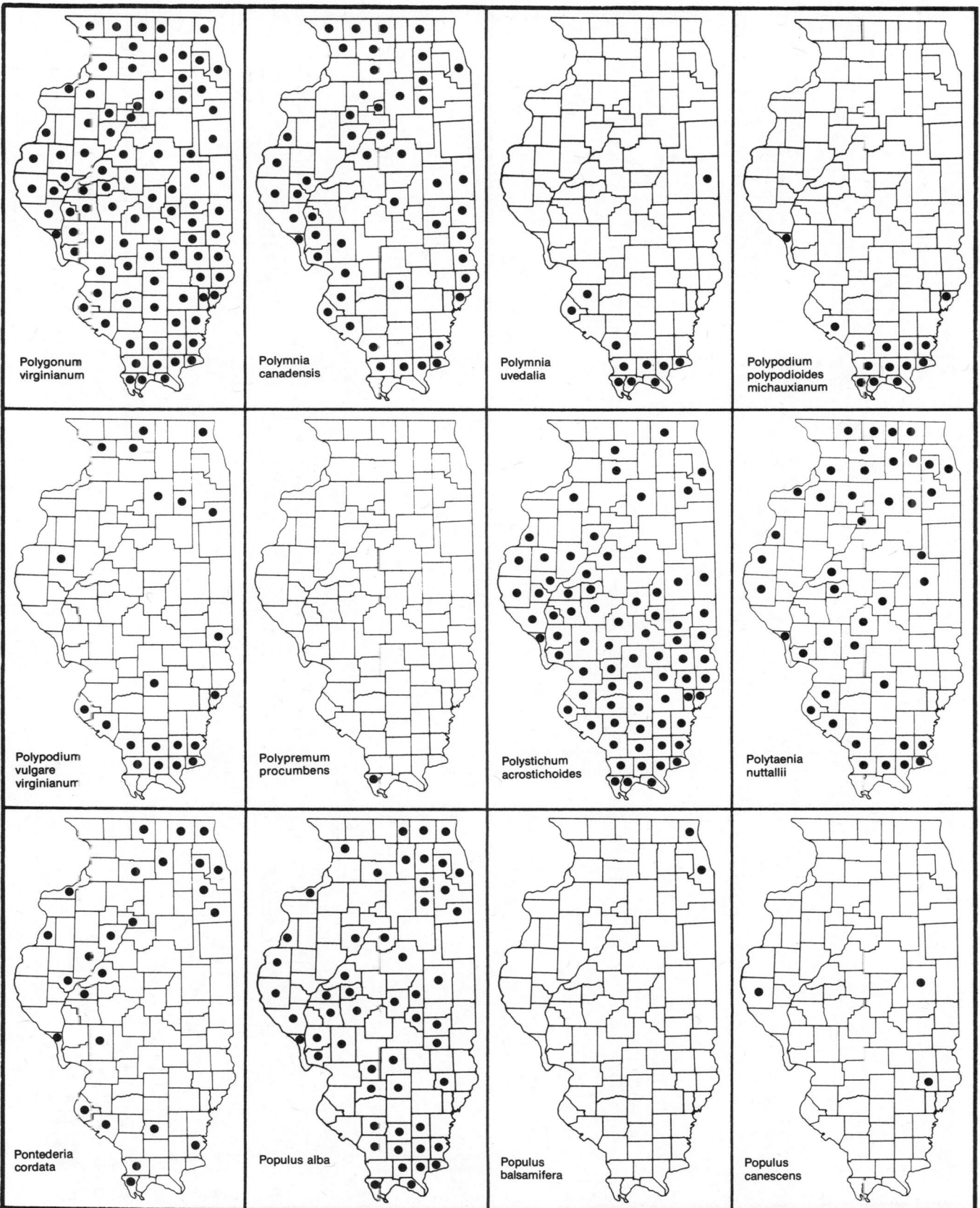

Polygonum
virginianum

Polymnia
canadensis

Polymnia
uvedalia

Polypodium
polypodioides
michauxianum

Polypodium
vulgare
virginianum

Polypremum
procumbens

Polystichum
acrostichoides

Polytaenia
nuttallii

Pontederia
cordata

Populus alba

Populus
balsamifera

Populus
canescens

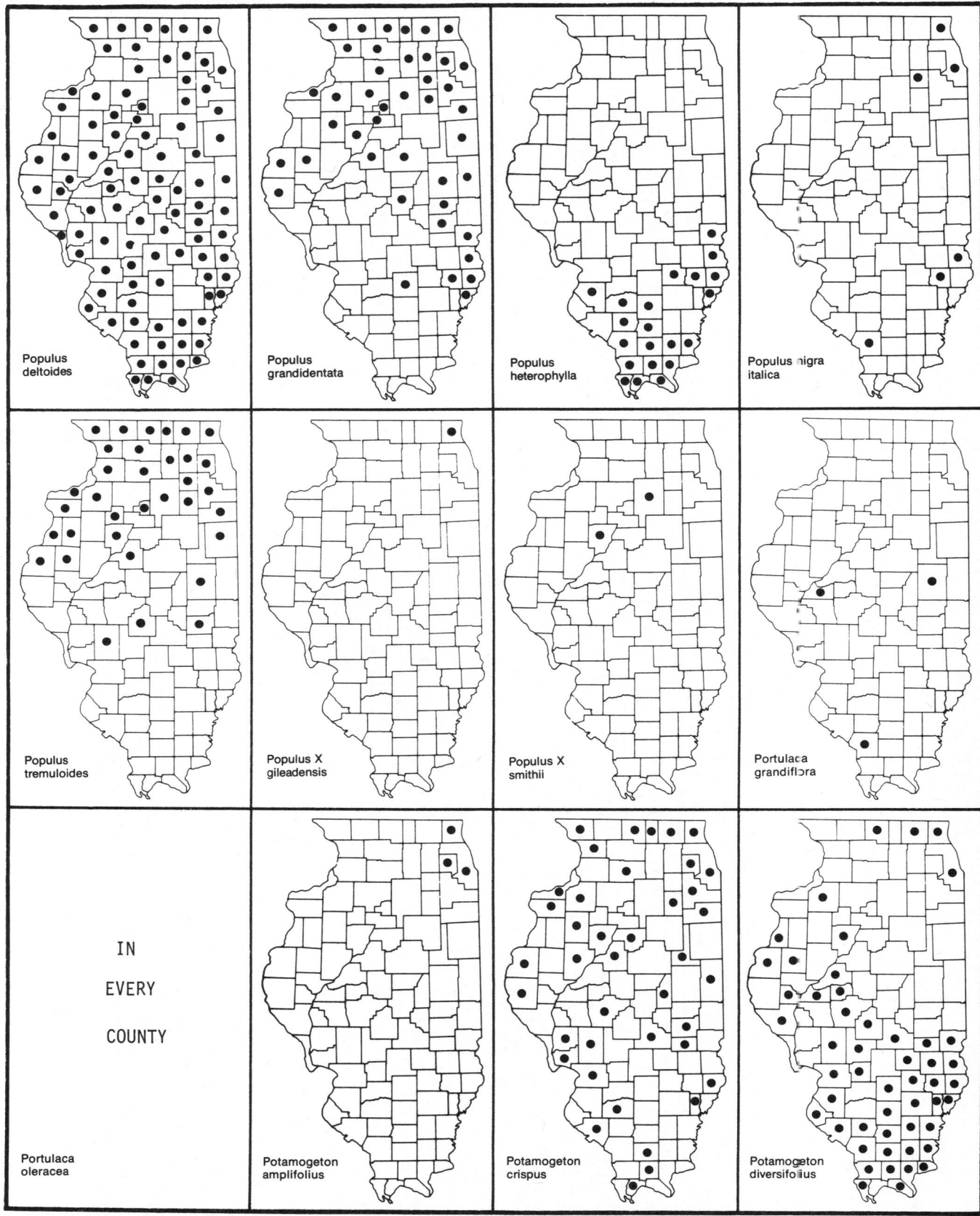

Populus
deltoides

Populus
grandidentata

Populus
heterophylla

Populus nigra
italica

Populus
tremuloides

Populus X
gileadensis

Populus X
smithii

Portulaca
grandiflora

IN

EVERY

COUNTY

Portulaca
oleracea

Potamogeton
amplifolius

Potamogeton
crispus

Potamogeton
diversifolius

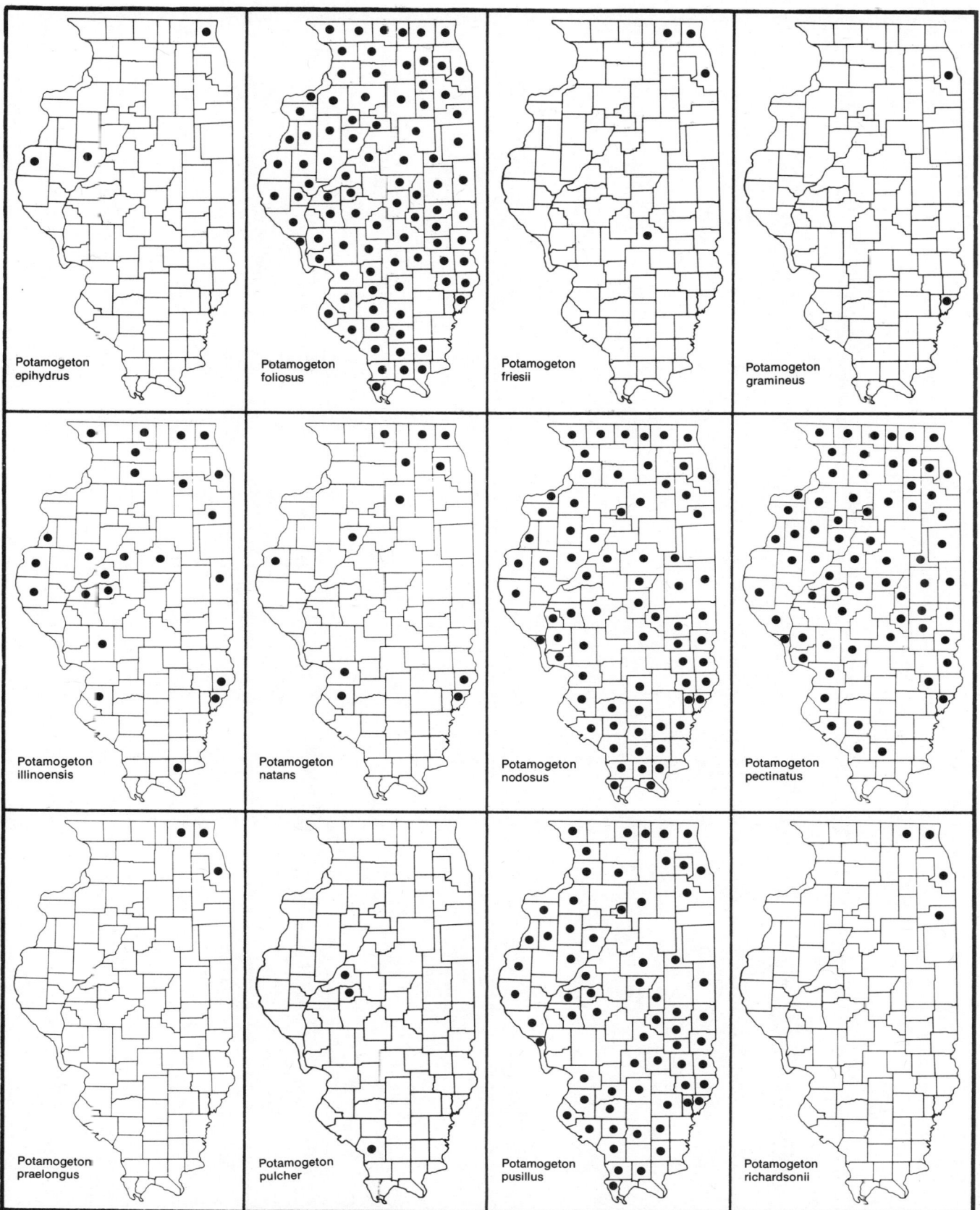

Potamogeton epihydrus

Potamogeton foliosus

Potamogeton friesii

Potamogeton gramineus

Potamogeton illinoensis

Potamogeton natans

Potamogeton nodosus

Potamogeton pectinatus

Potamogeton praelongus

Potamogeton pulcher

Potamogeton pusillus

Potamogeton richardsonii

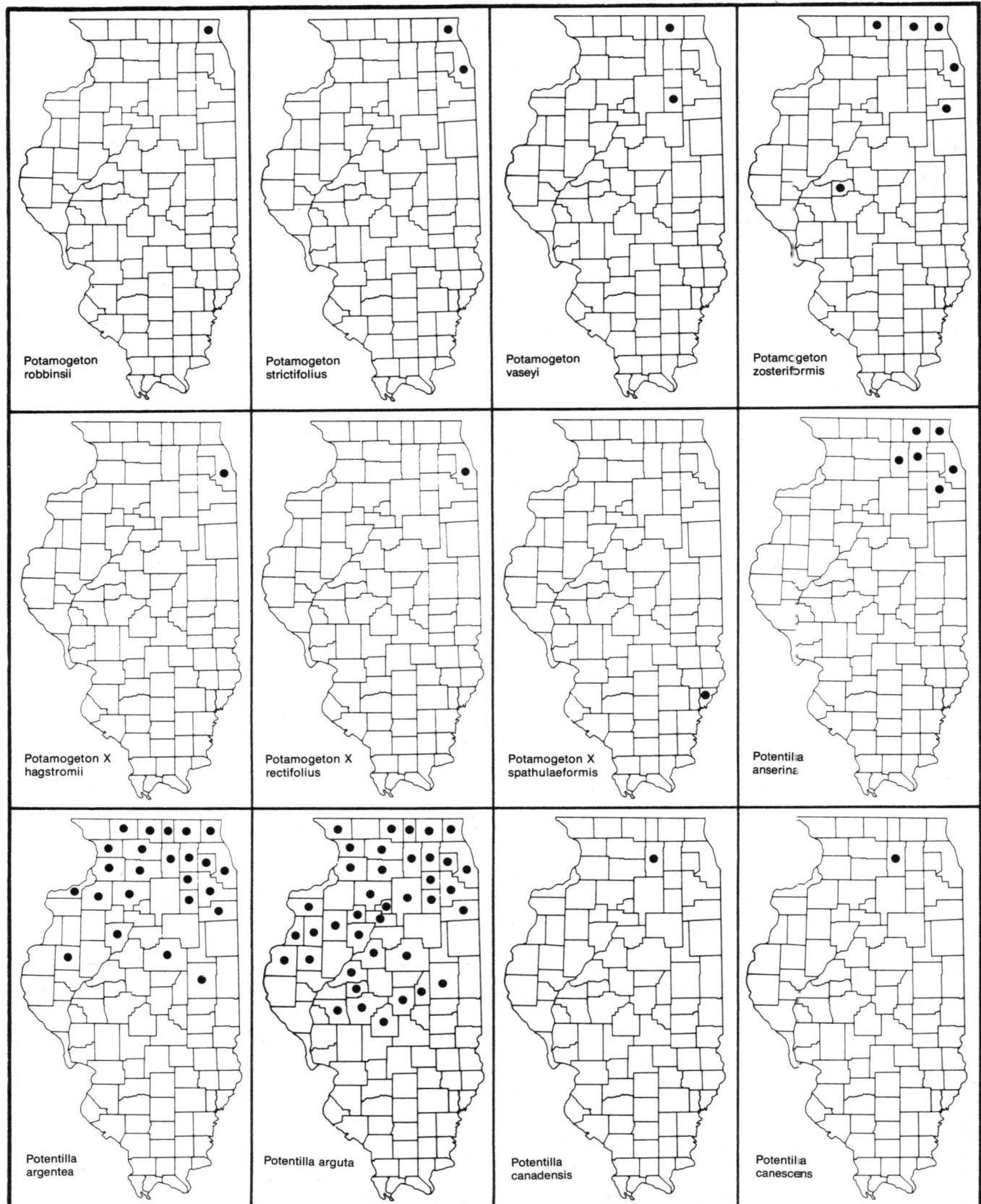

Potamogeton
robbinsii

Potamogeton
strictifolius

Potamogeton
vaseyi

Potamogeton
zosteriformis

Potamogeton X
hagstromii

Potamogeton X
rectifolius

Potamogeton X
spathulaeformis

Potentilla
anserina

Potentilla
argentea

Potentilla
arguta

Potentilla
canadensis

Potentilla
canescens

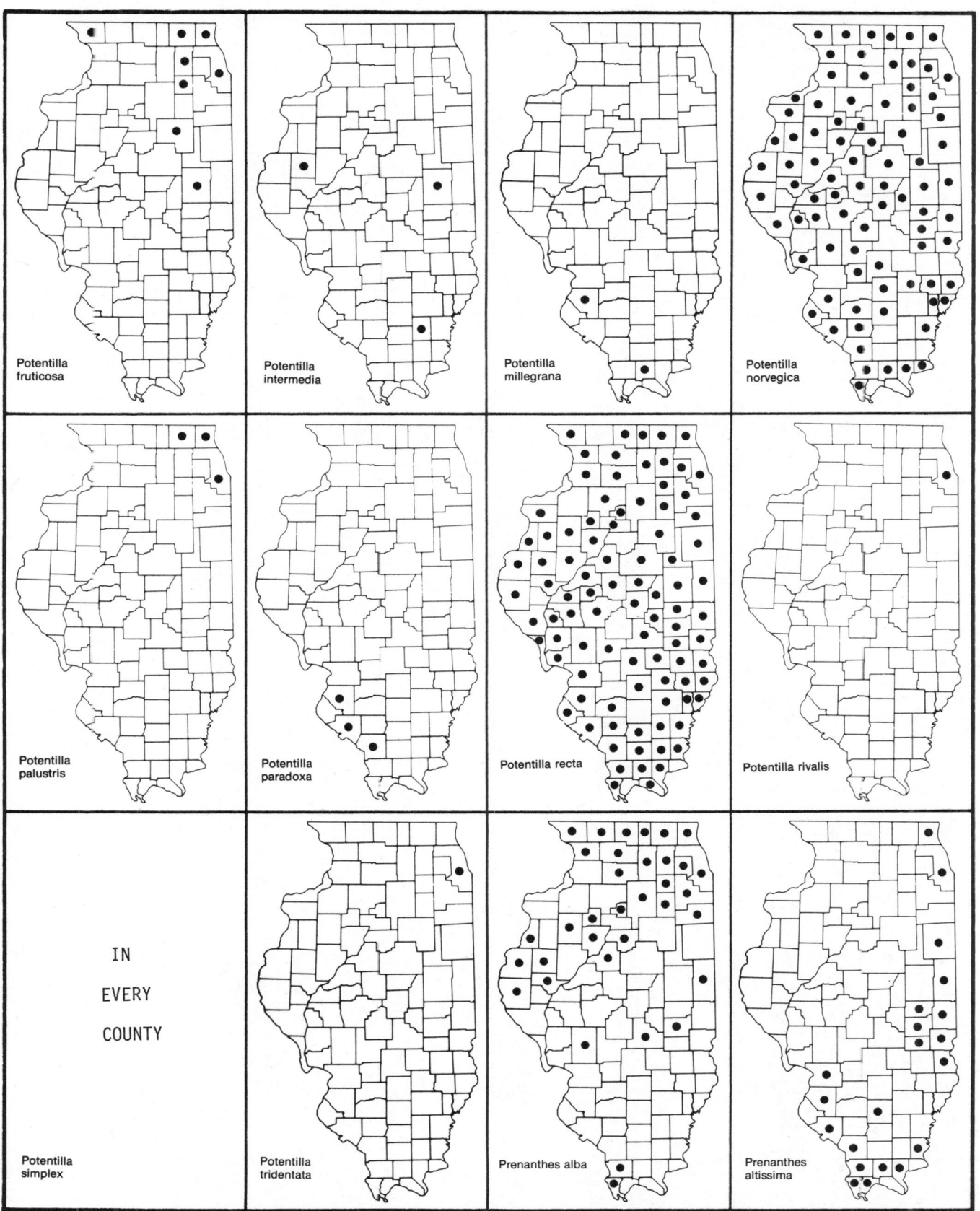

Potentilla
fruticosa

Potentilla
intermedia

Potentilla
millegrana

Potentilla
norvegica

Potentilla
palustris

Potentilla
paradoxa

Potentilla
recta

Potentilla
rivalis

IN

EVERY

COUNTY

Potentilla
simplex

Potentilla
tridentata

Prenanthes alba

Prenanthes
altissima

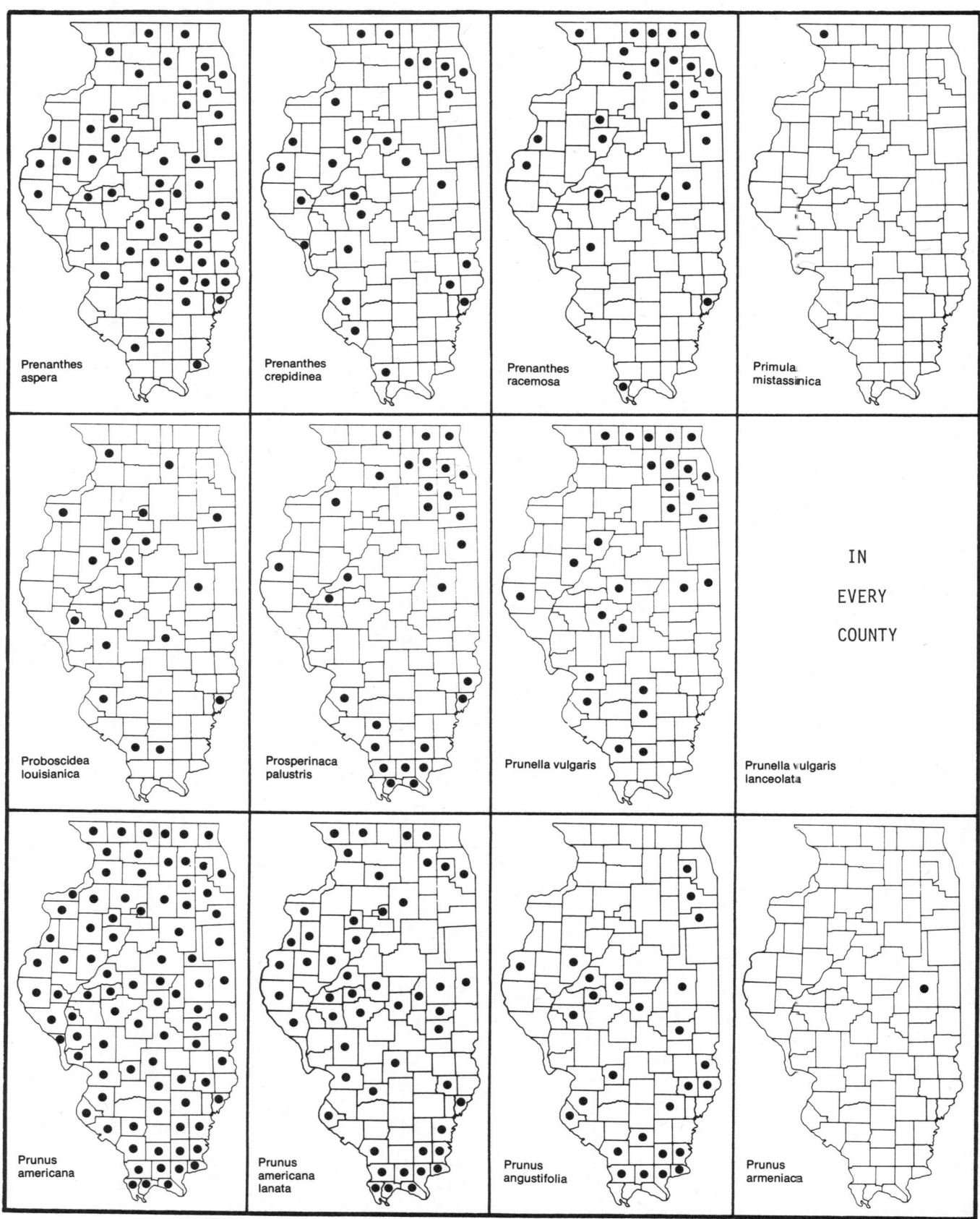

Prenanthes
aspera

Prenanthes
crepidinea

Prenanthes
racemosa

Primula
mistassinica

Proboscidea
louisianica

Prosperinaca
palustris

Prunella vulgaris

IN

EVERY

COUNTY

Prunella vulgaris
lanceolata

Prunus
americana

Prunus
americana
lanata

Prunus
angustifolia

Prunus
armeniaca

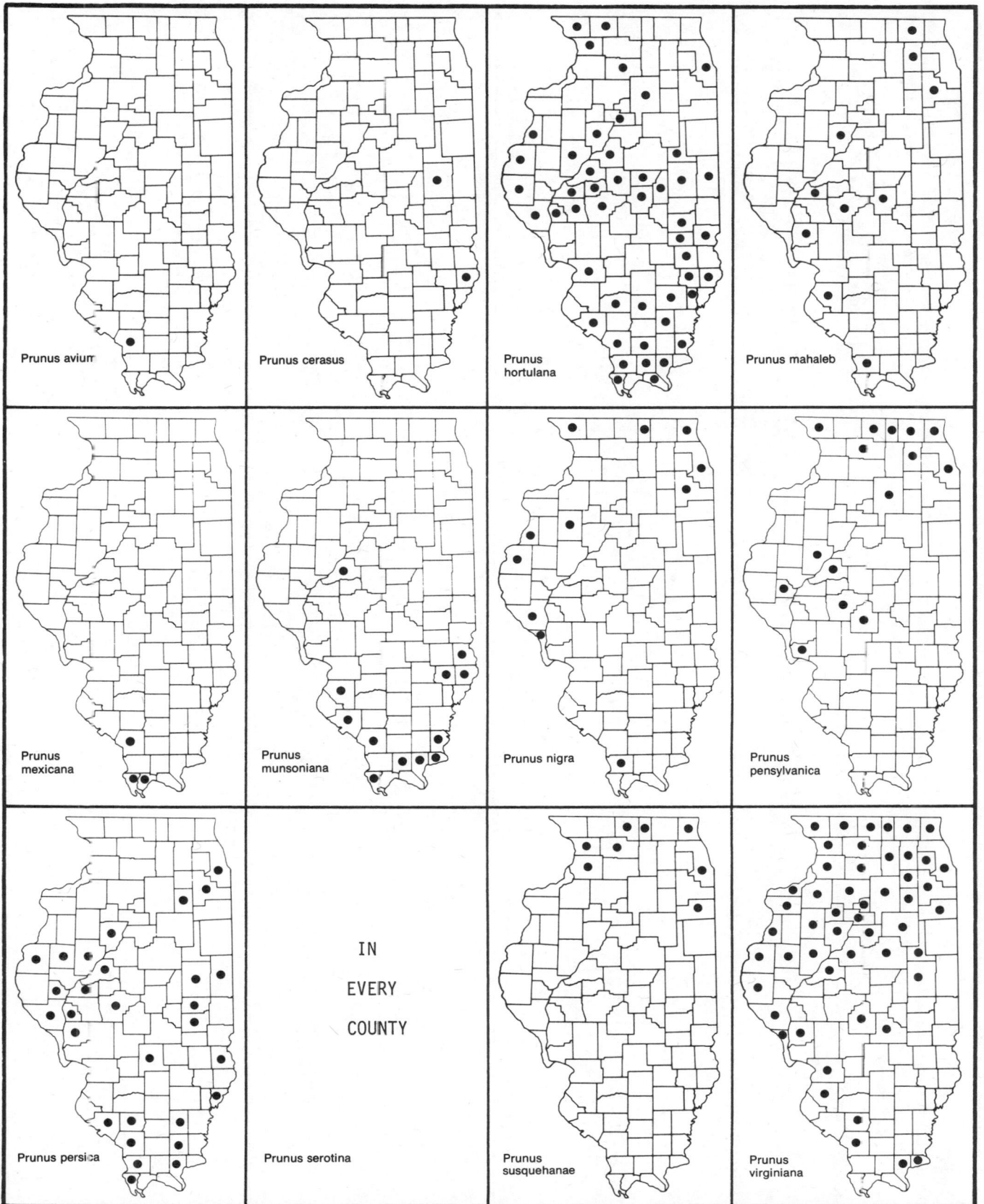

Prunus avium

Prunus cerasus

Prunus hortulana

Prunus mahaleb

Prunus mexicana

Prunus munsoniana

Prunus nigra

Prunus pensylvanica

Prunus persica

Prunus serotina

IN

EVERY

COUNTY

Prunus susquehanae

Prunus virginiana

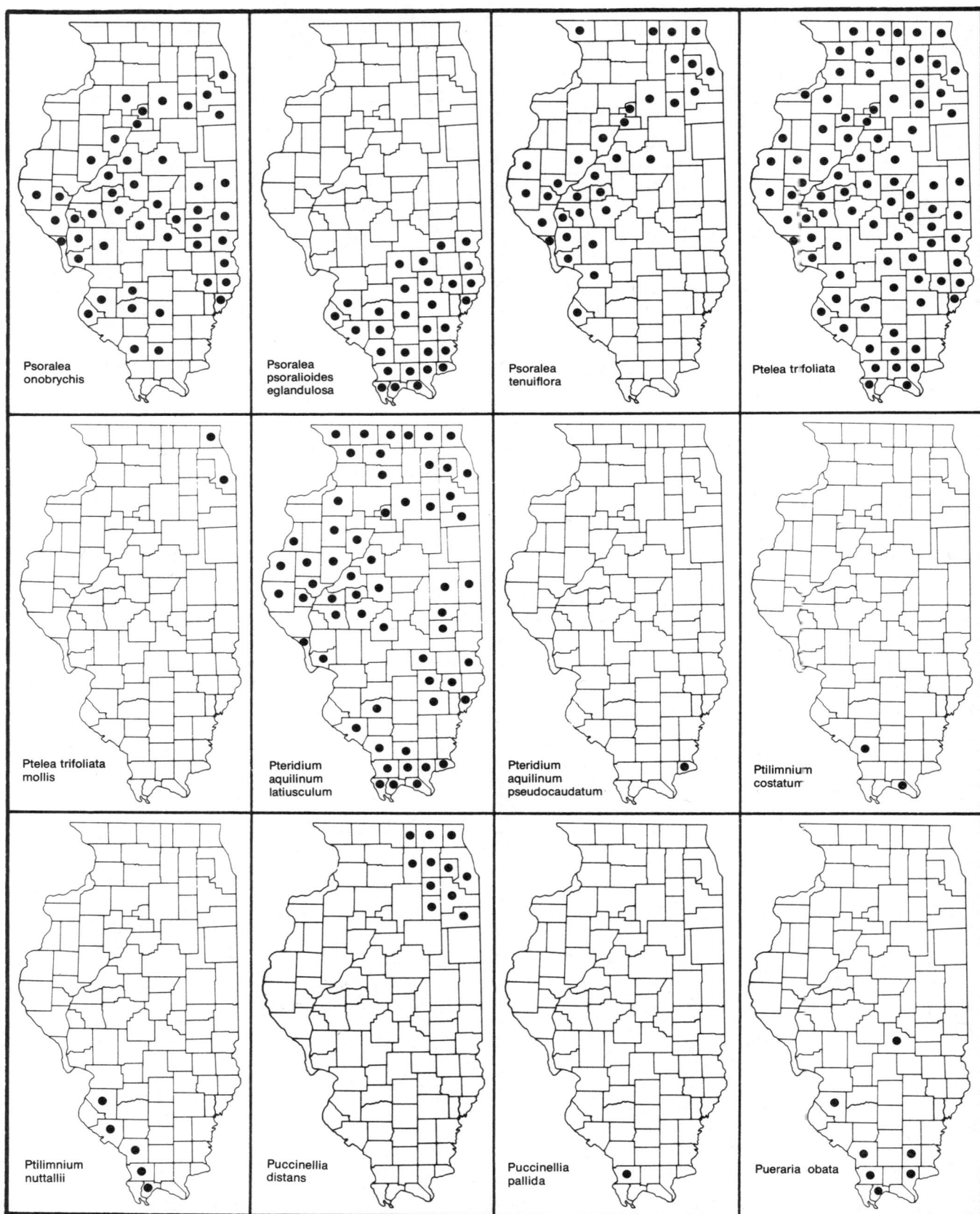

Psoralea
onobrychis

Psoralea
psoralioides
eglandulosa

Psoralea
tenuiflora

Ptelea trifoliata

Ptelea trifoliata
mollis

Pteridium
aquilinum
latiusculum

Pteridium
aquilinum
pseudocaudatum

Ptilimnium
costatum

Ptilimnium
nuttallii

Puccinellia
distans

Puccinellia
pallida

Pueraria obata

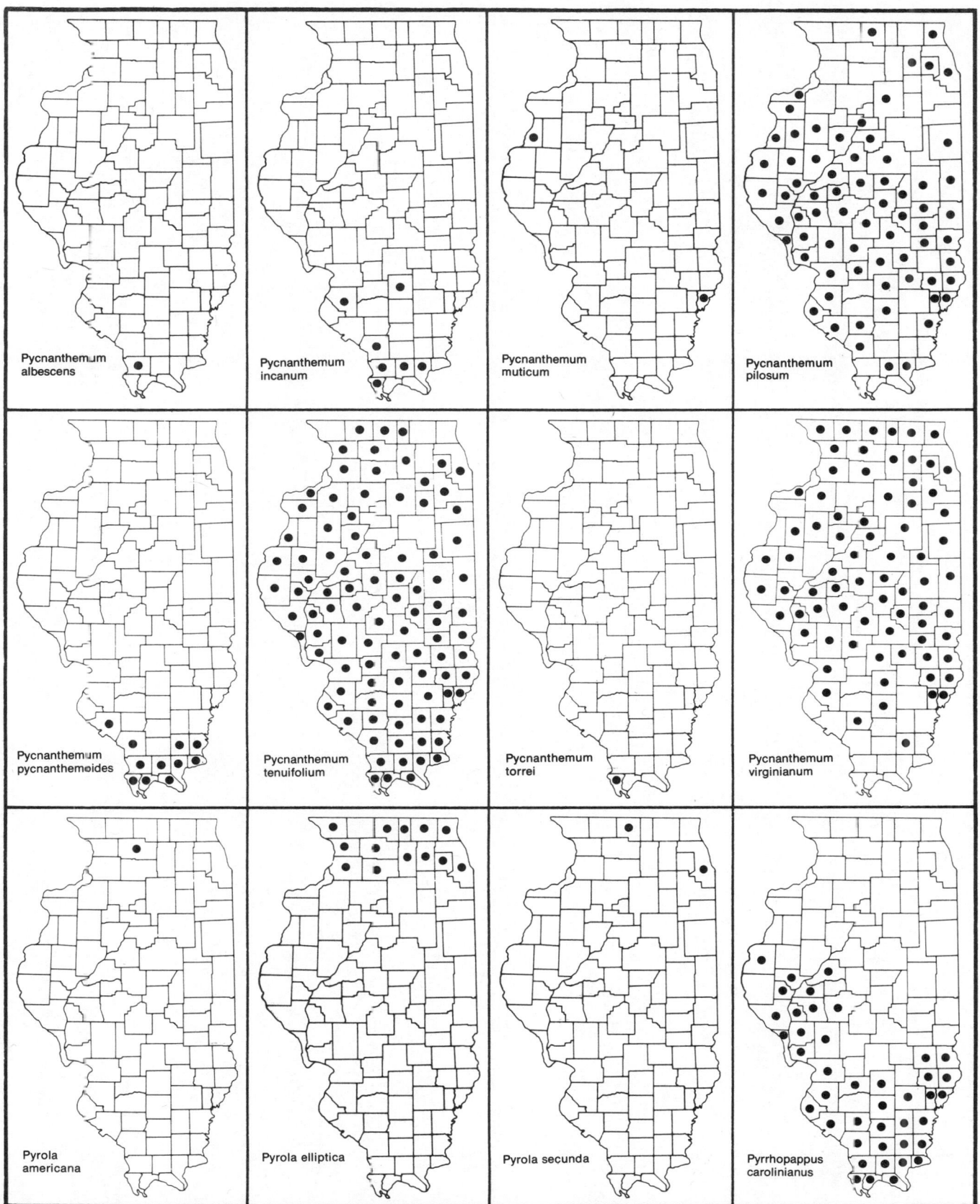

Pycnanthemum
albescens

Pycnanthemum
incanum

Pycnanthemum
muticum

Pycnanthemum
pilosum

Pycnanthemum
pycnanthemoides

Pycnanthemum
tenuifolium

Pycnanthemum
torrei

Pycnanthemum
virginianum

Pyrola
americana

Pyrola elliptica

Pyrola secunda

Pyrrhopappus
carolinianus

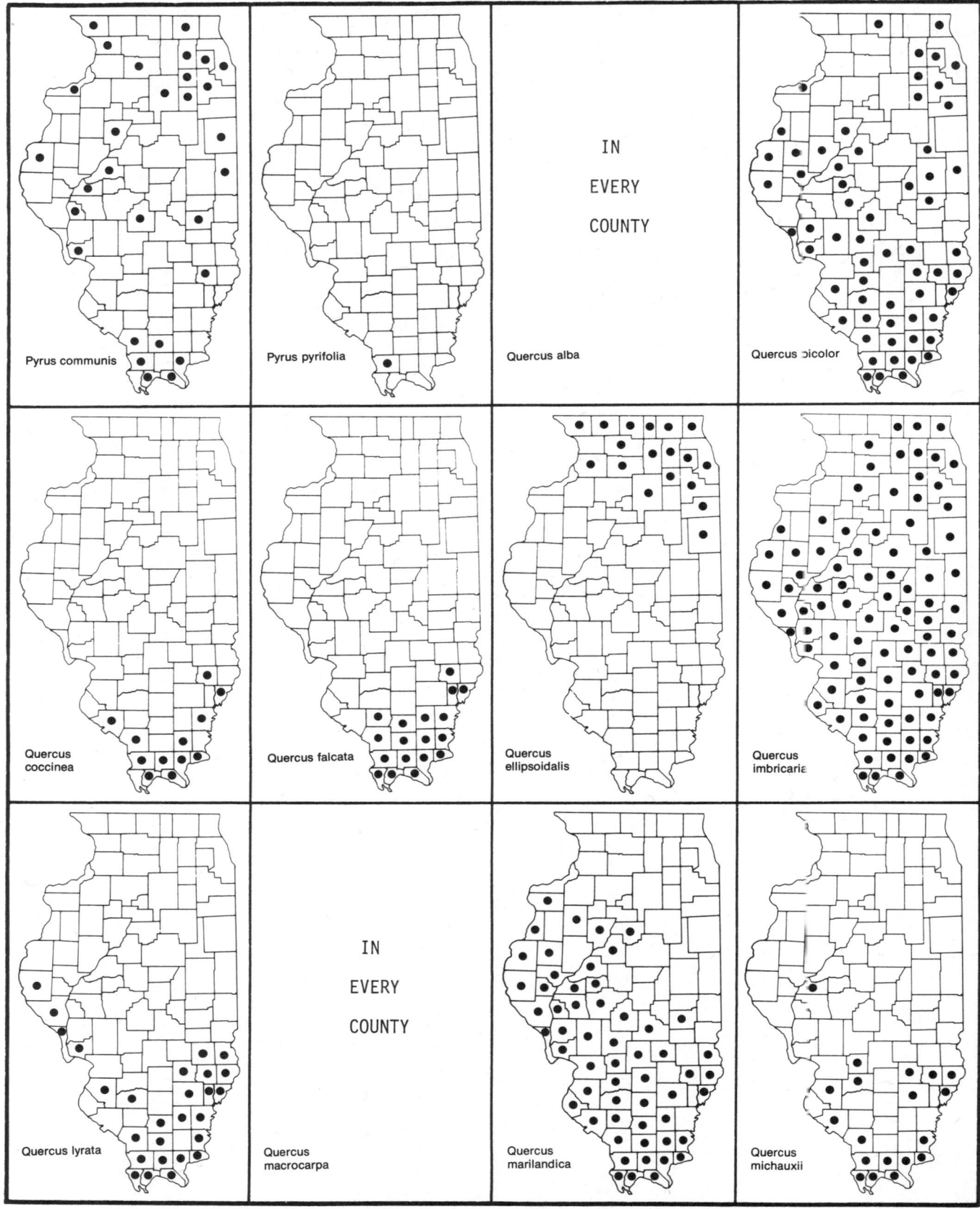

Pyrus communis

Pyrus pyrifolia

IN

EVERY

COUNTY

Quercus alba

Quercus bicolor

Quercus
coccinea

Quercus falcata

Quercus
ellipsoidalis

Quercus
imbricaria

Quercus lyrata

IN

EVERY

COUNTY

Quercus
macrocarpa

Quercus
marilandica

Quercus
michauxii

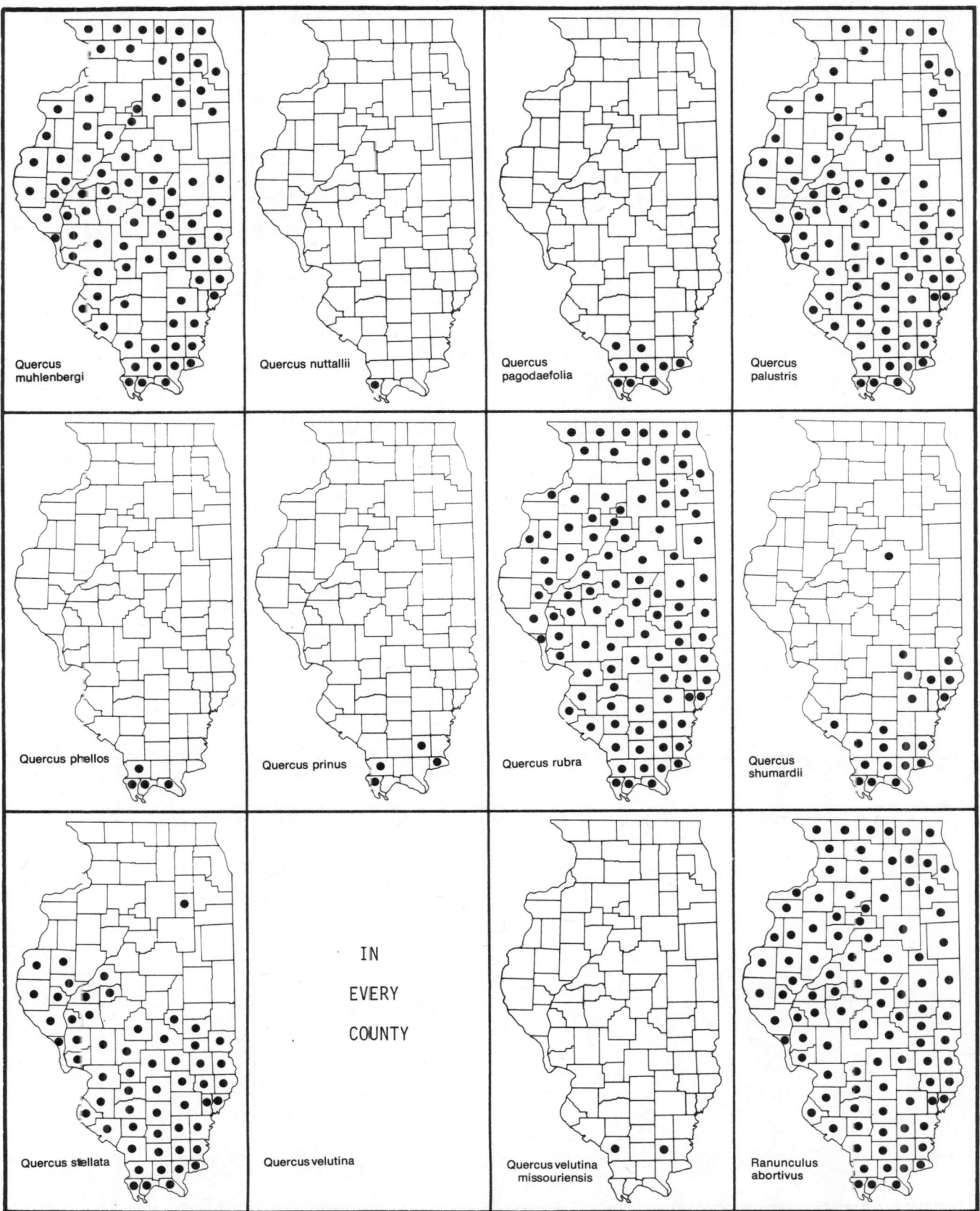

Quercus
muhlenbergi

Quercus
nuttallii

Quercus
pagodaefolia

Quercus
palustris

Quercus phellos

Quercus prinus

Quercus rubra

Quercus
shumardii

Quercus stellata

IN

EVERY

COUNTY

Quercus velutina

Quercus velutina
missouriensis

Ranunculus
abortivus

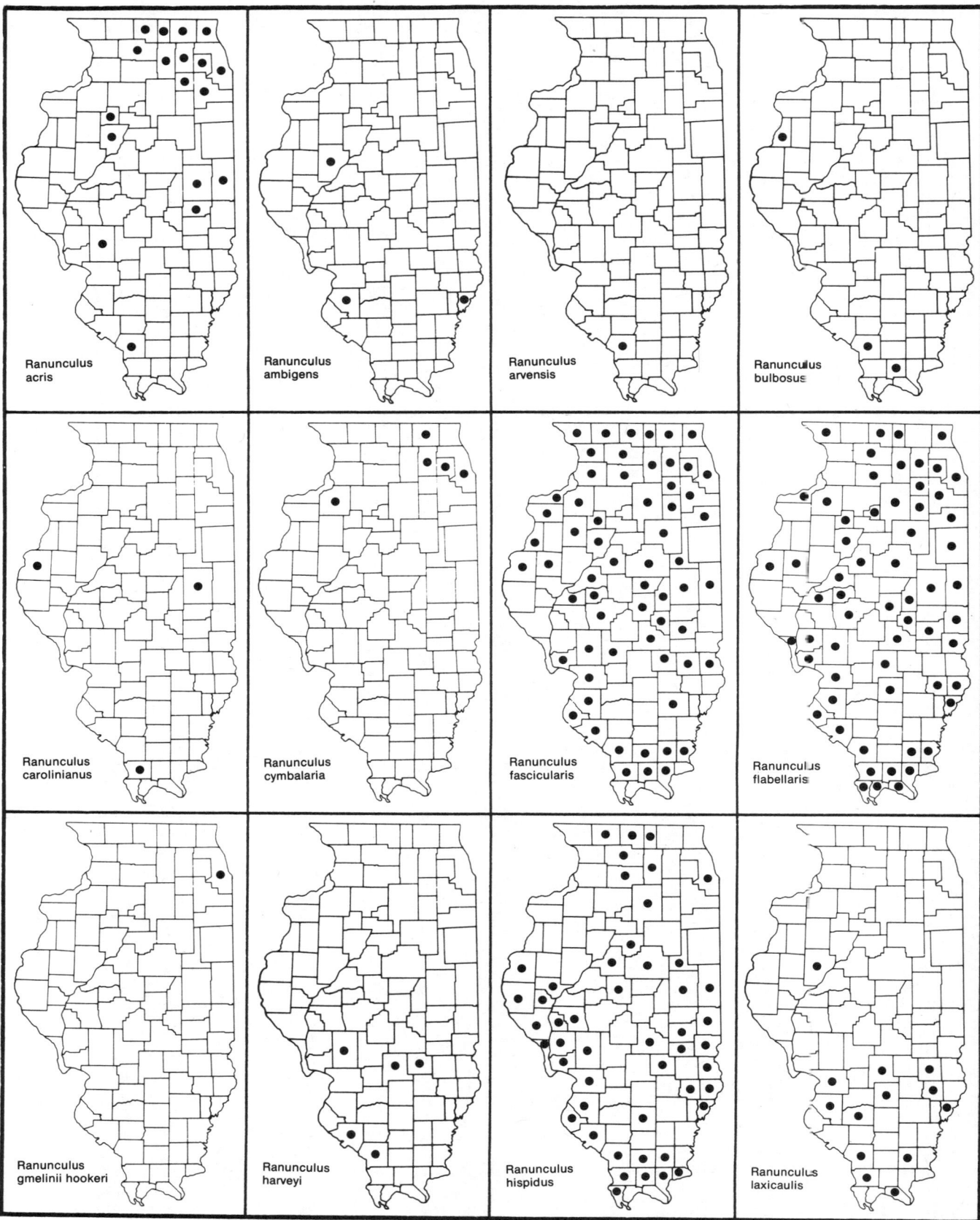

Ranunculus
acris

Ranunculus
ambigens

Ranunculus
arvensis

Ranunculus
bulbosus

Ranunculus
carolinianus

Ranunculus
cymbalaria

Ranunculus
fascicularis

Ranunculus
flabellaris

Ranunculus
gmelinii hookeri

Ranunculus
harveyi

Ranunculus
hispidus

Ranunculus
laxicaulis

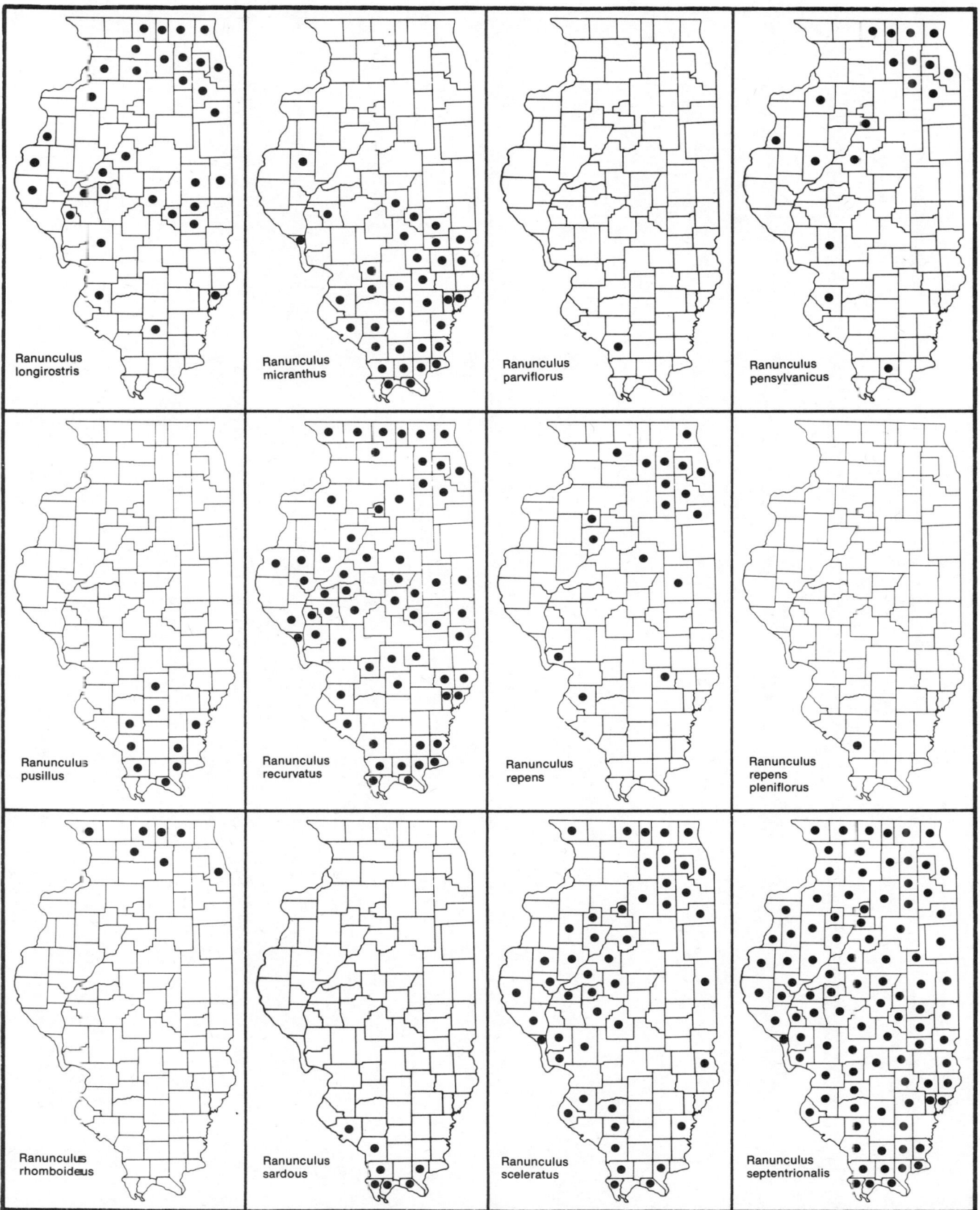

Ranunculus
longirostris

Ranunculus
micranthus

Ranunculus
parviflorus

Ranunculus
pensylvanicus

Ranunculus
pusillus

Ranunculus
recurvatus

Ranunculus
repens

Ranunculus
repens
pleniflorus

Ranunculus
rhomboideus

Ranunculus
sardous

Ranunculus
sceleratus

Ranunculus
septentrionalis

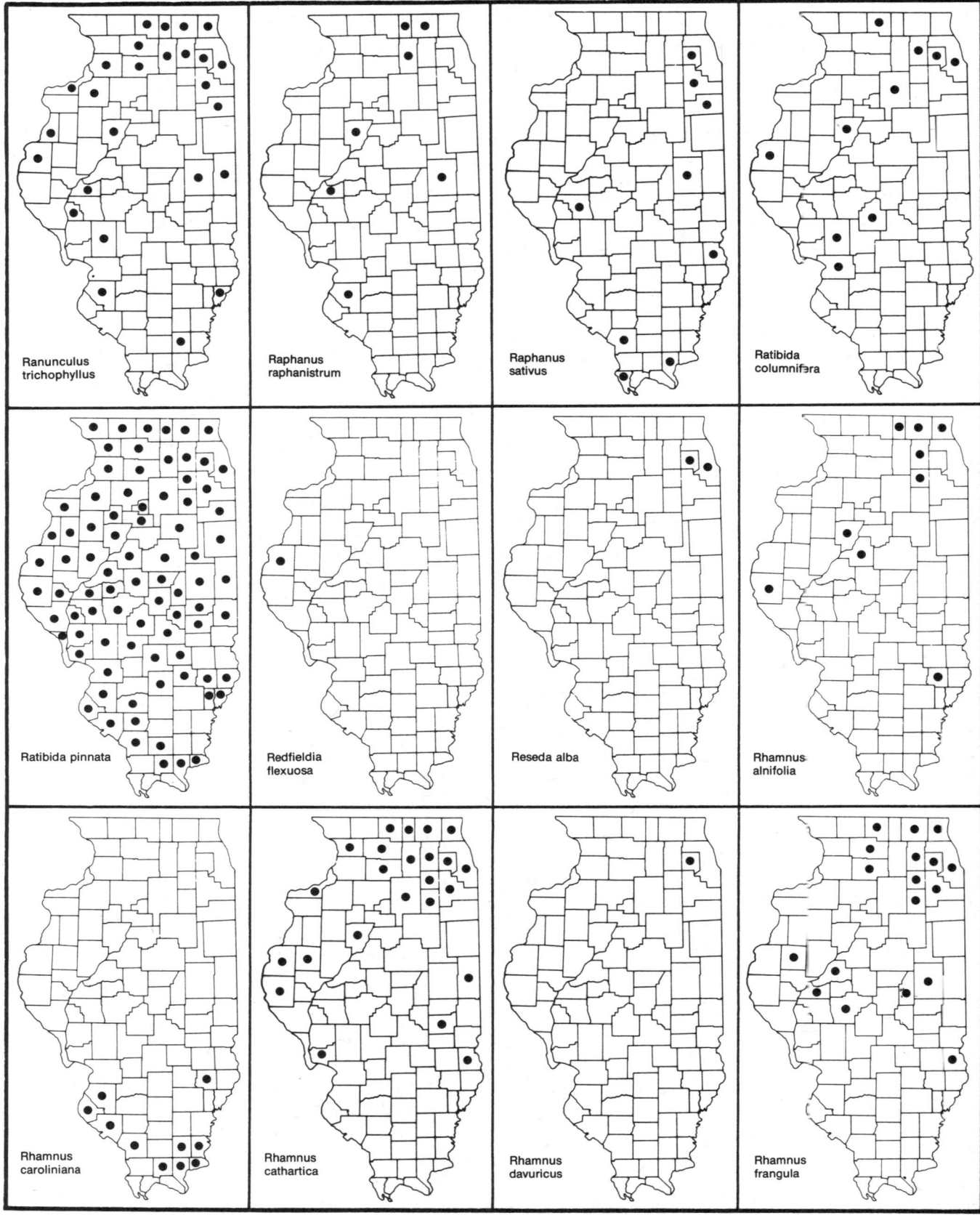

Ranunculus
trichophyllus

Raphanus
raphanistrum

Raphanus
sativus

Ratibida
columnifera

Ratibida pinnata

Redfieldia
flexuosa

Reseda alba

Rhamnus
alnifolia

Rhamnus
caroliniana

Rhamnus
cathartica

Rhamnus
davuricus

Rhamnus
frangula

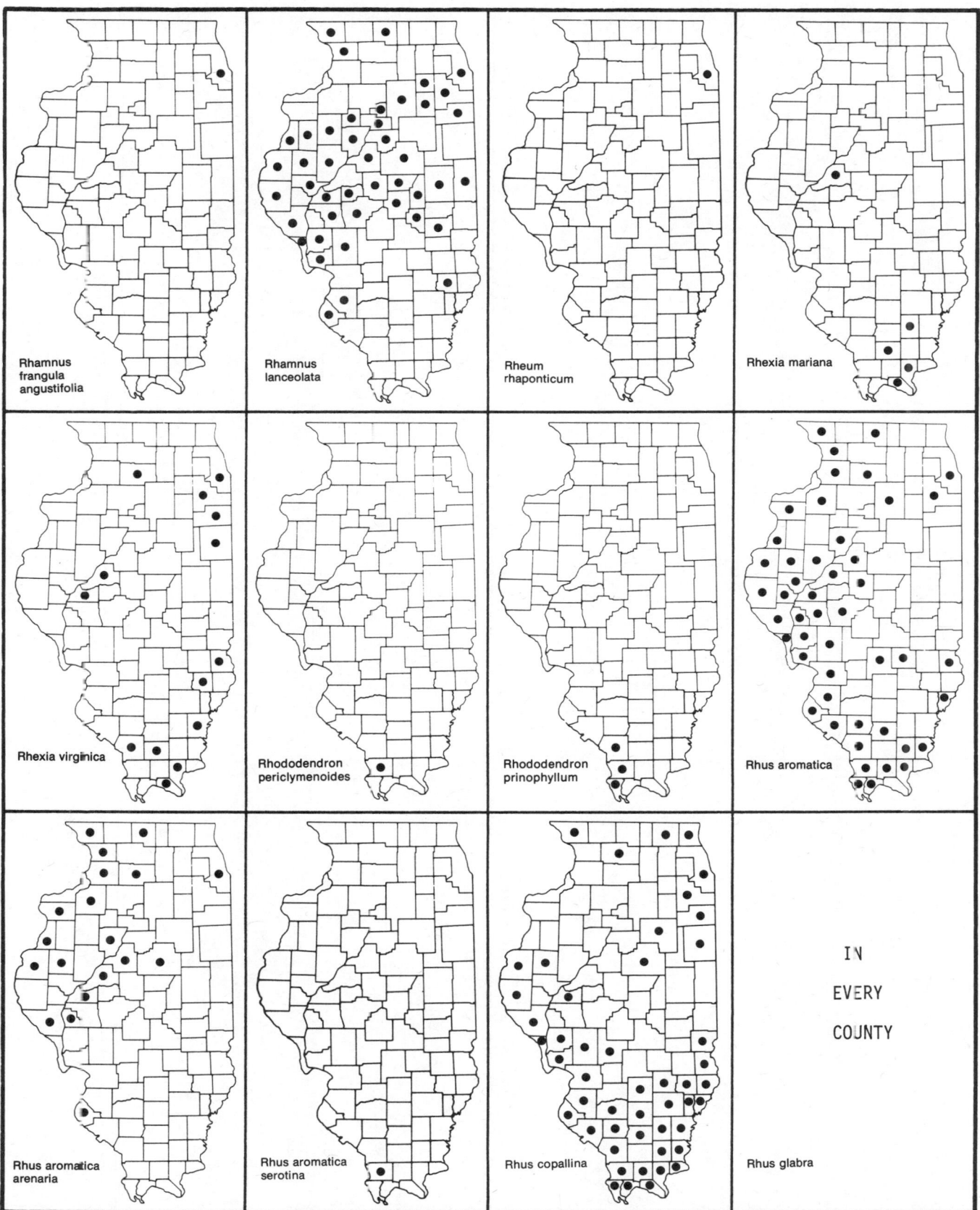

Rhamnus
frangula
angustifolia

Rhamnus
lanceolata

Rheum
rhaponticum

Rhexia mariana

Rhexia virginica

Rhododendron
periclymenoides

Rhododendron
prinophyllum

Rhus aromatica

Rhus aromatica
arenaria

Rhus aromatica
serotina

Rhus copallina

IN

EVERY

COUNTY

Rhus glabra

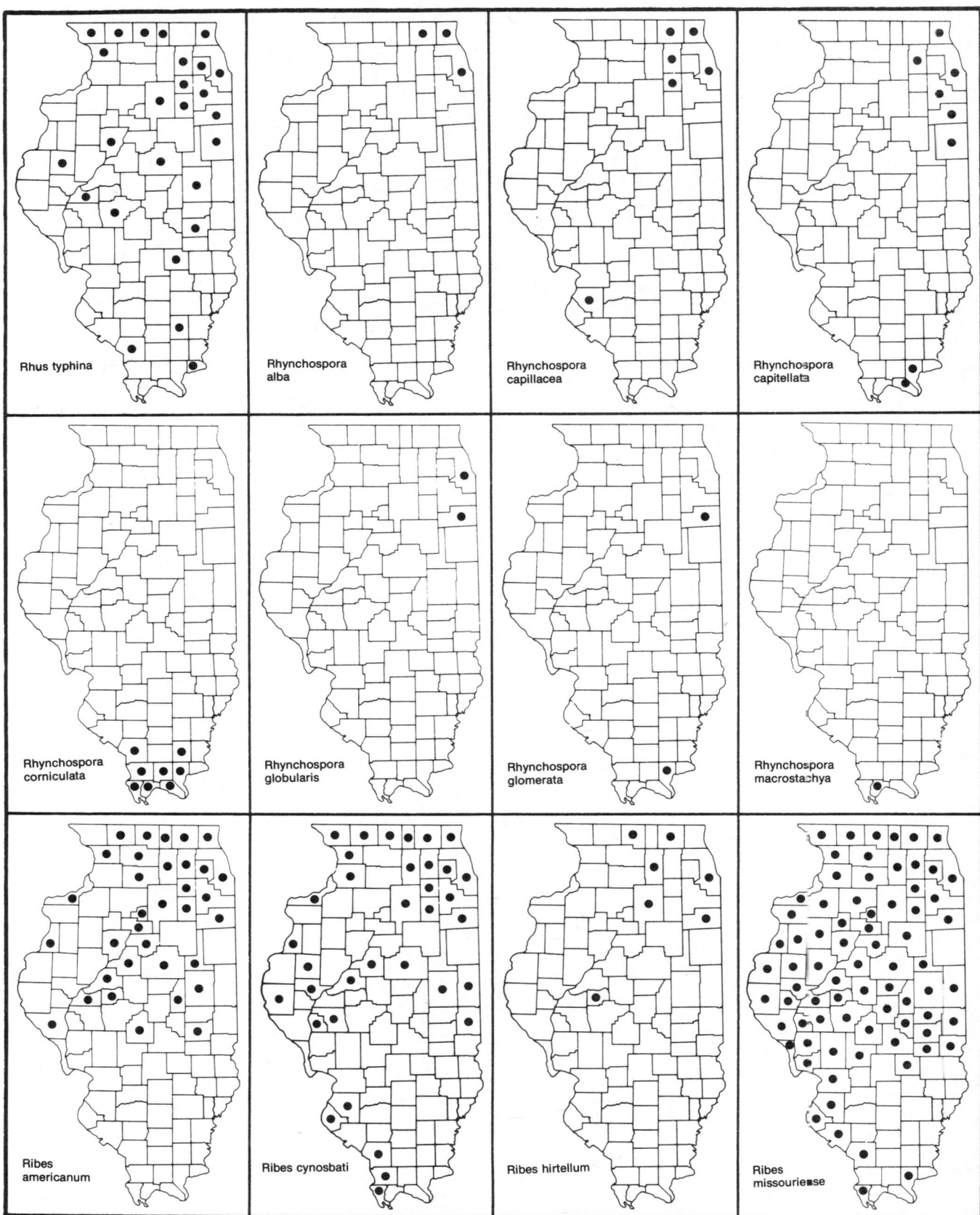

Rhus typhina

Rhynchospora
alba

Rhynchospora
capillacea

Rhynchospora
capitellata

Rhynchospora
corniculata

Rhynchospora
globularis

Rhynchospora
glomerata

Rhynchospora
macrostachya

Ribes
americanum

Ribes cynosbati

Ribes hirtellum

Ribes
missouriense

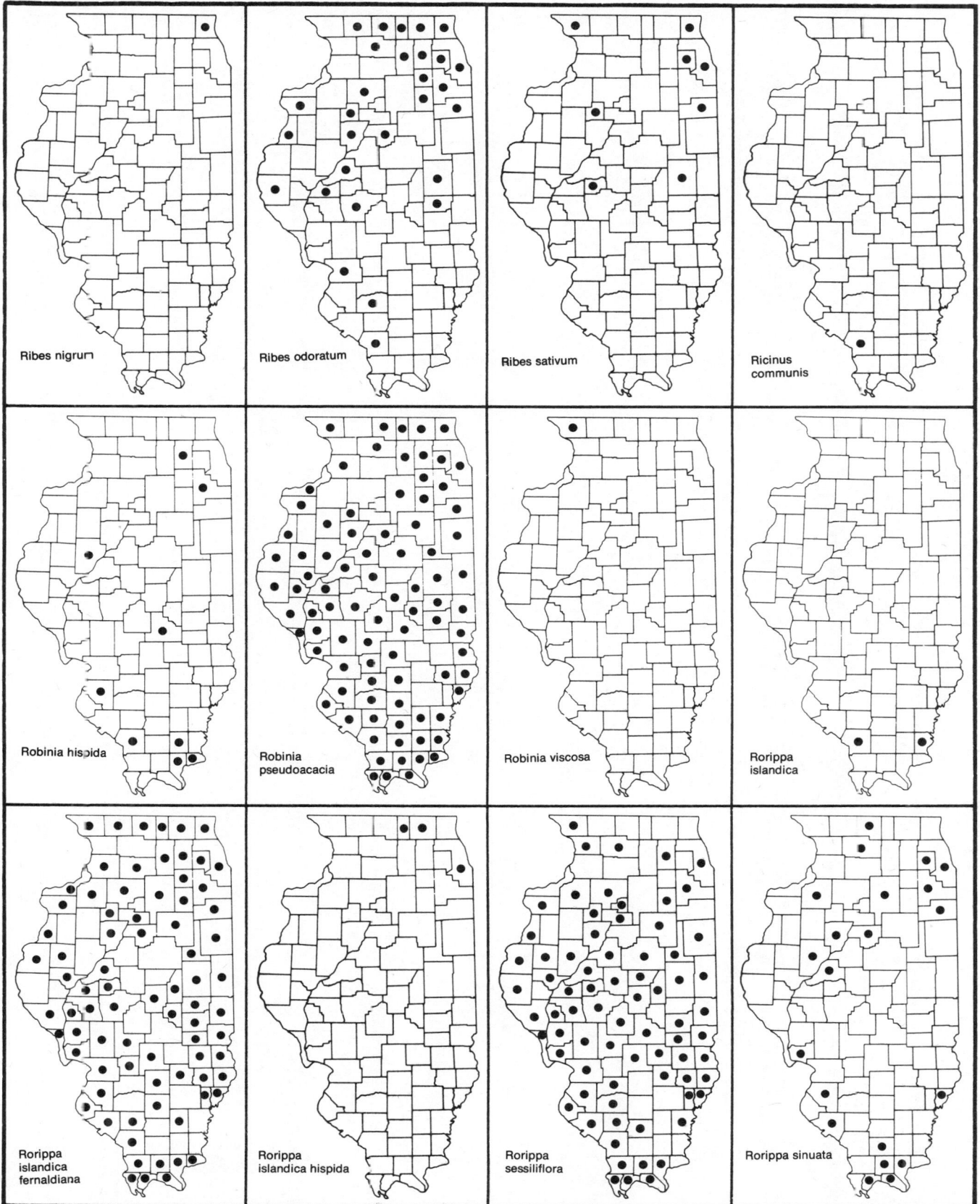

Ribes nigrum

Ribes odoratum

Ribes sativum

Ricinus
communis

Robinia hispida

Robinia
pseudoacacia

Robinia viscosa

Rorippa
islandica

Rorippa
islandica
fernaldiana

Rorippa
islandica hispida

Rorippa
sessiliflora

Rorippa sinuata

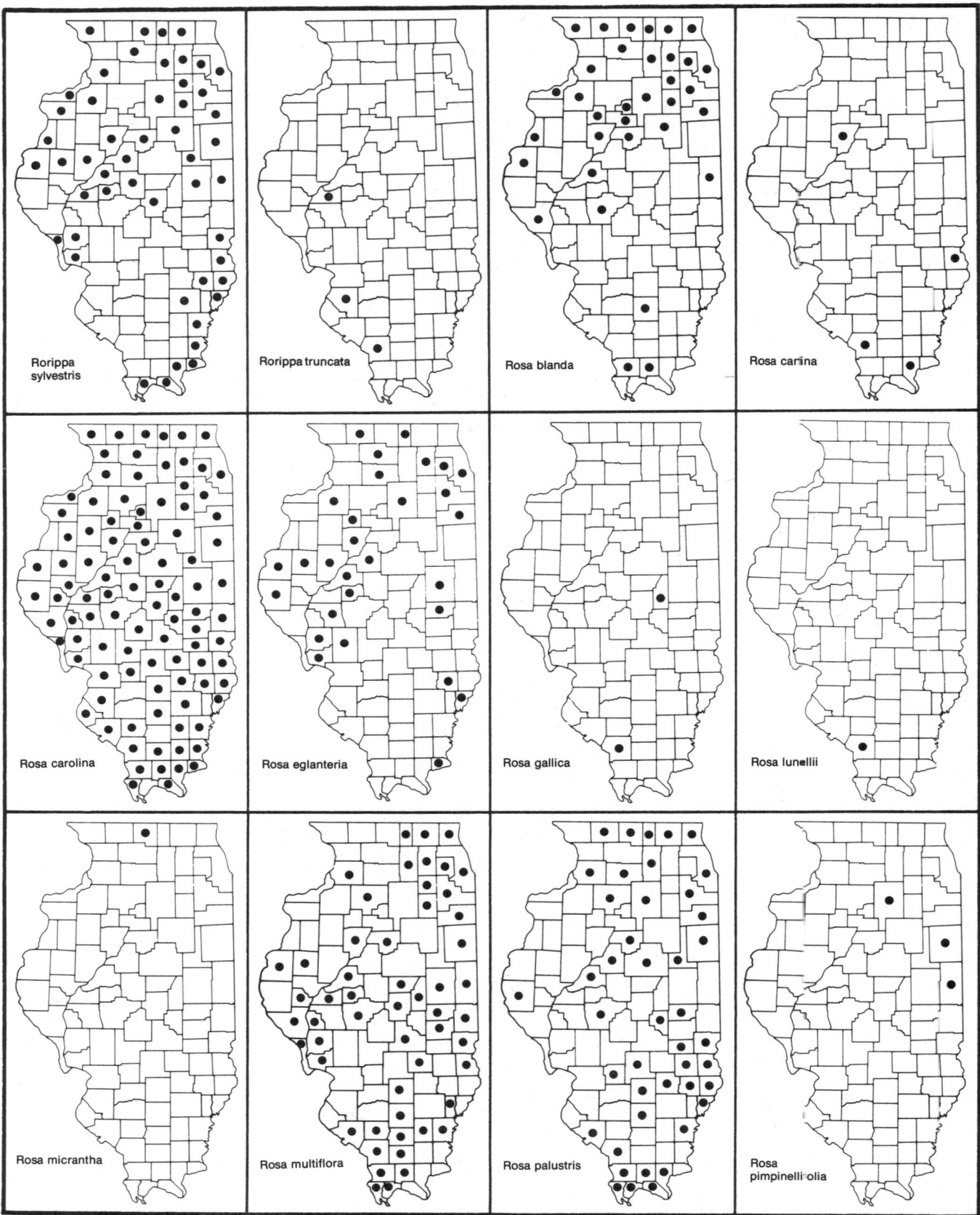

Rorippa
sylvestris

Rorippa truncata

Rosa blanda

Rosa carlina

Rosa carolina

Rosa eglanteria

Rosa gallica

Rosa lunellii

Rosa micrantha

Rosa multiflora

Rosa palustris

Rosa
pimpinellifolia

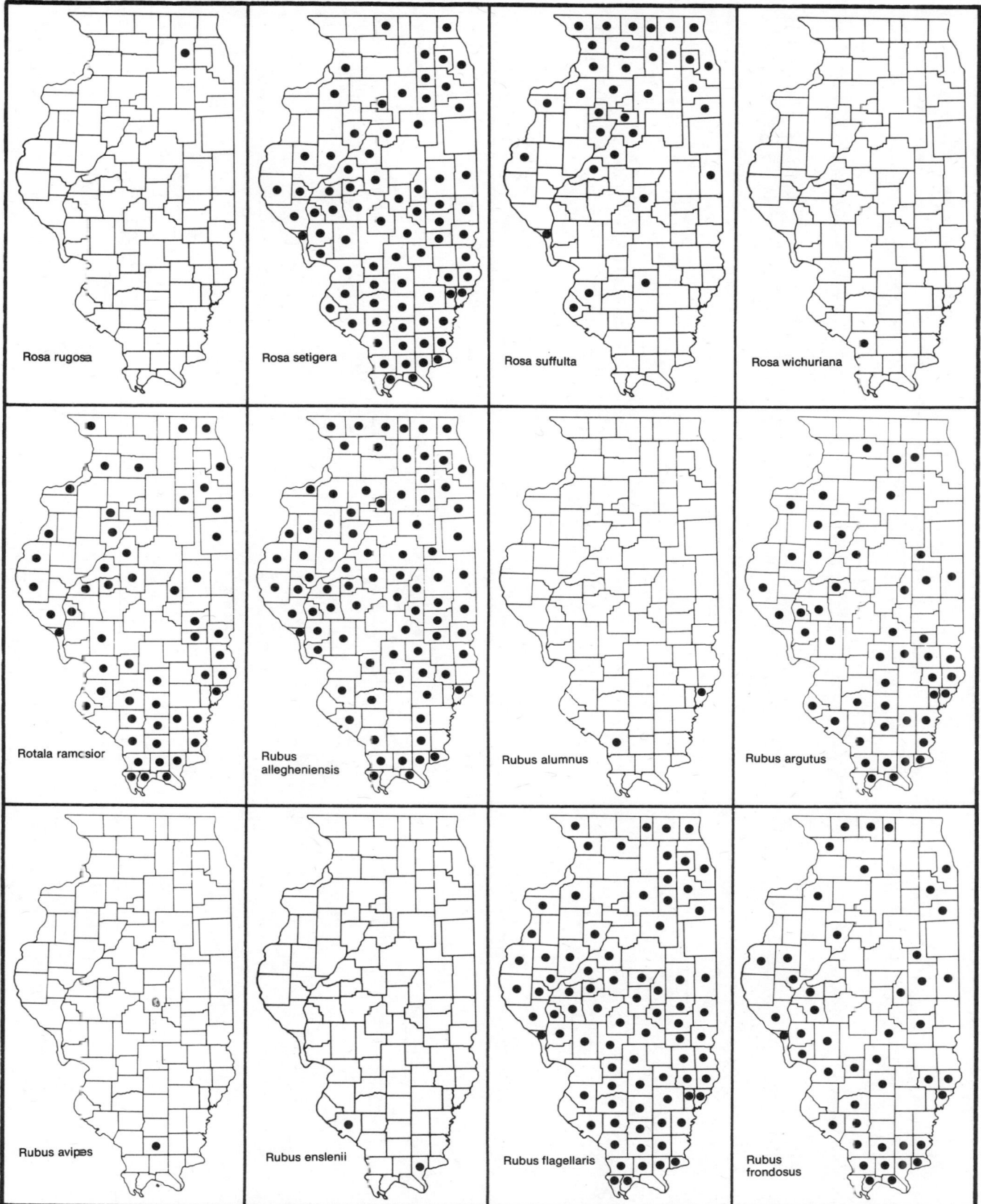

Rosa rugosa

Rosa setigera

Rosa suffulta

Rosa wichuriana

Rotala ramcsior

Rubus
allegheniensis

Rubus alumnus

Rubus argutus

Rubus avipes

Rubus enslenii

Rubus flagellaris

Rubus
frondosus

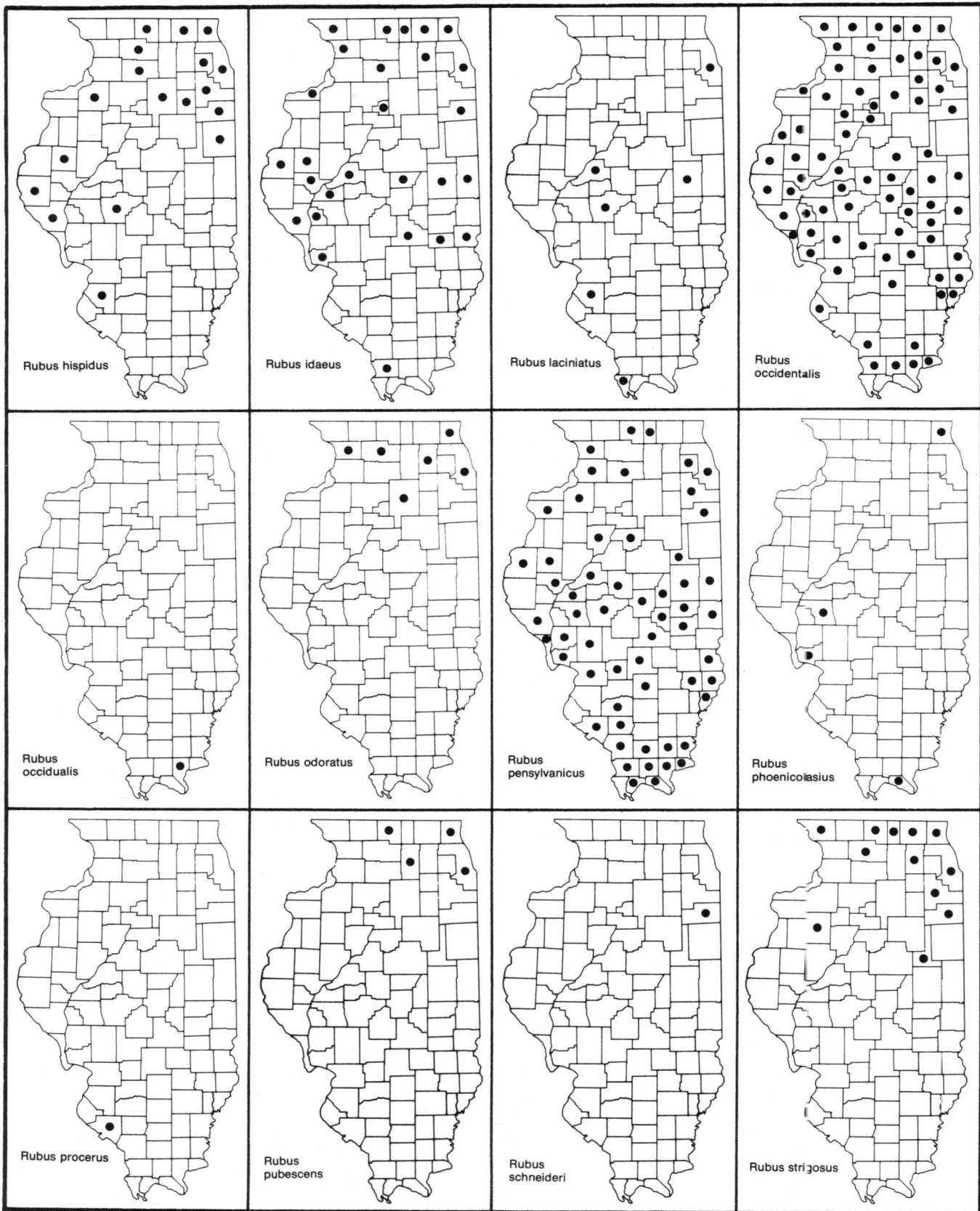

Rubus hispidus

Rubus idaeus

Rubus laciniatus

Rubus occidentalis

Rubus occidualis

Rubus odoratus

Rubus pensylvanicus

Rubus phoenicolasius

Rubus procerus

Rubus pubescens

Rubus schneideri

Rubus strigosus

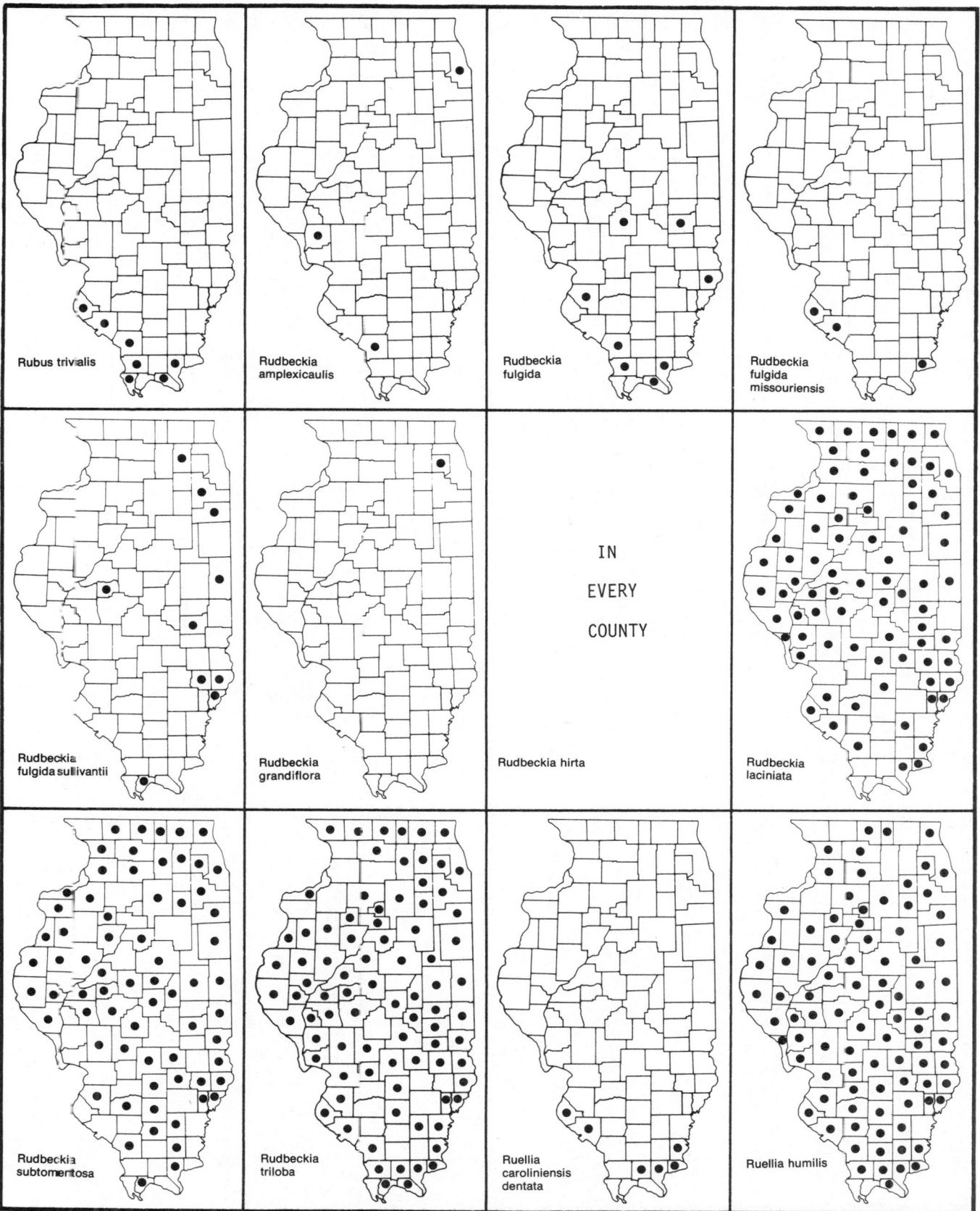

Rubus trivialis

Rudbeckia
amplexicaulis

Rudbeckia
fulgida

Rudbeckia
fulgida
missouriensis

Rudbeckia
fulgida sullivantii

Rudbeckia
grandiflora

IN

EVERY

COUNTY

Rudbeckia hirta

Rudbeckia
laciniata

Rudbeckia
subtomentosa

Rudbeckia
triloba

Ruellia
caroliniensis
dentata

Ruellia humilis

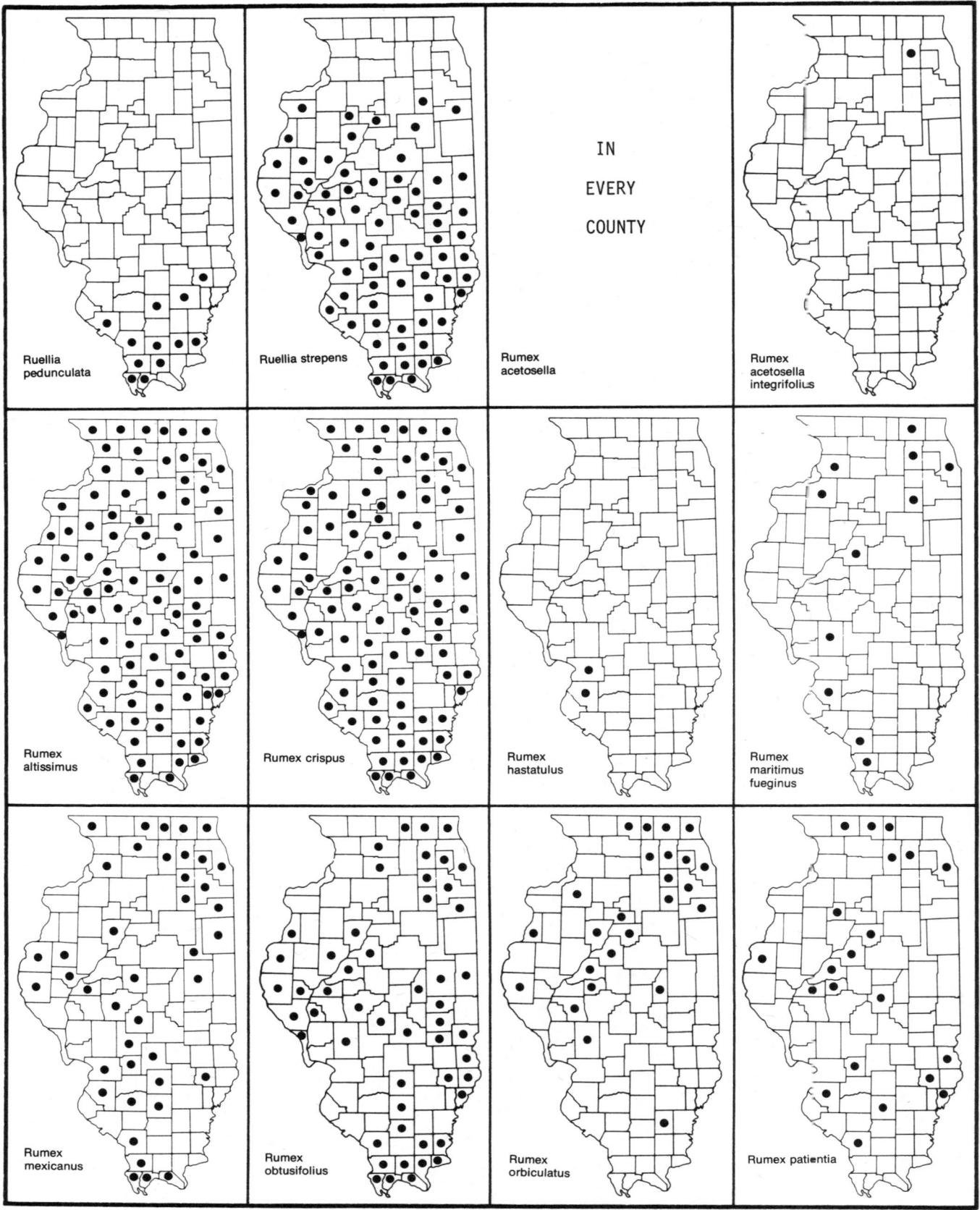

Ruellia
pedunculata

Ruellia strepens

Rumex
acetosella

IN

EVERY

COUNTY

Rumex
acetosella
integrifolius

Rumex
altissimus

Rumex
crispus

Rumex
hastatulus

Rumex
maritimus
fueginus

Rumex
mexicanus

Rumex
obtusifolius

Rumex
orbiculatus

Rumex patientia

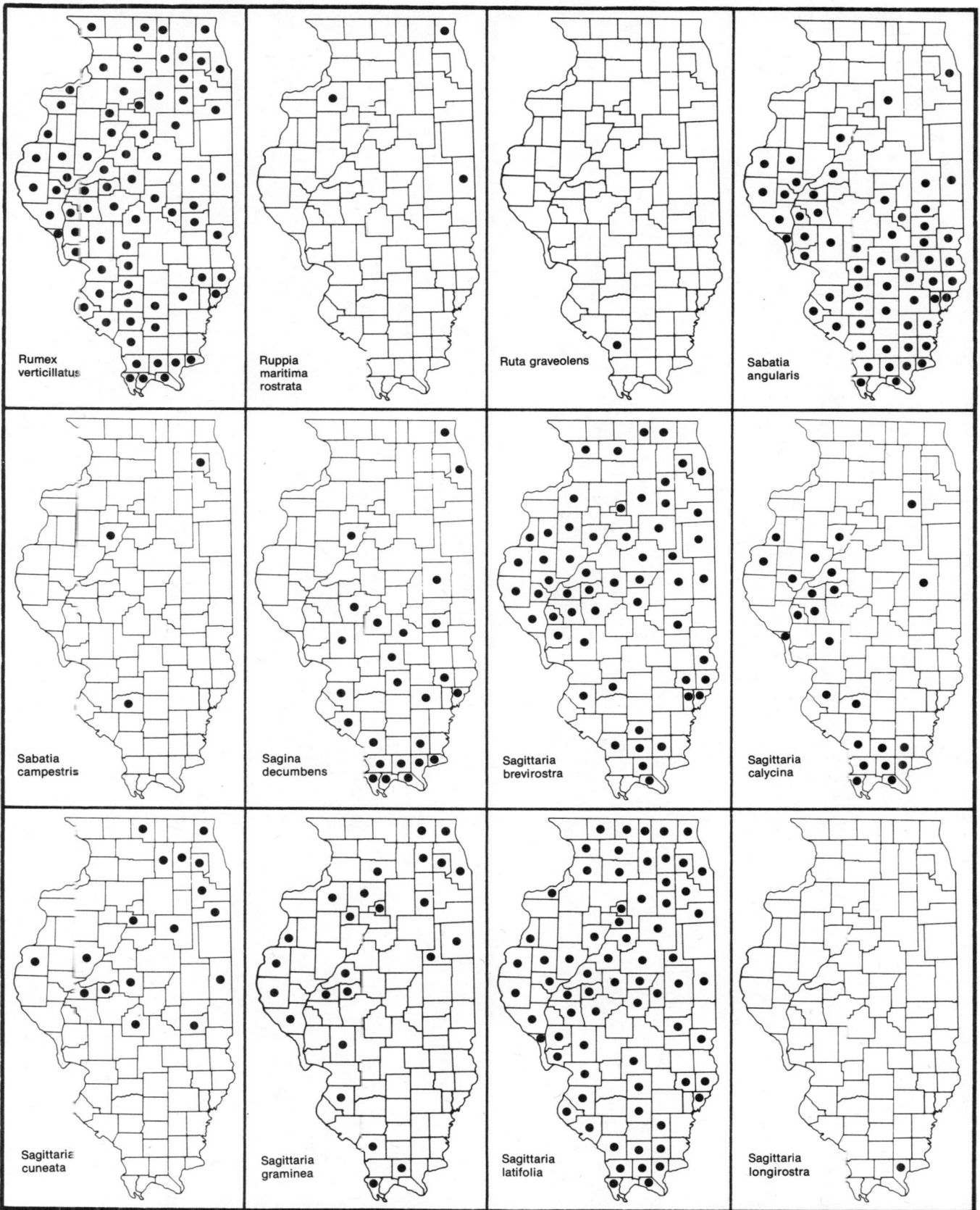

Rumex
verticillatus

Ruppia
maritima
rostrata

Ruta graveolens

Sabatia
angularis

Sabatia
campestris

Sagina
decumbens

Sagittaria
brevirostra

Sagittaria
calycina

Sagittaria
cuneata

Sagittaria
graminea

Sagittaria
latifolia

Sagittaria
longirostra

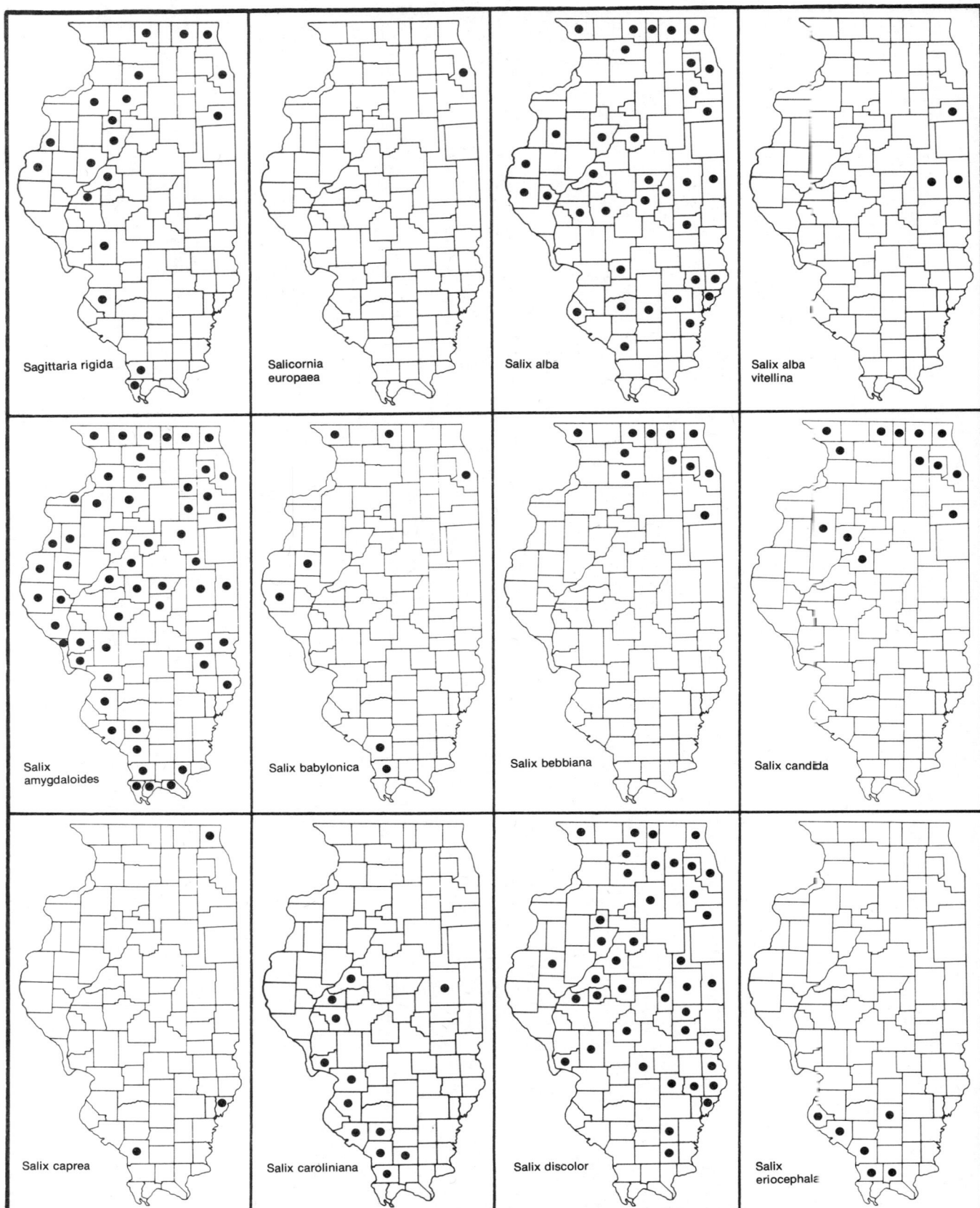

Sagittaria rigida

Salicornia europaea

Salix alba

Salix alba vitellina

Salix amygdaloides

Salix babylonica

Salix bebbiana

Salix candida

Salix caprea

Salix caroliniana

Salix discolor

Salix eriocephala

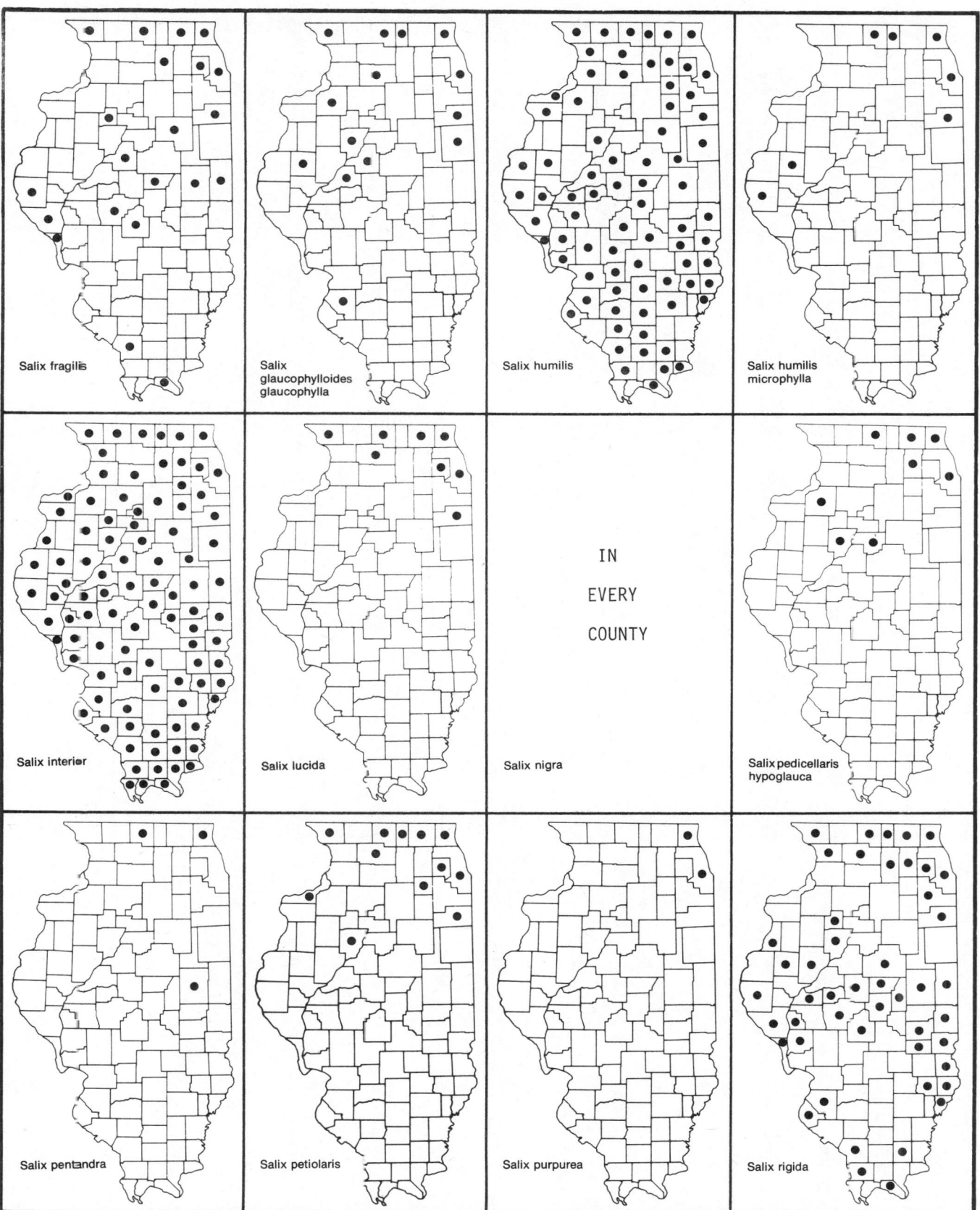

Salix fragilis

Salix glaucophylloides glaucophylla

Salix humilis

Salix humilis microphylla

Salix interior

Salix lucida

IN

EVERY

COUNTY

Salix nigra

Salix pedicellaris hypoglauca

Salix pentandra

Salix petiolaris

Salix purpurea

Salix rigida

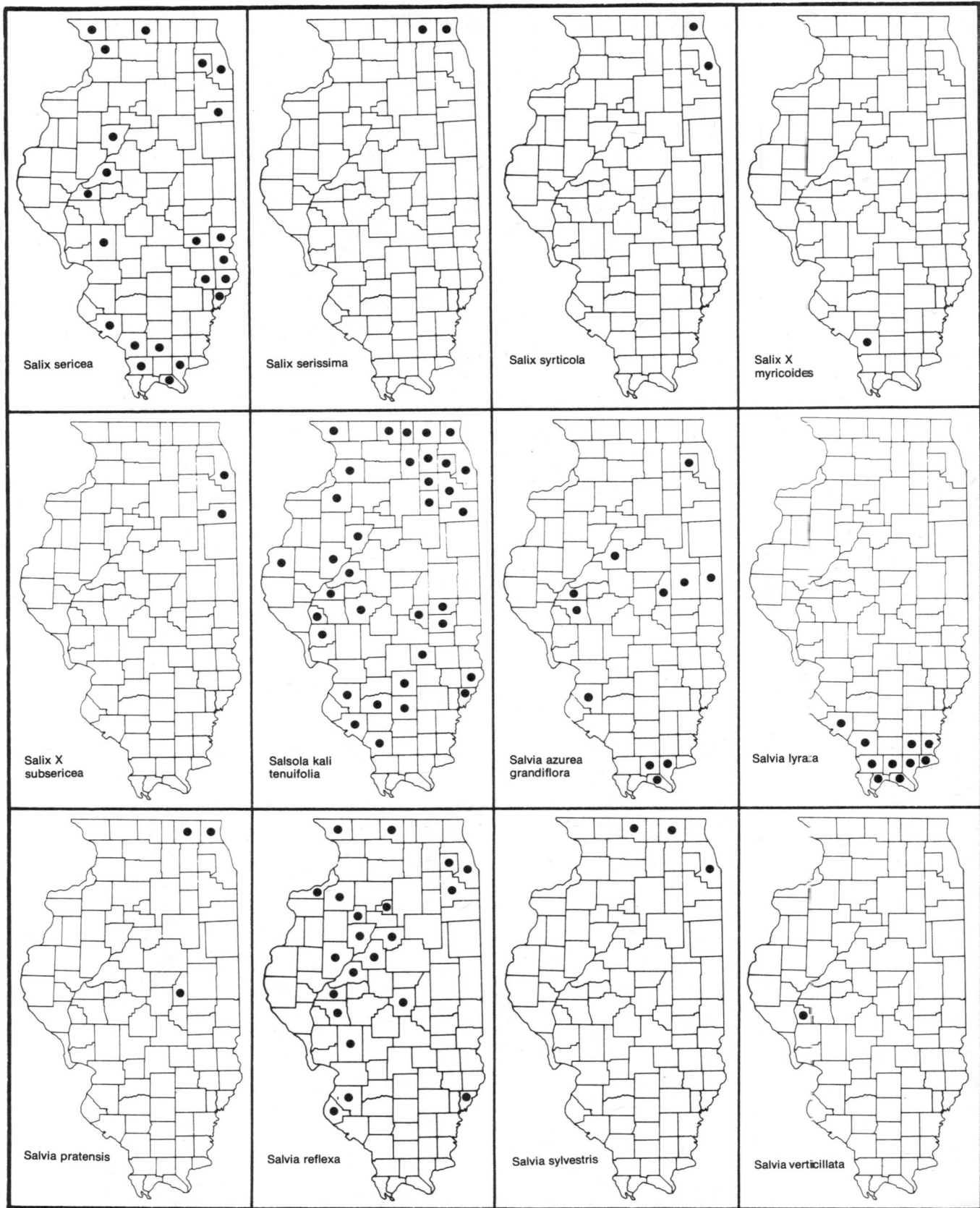

Salix sericea

Salix serissima

Salix syrticola

Salix X myricoides

Salix X subsericea

Salsola kali tenuifolia

Salvia azurea grandiflora

Salvia lyrata

Salvia pratensis

Salvia reflexa

Salvia sylvestris

Salvia verticillata

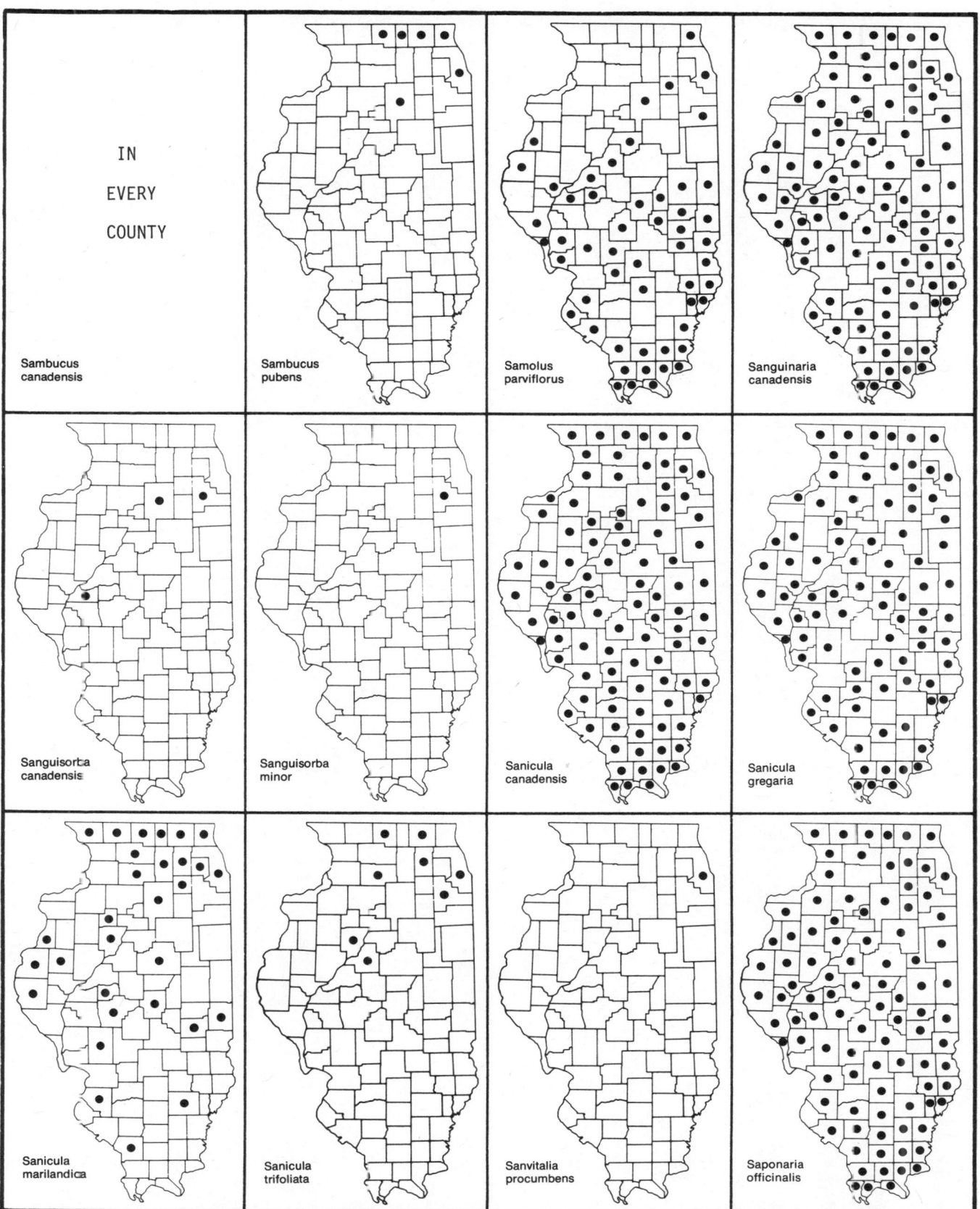

IN
EVERY
COUNTY

Sambucus
canadensis

Sambucus
pubens

Samolus
parviflorus

Sanguinaria
canadensis

Sanguisorba
canadensis

Sanguisorba
minor

Sanicula
canadensis

Sanicula
gregaria

Sanicula
marilandica

Sanicula
trifoliata

Sanvitalia
procumbens

Saponaria
officinalis

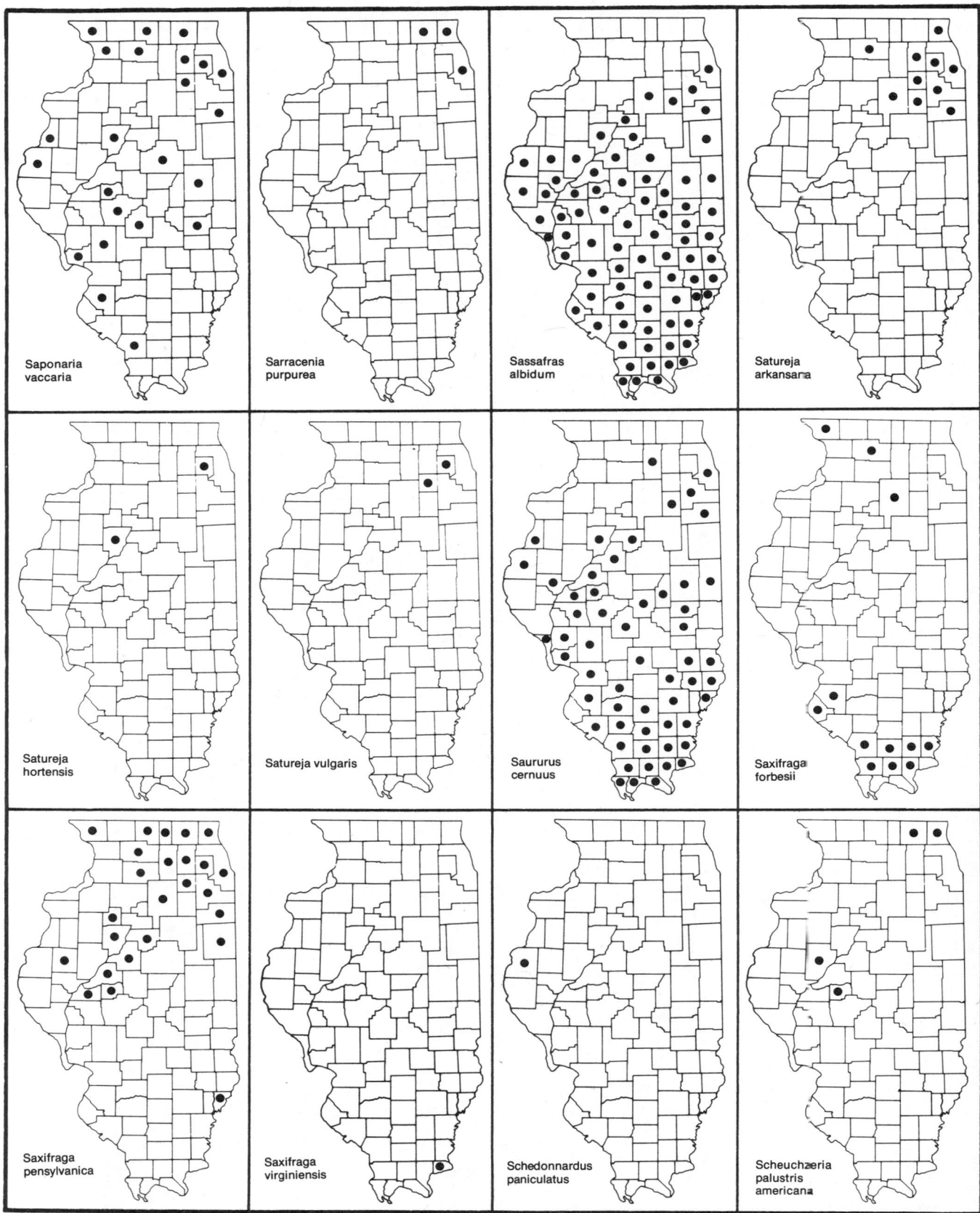

Saponaria
vaccaria

Sarracenia
purpurea

Sassafras
albidum

Satureja
arkansana

Satureja
hortensis

Satureja
vulgaris

Saururus
cernuus

Saxifraga
forbesii

Saxifraga
pensylvanica

Saxifraga
virginiensis

Schedonnardus
paniculatus

Scheuchzeria
palustris
americana

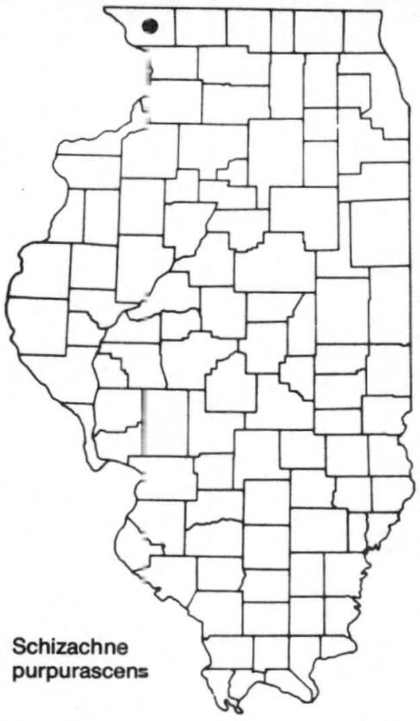

Schizachne purpurascens

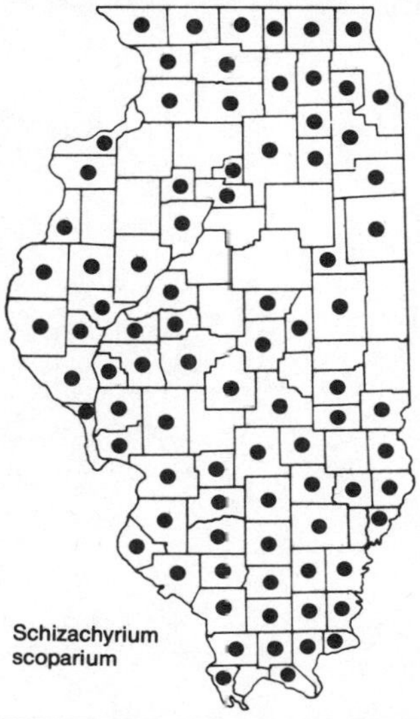

Schizachyrium scoparium

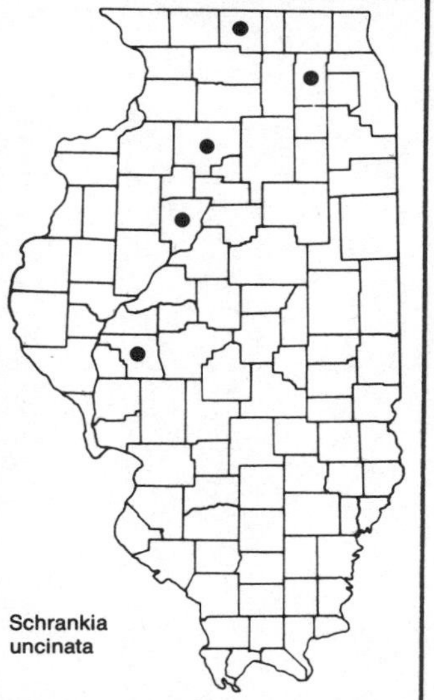

Schrankia uncinata

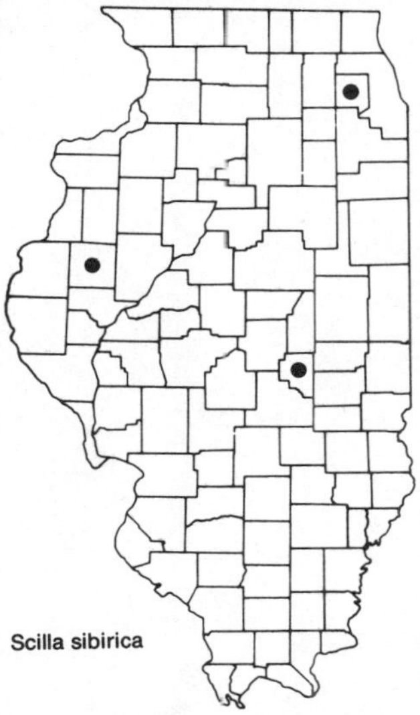

Scilla sibirica

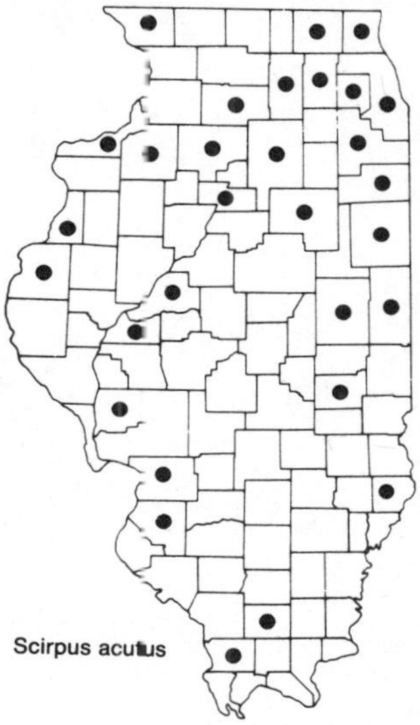

Scirpus acutus

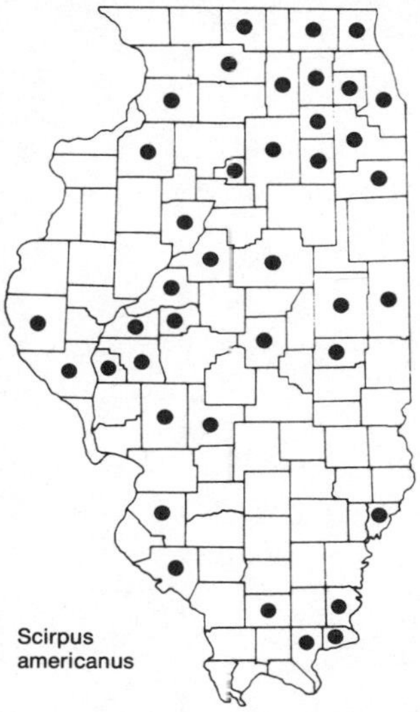

Scirpus americanus

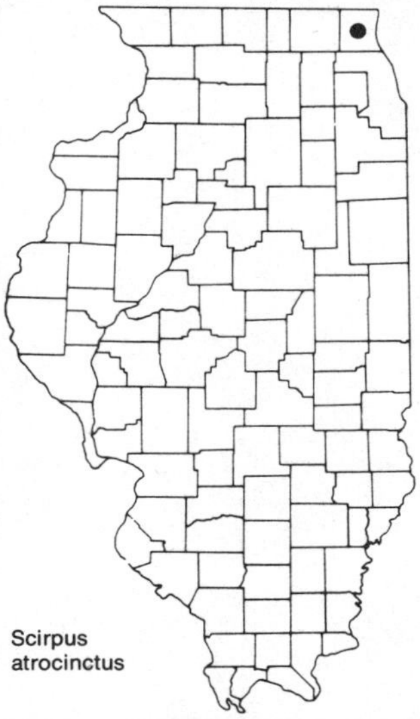

Scirpus atrocinctus

IN
EVERY
COUNTY

Scirpus atrovirens

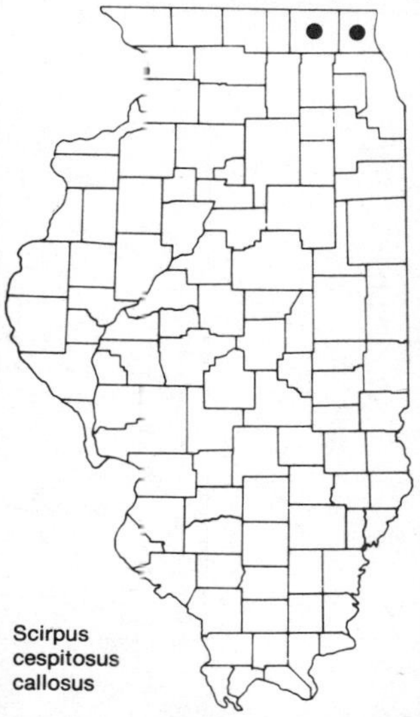

Scirpus cespitosus callosus

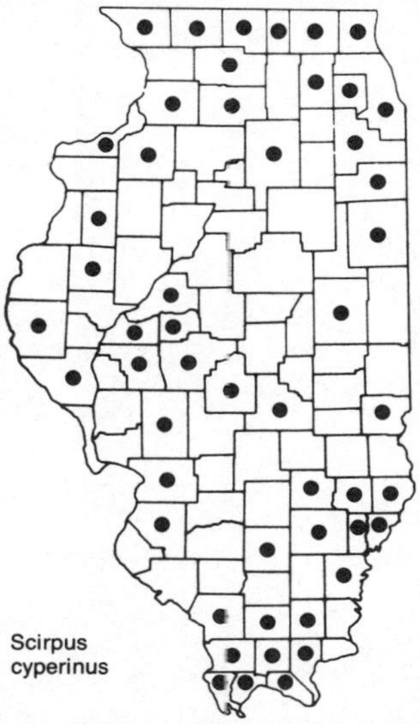

Scirpus cyperinus

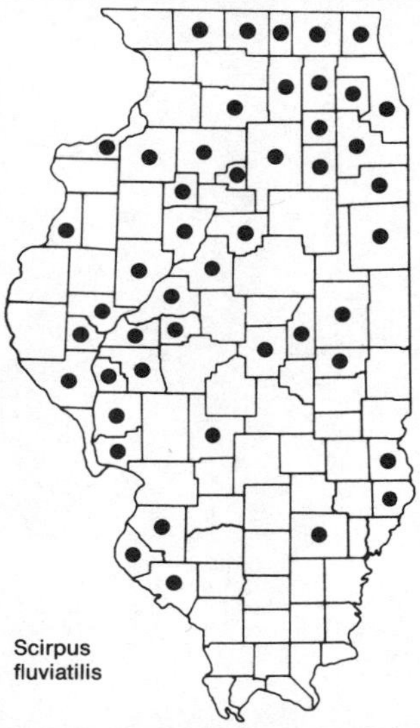

Scirpus fluviatilis

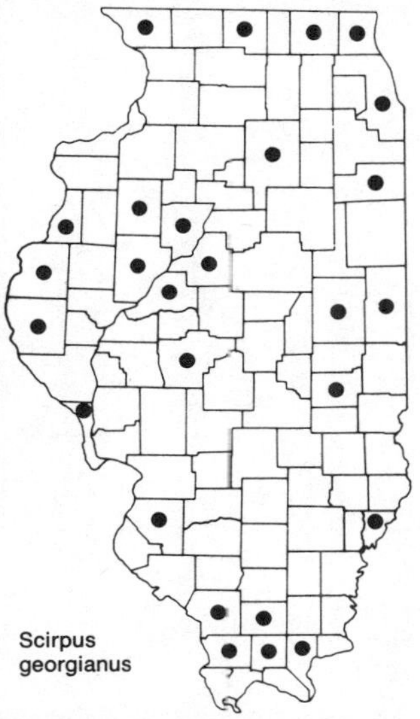

Scirpus georgianus

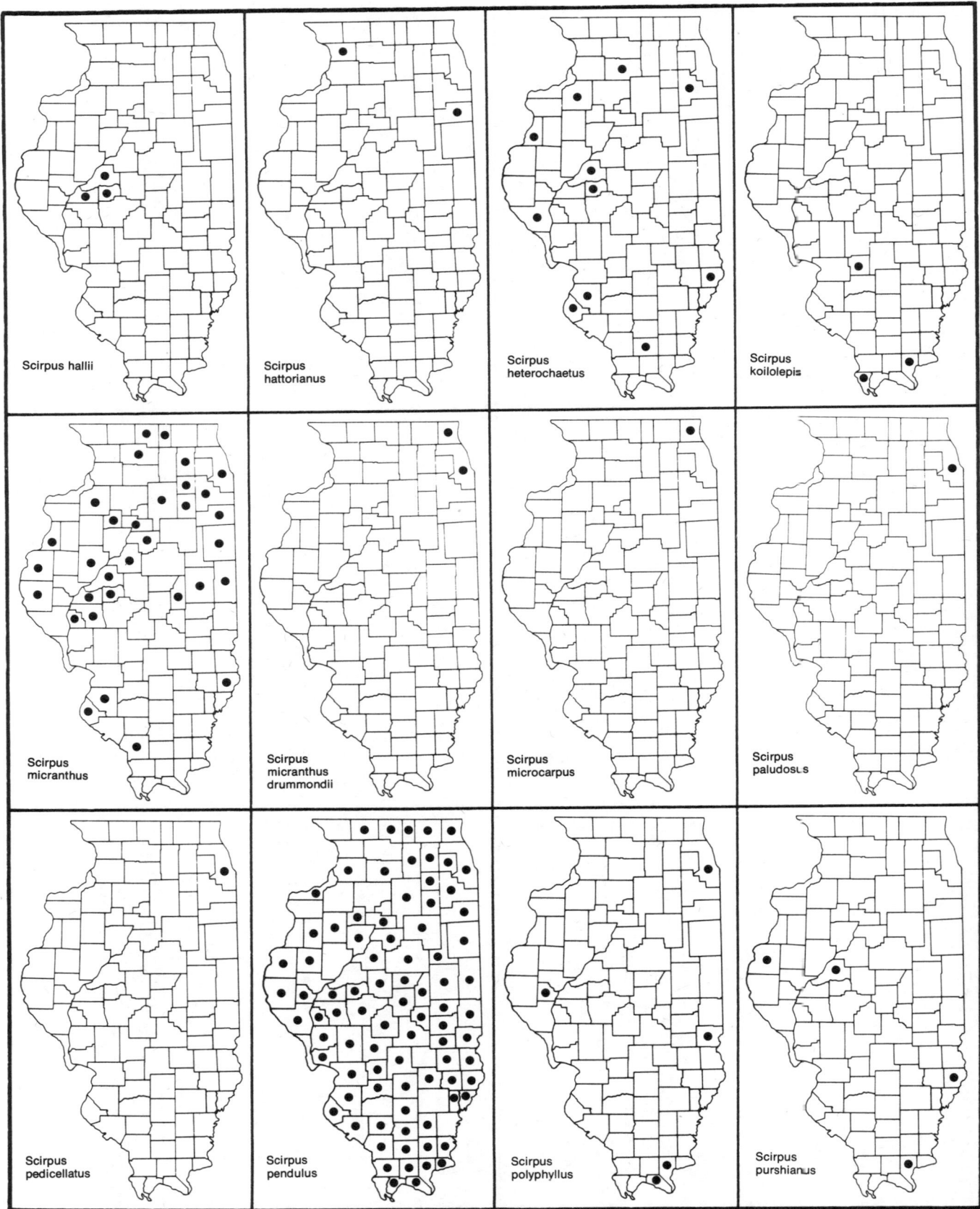

Scirpus hallii

Scirpus hattorianus

Scirpus heterochaetus

Scirpus koilolepis

Scirpus micranthus

Scirpus micranthus drummondii

Scirpus microcarpus

Scirpus paludosus

Scirpus pedicellatus

Scirpus pendulus

Scirpus polyphyllus

Scirpus purshianus

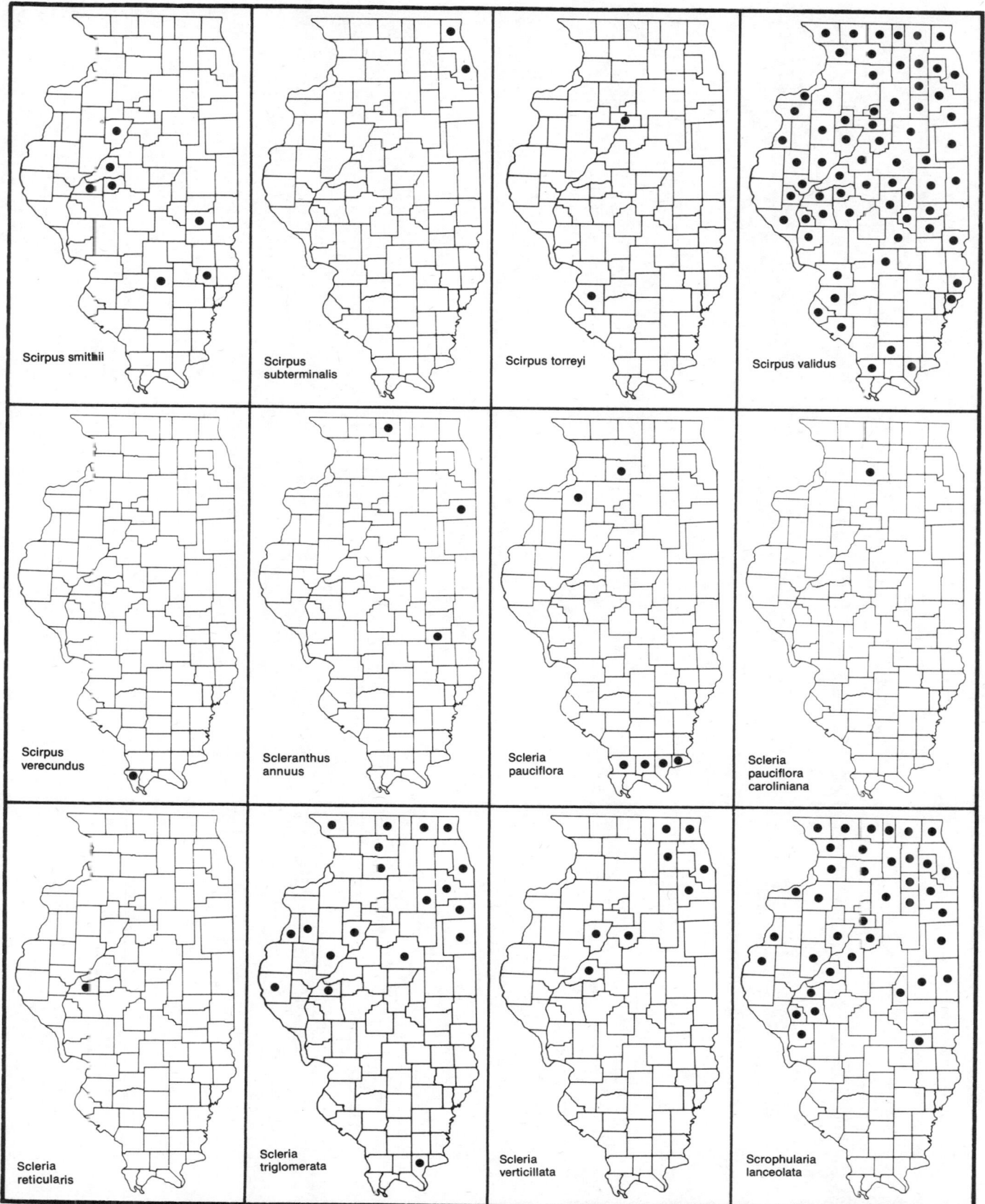

Scirpus smithii

Scirpus
subterminalis

Scirpus torreyi

Scirpus validus

Scirpus
verecundus

Scleranthus
annuus

Scleria
pauciflora

Scleria
pauciflora
caroliniana

Scleria
reticularis

Scleria
triglomerata

Scleria
verticillata

Scrophularia
lanceolata

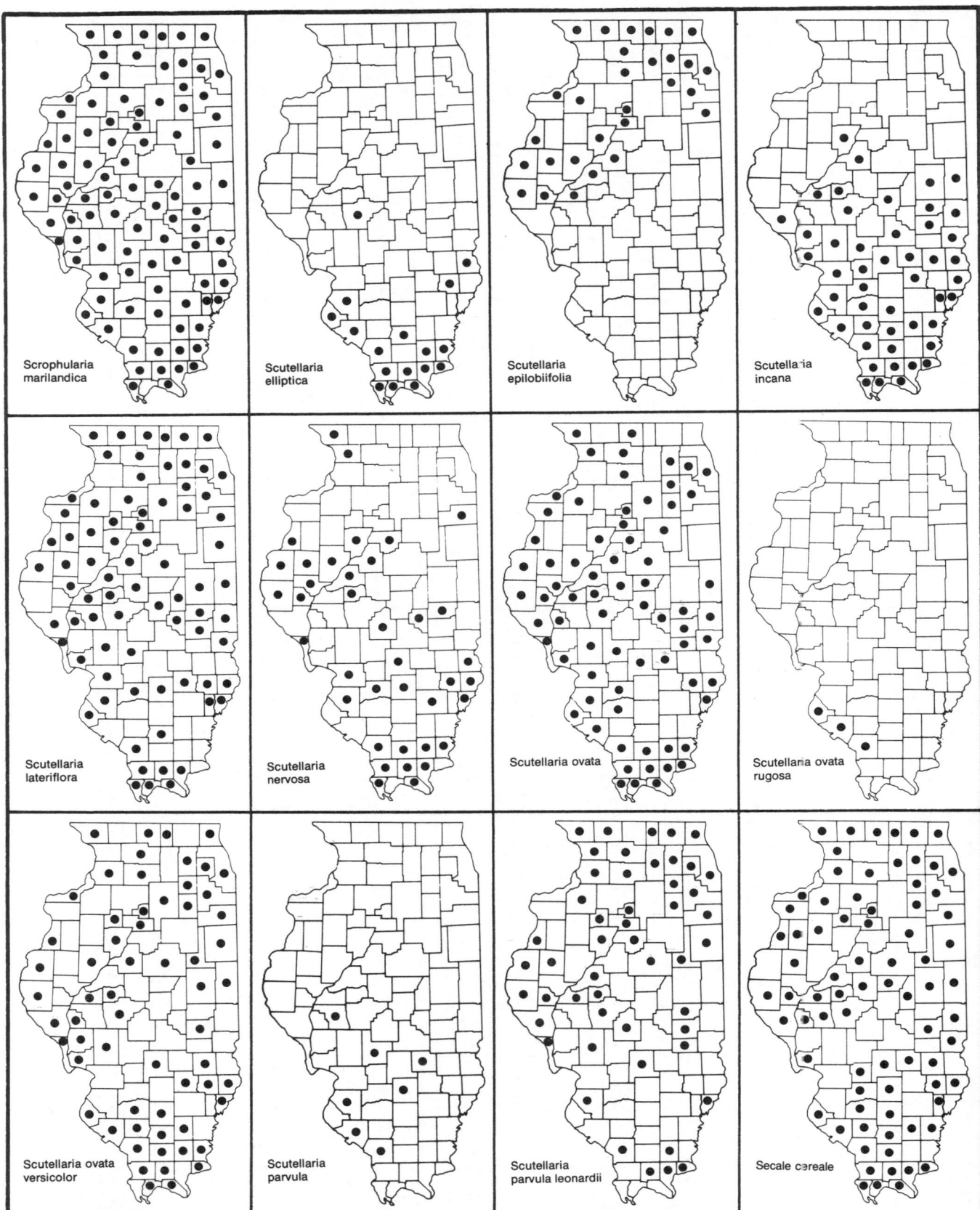

Scrophularia
marilandica

Scutellaria
elliptica

Scutellaria
epilobiifolia

Scutellaria
incana

Scutellaria
lateriflora

Scutellaria
nervosa

Scutellaria ovata

Scutellaria ovata
rugosa

Scutellaria ovata
versicolor

Scutellaria
parvula

Scutellaria
parvula leonardii

Secale cereale

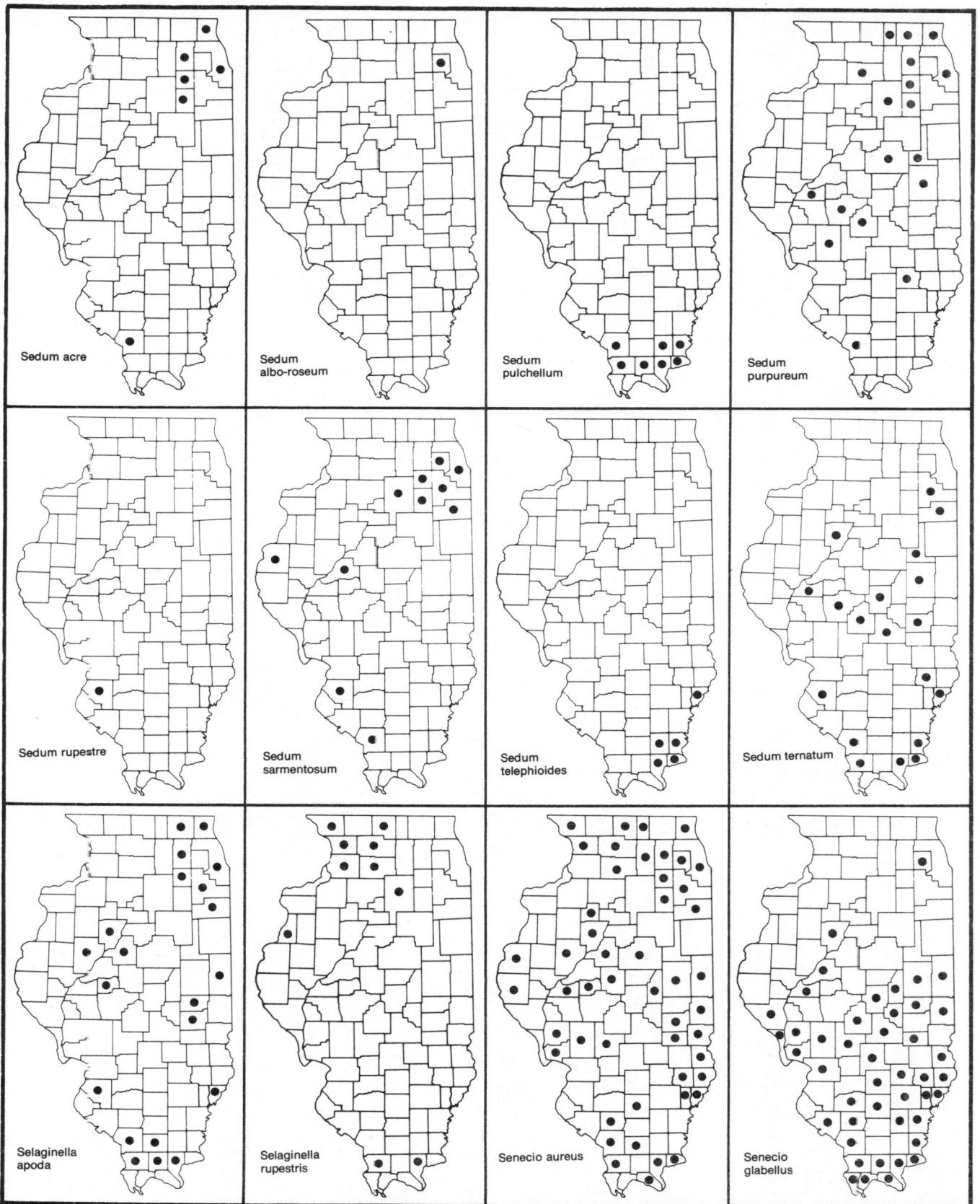

Sedum acre

Sedum
albo-roseum

Sedum
pulchellum

Sedum
purpureum

Sedum rupestre

Sedum
sarmentosum

Sedum
telephioides

Sedum ternatum

Selaginella
apoda

Selaginella
rupestris

Senecio aureus

Senecio
glabellus

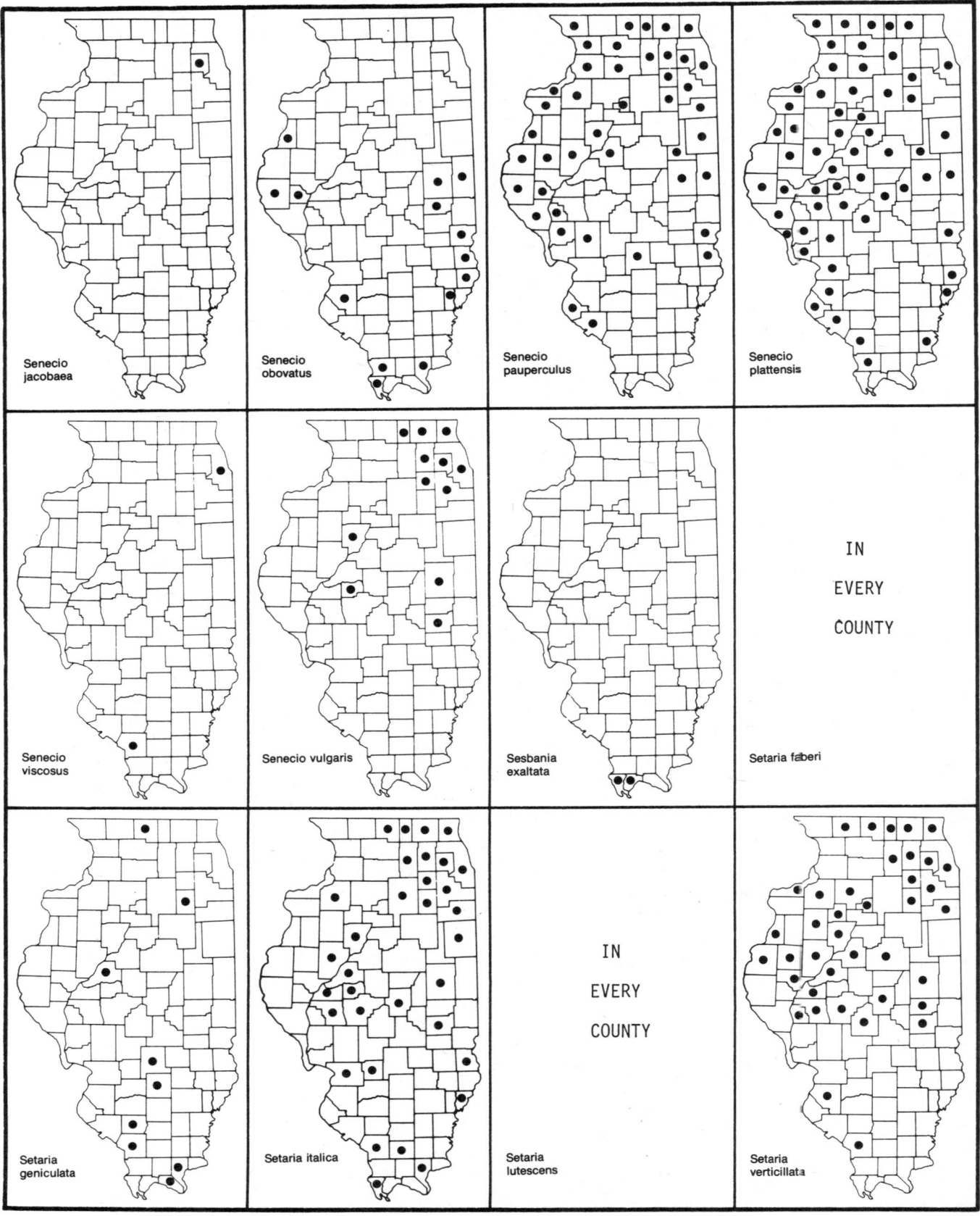

Senecio
jacobaea

Senecio
obovatus

Senecio
pauperculus

Senecio
plattensis

Senecio
viscosus

Senecio vulgaris

Sesbania
exaltata

Setaria faberi

IN

EVERY

COUNTY

Setaria
geniculata

Setaria italica

Setaria
lutescens

Setaria
verticillata

IN

EVERY

COUNTY

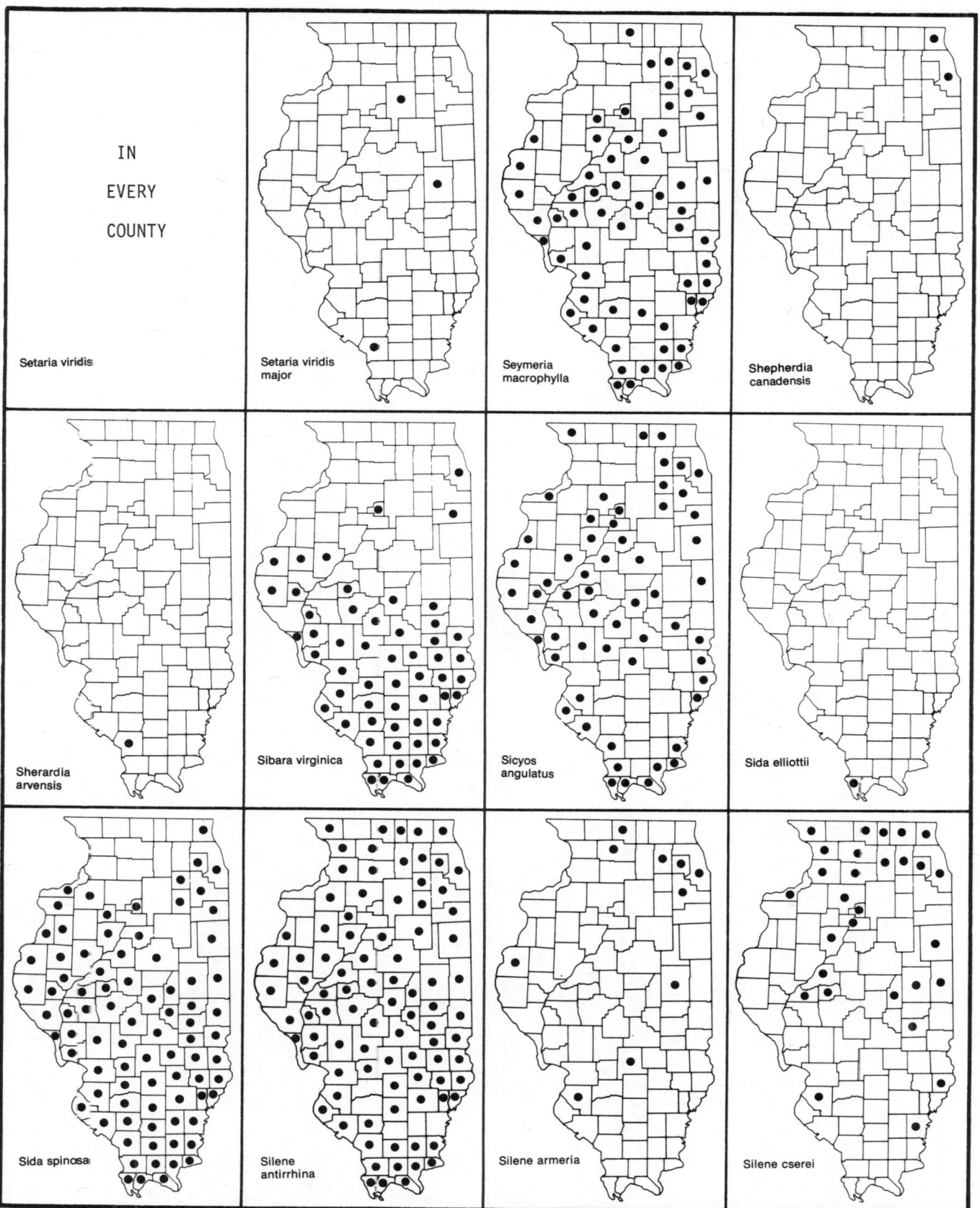

IN
EVERY
COUNTY

Setaria viridis

Setaria viridis
major

Seymeria
macrophylla

Shepherdia
canadensis

Sherardia
arvensis

Sibara virginica

Sicyos
angulatus

Sida elliottii

Sida spinosa

Silene
antirrhina

Silene armeria

Silene cserei

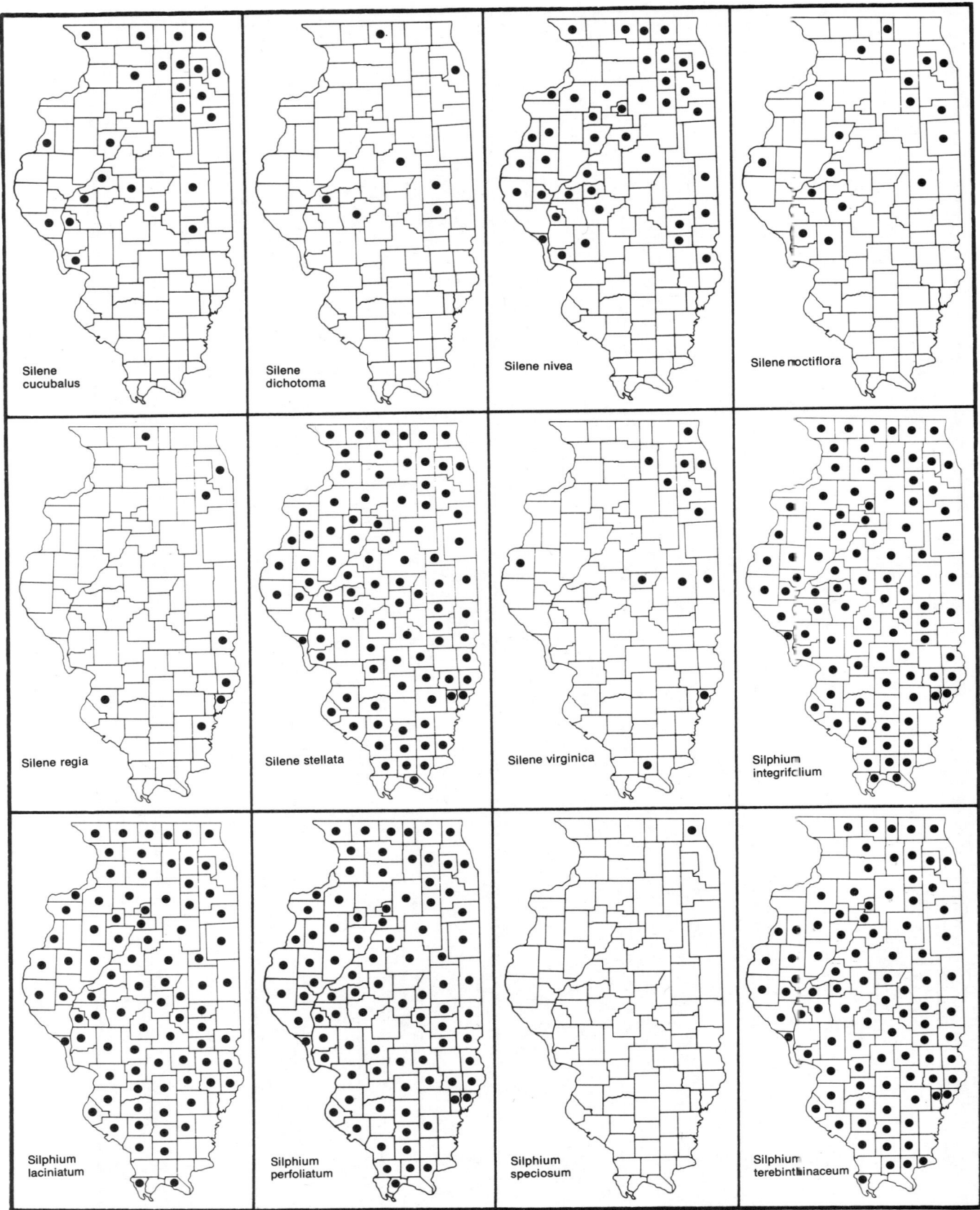

Silene
cucubalus

Silene
dichotoma

Silene nivea

Silene noctiflora

Silene regia

Silene stellata

Silene virginica

Silphium
integrifolium

Silphium
laciniatum

Silphium
perfoliatum

Silphium
speciosum

Silphium
terebinthinaceum

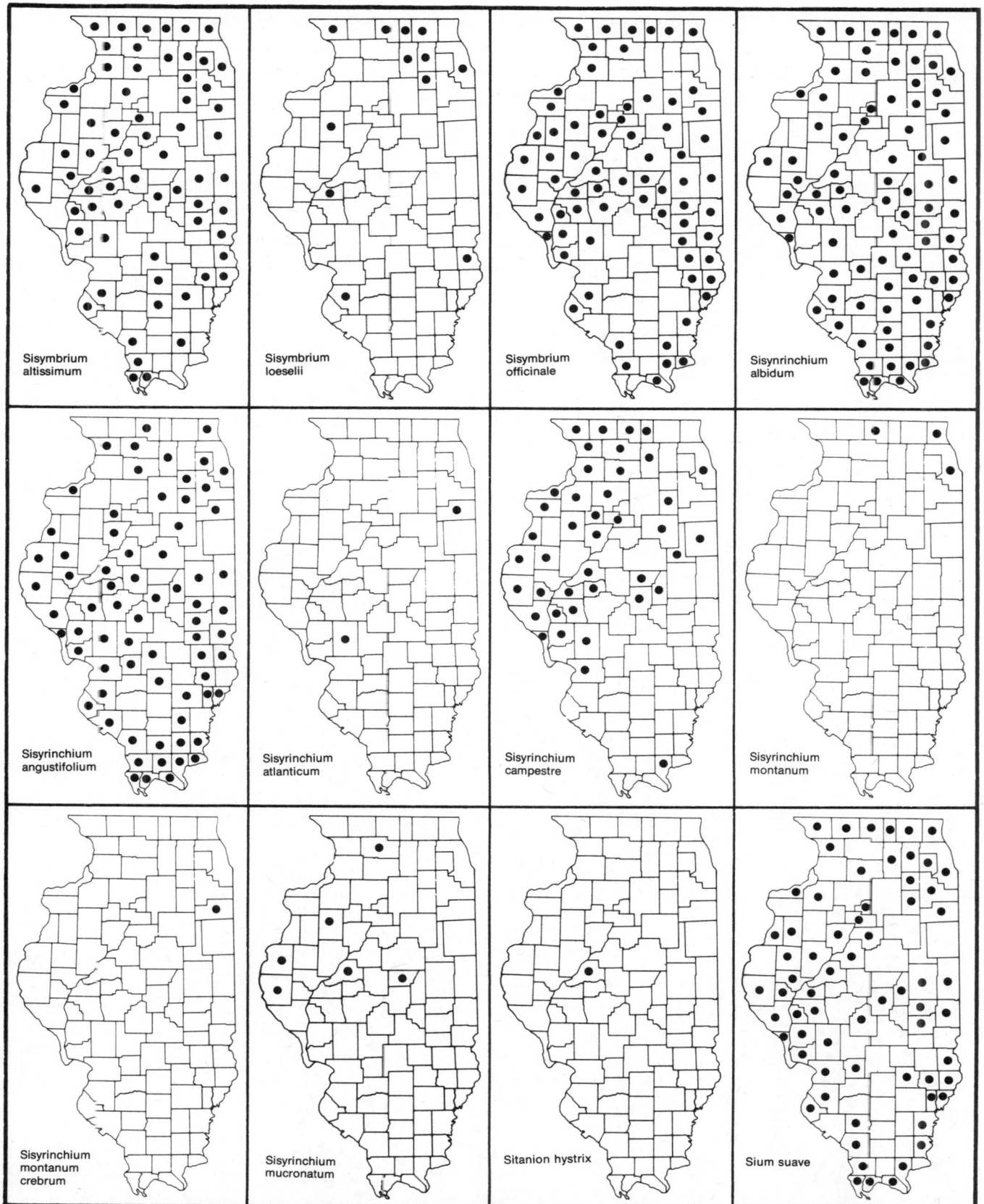

Sisymbrium
altissimum

Sisymbrium
loeselii

Sisymbrium
officinale

Sisynrinchium
albidum

Sisyrinchium
angustifolium

Sisyrinchium
atlanticum

Sisyrinchium
campestre

Sisynrinchium
montanum

Sisyrinchium
montanum
crebrum

Sisyrinchium
mucronatum

Sitanion hystrix

Sium suave

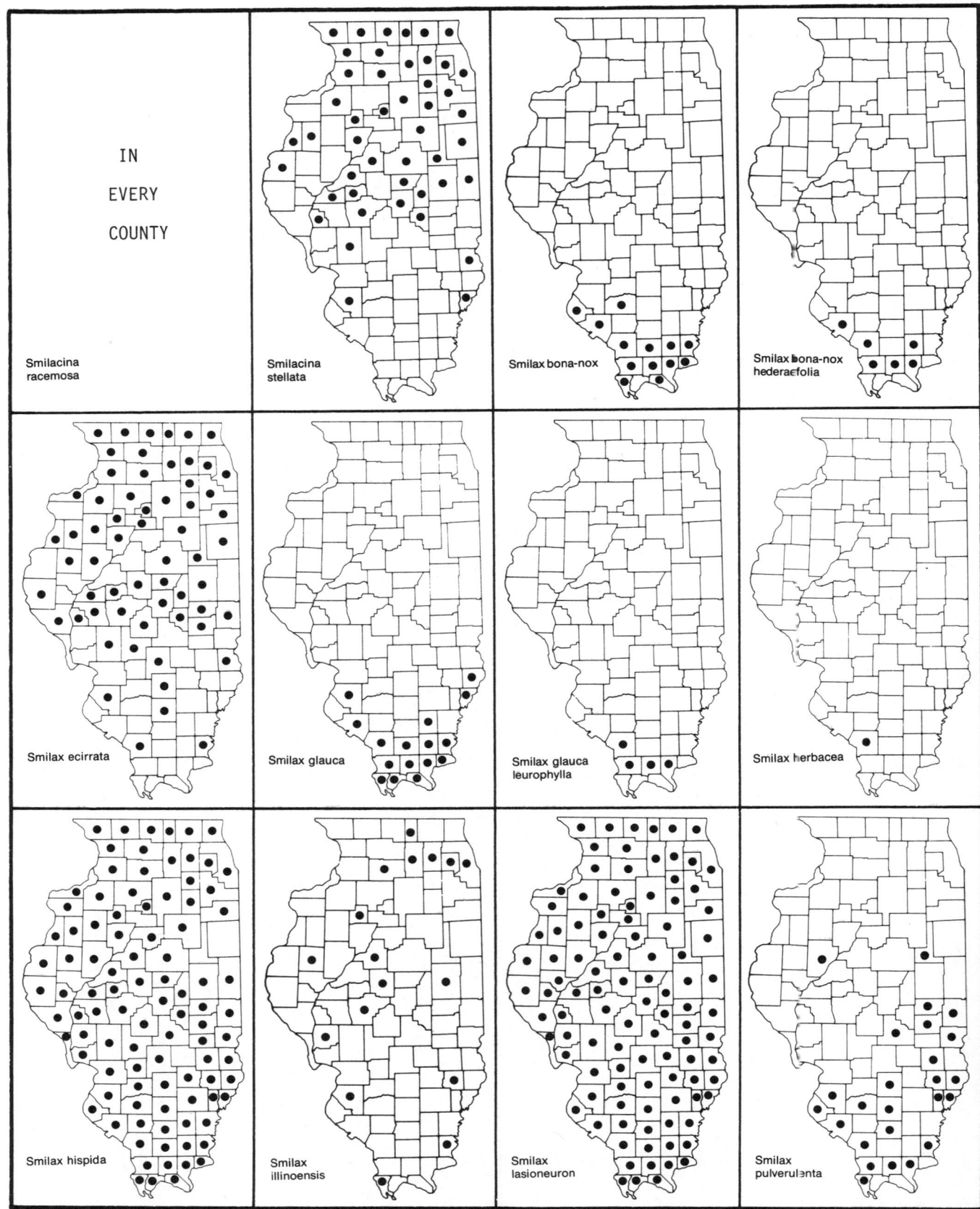

IN EVERY COUNTY

Smilacina racemosa

Smilacina stellata

Smilax bona-nox

Smilax bona-nox hederaefolia

Smilax ecirrata

Smilax glauca

Smilax glauca leurophylla

Smilax herbacea

Smilax hispida

Smilax illinoensis

Smilax lasioneuron

Smilax pulverulenta

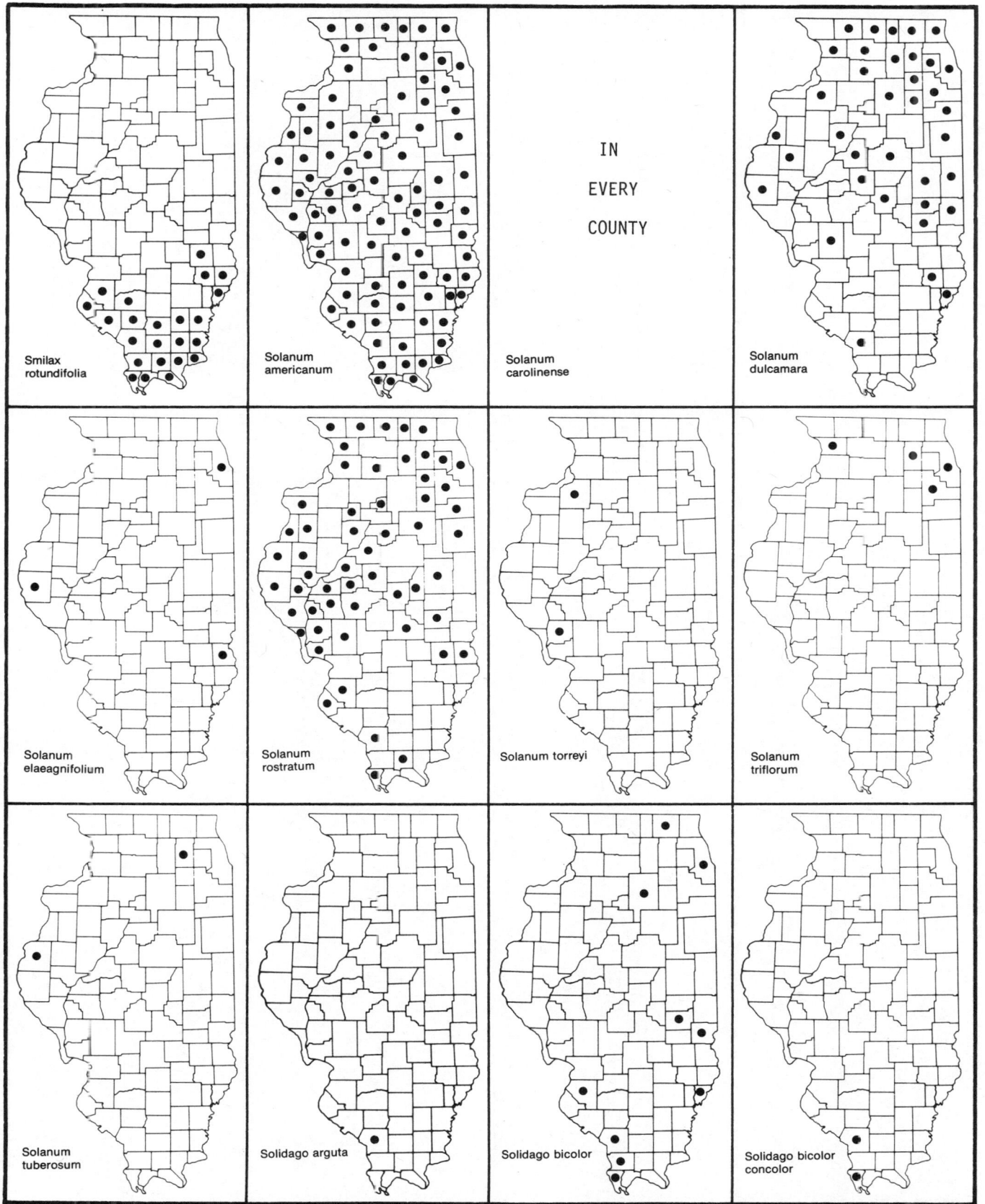

IN

EVERY

COUNTY

Smilax
rotundifolia

Solanum
americanum

Solanum
carolinense

Solanum
dulcamara

Solanum
elaeagnifolium

Solanum
rostratum

Solanum torreyi

Solanum
triflorum

Solanum
tuberosum

Solidago arguta

Solidago bicolor

Solidago bicolor
concolor

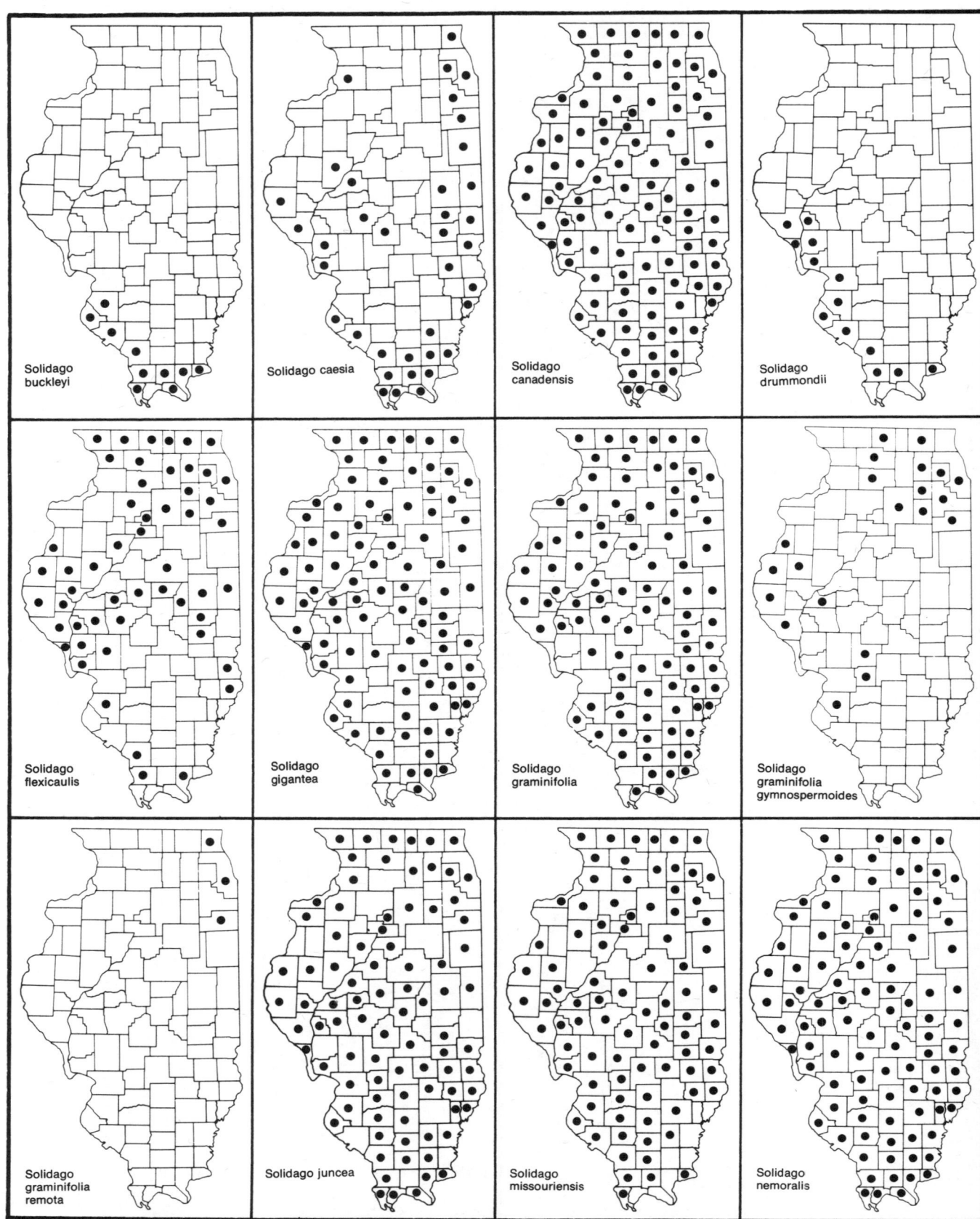

Solidago
buckleyi

Solidago caesia

Solidago
canadensis

Solidago
drummondii

Solidago
flexicaulis

Solidago
gigantea

Solidago
graminifolia

Solidago
graminifolia
gymnospermoides

Solidago
graminifolia
remota

Solidago juncea

Solidago
missouriensis

Solidago
nemoralis

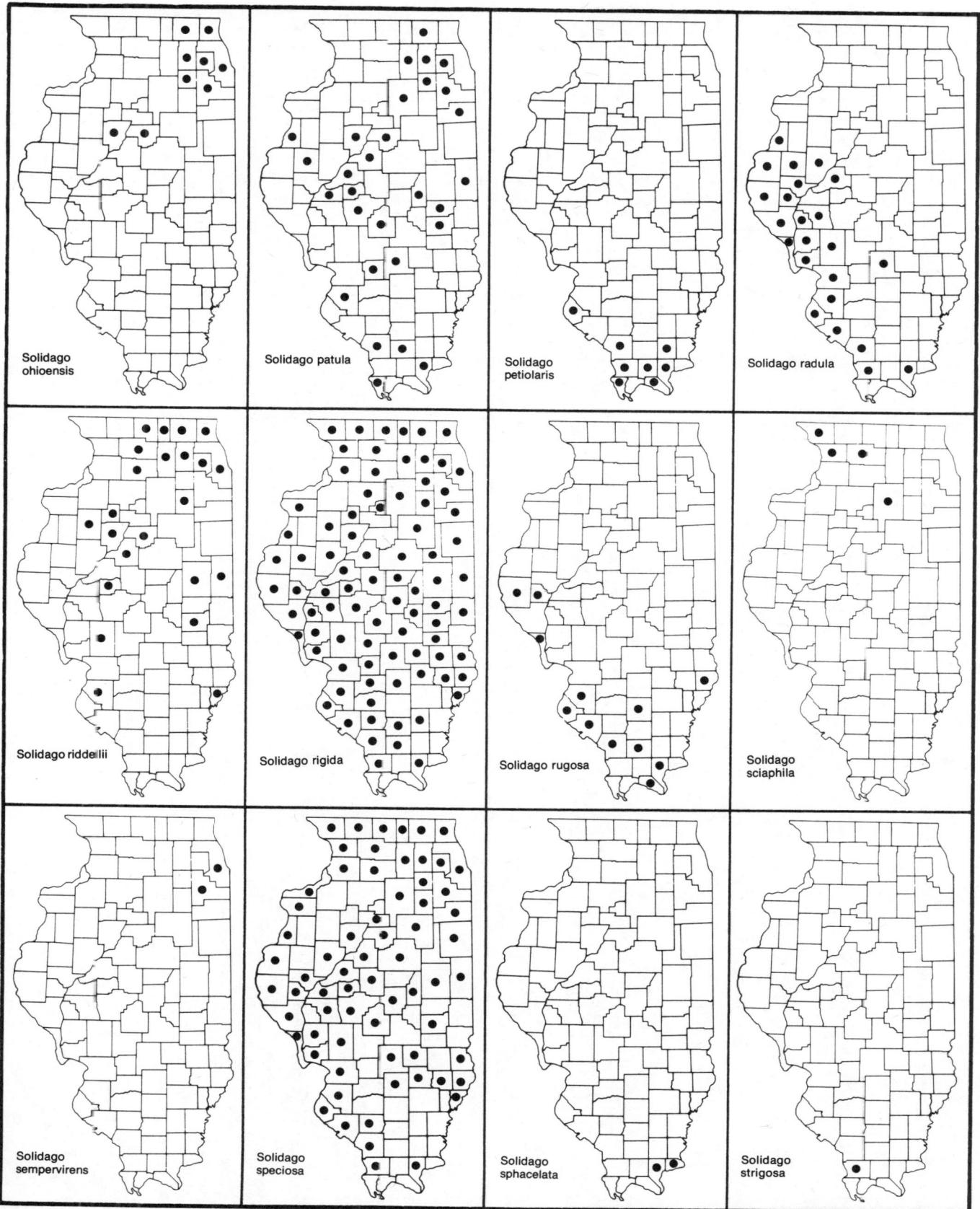

Solidago
ohioensis

Solidago patula

Solidago
petiolaris

Solidago radula

Solidago riddellii

Solidago rigida

Solidago rugosa

Solidago
sciaphila

Solidago
sempervirens

Solidago
speciosa

Solidago
sphacelata

Solidago
strigosa

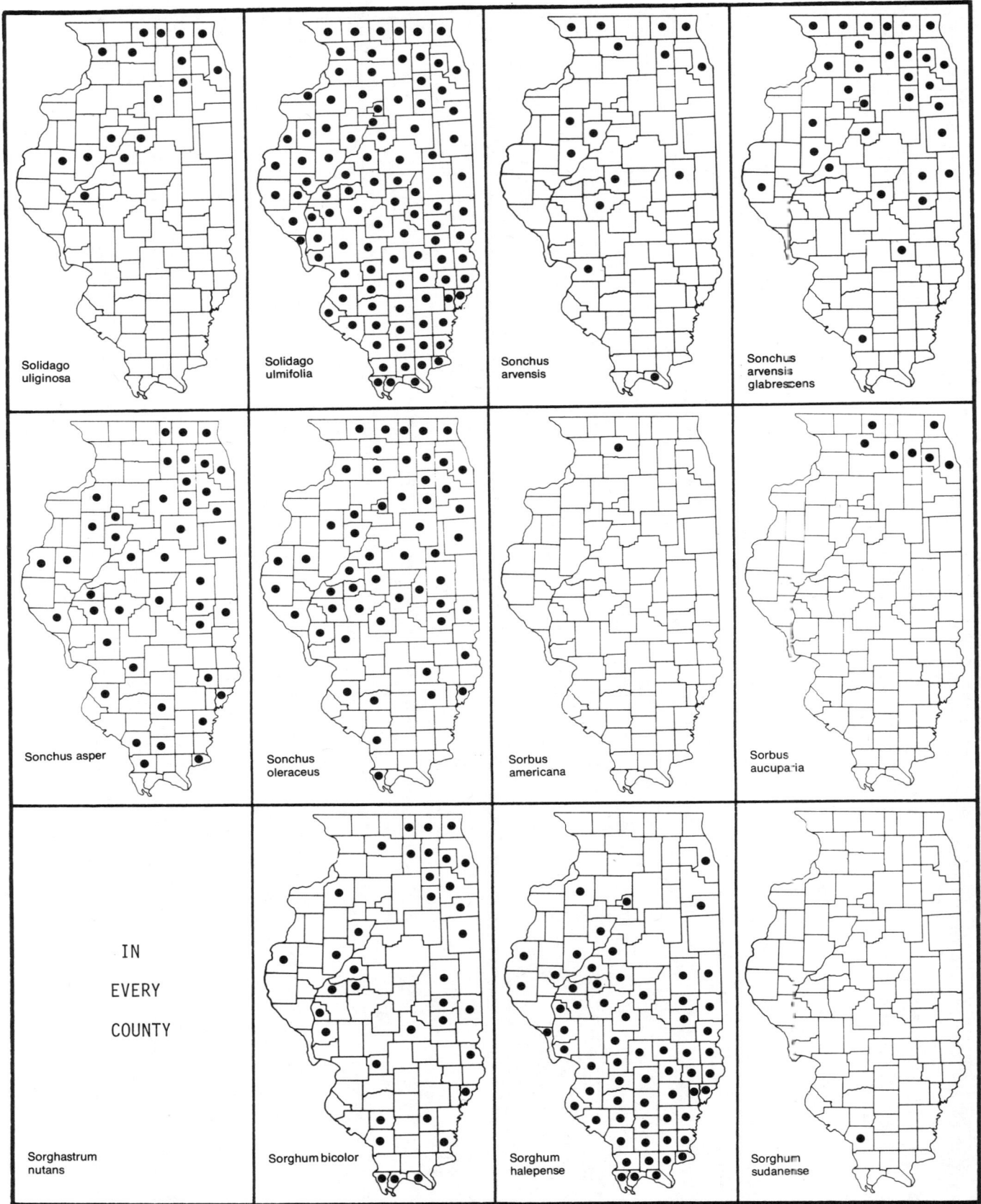

Solidago
uliginosa

Solidago
ulmifolia

Sonchus
arvensis

Sonchus
arvensis
glabrescens

Sonchus asper

Sonchus
oleraceus

Sorbus
americana

Sorbus
aucuparia

IN

EVERY

COUNTY

Sorghastrum
nutans

Sorghum bicolor

Sorghum
halepense

Sorghum
sudanense

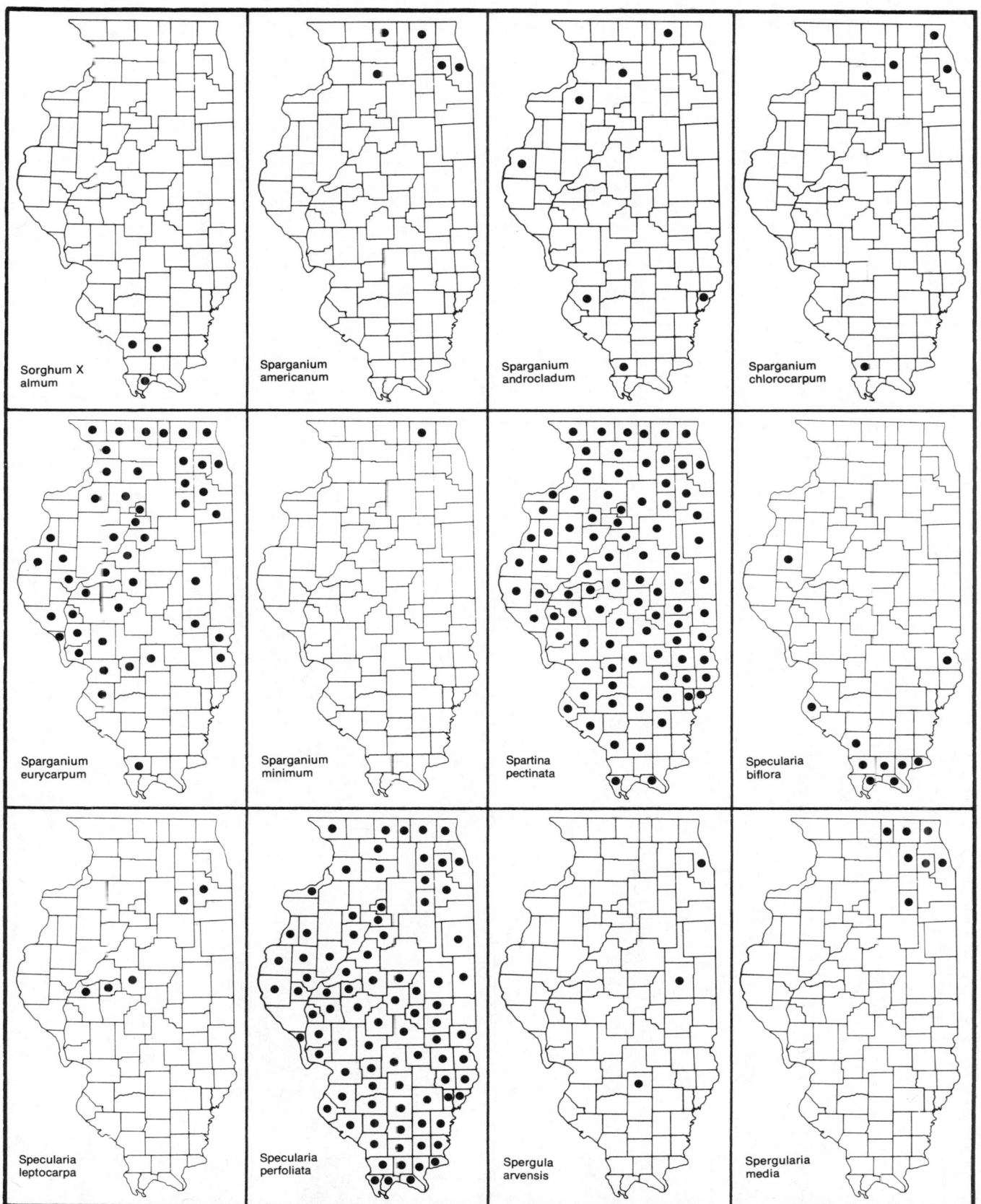

Sorghum X
almum

Sparganium
americanum

Sparganium
androcladum

Sparganium
chlorocarpum

Sparganium
eurycarpum

Sparganium
minimum

Spartina
pectinata

Specularia
biflora

Specularia
leptocarpa

Specularia
perfoliata

Spergula
arvensis

Spergularia
media

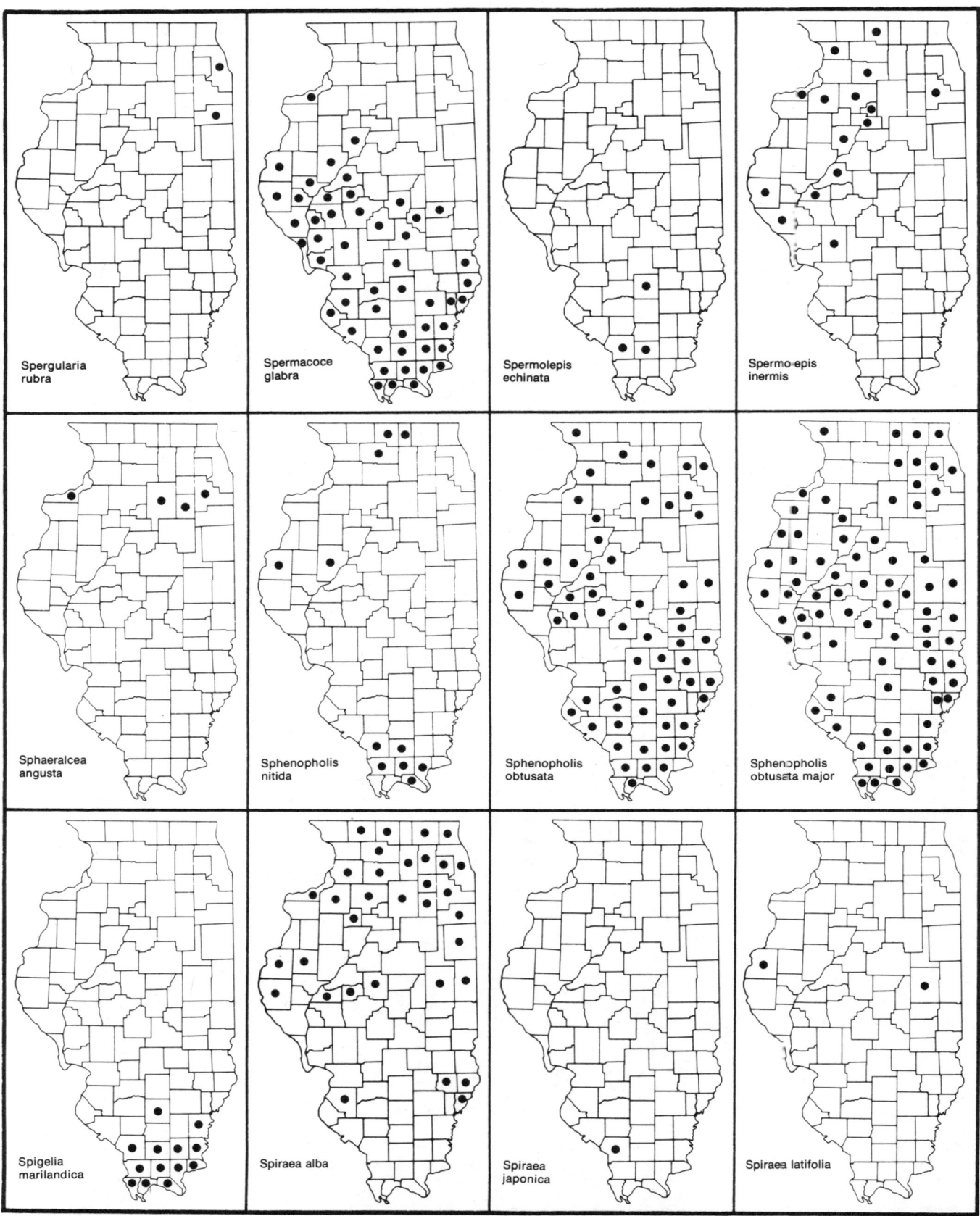

Spergularia
rubra

Spermacoce
glabra

Spermolepis
echinata

Spermoepis
inermis

Sphaeralcea
angusta

Sphenopholis
nitida

Sphenopholis
obtusata

Sphenopholis
obtusata major

Spigelia
marilandica

Spiraea alba

Spiraea
japonica

Spiraea latifolia

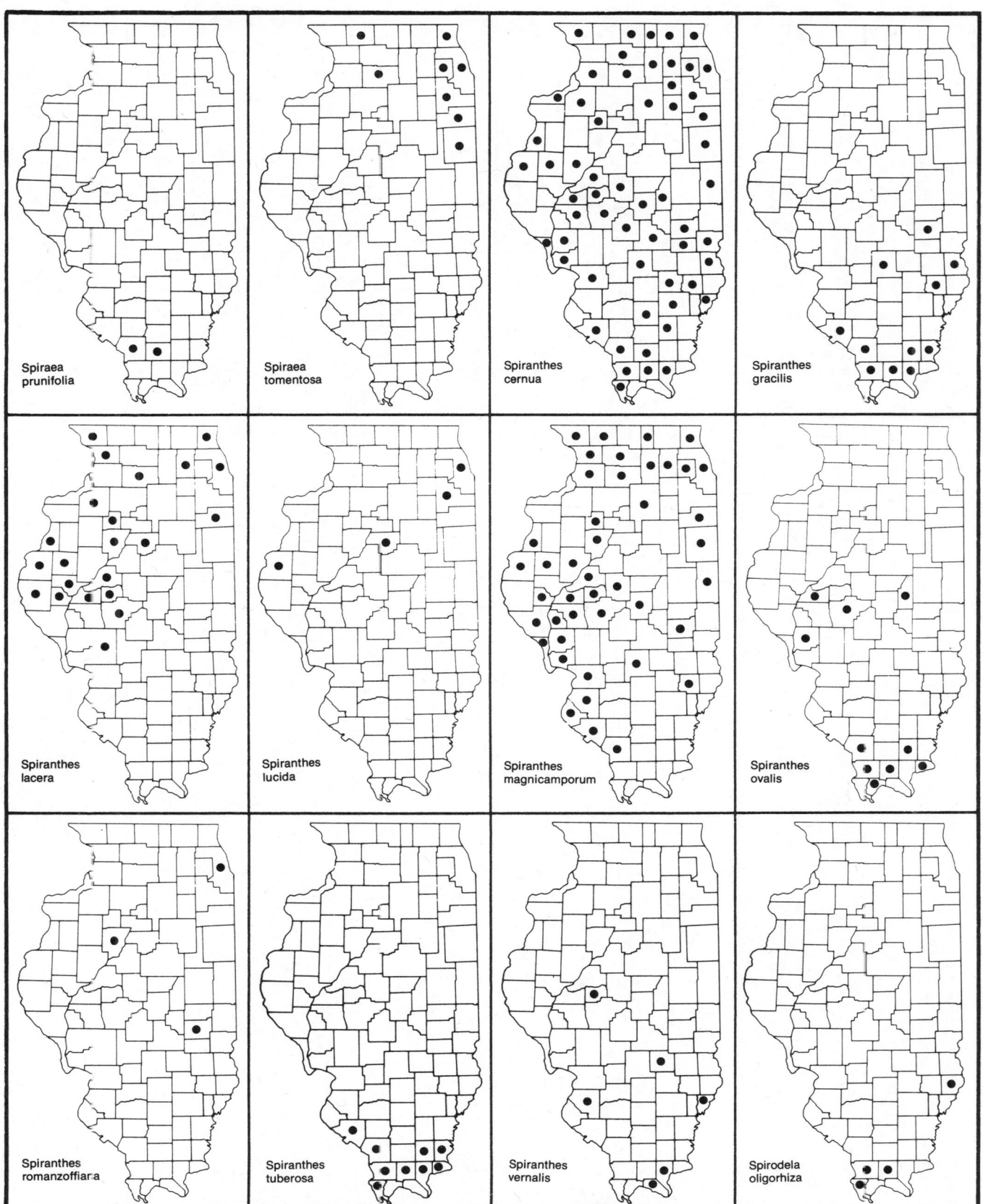

Spiraea
prunifolia

Spiraea
tomentosa

Spiranthes
cernua

Spiranthes
gracilis

Spiranthes
lacera

Spiranthes
lucida

Spiranthes
magnicamporum

Spiranthes
ovalis

Spiranthes
romanzoffiana

Spiranthes
tuberosa

Spiranthes
vernalis

Spirodela
oligorhiza

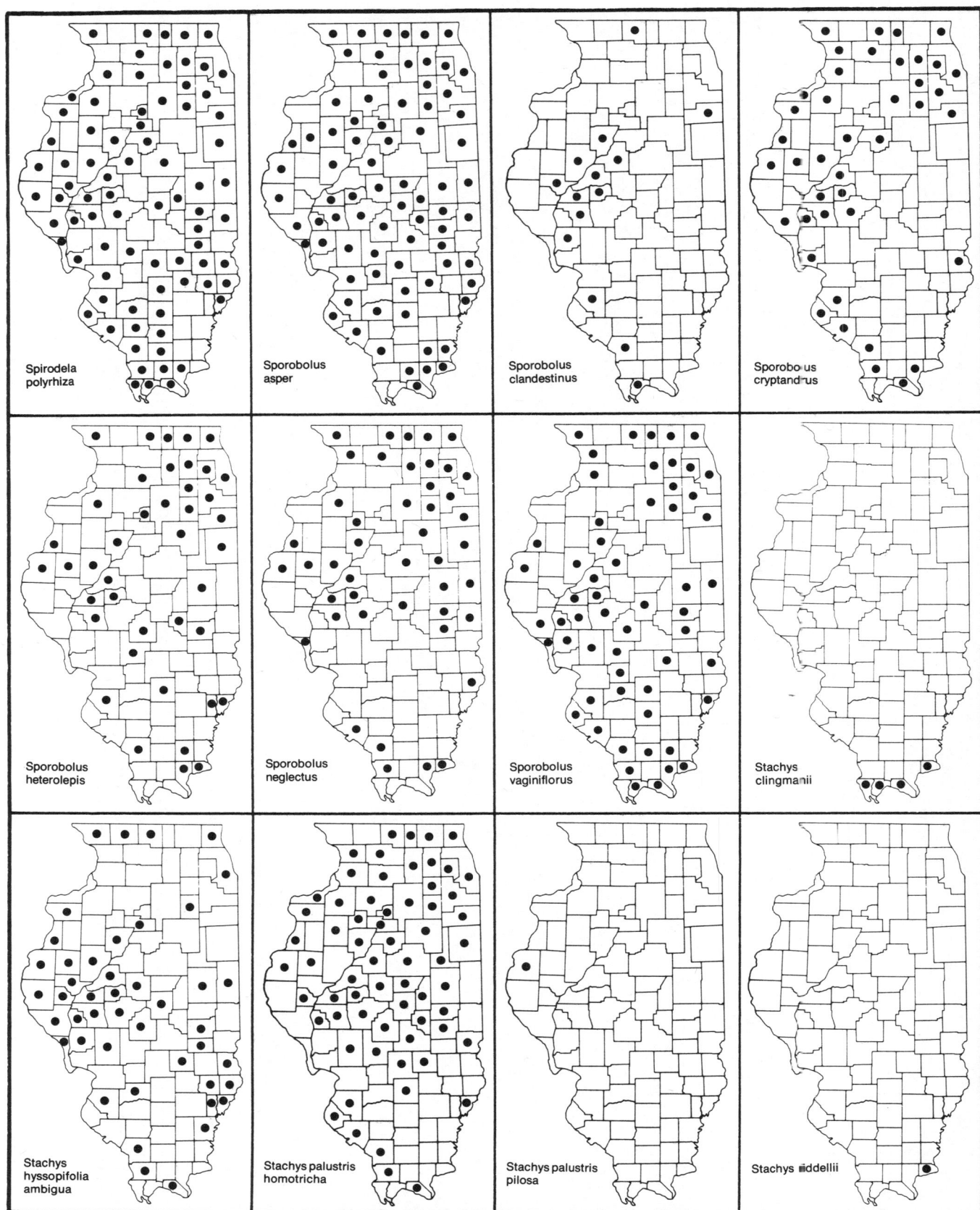

Spirodela
polyrhiza

Sporobolus
asper

Sporobolus
clandestinus

Sporobolus
cryptandrus

Sporobolus
heterolepis

Sporobolus
neglectus

Sporobolus
vaginiflorus

Stachys
clingmanii

Stachys
hyssopifolia
ambigua

Stachys palustris
homotricha

Stachys palustris
pilosa

Stachys riddellii

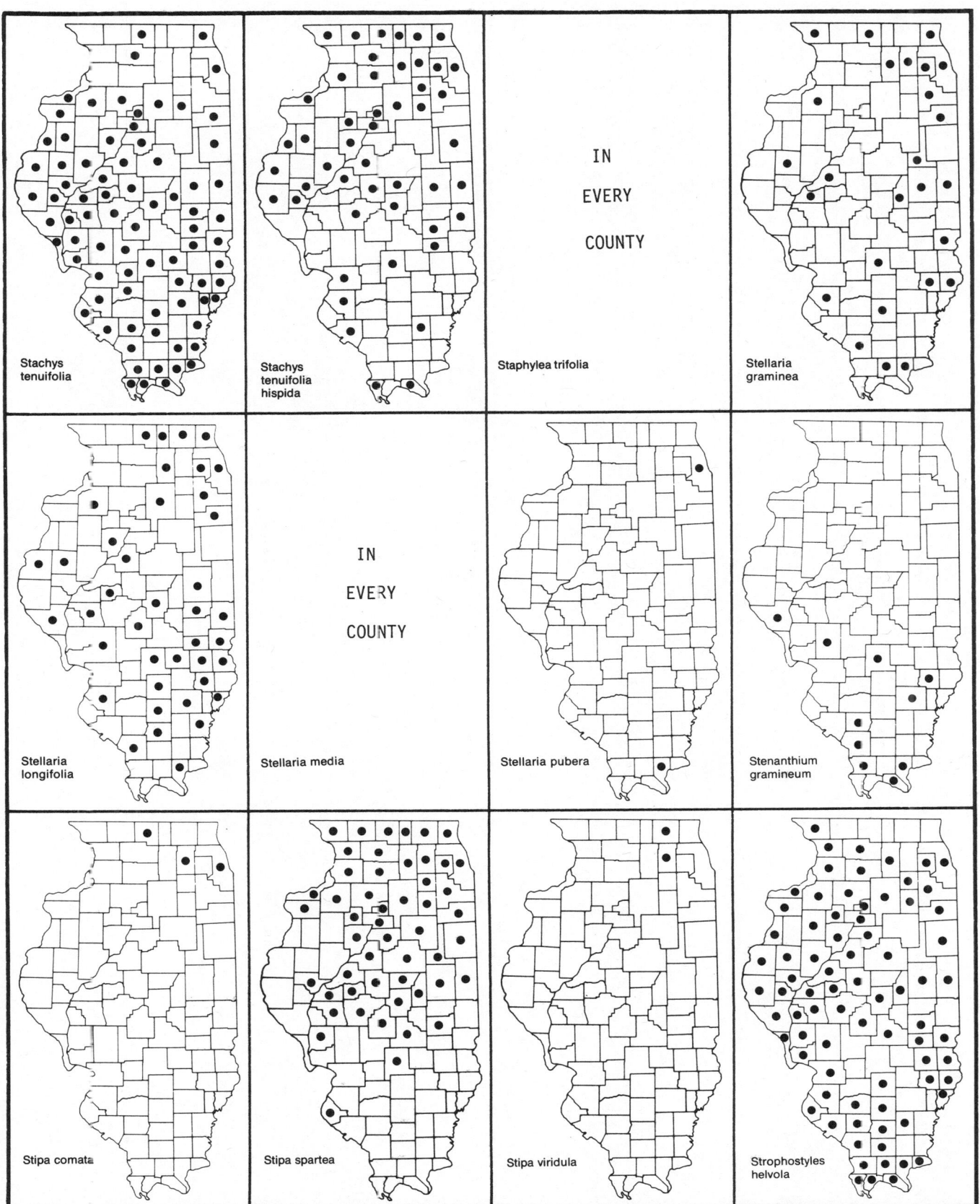

Stachys
tenuifolia

Stachys
tenuifolia
hispida

Staphylea trifolia

IN

EVERY

COUNTY

Stellaria
graminea

Stellaria
longifolia

Stellaria media

IN

EVERY

COUNTY

Stellaria pubera

Stenanthium
gramineum

Stipa comata

Stipa spartea

Stipa viridula

Strophostyles
helvola

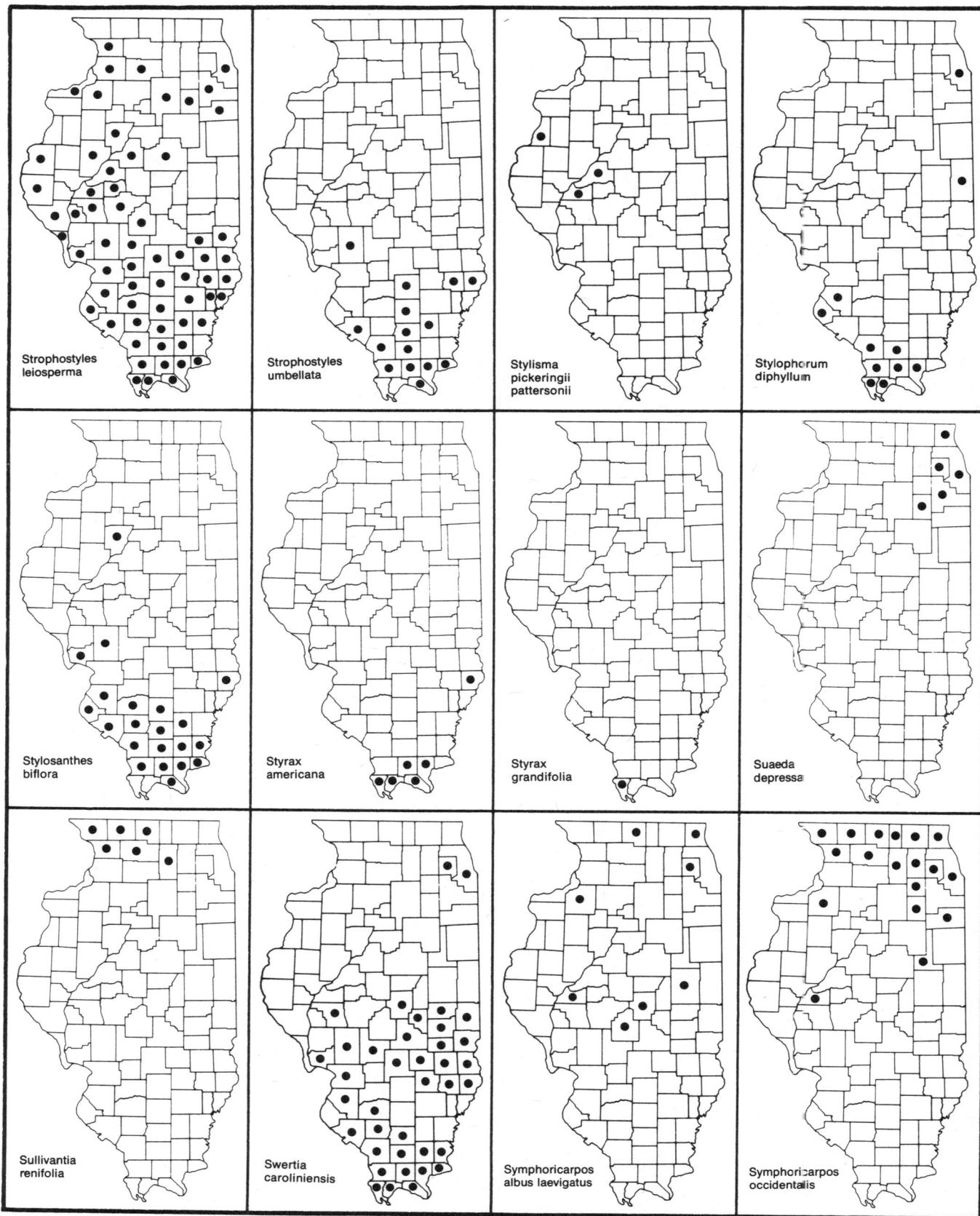

Strophostyles
leiosperma

Strophostyles
umbellata

Stylisma
pickeringii
pattersonii

Stylophorum
diphyllum

Stylosanthes
biflora

Styrax
americana

Styrax
grandifolia

Suaeda
depressa

Sullivantia
renifolia

Swertia
caroliniensis

Symphoricarpos
albus laevigatus

Symphoricarpos
occidentalis

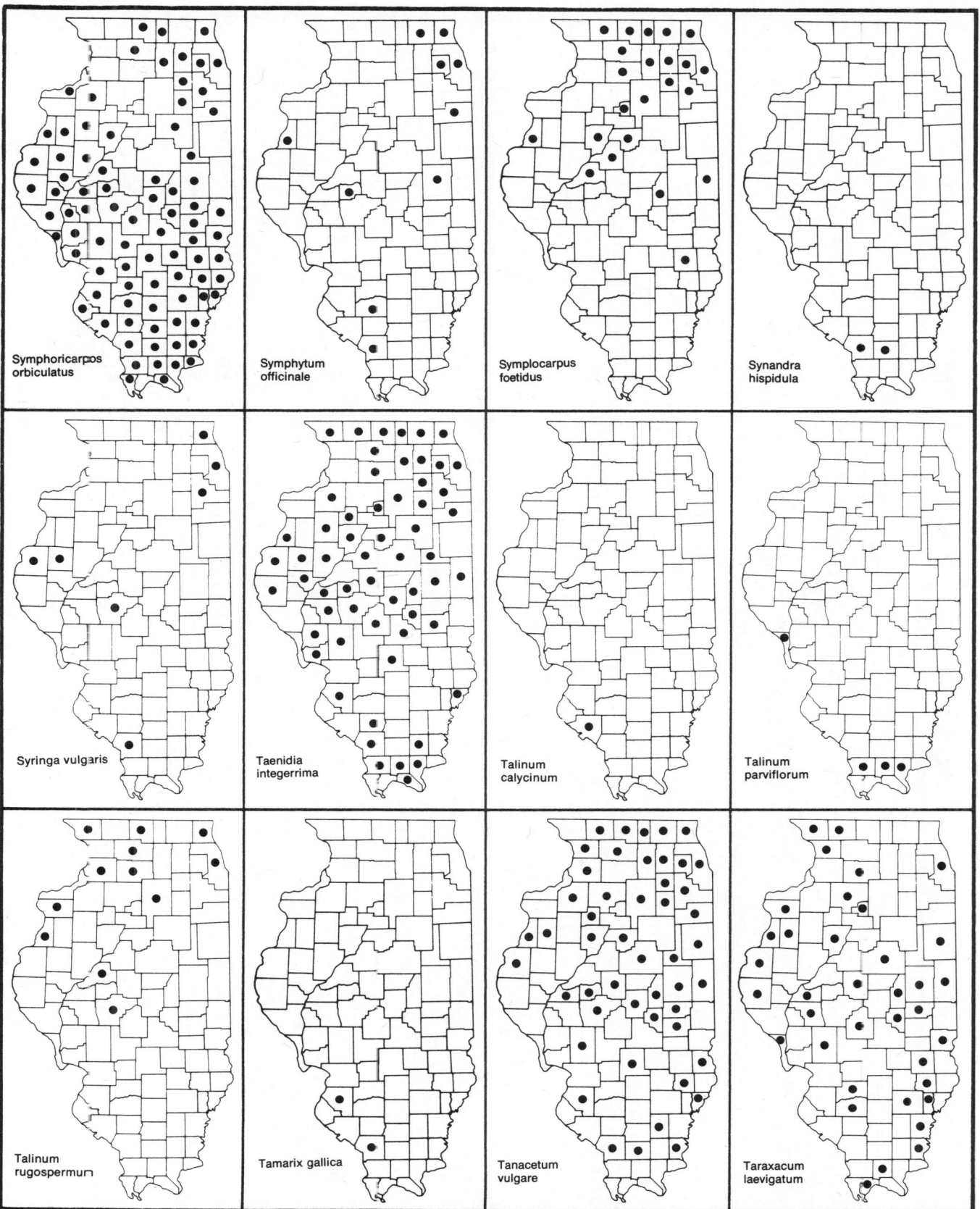

Symphoricarpos
orbiculatus

Symphytum
officinale

Symplocarpus
foetidus

Synandra
hispidula

Syringa vulgaris

Taenidia
integerrima

Talinum
calycinum

Talinum
parviflorum

Talinum
rugospermun

Tamarix gallica

Tanacetum
vulgare

Taraxacum
laevigatum

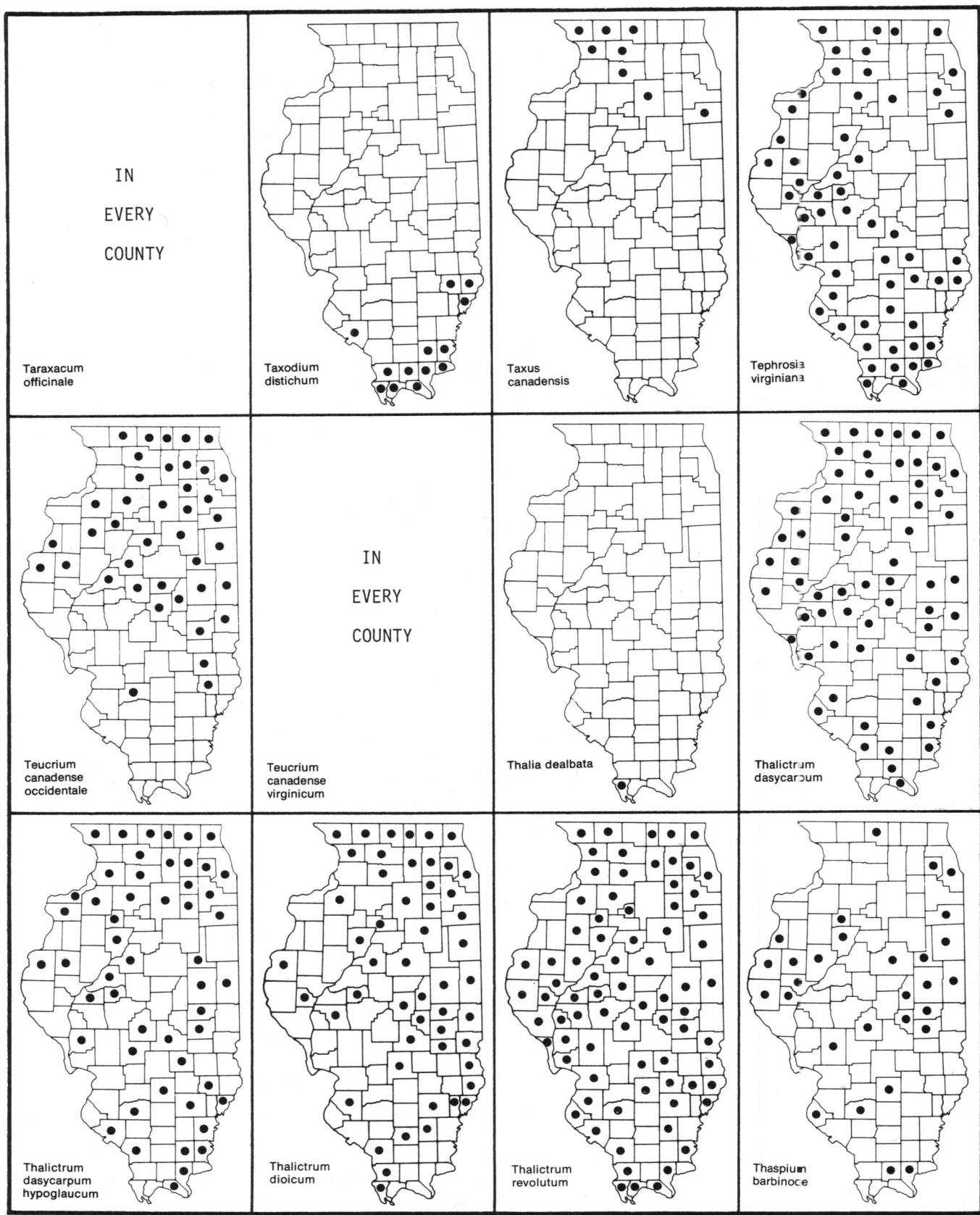

IN EVERY COUNTY

Taraxacum
officinale

Taxodium
distichum

Taxus
canadensis

Tephrosia
virginiana

Teucrium
canadense
occidentale

Teucrium
canadense
virginicum

IN EVERY COUNTY

Thalia dealbata

Thalictrum
dasycarpum

Thalictrum
dasycarpum
hypoglaucum

Thalictrum
dioicum

Thalictrum
revolutum

Thaspium
barbinode

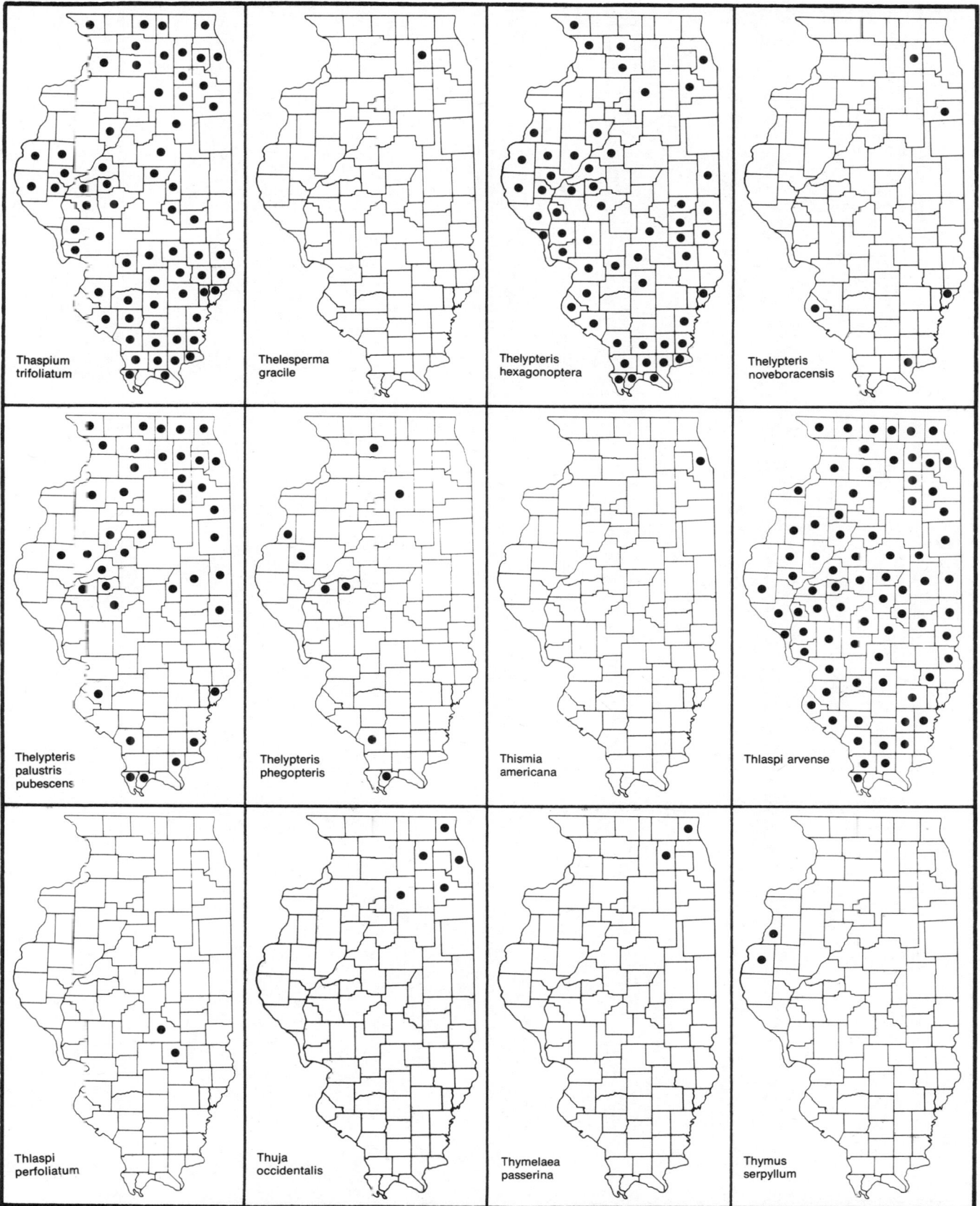

Thaspium
trifoliatum

Thelesperma
gracile

Thelypteris
hexagonoptera

Thelypteris
noveboracensis

Thelypteris
palustris
pubescens

Thelypteris
phegopteris

Thismia
americana

Thlaspi arvense

Thlaspi
perfoliatum

Thuja
occidentalis

Thymelaea
passerina

Thymus
serpyllum

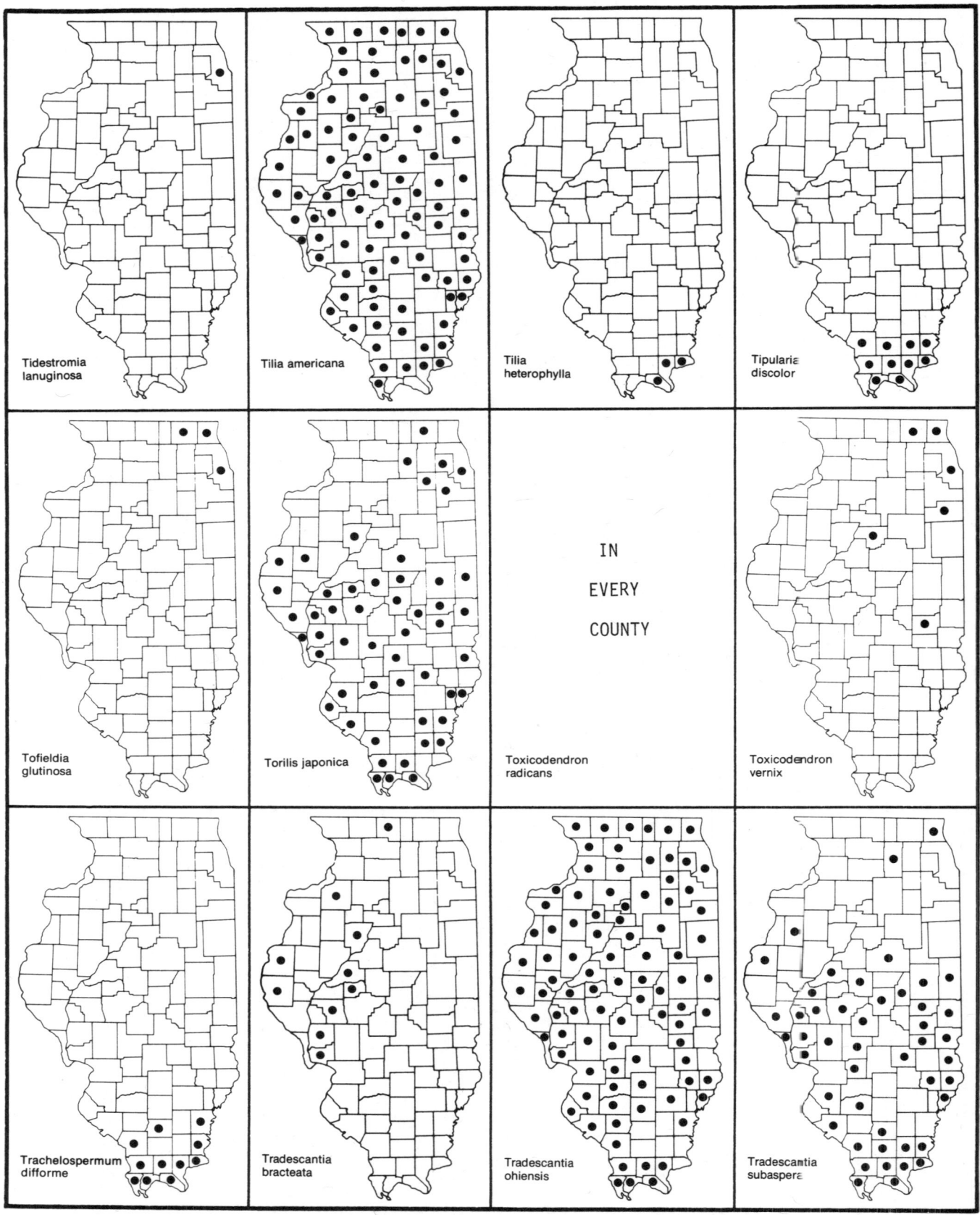

Tidestromia
lanuginosa

Tilia americana

Tilia
heterophylla

Tipularia
discolor

Tofieldia
glutinosa

Torilis japonica

Toxicodendron
radicans

IN

EVERY

COUNTY

Toxicodendron
vernix

Trachelospermum
difforme

Tradescantia
bracteata

Tradescantia
ohiensis

Tradescantia
subaspera

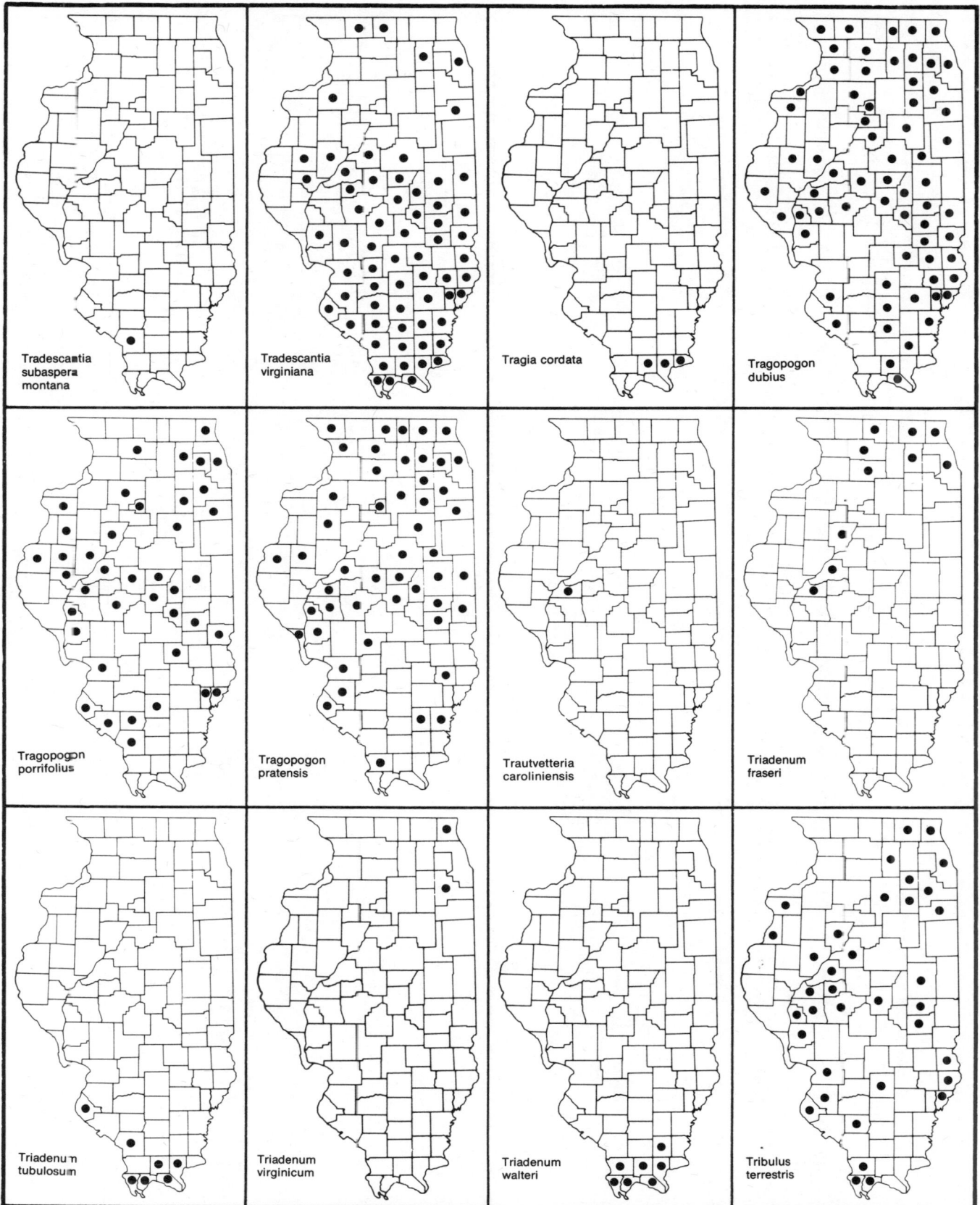

Tradescantia
subaspera
montana

Tradescantia
virginiana

Tragia cordata

Tragopogon
dubius

Tragopogon
porrifolius

Tragopogon
pratensis

Trautvetteria
caroliniensis

Triadenum
fraseri

Triadenum
tubulosum

Triadenum
virginicum

Triadenum
walteri

Tribulus
terrestris

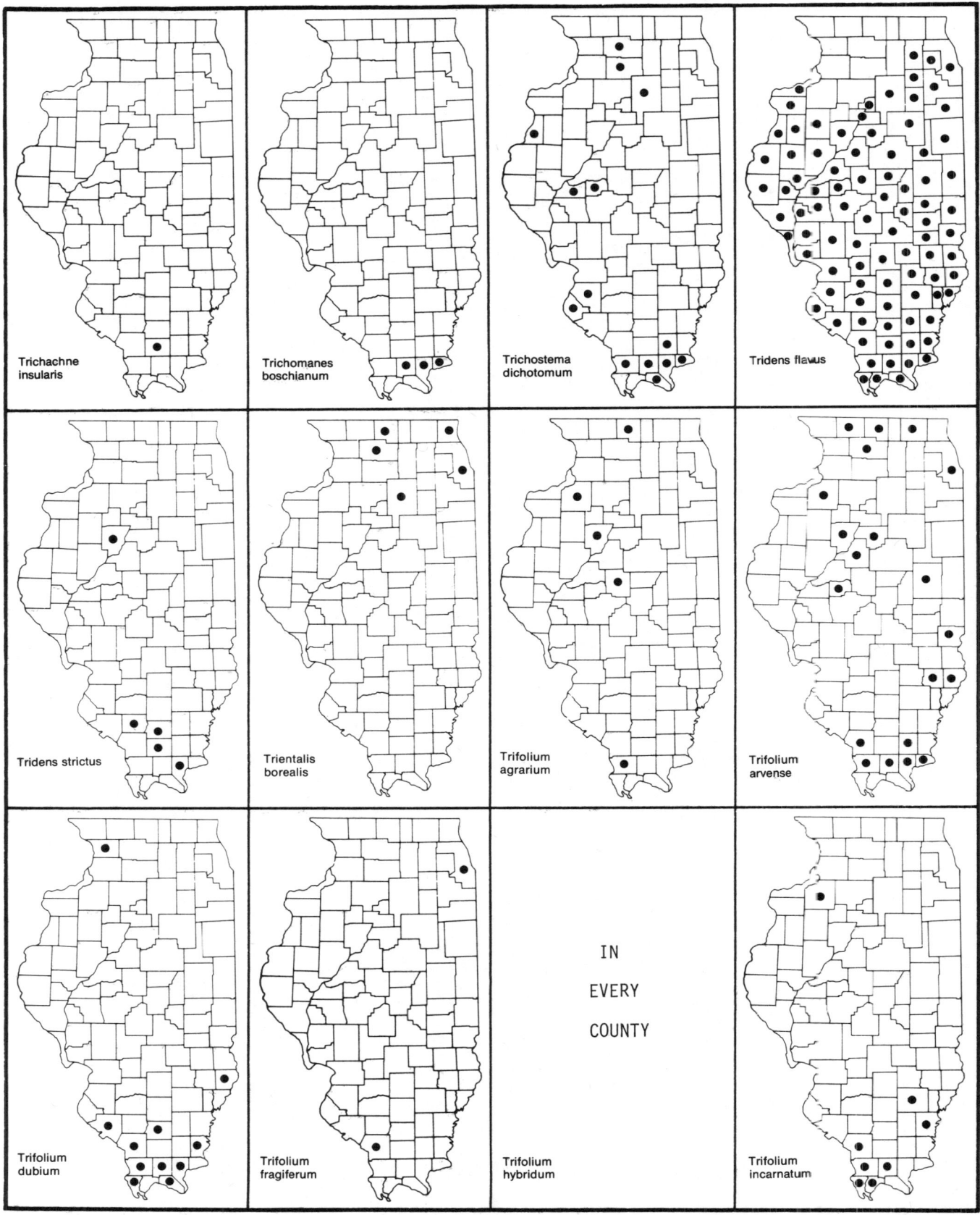

Trichachne
insularis

Trichomanes
boschianum

Trichostema
dichotomum

Tridens flavus

Tridens strictus

Trientalis
borealis

Trifolium
agrarium

Trifolium
arvense

Trifolium
dubium

Trifolium
fragiferum

Trifolium
hybridum

IN

EVERY

COUNTY

Trifolium
incarnatum

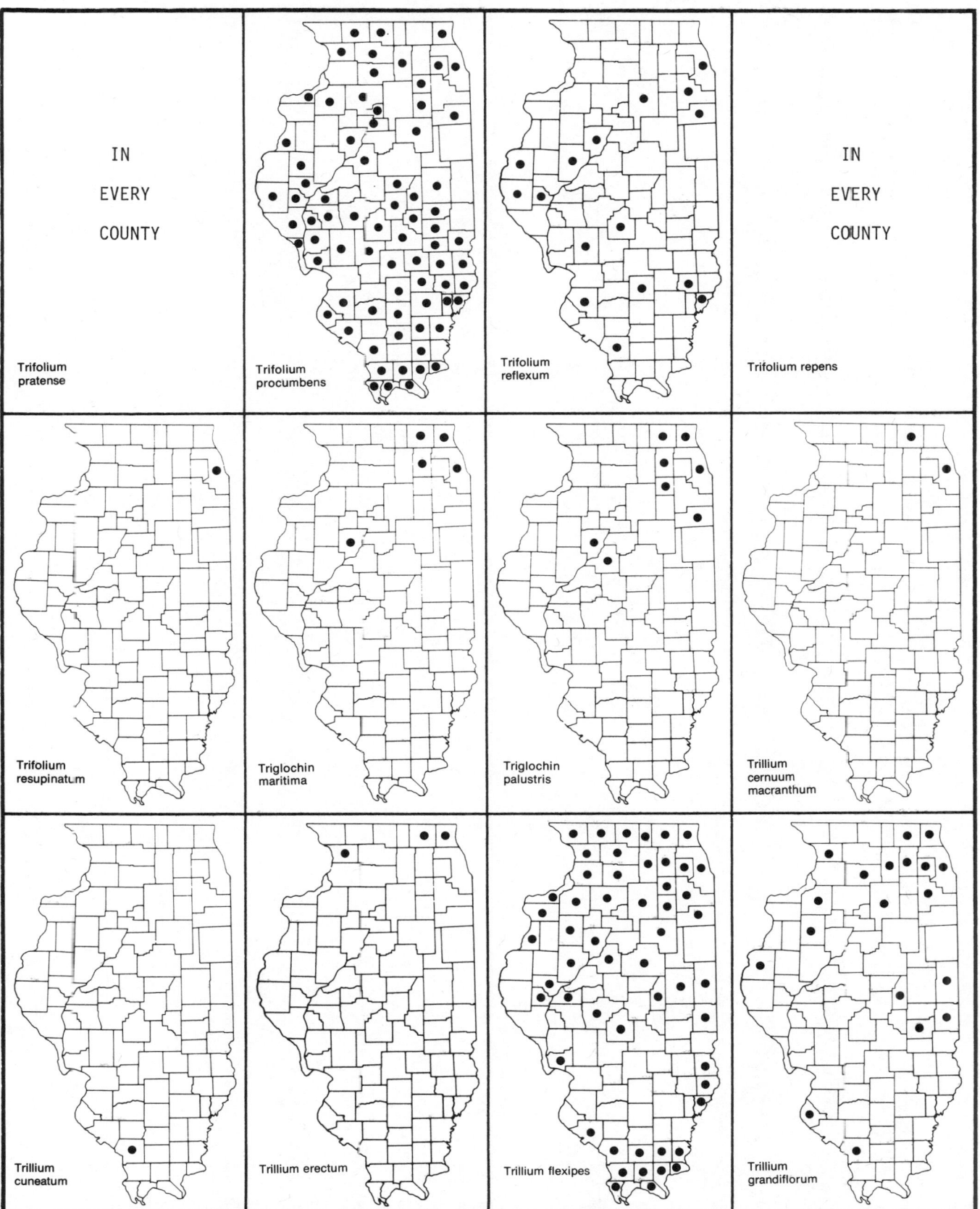

IN
EVERY
COUNTY

Trifolium
pratense

Trifolium
procumbens

Trifolium
reflexum

IN
EVERY
COUNTY

Trifolium repens

Trifolium
resupinatum

Triglochin
maritima

Triglochin
palustris

Trillium
cernuum
macranthum

Trillium
cuneatum

Trillium erectum

Trillium
flexipes

Trillium
grandiflorum

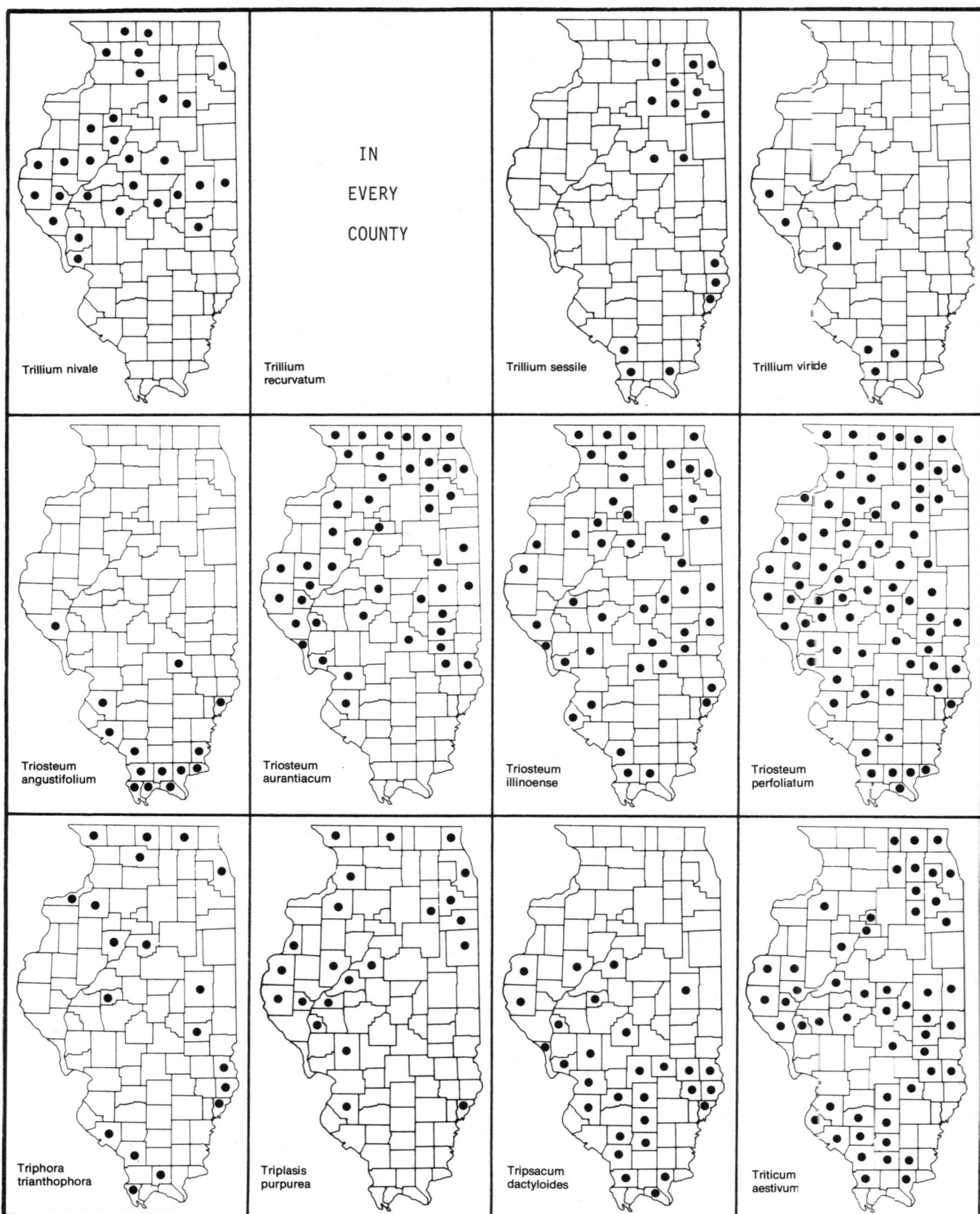

Trillium nivale

Trillium recurvatum

IN

EVERY

COUNTY

Trillium sessile

Trillium viride

Triosteum angustifolium

Triosteum aurantiacum

Triosteum illinoense

Triosteum perfoliatum

Triphora trianthophora

Triplasis purpurea

Tripsacum dactyloides

Triticum aestivum

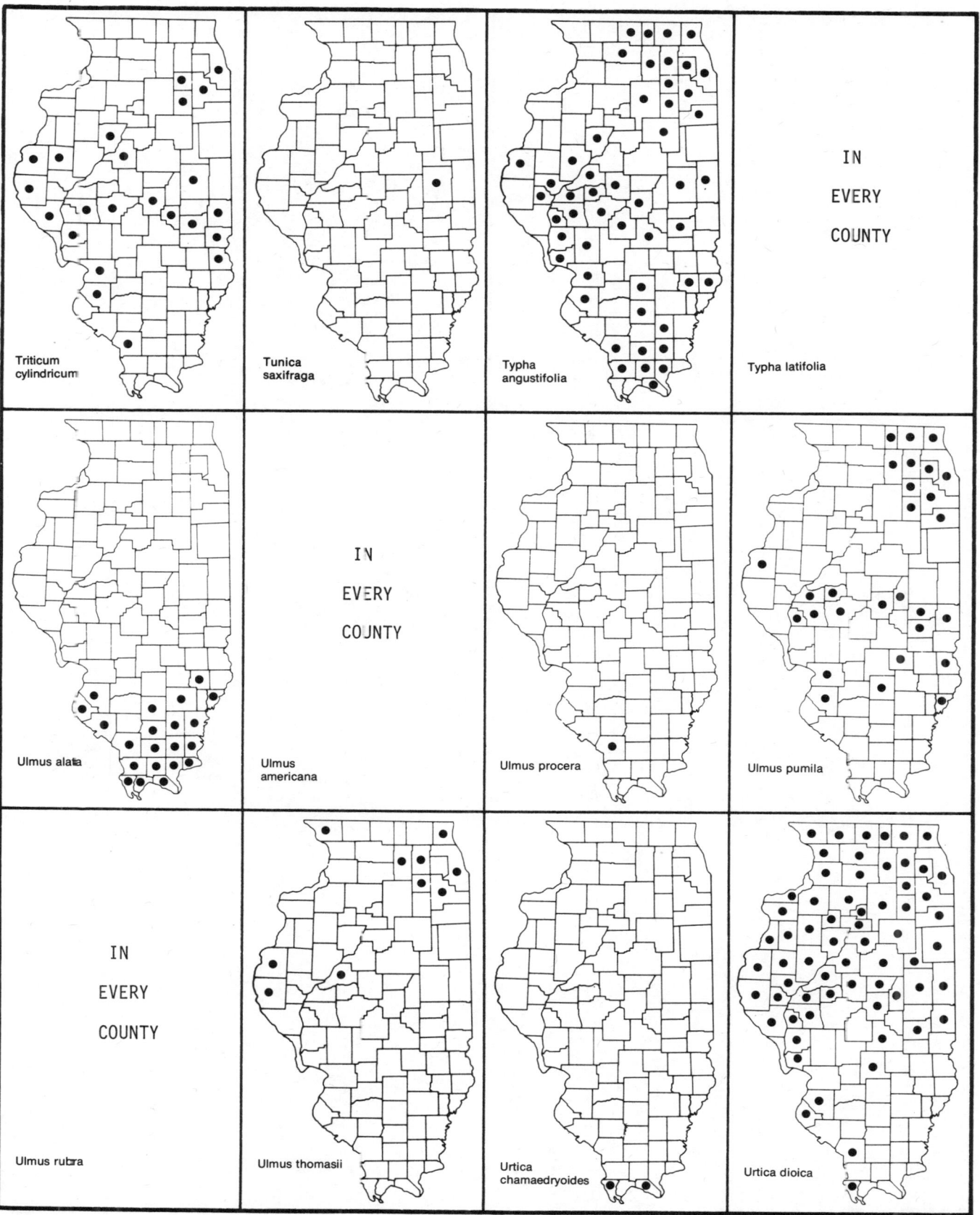

Triticum
cylindricum

Tunica
saxifraga

Typha
angustifolia

Typha latifolia

IN

EVERY

COUNTY

Ulmus alata

Ulmus
americana

IN

EVERY

COUNTY

Ulmus procera

Ulmus pumila

Ulmus rubra

IN

EVERY

COUNTY

Ulmus thomasii

Urtica
chamaedryoides

Urtica dioica

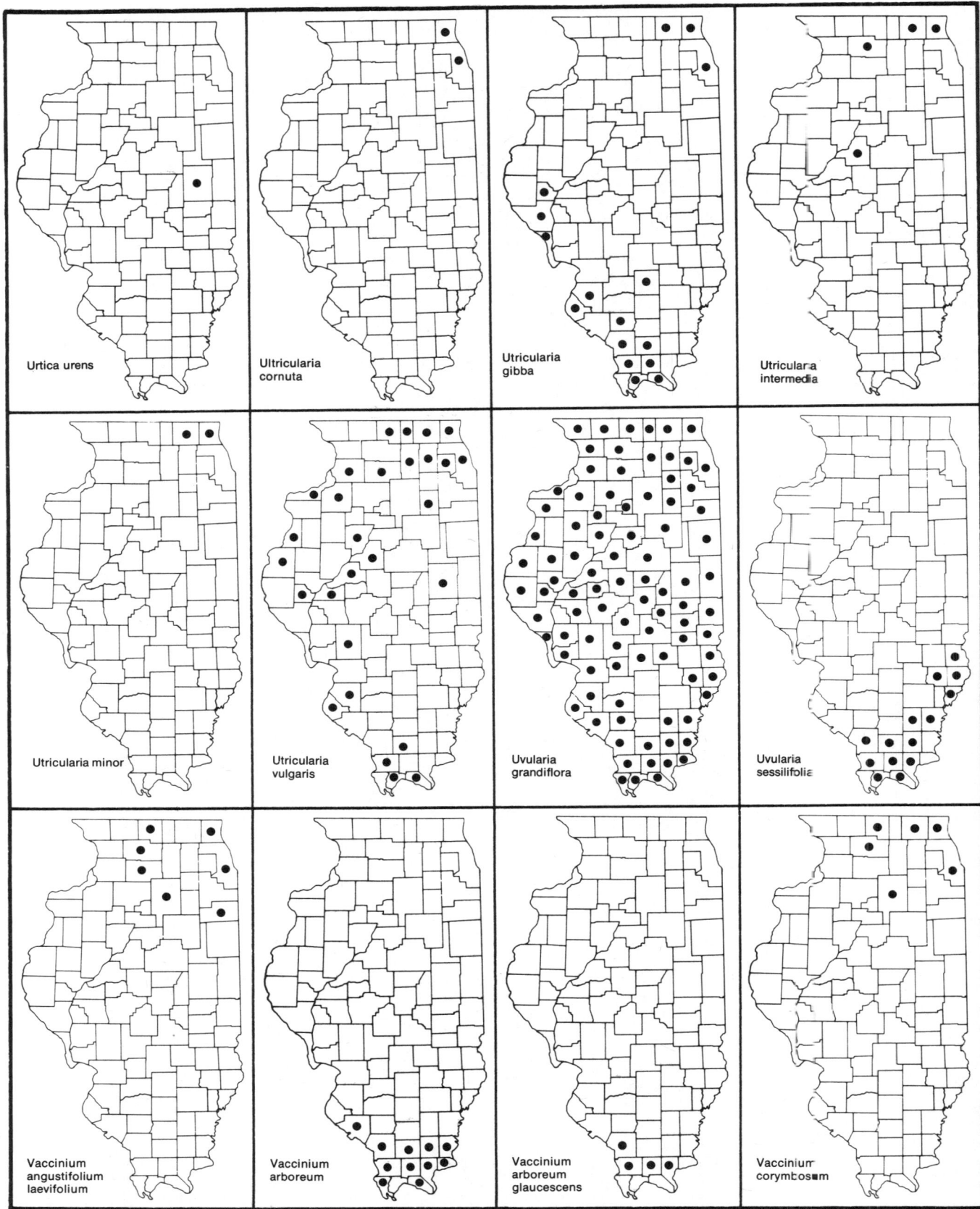

Urtica urens

Ultricularia
cornuta

Utricularia
gibba

Utricularia
intermedia

Utricularia minor

Utricularia
vulgaris

Uvularia
grandiflora

Uvularia
sessilifolia

Vaccinium
angustifolium
laevifolium

Vaccinium
arboreum

Vaccinium
arboreum
glaucescens

Vaccinium
corymbosum

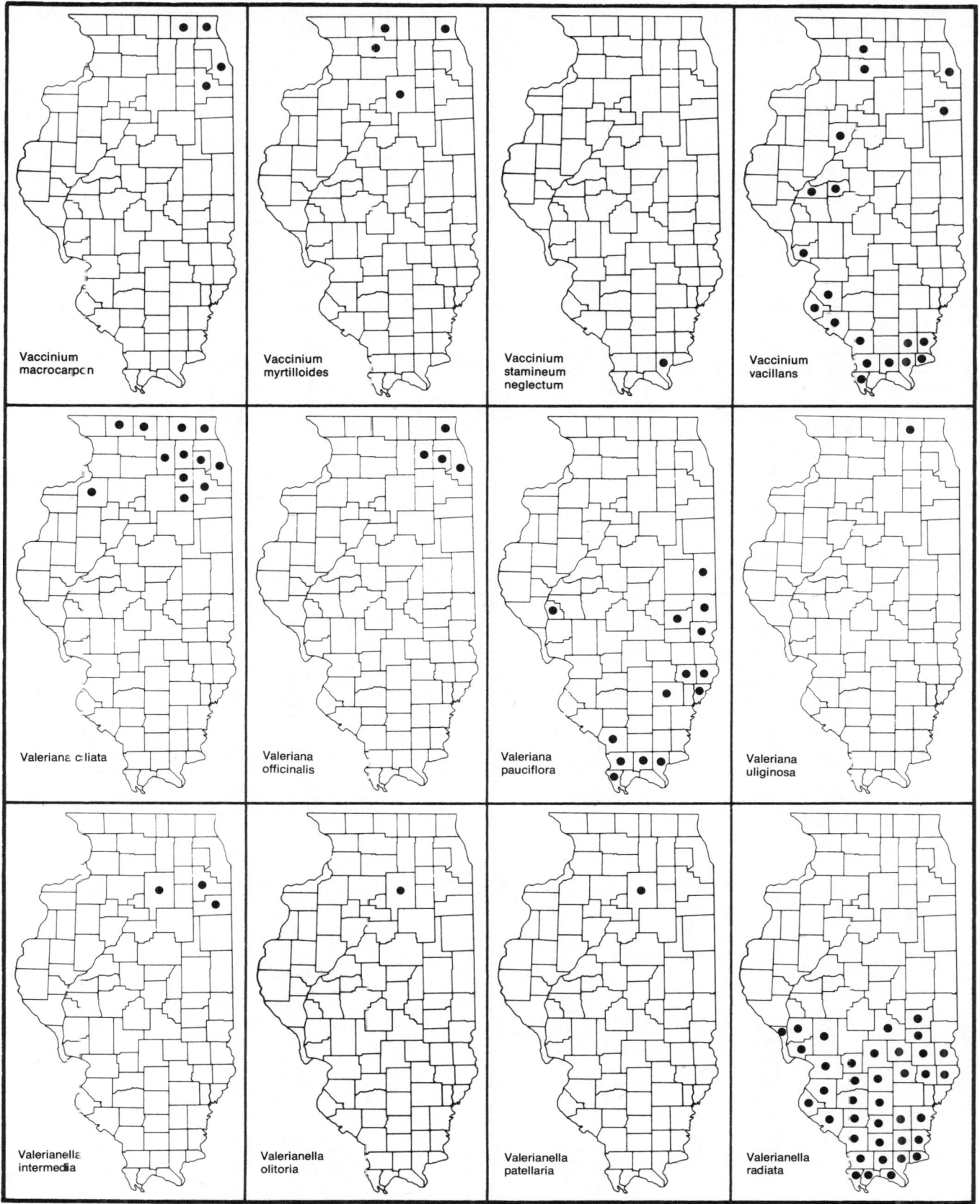

Vaccinium
macrocarpon

Vaccinium
myrtilloides

Vaccinium
stamineum
neglectum

Vaccinium
vacillans

Valeriana ciliata

Valeriana
officinalis

Valeriana
pauciflora

Valeriana
uliginosa

Valerianella
intermedia

Valerianella
olitoria

Valerianella
patellaria

Valerianella
radiata

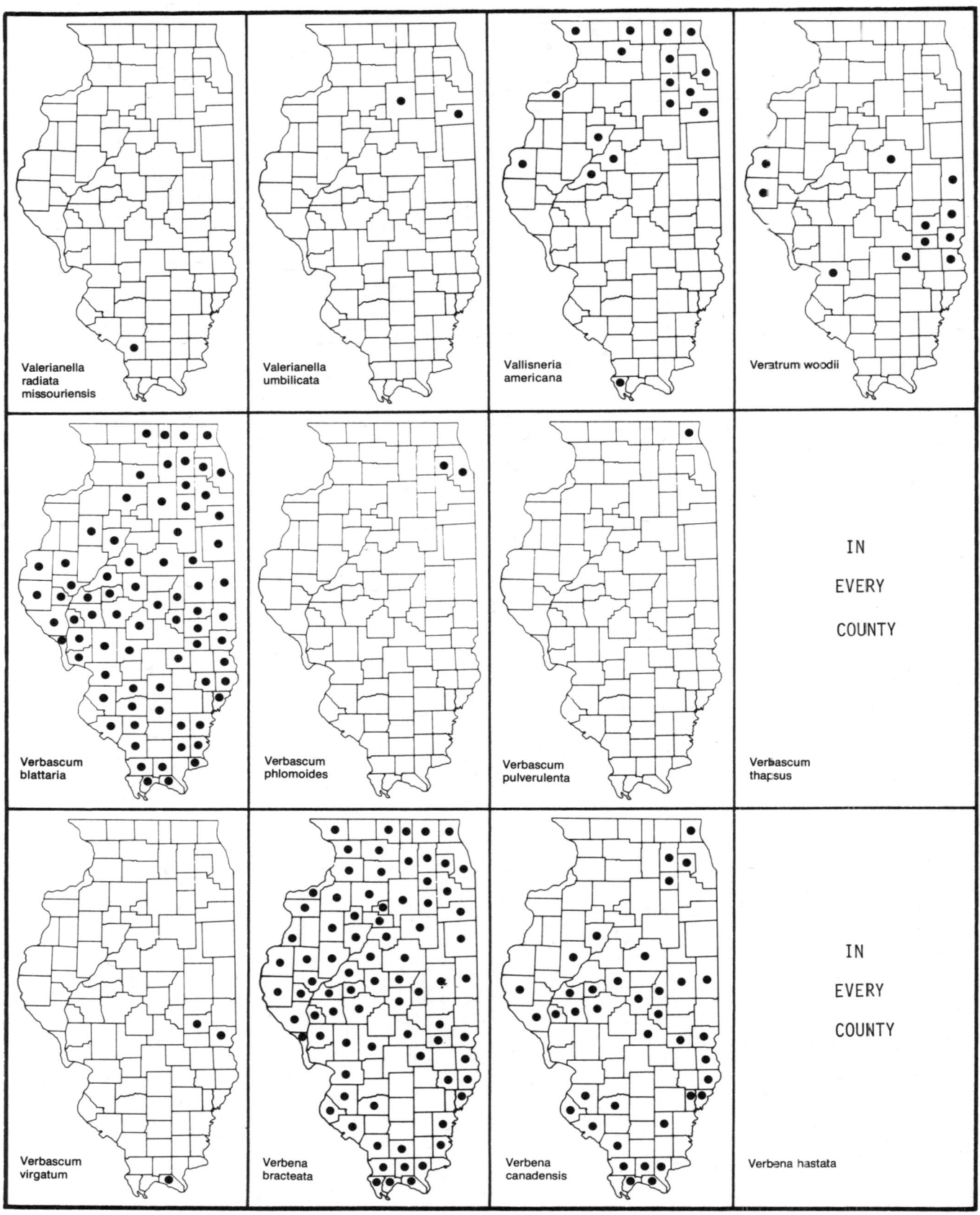

Valerianella
radiata
missouriensis

Valerianella
umbilicata

Vallisneria
americana

Veratrum woodii

Verbascum
blattaria

Verbascum
phlomoides

Verbascum
pulverulenta

Verbascum
thapsus

IN

EVERY

COUNTY

Verbascum
virgatum

Verbena
bracteata

Verbena
canadensis

Verbena hastata

IN

EVERY

COUNTY

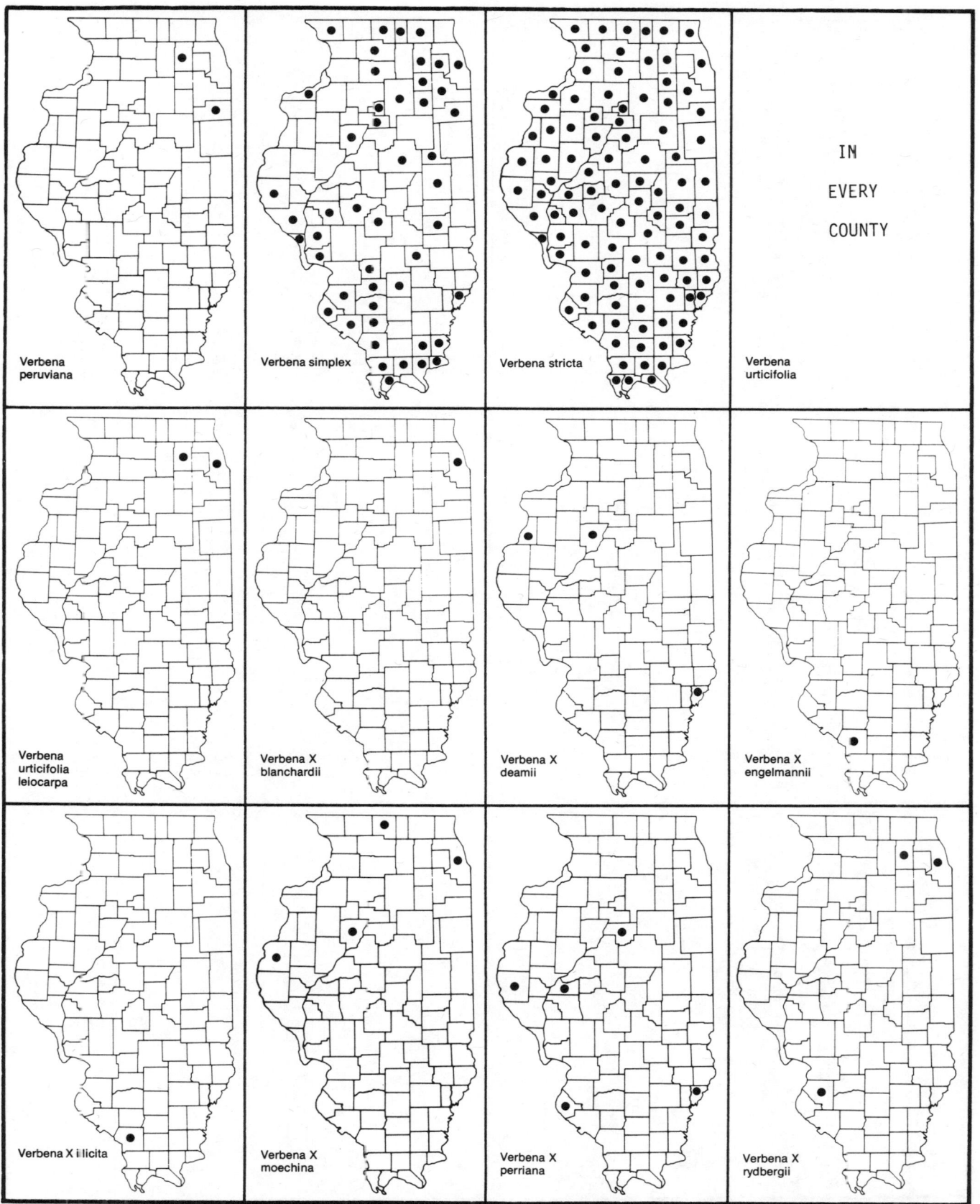

Verbena
peruviana

Verbena simplex

Verbena stricta

Verbena
urticifolia

IN

EVERY

COUNTY

Verbena
urticifolia
leiocarpa

Verbena X
blanchardii

Verbena X
deamii

Verbena X
engelmannii

Verbena X illicita

Verbena X
moechina

Verbena X
perriana

Verbena X
rydbergii

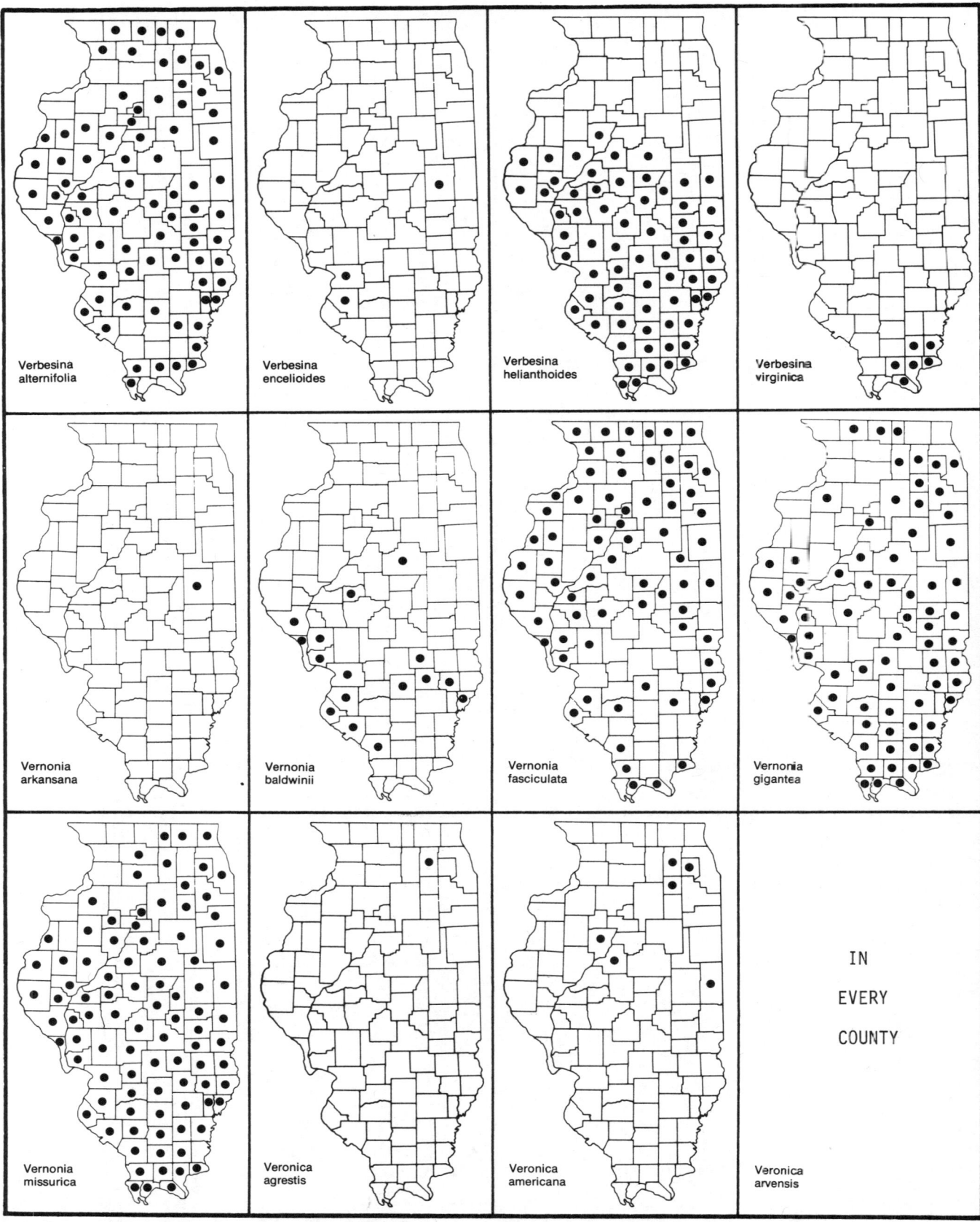

Verbesina
alternifolia

Verbesina
encelioides

Verbesina
helianthoides

Verbesina
virginica

Vernonia
arkansana

Vernonia
baldwinii

Vernonia
fasciculata

Vernonia
gigantea

Vernonia
missurica

Veronica
agrestis

Veronica
americana

Veronica
arvensis

IN

EVERY

COUNTY

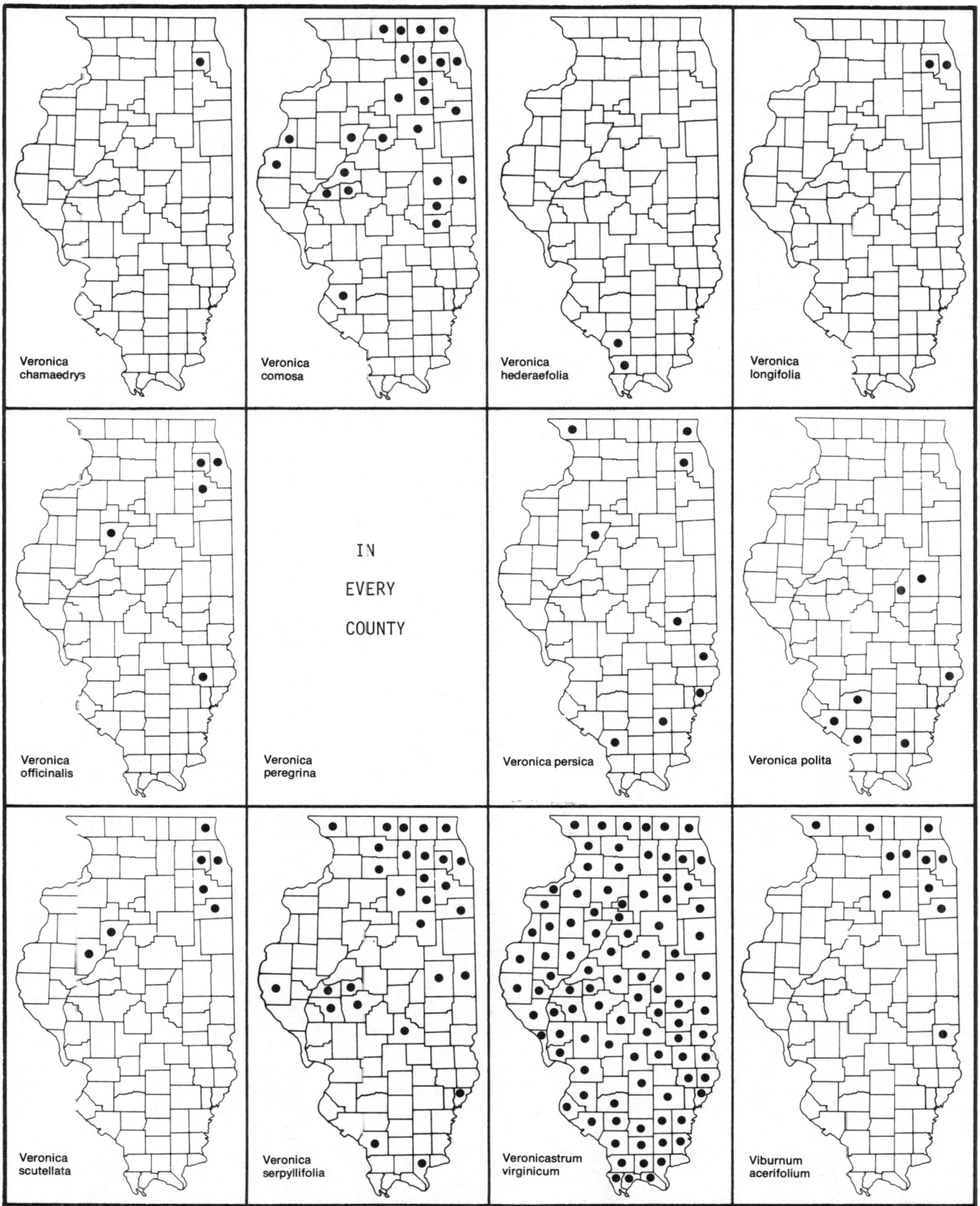

Veronica
chamaedrys

Veronica
comosa

Veronica
hederaefolia

Veronica
longifolia

Veronica
officinalis

Veronica
peregrina

IN

EVERY

COUNTY

Veronica
persica

Veronica
polita

Veronica
scutellata

Veronica
serpyllifolia

Veronicastrum
virginicum

Viburnum
acerifolium

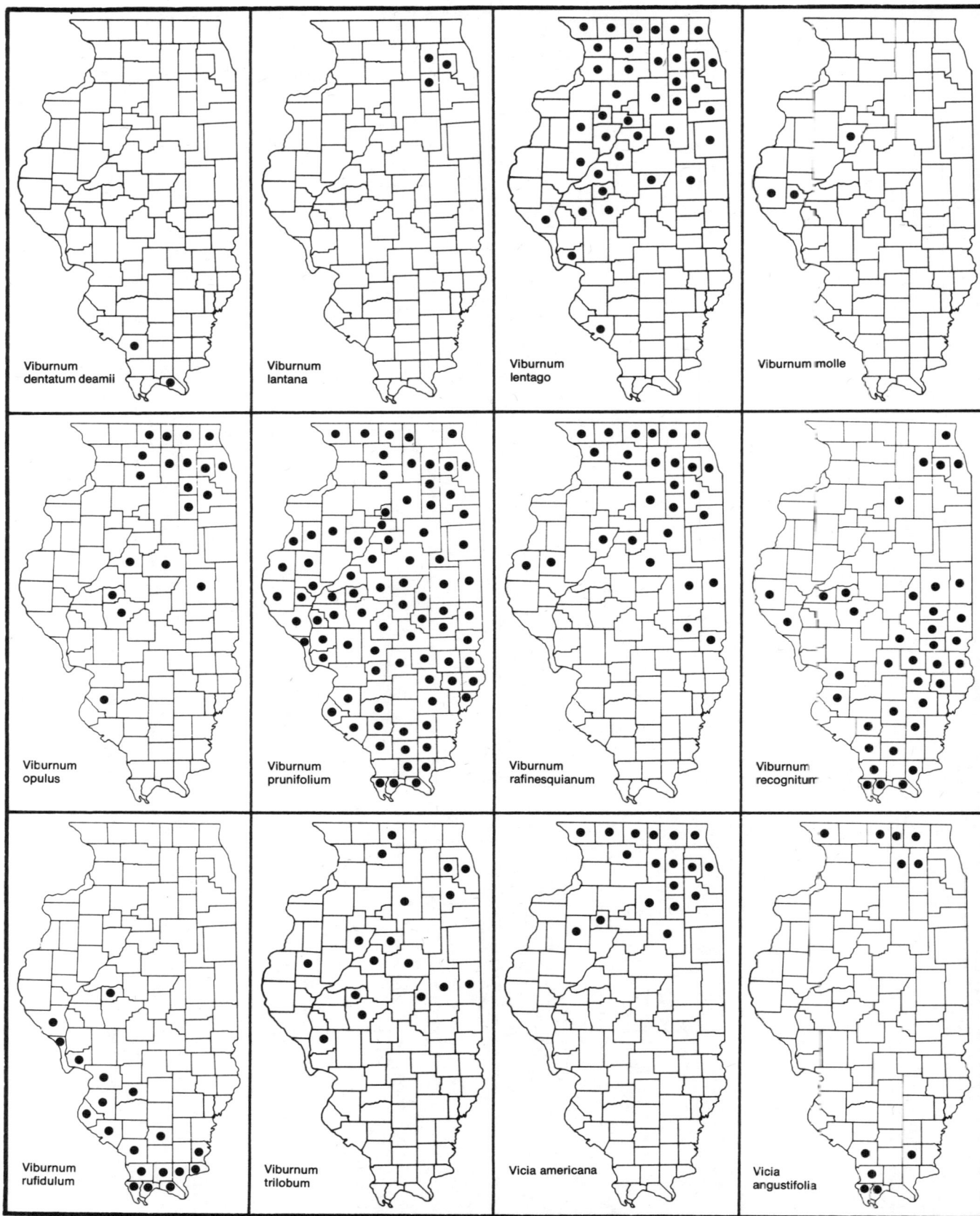

Viburnum
dentatum deamii

Viburnum
lantana

Viburnum
lentago

Viburnum molle

Viburnum
opulus

Viburnum
prunifolium

Viburnum
rafinesquianum

Viburnum
recognitum

Viburnum
rufidulum

Viburnum
trilobum

Vicia americana

Vicia
angustifolia

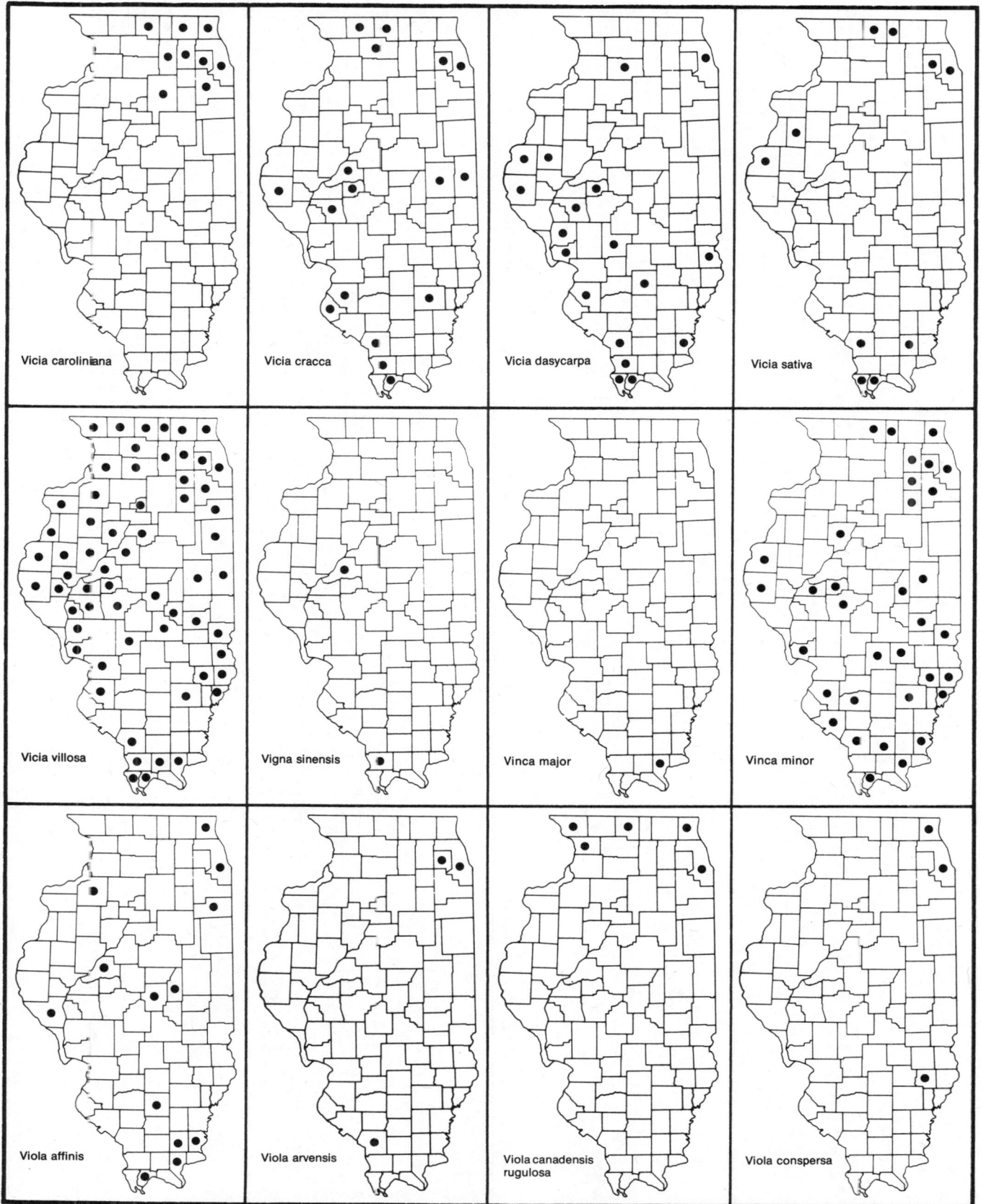

Vicia caroliniana

Vicia cracca

Vicia dasycarpa

Vicia sativa

Vicia villosa

Vigna sinensis

Vinca major

Vinca minor

Viola affinis

Viola arvensis

Viola canadensis
rugulosa

Viola conspersa

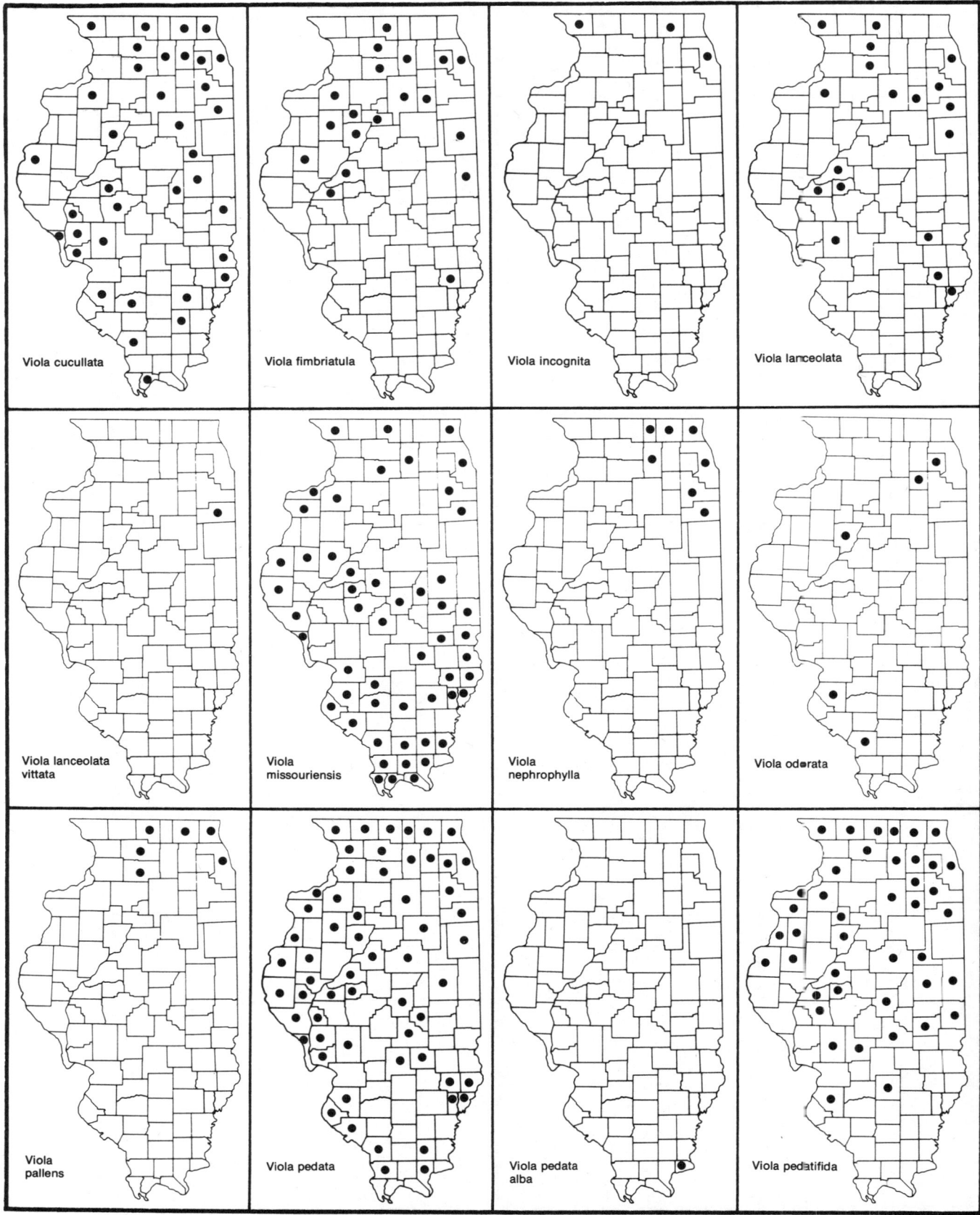

Viola cucullata

Viola fimbriatula

Viola incognita

Viola lanceolata

Viola lanceolata vittata

Viola missouriensis

Viola nephrophylla

Viola odorata

Viola pallens

Viola pedata

Viola pedata alba

Viola pedatifida

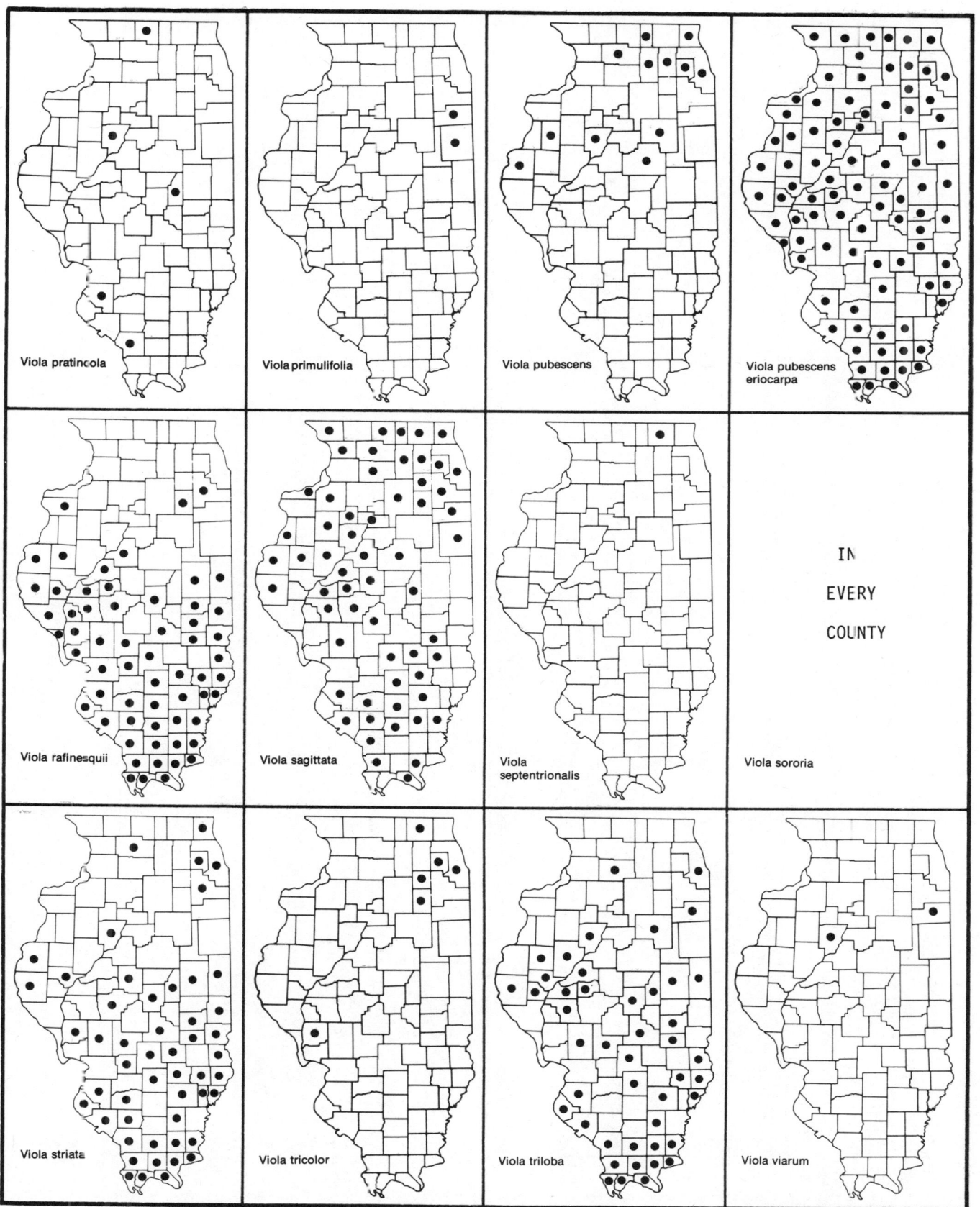

Viola pratincola

Viola primulifolia

Viola pubescens

Viola pubescens
eriocarpa

Viola rafinesquii

Viola sagittata

Viola
septentrionalis

IN
EVERY
COUNTY

Viola sororia

Viola striata

Viola tricolor

Viola triloba

Viola viarum

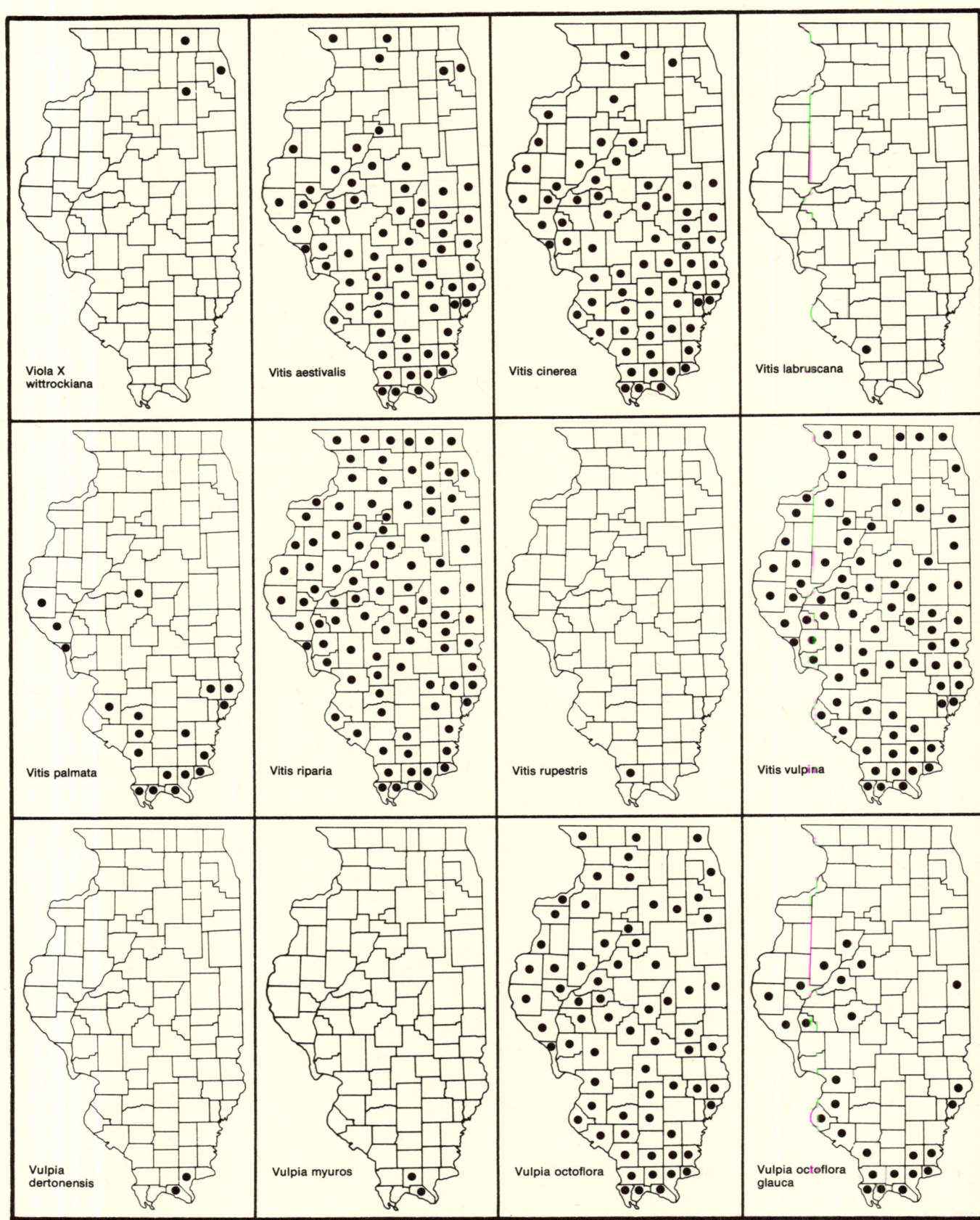

Viola X wittrockiana

Vitis aestivalis

Vitis cinerea

Vitis labruscana

Vitis palmata

Vitis riparia

Vitis rupestris

Vitis vulpina

Vulpia dertonensis

Vulpia myuros

Vulpia octoflora

Vulpia octoflora glauca

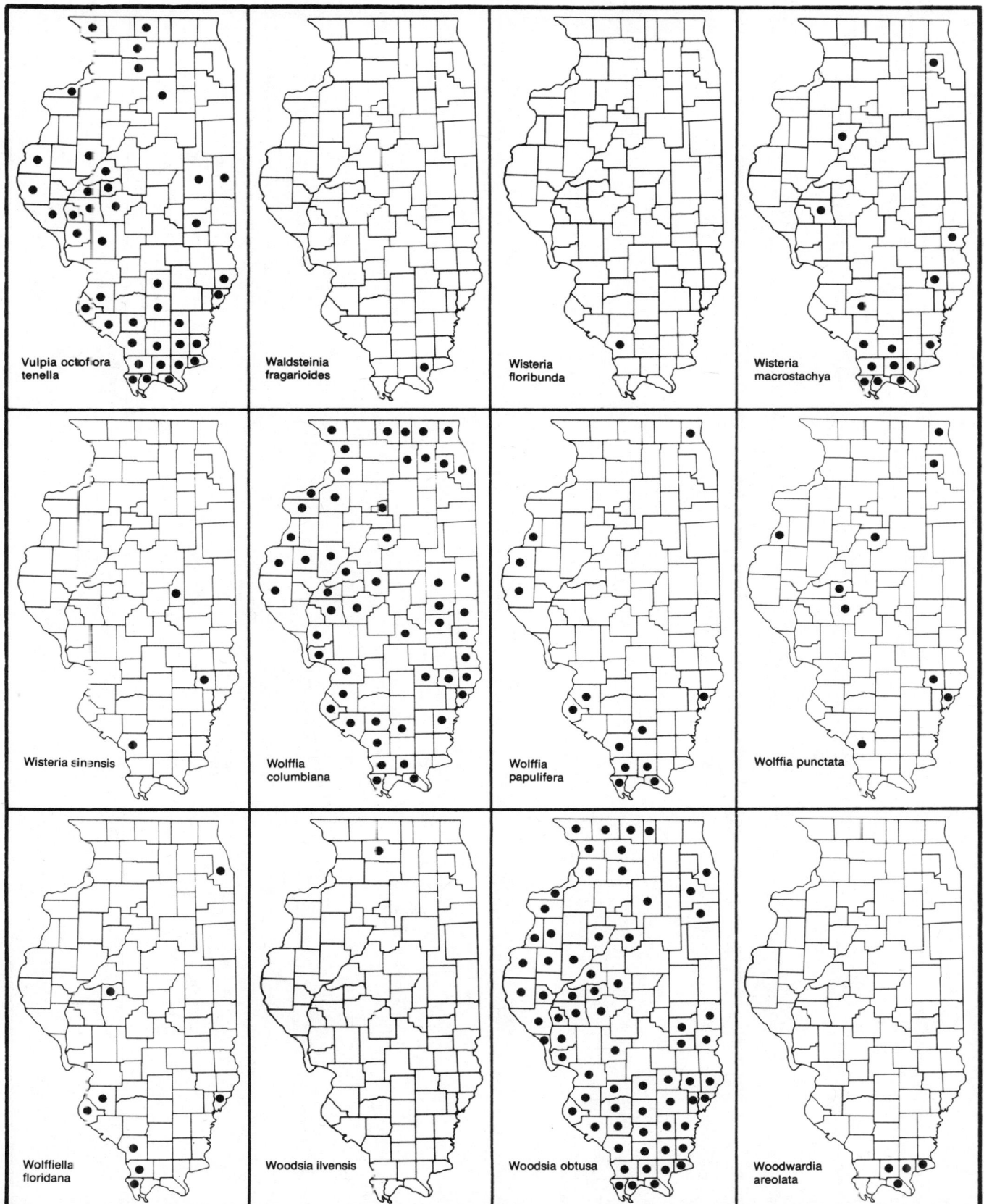

Vulpia octoflora tenella

Waldsteinia fragarioides

Wisteria floribunda

Wisteria macrostachya

Wisteria sinensis

Wolffia columbiana

Wolffia papulifera

Wolffia punctata

Wolffiella floridana

Woodsia ilvensis

Woodsia obtusa

Woodwardia areolata

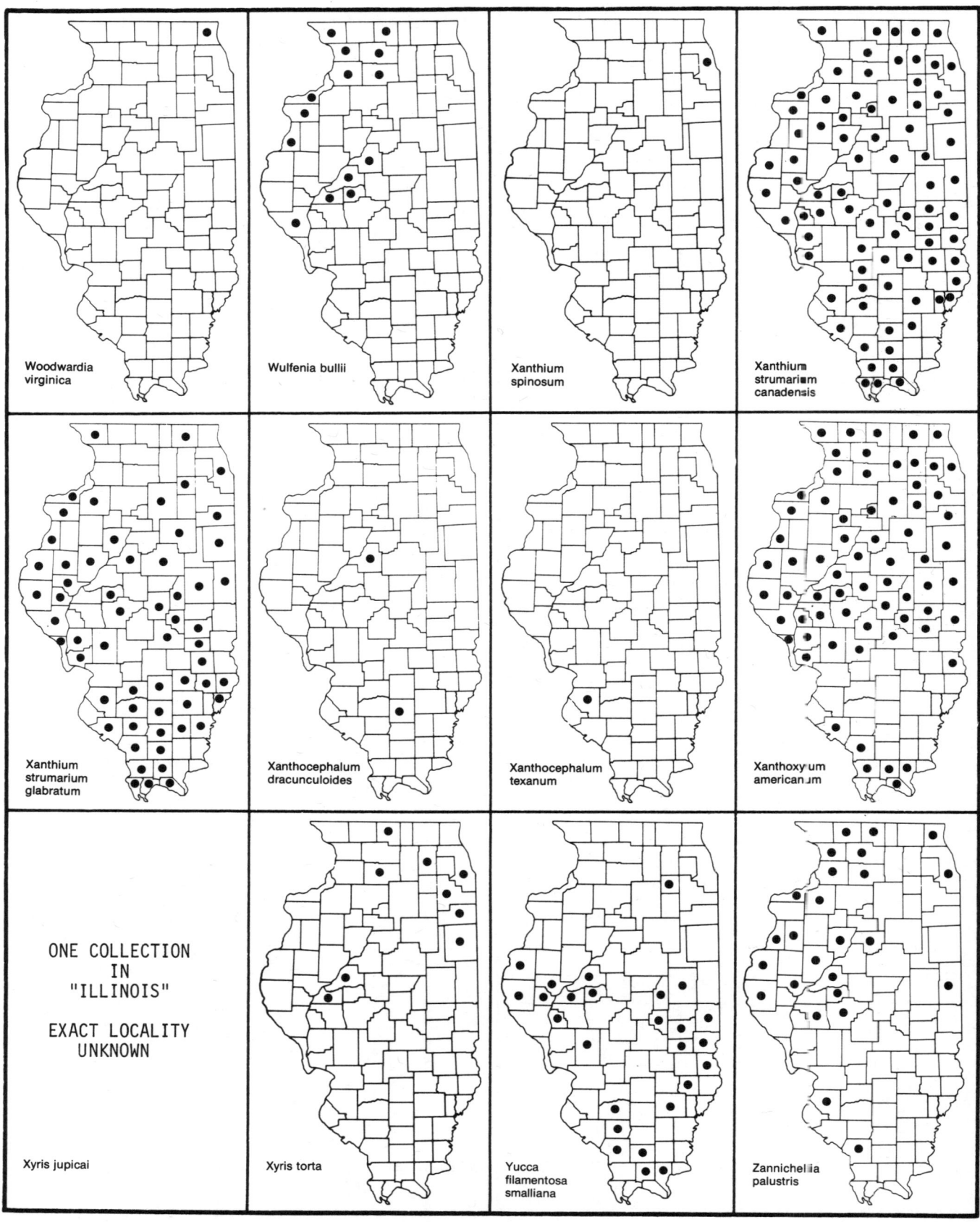

Woodwardia
virginica

Wulfenia bullii

Xanthium
spinosum

Xanthium
strumarium
canadensis

Xanthium
strumarium
glabratum

Xanthocephalum
dracunculoides

Xanthocephalum
texanum

Xanthoxylum
americanum

ONE COLLECTION
IN
"ILLINOIS"

EXACT LOCALITY
UNKNOWN

Xyris jupicai

Xyris torta

Yucca
filamentosa
smalliana

Zannichellia
palustris

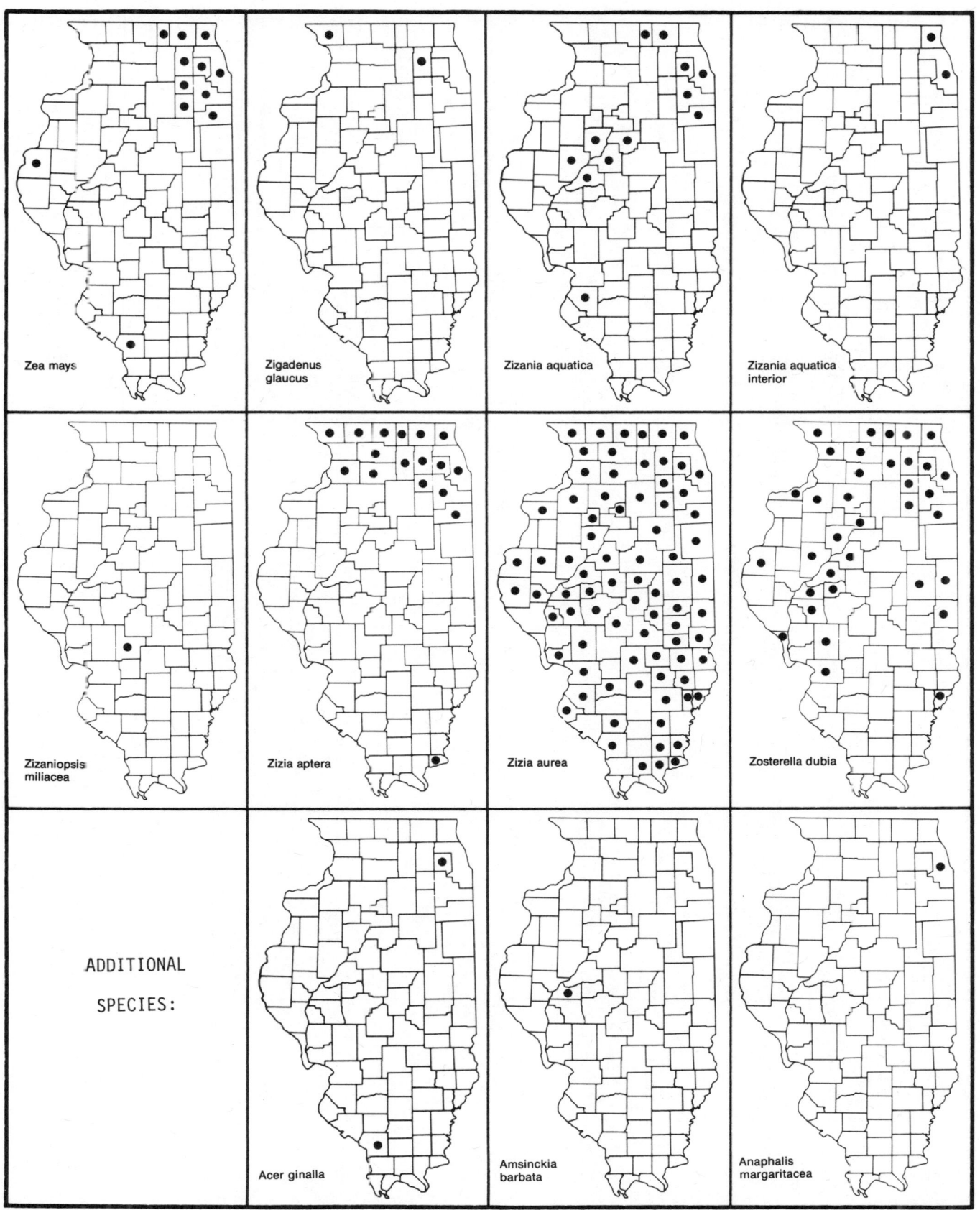

Zea mays

Zigadenus
glaucus

Zizania aquatica

Zizania aquatica
interior

Zizaniopsis
miliacea

Zizia aptera

Zizia aurea

Zosterella dubia

ADDITIONAL

SPECIES:

Acer ginalla

Amsinckia
barbata

Anaphalis
margaritacea

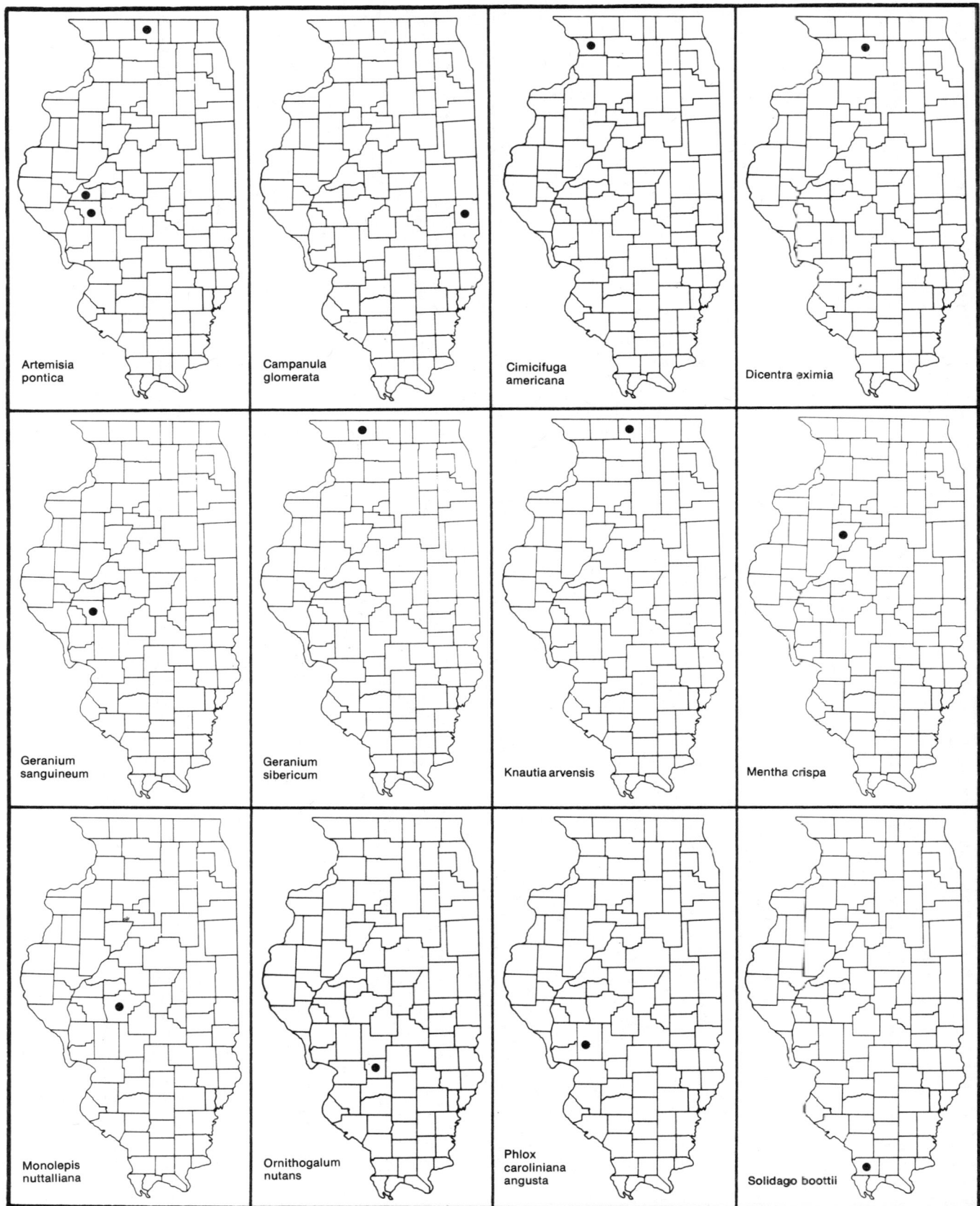

Artemisia
pontica

Campanula
glomerata

Cimicifuga
americana

Dicentra eximia

Geranium
sanguineum

Geranium
sibericum

Knautia arvensis

Mentha crispa

Monolepis
nuttalliana

Ornithogalum
nutans

Phlox
caroliniana
angusta

Solidago boottii

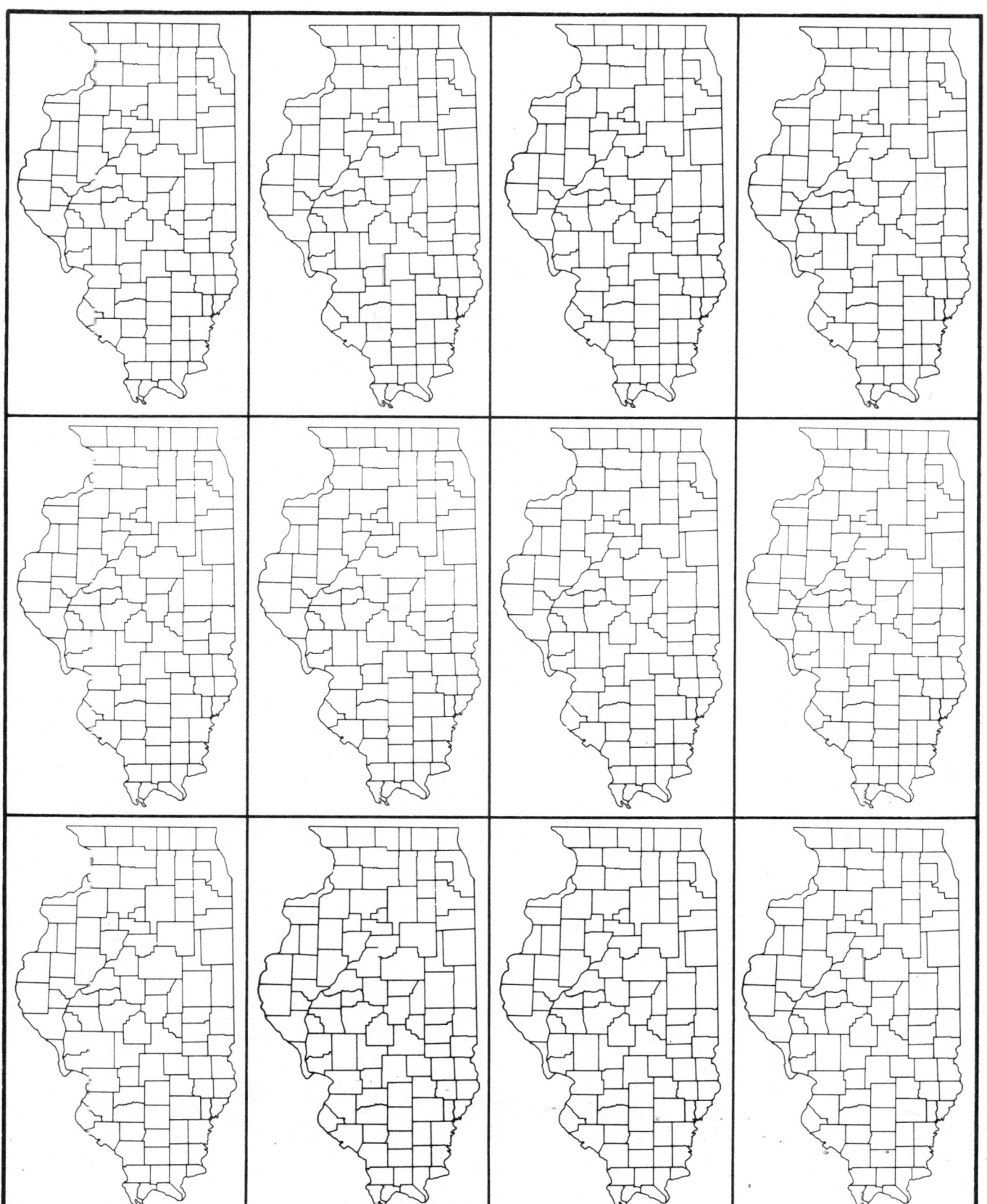

Synonymy

The nomenclature used in this book follows Mohlenbrock (1975). Older floras, such as those by Fernald (1950), Gleason (1952), and Jones (1963), differ in the nomenclature used for certain taxa. In order to equate this earlier nomenclature with that used in this book, the following list of equivalent names is given. The name to the left is that used by either Fernald, Gleason, or Jones, while the name to the right of the equal sign is the one used in this work.

Acer drummondii H. & A. = A. rubrum L. var. drummondii (H. & A.) Sarg.

Acerates angustifolia (Nutt.) Decne. = Asclepias stenophylla Gray

Acerates hirtella Pennell = Asclepias hirtella (Pennell) Woodson

Acerates lanuginosa (Nutt.) Decne. = Asclepias lanuginosa Nutt.

Acerates viridiflora (Raf.) Eat. = Asclepias viridiflora Raf.

Achillea lanulosa Nutt. = A. millefolium L. ssp. lanulosa (Nutt.) Piper

Acnida altissima (Riddell) Riddell = Amaranthus ambigens Standl.

Acnida subnuda (S. Wats.) Standl. = Amaranthus tuberculatus (Moq.) Sauer

Acnida tamariscina (Nutt.) Wood = Amaranthus tamariscinus Nutt.

Actaea alba (L.) Mill. = A. pachypoda Ell.

Actinea herbacea (Greene) B. L. Robins. = Hymenoxys acaulis (Pursh) Parker

Actinomeris alternifolia (L.) DC. = Verbesina alternifolia (L.) Britt.

Aegilops cylindrica Host = Triticum cylindricum (Host) Ces.

Aesculus pavia L. = A. discolor Pursh

Agave virginica L. = Polianthes virginica (L.) Shinners

Agoseris cuspidata (Pursh) D. Dietr. = Microseris cuspidata (Pursh) Sch.-Bip.

Agrimonia mollis (Torr. & Gray) Britt. = A. pubescens Wallr.

Agropyron dasystachyum (Hook.) Scribn. = A. smithii Rydb. var. molle (Scribn. & Smith) Jones

Agropyron molle (Scribn. & Smith) Rydb. = A. smithii Rydb. var. molle (Scribn. & Smith) Jones

Agrostis stolonifera L. = A. alba L.

Alisma triviale Pursh = A. plantago-aquatica L.

Allocarya figurata Piper = Plagiobothrys hirtus (Greene) I. M. Johnston var. figuratus (Piper) I. M. Johnston

Amaranthus arenicola I. M. Johnston = A. torreyi (Gray) Benth.

Ambrosia coronopifolia Torr. & Gray = A. psilostachya DC.

Amelanchier spicata K. Koch = A. humilis Weg.

Ampelamus albidus (Nutt.) Britt. = Cynanchum laeve (Michx.) Pers.

Amphicarpa comosa (L.) G. Don = A. bracteata (L.) Fern. var. comosa (L.) Fern.

Anacharis canadensis (Michx.) Rich. = Elodea canadensis Michx.

Anacharis densa (Planch.) Vict. = Elodea densa (Planch.) Caspary

Anacharis nuttallii Planch. = Elodea nuttallii (Planch.) St. John

Andropogon furcatus Muhl. = A. gerardii Vitman

Andropogon saccharoides Sw. = Bothriochloa saccharoides (Sw.) Rydb.

Andropogon scoparius Michx. = Schizachyrium scoparium (Michx.) Nash

Anemone ludoviciana Nutt. = A. patens L.

Antennaria campestris Rydb. = A. neglecta Greene

Antennaria fallax Greene = A. plantaginifolia (L.) Richards var. ambigens (Greene) Cronq.

Antennaria munda Fern. = A. plantaginifolia (L.) Richards var. ambigens (Greene) Cronq.

Antennaria neodioica Greene = A. neglecta Greene var. attenuata (Fern.) Cronq.

Antennaria parlinii Fern. = A.

plantaginifolia (L.) Richards var.
arnoglossa (Greene) Cronq.
Apocynum pubescens R. Br. = A.
cannabinum L. var. pubescens
(Mitchell) A. DC.
Arabis confinis S. Wats. = A.
drummondii Gray
Arabis perstellata E. L. Br. = A. shortii
(Fern.) Gl
Arabis pycnocarpa Hopkins = A.
hirsuta (L.) Scop. var. pycnocarpa
(Hopkins) Rollins
Arabis virginica (L.) Poir. = Sibara
virginica (L.) Rollins
Argemone intermedia Sweet = A.
albiflora Hornem.
Arisaema atrorubens (Ait.) Blume = A.
triphyllum (L.) Schott
Aristolochia nashii Kearney = A.
serpentaria L. var. hastata (Nutt.)
Duchartre
Armoracia rusticana (Lam.) Gaertn. =
A. lapathifolia Gilib.
Artemisia caudata Michx. = A.
campestris L.
Artemisia dracunculoides Pursh = A.
dracunculus L.
Artemisia glauca Pall. = A.
dracunculus L.
Artemisia gnaphalodes Nutt. = A.
ludoviciana Nutt. var. gnaphalodes
(Nutt.) Torr. & Gray
Asarum acuminatum (Ashe) Bickn. =
A. canadense L.
Asarum reflexum Bickn. = A.
canadense L. var. reflexum (Bickn.)
Robins.
Asclepias angustifolia Nutt. = A.
stenophylla Gray
Asclepiodora viridis (Walt.) Gray =
Asclepias viridis Walt.
Ascyrum multicaule Michx. = A.
hypericoides L. var. multicaule
(Michx.) Fern.
Asplenium cryptolepis Fern. = A.
ruta-muraria L.
Asplenosorus X ebenoides (R. R. Scott)
Wherry = Asplenium X ebenoides R.
R. Scott
Aster chasei G. N. Jones = A. schreberi
Nees
Aster drummondii Lindl. = A.
sagittifolius Wedem. var.
drummondii (Lindl.) Shinners
Aster exiguus (Fern.) Rydb. = A.
ericoides L. var. prostratus (Ktze.)
Blake
Aster longifolius Lam. = A. novi-belgii
L.
Aster lucidulus (Gray) Wieg. = A.
puniceus L. var. lucidulus Gray
Aster pringei (Gray) Britt. = A. pilosus
Willd. var. pringlei (Gray) Blake
Astragalus crassicarpus Nutt. = A.
trichocalyx Nutt.

Astragalus mexicanus A. DC. = A.
trichocalyx Nutt.
Athyrium angustum (Willd.) Presl = A.
filix-femina (L.) Roth
Atriplex hastata L. = A. patula L.
Aureolaria flava (L.) Farw. = Gerardia
flava L.
Aureolaria grandiflora (Benth.) Pennell
= Gerardia grandiflora Benth.
Aureolaria pedicularia (L.) Raf. =
Gerardia pedicularia L.

Berula erecta Cowles = B. pusilla
(Nutt.) Fern.
Berula incisa (Torr.) G. N. Jones = B.
pusilla (Nutt.) Fern.
Besseya bullii (Eat.) Rydb. = Wulfenia
bullii (Eat.) Barnh.
Bidens polylepis Blake = B. aristosa L.
var. retrorsa (Sherff) Wunderlin
Bidens tripartita L. = B. connata Muhl.
Boehmeria drummondiana Wedd. = B.
cylindrica (L.) Sw. var.
drummondiana Wedd.
Boltonia decurrens (Torr. & Gray) Wood
= B. asteroides (L.) L'Hér. var.
decurrens (Torr. & Gray) Engelm.
Boltonia interior (Fern. & Grisc.) G. N.
Jones = B. diffusa Ell.
Boltonia latisquama Gray = B.
asteroides (L.) L'Hér.
Boltonia recognita (Fern. & Grisc.)
G. N. Jones = B. asteroides (L.)
L'Hér.
Botrychium obliquum Muhl. = B.
dissectum Spreng. var. obliquum
(Muhl.) Clute
Brassica alba (L.) Rabenh. = B. hirta
Moench
Brassica campestris L. = B. rapa L.
Breweria pickeringii (Torr.) Gray =
Stylisma pickeringii (Torr.) Gray
Bromus catharticus Vahl = B.
willdenovii Kunth
Bromus latiglumis (Shear) Hitchcock =
B. purgans L.

Callitriche deflexa A. Br. = C. terrestris
Raf.
Calopogon pulchellus (Salisb.) R.
Br. = C. tuberosus (L.) BSP.
Campanula intercedens Witasek = C.
rotundifolia L.
Camptosorus rhizophyllus (L.) Link =
Asplenium rhizophyllum L.
Cardamine arenicola Britt. = C.
parviflora L. var. arenicola (Britt.)
O. E. Schulz
Carex aquatilis Wahl. = C. substricta
(Kükenth.) Mack.
Carex brachyglossa Mack. = C.
annectens Bickn. var. xanthocarpa
(Bickn.) Wieg.
Carex complanata Torr. & Hook. = C.
hirsutella Mack.

Carex eleocharis Bailey = C.
stenophylla Wahlenb. var. enervis
(C. A. Mey.) Kükenth.
Carex flava L. = C. cryptolepis Mack.
Carex muricata L. = C. sterilis Willd.
Celtis pumila Pursh = C. tenuifolia
Nutt.
Cenchrus pauciflorus Benth. = C.
longispinus (Hack.) Fern.
Centaurea vochinensis Bernh. = C.
dubia Suter
Cerastium arvense L. = C. velutinum
Raf.
Cheilanthes vestita (Spreng.) Sw. = C.
lanosa (Michx.) D. C. Eaton
Chenopodium hybridum L. = C.
gigantospermum Aellen
Chenopodium paganum Reich. = C.
bushianum Aellen
Chimaphila corymbosa Pursh = C.
umbellata (L.) Bart. var. cisatlantica
Blake
Chrysopsis camporum Greene =
Heterotheca villosa (Pursh) Shinners
Chrysopsis villosa (Pursh) Nutt. =
Heterotheca villosa (Pursh) Shinners
Cimicifuga cordifolia Pursh = C.
rubifolia Kearney
Circaea latifolia Hill = C. quadrisulcata
(Maxim.) Franch. & Sav.
Cirsium hillii (Canby) Fern. = C.
pumilum (Nutt.) Spreng.
Cirsium setosum (Willd.) Bieb. = C.
arvense (L.) Scop.
Cleome speciosissima Deppe = C.
hassleriana Chod.
Clinopodium arkansanum (Nutt.) G. N.
Jones = Satureja arkansana (Nutt.)
Briq.
Comandra umbellata (L.) Nutt. = C.
richardsiana Fern.
Convolvulus americanus (Sims)
Greene = Calystegia sepium (L.) R.
Br. var. americana (Sims) Mohlenbr.
Convolvulus fraterniflorus Mack. &
Bush = Calystegia sepium (L.) R. Br.
var. fraterniflora (Mack. & Bush)
Shinners
Convolvulus japonicus Thunb. =
Calystegia pubescens Lindl.
Convolvulus pellitus Ledeb. =
Calystegia pubescens Lindl.
Convolvulus repens L. = Calystegia
sepium (L.) R. Br. var. repens (L.)
Mohlenbr.
Convolvulus sepium L. = Calystegia
sepium (L.) R. Br. var. americana
(Sims) Mohlenbr.
Convolvulus spithamaeus L. =
Calystegia spithamaea (L.) Pursh
Conyza canadensis (L.) Cronq. =
Erigeron canadensis L.
Conyza ramosissima Cronq. =
Erigeron divaricatus Michx.
Cornus baileyi Coult. & Evans = C.

stolonifera Michx. var. baileyi (Coult. & Evans) Drescher

Crataegus coccinea L. = C. holmesiana Ashe

Crataegus disperma Ashe = C. cuneiformis (Marsh.) Egglest.

Crataegus macrosperma Ashe = C. lucorum Sarg.

Crataegus margaretta Ashe = C. faxonii Sarg.

Crataegus putnamiana Sarg. = C. corusca Sarg.

Cristatella jamesii Torr. & Gray = Polanisia jamesii (Torr. & Gray) Iltis

Cubelium concolor (Forst.) Raf. = Hybanthus concolor (Forst.) Spreng.

Cynosciadium pinnatum DC. = Limnosciadium pinnatum (DC.) Math. & Const.

Cyperus inflexus Muhl. = C. aristatus Rottb.

Cyperus X mesochorus Geise = C. filiculmis Vahl

Cyperus tenuifolius (Steud.) Dandy = C. densicaespitosus Mattf. & Kükenth.

Cyperus virens Gray = C. pseudovegetus Steud.

Cypripedium parviflorum Salisb. = C. calceolus L.

Dasistoma macrophylla (Nutt.) Raf. = Seymeria macrophylla Nutt.

Datura meteloides DC. = D. innoxia Mill.

Descurainia brachycarpa (Richards.) O. E. Schulz = D. pinnata (Walt.) Britt. var. brachycarpa (Richards.) Fern.

Desmodium glabellum (Michx.) DC. = D. dillenii Darl.

Desmodium longifolium (Torr. & Gray) Smythe = D. cuspidatum (Muhl.) Loud. var. longifolium (Torr. & Gray) Schub.

Desmodium perplexum Schub. = D. dillenii Darl.

Dianthera americana L. = Justicia americana (L.) Vahl

Diplachne acuminata Nash = Leptochloa acuminata (Nash) Mohlenbr.

Diplachne fascicularis (Lam.) Beauv. = Leptochloa fascicularis (Lam.) Gray

Diplachne halei Nash = Leptochloa panicoides (Presl) Hitchc.

Dodecatheon radicatum Greene = D. amethystinum Fassett

Dracocephalum formosius (Lunell) Rydb. = Physostegia speciosa (Sweet) Sweet

Dracocephalum intermedium Nutt. = Physostegia intermedia (Nutt.) Engelm. & Gray

Dracocephalum nuttallii Britt. = Physostegia parviflora Nutt.

Dryopteris austriaca (Jacq.) Woynar = D. X triploidea Wherry

Dryopteris disjuncta (Ledeb.) C. V. Morton = Gymnocarpium dryopteris (L.) Newm.

Dryopteris hexagonoptera (Michx.) Christens. = Thelypteris hexagonoptera (Michx.) Weatherby

Dryopteris noveboracensis (L.) Gray = Thelypteris noveboracensis (L.) Nieuwl.

Dryopteris phegopteris (L.) Christens. = Thelypteris phegopteris (L.) Slosson

Dryopteris spinulosa (O. F. Muell.) Watt = D. carthusiana (Villars) H. P. Fuchs

Dryopteris thelypteris (L.) Gray = Thelypteris palustris Schott var. pubescens (Laws.) Fern.

Echinochloa microstachya (Wieg.) Rydb. = E. pungens (Poir.) Rydb. var. microstachya (Wieg.) Mohl.

Echinodorus parvulus Engelm. = E. tenellus (Mart.) Buchenau

Echinodorus rostratus (Nutt.) Engelm. = E. berteroi (Spreng.) Fassett

Elatine triandra Schk. = E. brachysperma Gray

Eleocharis calva (Gray) Torr. = E. erythropoda Steud.

Eleocharis compressa Sull. = E. elliptica Kunth var. compressa (Sull.) Drap. & Mohlenbr.

Eleocharis engelmannii Steud. = E. obtusa (Willd.) Schult. var. detonsa (Gray) Drap. & Mohlenbr.

Eleocharis geniculata (L.) Roem. & Schultes = E. caribaea (Rottb.) Blake

Eleocharis ovata (Roth) Roem. & Schultes = E. obtusa (Willd.) Schult. var. ovata (Roth) Drap. & Mohlenbr.

Eleocharis palustris (L.) Roem. & Schultes = E. smallii Britt.

Elodea occidentalis (Pursh) St. John = E. nuttallii (Planch.) St. John

Elymus macounii Vasey = Agrohordeum X macounii (Vasey) Lepage

Elymus mollis Trin. = E. arenarius L.

Epilobium glandulosum Lehm. = E. adenocaulon Haussk.

Epipactis latifolia (L.) Crantz = E. helleborine (L.) Crantz

Eragrostis megastachya (Koel.) Link = E. cilianensis (All.) Mosher

Eragrostis mexicana (Hornem.) Link = E. neomexicana Vasey

Erysimum arkansanum Nutt. = E. capitatum (Dougl.) Greene

Erysimum asperum (Nutt.) DC. = E. capitatum (Dougl.) Greene

Eulalia viminea (Trin.) Kuntze = Microstegium vimineum (Trin.) A. Camus

Euphorbia dentata Michx. = Poinsettia dentata (Michx.) Kl. & Garcke

Euphorbia dcryosperma Fisch. & Mey. = E. spathulata Lam.

Euphorbia geyeri Engelm. = Chamaesyce geyeri (Engelm.) Small

Euphorbia gyptosperma Engelm. = Chamaesyce glyptosperma (Engelm.) Small

Euphorbia heterophylla L. = Poinsettia cyanthophora (Murr.) Kl. & Garcke

Euphorbia humistrata Engelm. = Chamaesyce humistrata (Engelm.) Small

Euphorbia maculata L. = Chamaesyce maculata (L.) Small

Euphorbia polygonifolia L. = Chamaesyce polygonifolia (L.) Small

Euphorbia serpens HBK. = Chamaesyce serpens (HBK.) Small

Euphorbia serpyllifolia Pers. = Chamaesyce serpyllifolia (Pers.) Small

Euphorbia supina Raf. = Chamaesyce supina (Raf.) Moldenke

Euphorbia vermiculata Raf. = Chamaesyce vermiculata (Raf.) House

Fagopyrum sagittatum Gilib. = F. esculentum Moench

Festuca dertonensis (All.) Aschers. & Graebn. = Vulpia dertonensis All.

Festuca elatior L. = F. pratensis Huds.

Festuca myuros L. = Vulpia myuros (L.) K. Gmel.

Festuca octoflora Walt. = Vulpia octoflora (Walt.) Rydb.

Fimbristylis annua (All.) Roem. & Schultes = F. baldwiniana (Schult.) Torr.

Fimbristylis caroliniana (Lam.) Fern. = F. puberula (Michx.) Vahl var. drummondi (Boeckl.) Ward

Fimbristylis drummondii Boeckl. = F. puberula (Michx.) Vahl var. drummondi (Boeckl.) Ward

Franseria discolor Nutt. = Ambrosia tomentosa Nutt.

Frasera caroliniensis Walt. = Swertia caroliniensis (Walt.) Kuntze

Fraxinus lanceolata Borkh. = F. pennsylvanica Marsh. var. subintegerrima (Vahl) Fern.

Froelichia campestris Small = F. floridana (Nutt.) Moq. var. campestris (Small) Fern.

Gaillardia lanceolata Michx. = G. aestivalis (Walt.) Rock

Gaillardia lutea Greene = G. aestivalis (Walt.) Rock

Gentiana flavida Gray = G. alba Gray

Gentiana puberula Michx. = G. puberulenta Pringle

Geum aleppicum Jacq. = G. strictum Ait.

Glechoma heterophylla Waldst. & Kit. = G. hederacea L. var. micrantha Moricand

Glyceria pallida (Torr.) Trin. = Puccinellia pallida (Torr.) Clausen

Gonolobus decipiens (Alex.) Perry = Matelea decipiens (Alex.) Woodson

Gonolobus gonocarpos (Walt.) Perry = Matelea gonocarpa (Walt.) Shinners

Gonolobus obliquus (Jacq.) Schultes = Matelea obliqua (Jacq.) Woodson

Gutierrezia dracunculoides (DC.) Blake = Xanthocephalum dracunculoides (DC.) Shinners

Habenaria bracteata (Muhl.) R. Br. = H. viridis (L.) R. Br. var. bracteata (Muhl.) Gray

Habenaria huronensis (Nutt.) Spreng. = H. hyperborea (L.) R. Br. var. huronensis (Nutt.) Farw.

Helenium rudiflorum Nutt. = H. flexuosum Raf.

Helenium tenuifolium Nutt. = H. amarum (Raf.) Rock

Heleochloa schoenoides (L.) Host = Crypsis schoenoides (L.) Lam.

Helianthus laetiflorus Pers. = H. rigidus (Cass.) Desf.

Helianthus tomentosus Michx. = H. tuberosus L. var. subcanescens Gray

Hemerocallis flava L. = H. lilio-asphodelus L.

Hemicarpha drummondii Nees = Scirpus micranthus Vahl var. drummondii (Nees) Mohlenbr.

Hemicarpha micrantha (Vahl) Pax = Scirpus micranthus Vahl

Hepatica acutiloba DC. = H. nobilis Schreb. var. acuta (Pursh) Steyerm.

Hepatica americana (DC.) Ker = H. nobilis Schreb. var. obtusa (Pursh) Steyerm.

Heracleum lanatum Michx. = H. maximum Bartr.

Heranthera dubia (Jacq.) MacM. = Zosterella dubia (Jacq.) Small

Heterotheca subaxillaris (Lam.) Britt. & Rusby = H. latifolia Buckl.

Heuchera americana L. = H. hirsuticaulis (Wheelock) Rydb.

Hibiscus moscheutos L. = H. palustris L.

Hosta japonica (Thunb.) Voss = H. lancifolia (Thunb.) Engl.

Houstonia canadensis Willd. = H. longifolia Gaertn. var. ciliolata (Torr.) Torr.

Houstonia lanceolata (Poir.) Britt. = H. purpurea L. var. calycosa Gray

Houstonia patens Ell. = H. pusilla Schoepf

Houstonia tenuifolia Nutt. = H. longifolia Gaertn. var. tenuifolia (Nutt.) Wood

Humulus americanus Nutt. = H. lupulus L.

Hydrolea affinis Gray = H. uniflora Raf.

Hypericum ascyron L. = H. pyrimidatum Ait.

Hypericum prolificum L. = H. spathulatum (Spach) Steud.

Hypericum pseudomaculatum Bush = H. punctatum Lam. var. pseudomaculatum (Bush) Fern.

Hypericum tubulosum Walt. = Triadenum tubulosum (Walt.) Gl.

Hystrix patula Moench = Elymus hystrix L.

Iris virginica L. = I. shrevei Small

Juncus richardsonianus Schult. = J. alpinus Vill. var. rariflorus Hartm.

Juniperus canadensis Burgsdorf = J. communis L.

Jussiaea diffusa Forsk. = J. repens L. var. glabrescens Ktze.

Justicia humilis Michx. = J. ovata (Walt.) Lindau

Koeleria cristata Pers. = K. macrantha (Ledeb.) Spreng.

Kuhnia eupatorioides L. = Brickellia eupatorioides (L.) Shinners

Lactuca pulchella (Pursh) DC. = L. tatarica (L.) C. A. Mey.

Lactuca scariola L. = L. serriola L.

Lappula myosotis Moench = L. echinata Gilib.

Lappula occidentalis (S. Wats.) Greene = L. redowskii (Hornem.) Greene var. occidentalis (S. Wats.) Rydb.

Lathyrus japonicus L. = L. maritimus (L.) Bigel.

Lechea leggettii Britt. & Holl. = L. pulchella Raf.

Lechea villosa Ell. = L. mucronata Raf.

Leucospora multifida (Michx.) Nutt. = Conobea multifida (Michx.) Benth.

Lilium tigrinum L. = L. lancifolium Thunb.

Lilium umbellatum Pursh = L. philadelphicum L. var. andinum (Nutt.) Ker

Linaria texana Scheele = L. canadensis (L.) Dumort. var. texana (Scheele) Pennell

Lithospermum croceum Fern. = L. caroliniense (Walt.) MacM.

Lobelia leptostachys A. DC. = L. spicata Lam. var. leptostachys (A. DC.) Mack. & Bush

Lophotocarpus calycinus (Engelm.) J. G. Sm. = Sagittaria calycina Engelm.

Lotus americanus (Nutt.) Bisch. = Hosackia americana (Nutt.) Piper

Lotus purshianus Clem. & Clem. = Hosackia americana (Nutt.) Piper

Luzula echinata (Small) F. J. Herm. = L. multiflora (Retz.) Lejeune var. echinata (Small) Mohl.

Luzula saltuensis Fern. = L. acuminata Raf.

Lycopodium obscurum L. = L. dendroideum Michx.

Malaxis brachypoda (Gray) Fern. = M. monophylla (L.) Sw. var. brachypoda (Gray) F. Morris

Malva crispa L. = M. verticillata L. var. crispa L.

Malvastrum angustum Gray = Sphaeralcea angusta (Gray) Fern.

Mecardonia acuminata (Walt.) Small = Bacopa acuminata (Walt.) Robins.

Melanthera hastata Michx. = M. nivea (L.) Small

Megalodonta beckii (Torr.) Greene = Bidens beckii Torr.

Mentha canadensis L. = M. arvensis L. var. villosa (Benth.) S. R. Stewart

Mimulus geyeri Torr. = M. glabratus HBK.

Moldavica parviflora (Nutt.) Adans. = Dracocephalum parviflorum Nutt.

Monarda russeliana Nutt. = M. bradburiana Beck

Monotropa lanuginosa Michx. = M. hipopithys L.

Muhlenbergia brachyphylla Bush = M. bushii Pohl

Muhlenbergia glabriflora Scribn. = M. glabrifloris Scribn.

Muscari racemosum (L.) Mill. = M. atlanticum Boiss. & Reut.

Myosotis macrosperma Engelm. = M. virginica (L.) BSP. var. macrosperma (Engelm.) Fern.

Myosotis verna Nutt. = M. virginica (L.) BSP.

Myrica asplenifolia L. = Comptonia peregrina (L.) Coult.

Neobeckia aquatica (Eat.) Greene = Armoracia aquatica (Eat.) Wieg.

Nuphar advena (Ait.) Ait. f. = N. luteum L. ssp. macrophyllum (Small) Beal

Nuphar variegatum Engelm. = N. luteum L. ssp. variegatum (Engelm.) Beal

Oenothera albicaulis Nutt. = O. nuttallii Sweet

Oenothera parviflora L. = O. cruciata Nutt.

Oenothera strigosa (Rydb.) Mack. & Bush = O. biennis L. var. canescens Torr. & Gray

Opuntia humifusa Raf. = O. compressa (Salisb.) Macbr.

Opuntia rafinesquii Engelm. = O. compressa (Salisb.) Macbr.

Oxalis cymosa Small = O. stricta L.

Oxalis europaea Jord. = O. stricta L.

Oxybaphus albidus (Walt.) Sweet = Mirabilis albida (Walt.) Heimerl

Oxybaphus hirsutus (Pursh) Sweet = Mirabilis hirsuta (Pursh) MacM.

Oxybaphus linearis (Pursh) Robins. = Mirabilis linearis (Pursh) Heimerl

Oxybaphus nyctagineus (Michx.) Sweet = Mirabilis nyctaginea (Michx.) MacM.

Oxycoccus macrocarpus (Ait.) Pursh = Vaccinium macrocarpon Ait.

Panicum agrostoides Spreng. = P. rigidulum Bosc

Panicum helleri Nash = P. oligosanthes Schult. var. helleri (Nash) Fern.

Panicum implicatum Scribn. = P. lanuginosum Ell. var. implicatum (Scribn.) Fern.

Panicum lindheimeri Nash = P. lanuginosum Ell. var. lindheimeri (Nash) Fern.

Panicum pseudopubescens Nash = P. villosissimum Nash var. pseudopubescens (Nash) Fern.

Panicum scribnerianum Nash = P. oligosanthes Schult. var. scribnerianum (Nash) Fern.

Panicum tennesseense Ashe = P. lanuginosum Ell.

Panicum xalapense HBK. = P. laxiflorum Lam.

Parthenocissus inserta (Kerner) K. Fritsch = P. vitacea (Knerr) Hitchc.

Paspalum circulare Nash = P. laeve Michx.

Paspalum geminum Nash = P. pubiflorum Rupr. var. glabrum (Vasey) Vasey

Paspalum glabratum (Engelm.) Mohr = P. floridanum Michx.

Paspalum pubescens Muhl. = P. ciliatifolium Michx.

Paspalum stramineum Nash = P. ciliatifolium Michx.

Phragmites communis Trin. = P. australis Trin.

Phyla cuneifolia (Torr.) Greene = Lippia cuneifolia (Torr.) Standl.

Phyla lanceolata (Michx.) Greene = Lippia lanceolata Michx.

Poinsettia heterophylla (L.) Kl. & Garcke = P. cyanthophora (Murr.) Kl. & Garcke

Polanisia graveolens Raf. = P. dodecandra (L.) DC.

Polanisia trachysperma Torr. & Gray = P. dodecandra (L.) DC. ssp. trachysperma (Torr. & Gray) Iltis

Polygala ambigua Nutt. = P. verticillata L. var. ambigua (Nutt.) Wood

Polygonum buxiforme Small = P. aviculare L. var. littorale (Link) W. D. J. Koch

Polygonum dumetorum L. = P. cristatum Engelm. & Gray

Polygonum fluitans Eaton = P. amphibium L. var. stipulaceum (Coleman) Fern.

Polygonum longisetum DeBruyn = P. cespitosum Blum var. longisetum (DeBruyn) Steward

Polygonum natans Eaton = P. amphibium L. var. stipulaceum (Coleman) Fern.

Polygonum ramosissimum Michx. = P. exsertum Small

Polygonum tomentosum Schrank = P. scabrum Moench

Polypodium virginianum L. = P. vulgare L.

Populus X jackii Sarg. = P. balsamifera L.

Potamogeton americanus Cham. & Schlecht. = P. nodosus Poir.

Potamogeton berchtoldii Fieber = P. pusillus L.

Potentilla monspeliensis L. = P. norvegica L.

Prunus lanata (Sudw.) Mack. & Bush = P. americana Marsh. var. lanata Sudw.

Prunus pumila L. = P. susquehanae Willd.

Pteretis pensylvanica (Willd.) Fern. = Matteuccia struthiopteris (L.) Todaro

Pycnanthemum flexuosum [Walt.] BSP. = P. tenuifolium Schrad.

Pyrola rotundifolia L. = P. americana Sweet

Pyrus americana (Marsh.) DC. = Sorbus americana (Marsh.) DC.

Pyrus angustifolia Ait. = Malus angustifolia (Ait.) Michx.

Pyrus aucuparia (L.) Gaertn. = Sorbus aucuparia L.

Pyrus coronaria L. = Malus coronaria (L.) Mill.

Pyrus floribunda Lindl. = Aronia prunifolia (Marsh.) Rehd.

Pyrus ioensis (Wood) Bailey = Malus ioensis (Wood) Britt.

Pyrus malus L. = Malus pumila Mill.

Pyrus melanocarpa (Michx.) Willd. = Aronia melanocarpa (Michx.) Ell.

Pyrus X soulardii Bailey = Malus X soulardii (Bailey) Britt.

Quamoclit coccinea (L.) Moench = Ipomoea coccinea L.

Ranunculus aquatilis L. = R. trichophyllus Chaix

Ranunculus circinatus Sibth. = R. longirostris Godr.

Ranunculus purshii Richards = R.

gmelinii DC. var. hookeri (D. Don) L. Benson

Rhododendron nudiflorum (L.) Torr. = R. periclymenoides (Michx.) Shinners

Rhododendron roseum (Loisel.) Rehd. = R. prinophyllum (Small) Millais

Rhus arenaria (Greene) G. N. Jones = R. aromatica Ait. var. arenaria (Greene) Fern.

Rhus radicans L. = Toxicodendron radicans (L.) Kuntze

Rhus trilobata Nutt. = R. aromatica Ait. var. arenaria (Greene) Fern.

Rhus vernix L. = Toxicodendron vernix (L.) Kuntze

Rorippa hispida (Desv.) Britt. = R. islandica (Oeder) Borbas var. hispida (Desv.) Britt.

Rorippa obtusa (Nutt.) Britt. = R. truncata (Jepson) Stuckey

Rosa arkansana Porter = R. suffulta Greene

Rubus ostryifolius Rydb. = R. argutus Link

Rosa spinosissima L. = R. pimpinellifolia L.

Rubus setosus Bigel. = R. schneideri Bailey

Rudbeckia missouriensis Engelm. = R. fulgida Ait. var. missouriensis (Engelm.) Cronq.

Rudbeckia serotina Nutt. = R. hirta L.

Rudbeckia sullivantii Boynt. & Beadle = R. fulgida Ait. var. sullivantii (Boynt. & Beadle) Cronq.

Rumex fueginus Phil. = R. maritimus L. var. fueginus (Phil.) Dusén

Rumex triangulivalvis (Danser) Rech. f. = R. mexicanus Meisn.

Sagittaria australis (J. G. Sm.) Small = S. longirostra (Micheli) J. G. Sm.

Sagittaria engelmanniana J. G. Sm. = S. brevirostra Mack. & Bush

Sagittaria montevidensis Cham. & Schlecht. = S. calycina Engelm.

Salix cordata Michx. = S. syrticola Fern.

Salix gracilis Anderss. = S. petiolaris Sm.

Salix tristis Ait. = S. humilis Marsh. var. microphylla (Anderss.) Fern.

Salix vitellina L. = S. alba L. var. vitellina (L.) Stokes

Salsola pestifer Nels. = S. kali L. var. tenuifolia Tausch

Salvia pitcheri Torr. = S. azurea Lam. var. grandiflora Benth.

Satureja glabella (Michx.) Briq. = S. arkansana (Nutt.) Briq.

Scheuchzeria americana (Fern.) G. N. Jones = S. palustris L. var. americana Fern.

Schrankia nuttallii (DC.) Standl. = S. uncinata Willd.

Scirpus eriophorum Michx. = S. cyperinus (L.) Kunth

Scirpus lineatus Michx. = S. pendulus Muhl.

Scirpus maritimus L. = S. paludosus A. Nels.

Scirpus rubrotinctus Fern. = S. microcarpus Presl

Scutellaria australis (Fassett) Epling = S. parvula Michx. var. australis Fassett

Scutellaria galericulata L. = S. epilobiifolia Muhl.

Scutellaria leonardii Epling = S. parvula Michx. var. leonardii (Epling) Fern.

Sedum telephium L. = S. purpureum (L.) Link

Sedum triphyllum (Haw.) S. F. Gray = S. purpureum (L.) Link

Serinia oppositifolia (Raf.) Kuntze = Krigia oppositifolia Raf.

Setaria glauca (L.) Beauv. = S. lutescens (Weigel) Hubb

Sisyrinchium bermudiana L. = S. angustifolium Mill.

Sisyrinchium graminoides Bickn. = S. angustifolium Mill.

Smilax tamnoides L. = S. hispida Muhl.

Solanum nigrum L. = S. americanum Mill.

Solidago altissima L. = S. canadensis L.

Solidago glaberrima Martens = S. missouriensis Nutt.

Solidago gymnospermoides (Greene) Fern. = S. graminifolia (L.) Salisb. var. gymnospermoides (Greene) Croat

Solidago hirtella (Greene) Bush = S. graminifolia (L.) Salisb.

Solidago hispida Muhl. = S. bicolor L. var. concolor Torr. & Gray

Solidago latifolia L. = S. flexicaulis L.

Solidago media (Greene) Bush = S. graminifolia (L.) Salisb.

Solidago remota (Greene) Friesner = S. graminifolia (L.) Salisb. var. remota (Greene) Harris

Sonchus uliginosus Bieb. = S. arvensis L. var. glabrescens Guenth., Gram., & Winn

Sorghum vulgare Pers. = S. bicolor (L.) Moench

Sphenopholis intermedia (Rydb.) Rydb. = S. obtusata (Michx.) Scribn. var. major (Torr.) Erdman

Spiranthes grayi Ames = S. tuberosa Raf.

Stachys arenicola Britt. = S. plaustris L. var. homotricha Fern.

Stachys aspera Michx. = S. hyssopfolia Michx. var. ambigua Gray

Stachys hispida Pursh = S. tenuifolia Willd. var. hispida (Pursh) Fern.

Stellaria aquatica (L.) Scop. = Myosoton aquaticum (L.) Moench

Stylisma pattersonii (Fern. & Schub.) G. N. Jones = S. pickeringii (Torr.) Gray

Symphoricarpos rivularis Suksd. = S. albus (L.) Blake

Synthyris bullii (Eat.) Heller = Wulfenia bullii (Eat.) Barnh.

Taraxacum erythrospermum Andrz. = T. laevigatum (Willd.) DC.

Teucrium occidentale Gray = T. canadense L. var. occidentale (Gray) McClintock & Epling

Thalictrum hypoglaucum Rydb. = T. dasycarpum Fisch. & Lall. var. hypoglaucum (Rydb.) Boivin

Tilia neglecta Spach = T. americana L. var. neglecta (Spach) Fosberg

Tomanthera auriculata (Michx.) Raf. = Gerardia auriculata Michx.

Torilis arvensis Link = T. japonica (Houtt.) DC.

Tovara virginiana (L.) Raf. = Polygonum virginianum L.

Tragopogon major Jacq. = T. dubius Scop.

Trifolium agrarium L. = T. aureum Pollich

Trifolium procumbens L. = T. campestre Schreb.

Trillium gleasonii Fern. = T. flexipes Raf.

Triodia flava (L.) Smyth = Tridens flavus (L.) Hitchcock

Triodia stricta (Nutt.) Benth. = Tridens strictus (Nutt.) Nash

Uniola latifolia Michx. = Chasmanthium latifolium (Michx.) Yates

Urtica gracilis Ait. = U. dioica L.

Urtica procera Muhl. = U. dioica L.

Vaccinium neglectum (Small) Fern. = V. stamineum L.

Valeriana edulis Nutt. = V. ciliata Torr. & Gray

Vernonia altissima Nutt. = V. gigantea (Walt.) Trel.

Vernonia crinita Raf. = V. arkansana DC.

Veronica catenata Pennell = V. comosa Richter

Veronica didyma Tenore = V. polita Fries

Veronica salina Schur. = V. comosa Richter

Viola blanda Willd. = V. macloskeyi Lloyd ssp. pallens (Banks) Brainerd

Viola eriocarpa Schw. = V. pubescens Ait. var. eriocarpa (Schwein.) Russell

Viola falcata Greene = V. triloba Schwein.

Viola kitaibeliana Roem. & Schultes = V. rafinesquii Greene

Vitis lincecumii Buckl. = V. aestivalis Michx.

Viola pallens (Banks) Brainerd = V. macloskeyi Lloyd ssp. pallens (Banks) Brainerd

Viola pennsylvanica Michx. = V. pubescens Ait. var. eriocarpa (Schwein.) Russell

Viola priceana Pollard = V. pratincola Greene f. albiflora (Glover) Mohl.

Viola rugulosa Greene = V. canadensis L. var. rugulosa (Greene) C. L. Hitchcock

Viola tricolor L. = V. X wittrockiana Hort.

Viola vittata Greene = V. lanceolata L. var. vittata (Greene) Russell

Xanthium chasei Fern. = X. strumarium L. var. glabratum (DC.) Cronq.

Xanthium chinense Mill. = X. strumarium L. var. glabratum (DC.) Cronq.

Xanthium commune Britt. = X. strumarium L. var. canadensis (Mill.) Torr. & Gray

Xanthium italicum Moretti = X. strumarium L. var. canadensis (Mill.) Torr. & Gray

Xanthium pennsylvanicum Mill. = X. strumarium L. var. glabratum (DC.) Cronq.

Xyris caroliniana Walt. = X. jupicai L. Rich

Phylogenetic List of Illinois Vascular Plants

Equisetaceae

Equisetum pratense Ehrh.
Equisetum arvense L.
Equisetum scirpoides Michx.
Equisetum variegatum Schleich.
Equisetum X trachyodon A. Br.
Equisetum X nelsonii (A. A. Eaton)
  Schaffner
Equisetum palustre L.
Equisetum X litorale Kuhl.
Equisetum hyemale L. var. affine
  (Engelm.) A. A. Eaton
Equisetum laevigatum A. Br.
Equisetum X ferrissii Clute
Equisetum fluviatile L.

Lycopodiaceae

Lycopodium porophilum Lloyd &
  Underw.
Lycopodium lucidulum Michx.
Lycopodium lucidulum Michx. var.
  tryonii Mohlenbrock
Lycopodium adpressum Lloyd &
  Underw.
Lycopodium inundatum L.
Lycopodium dendroideum Michx.
Lycopodium flabelliforme (Fern.)
  Blanch.
Lycopodium X habereri House
Lycopodium clavatum L.
Lycopodium clavatum L. var.
  megastachyon Grev. & Hook.

Selaginellaceae

Selaginella apoda (L.) Fern.
Selaginella rupestris (L.) Spring.

Isoetaceae

Isoetes melanopoda Gay & Dur.
Isoetes engelmannii A. Br.

Ophioglossaceae

Botrychium multifidum (Gmel.) Rupr.
  ssp. silaifolium (Presl) Clausen
Botrychium dissectum Spreng.
Botrychium dissectum Spreng. var.
  obliquum (Muhl.) Clute
Botrychium biternatum (Sav.) Underw.
Botrychium simplex E. Hitchc.
Botrychium matricariaefolium A. Br.

Botrychium virginianum (L.) Sw.

Ophioglossum vulgatum L. var.
  pycnostichum Fern.
Ophioglossum vulgatum L. var.
  pseudopodum (Blake) Farw.
Ophioglossum engelmannii Prantl

Osmundaceae

Osmunda regalis L. var. spectabilis
  (Willd.) Gray
Osmunda cinnamomea L.
Osmunda claytoniana L.

Hymenophyllaceae

Trichomanes boschianum Sturm

Polypodiaceae

Dennstaedtia punctilobula (Michx.)
  Moore

Adiantum pedatum (Tourn.) L.

Pteridium aquilinum (L.) Kuhn var.
  latiusculum (Desv.) Underw.
Pteridium aquilinum (L.) Kuhn var.
  pseudocaudatum (Clute) Heller

Cryptogramma stelleri (S. G. Gmel.)
  Prantl

Pellaea atropurpurea (L.) Link
Pellaea glabella Mett.
Cheilanthes feei Moore
Cheilanthes Lanosa(Michx.) D. C.
  Eaton
Polypodium vulgare L. var. virginianum
  (L.) Eaton
Polypodium polypodioides (L.) Watt
  var. michauxianum Weatherby
Polystichum acrostichoides (Michx.)
  Schott

Onoclea sensibilis L.

Matteuccia struthiopteris (L.) Todaro

Gymnocarpium dryopteris (L.) Newm.
Thelypteris phegopteris (L.) Slosson
Thelypteris hexagonoptera (Michx.)
  Weatherby

Thelypteris noveboracensis (L.)
  Nieuwl.
Thelypteris palustris Schott var.
  pubescens (Laws.) Fern.

Dryopteris carthusiana (Villars) H. P.
  Fuchs
Dryopteris X boottii (Tuckerm.)
  Underw.
Dryopteris intermedia (Muhl.) Gray
Dryopteris X triploidea Wherry
Dryopteris cristata (L.) Gray
Dryopteris X clintoniana (D. C. Eaton)
  Dowell
Dryopteris goldiana (Hook.) Gray
Dryopteris celsa (Wm. Palmer) Small
Dryopteris X neo-wherryi W. H. Wagner
Dryopteris marginalis (L.) Gray

Woodwardia virginica (L.) Sm.
Woodwardia areolata (L.) Moore

Athyrium pycnocarpon (Spreng.)
  Tidestrom
Athyrium thelypterioides (Michx.)
  Desv.
Athyrium filix-femina (L.) Roth var.
  rubellum Gilib.
Athyrium filix-femina (L.) Roth var.
  asplenioides (Michx.) Farw.

Asplenium rhizophyllum L.
Asplenium ruta-muraria L.
Asplenium pinnatifidum Nutt.
Asplenium X gravesii Maxon
Asplenium X trudellii Wherry
Asplenium X kentuckiense McCoy
Asplenium trichomanes X Asplenium
  pinnatifidum
Asplenium bradleyi D. C. Eaton
Asplenium X ebenoides R. R. Scott
Asplenium trichomanes L.
Asplenium resiliens Kunze
Asplenium platyneuron (L.) Oakes

Woodsia obtusa (Spreng.) Torr.
Woodsia ilvensis (L.) R. Br.

Cystopteris bulbifera (L.) Bernh.
Cystopteris fragilis (L.) Bernh.
Cystopteris fragilis (L.) Bernh. var.
  protrusa Weatherby
Cystopteris fragilis (L.) Bernh. var.
  mackayi Laws.
Cystopteris X tennesseensis Shaver

Marsileaceae

Marsilea quadrifolia L.

Salviniaceae

Azolla caroliniana Willd.
Azolla mexicana Presl

Taxaceae

Taxus canadensis Marsh.

Pinaceae

Pinus strobus L.
Pinus echinata Mill.
Pinus rigida Mill.
Pinus taeda L.
Pinus pungens Lamb.
Pinus banksiana Lamb.
Pinus sylvestris L.
Pinus resinosa Ait.

Larix laricina (DuRoi) K. Koch
Larix decidua Mill.

Taxodiaceae

Taxodium distichum (L.) Rich.

Cupressaceae

Thuja occidentalis L.

Juniperus communis L.
Juniperus communis L. var. depressa
  Pursh
Juniperus horizontalis Moench
Juniperus virginiana L.

Typhaceae

Typha latifolia L.
Typha angustifolia L.

Sparganiaceae

Sparganium minimum (Hartm.) Fries
Sparganium chlorocarpum Rydb.
Sparganium androcladum (Engelm.)
  Morong
Sparganium americanum Nutt.
Sparganium eurycarpum Engelm.

Ruppiaceae

Ruppia maritima L. var. rostrata Agardh

Zannichelliaceae

Zannichellia palustris L.

Najadaceae

Najas flexilis (Willd.) Rostk. & Schmidt
Najas guadalupensis (Spreng.)
  Magnus
Najas gracillima (A. Br.) Magnus
Najas minor All.
Najas marina L.

Potamogetonaceae

Potamogeton pectinatus L.
Potamogeton robbinsii Oakes
Potamogeton crispus L.
Potamogeton praelongus Wulfen
Potamogeton richardsonii (Benn.)
  Rydb.
Potamogeton zosteriformis Fern.
Potamogeton foliosus Raf.
Potamogeton friesii Rupr.
Potamogeton strictifolius Benn.
Potamogeton pusillus L.
Potamogeton vaseyi Robbins

Potamogeton natans L.
Potamogeton diversifolius Raf.
Potamogeton epihydrus Raf.
Potamogeton amplifolius Tuckerm.
Potamogeton pulcher Tuckerm.
Potamogeton nodosus Poir.
Potamogeton illinoensis Morong
Potamogeton X hagstromii Benn.
Potamogenton gramineus L.
Potamogeton X spathulaeformis
  (Robbins) Morong
Potamogeton X rectifolius Benn.

Juncaginaceae

Triglochin maritima L.
Triglochin palustris L.

Scheuchzeria palustris L. var.
  americana Fern.

Alismaceae

Echinodorus tenellus (Mart.) Buchenau
  var. parvulus (Engelm.) Fassett
Echinodorus berteroi (Spreng.) Fassett
  var. lanceolatus (Wats. & Coult.)
  Fassett
Echinodorus cordifolius (L.) Griseb.

Sagittaria calycina Engelm.
Sagittaria rigida Pursh
Sagittaria graminea Michx.
Sagittaria cuneata Sheld.
Sagittaria brevirostra Mack. & Bush
Sagittaria longirostra (Micheli)
  J. G. Sm.
Sagittaria latifolia Willd.

Alisma plantago-aquatica L. var.
  americanum Roem. & Schultes
Alisma subcordatum Raf.

Butomaceae

Butomus umbellatus L.

Hydrocharitaceae

Elodea densa (Planch.) Caspary
Elodea canadensis Michx.
Elodea nuttallii (Planch.) St. John

Vallisneria americana Michx.

Limnobium spongia (Bosc) Steud.

Poaceae

Bromus sterilis L.
Bromus tectorum L.
Bromus marginatus Nees
Bromus nottowayanus Fern.
Bromus kalmii Gray
Bromus willdenovii Kunth
Bromus secalinus L.
Bromus brizaeformis Fisch. & Mey.
Bromus mollis L.
Bromus racemosus L.
Bromus commutatus Schrad.

Bromus arvensis L.
Bromus japonicus Thunb.
Bromus inermis Leyss.
Bromus erectus Huds.
Bromus purgans L.
Bromus pubescens Muhl.
Bromus ciliatus L.
Bromus squarrosus L.
Bromus carinatus L.

Vulpia octoflora (Walt.) Rydb.
Vulpia octoflora (Walt.) Rydb. var.
  tenella (Willd.) Fern.
Vulpia octoflora (Walt.) Rydb. var.
  glauca (Nutt.) Fern.
Vulpia myuros (L.) K. Gmel.
Vulpia dertonensis All.

Festuca capillata Lam.
Festuca ovina L. var. duriuscula (L.)
  Koch
Festuca rubra L.
Festuca pratensis Huds.
Festuca arundinacea Schreb.
Festuca obtusa Biehler
Festuca paradoxa Desv.

Lolium temulentum L.
Lolium multiflorum Lam.
Lolium perenne L.

Puccinellia distans (L.) Parl.
Puccinellia pallida (Torr.) Clausen

Poa annua L.
Poa chapmaniana Scribn.
Poa autumnalis Muhl.
Poa arachnifera Torr.
Poa pratensis L.
Poa angustifolia L.
Poa compressa L.
Poa languida Hitchc.
Poa trivialis L.
Poa alsodes Gray
Poa paludigena Fern. & Wieg.
Poa nemoralis L.
Poa palustris L.
Poa wolfii Scribn.
Poa sylvestris Gray
Poa bulbosa L.

Briza maxima L.

Pennisetum alopecuroides (L.) Spreng.

Dactylis glomerata L.

Koeleria macrantha (Ledeb.) Spreng.

Sphenopholis obtusata (Michx.)
  Scribn.
Sphenopholis obtusata (Michx.)
  Scribn. var. major (Torr.) Erdman
Sphenopholis nitida (Biehler) Scribn.

Aira caryophyllaea L.

Deschampsia cespitosa (L.) Beauv.

Avena fatua L.
Avena sativa L.

Arrhenatherum elatius (L.) Presl

Holcus lanatus L.

Calamagrostis canadensis (Michx.)
  Beauv.
Calamagrostis canadensis (Michx.)
  Beauv. var. macouniana (Vasey)
  Stebbins
Calamagrostis inexpansa Gray var.
  brevior (Vasey) Stebbins
Calamagrostis epigeios (L.) Roth

Ammophila breviligulata Fern.

Agrostis elliottiana Schult.
Agrostis hyemalis (Walt.) BSP.
Agrostis scabra Willd.
Agrostis perennans (Walt.) Tuckerm.
Agrostis alba L.
Agrostis alba L. var. palustris (Huds.)
  Pers.
Agrostis tenuis Sibth.
Agrostis interrupta L.

Cinna arundinacea L.
Cinna latifolia (Trev.) Griseb.

Anthoxanthum odoratum L.
Anthoxanthum aristatum Boiss.

Hierochloë odorata (L.) Beauv.

Phalaris arundinacea L.
Phalaris canariensis L.

Alopecurus pratensis L.
Alopecurus aequalis Sobol
Alopecurus carolinianus Walt.

Phleum pratense L.

Milium effusum L.

Beckmannia syzigachne (Steud.) Fern.

Elymus arenarius L.
Elymus hystrix L.
Elymus hystrix L. var. bigeloviana
  (Fern.) Mohlenbrock
Elymus virginicus L.
Elymus virginicus L. var. submuticus
  Hook.
Elymus virginicus L. var. glabriflorus
  (Vasey) Bush
Elymus riparius Wiegand
Elymus villosus Muhl.
Elymus canadensis L.

Sitanion hystrix (Nutt.) J. G. Sm.

Hordeum pusillum Nutt.
Hordeum brachyantherum Nevski
Hordeum jubatum L.
Hordeum vulgare L.
Hordeum vulgare L. var. trifurcatum
  (Schlecht.) Alefeld
Hordeum X montanense Scribn.
Hordeum geniculatum

Agrohordeum X macounii (Vasey)
  Lepage

Agropyron desertorum (Fisch.) Schult.
Agropyron cristatum (L.) Gaertn.
Agropyron subsecundum (Link)
  Hitchcock
Agropyron trachycaulum (Link) Malte
Agropyron repens (L.) Beauv.
Agropyron smithii Rydb.
Agropyron smithii Rydb. var. molle
  (Scribn. & Smith) Jones

Triticum aestivum L.
Triticum cylindricum (Host) Ces.

Secale cereale L.

Melica mutica Walt.
Melica nitens (Scribn.) Nutt.

Glyceria borealis (Nash) Batchelder
Glyceria septentrionalis Hitchcock
Glyceria arkansana Fern.
Glyceria canadensis (Michx.) Trin.
Glyceria striata (Lam.) Hitchcock
Glyceria striata (Lam.) Hitchcock var.
  stricta (Scribn.) Fern
Glyceria grandis S. Wats.

Schizachne purpurascens (Torr.)
  Swallen

Stipa viridula Trin.
Stipa comata Trin. & Rupr.
Stipa spartea Trin.

Oryzopsis racemosa (J. E. Smith)
  Ricker
Oryzopsis asperifolia Michx.
Oryzopsis pungens (Torr.) Hitchcock

Brachyelytrum erectum (Schreb.)
  Beauv.

Diarrhena americana Beauv.
Diarrhena americana Beauv. var.
  obovata Gleason

Digitaria sanguinalis (L.) Scop.
Digitaria sanguinalis (L.) Scop. var.
  ciliaris (Retz.) Parl.
Digitaria ischaemum (Schreb.) Muhl.
Digitaria filiformis (L.) Koel.
Digitaria villosa (Walt.) Pers.

Trichachne insularis (L.) Nees

Leptoloma cognatum (Schult.) Chase

Eriochloa villosa (Thunb.) Kunth
Eriochloa contracta Hitchcock
Eriochloa gracilis (Fourn.) Hitchcock

Paspalum dissectum (L.) L.
Paspalum fluitans (Ell.) Kunth
Paspalum pubiflorum Rupr. var.
  glabrum (Vasey) Vasey
Paspalum floridanum Michx.
Paspalum laeve Michx.
Paspalum lentiferum Lam.
Paspalum ciliatifolium Michx.
Paspalum bushii Nash

Panicum dichotomiflorum Michx.
Panicum dichotomiflorum Michx. var.
  geniculatum (Muhl.) Fern.
Panicum dichotomiflorum Michx. var.
  puritanorum Svenson
Panicum flexile (Gattinger) Scribn.
Panicum gattingeri Nash
Panicum philadelphicum Bernh.
Panicum capillare L.
Panicum capillare L. var. occidentale
  Rydb.
Panicum miliaceum L.
Panicum virgatum L.
Panicum rigidulum Bosc
Panicum rigidulum Bosc var.
  condensum (Nash) Mohlenbrock
Panicum stipitatum Nash
Panicum longifolium Torr.
Panicum anceps Michx.
Panicum nians Ell.
Panicum depauperatum Muhl.
Panicum perlongum Nash
Panicum linearifolium Scribn.
Panicum linearifolium Scribn. var.
  werneri (Scribn.) Fern.
Panicum laxiflorum Lam.
Panicum microcarpon Muhl.
Panicum nitidum Lam.
Panicum boreale Nash
Panicum dichotomum L.
Panicum dichotomum L. var.
  barbulatum (Michx.) Wood
Panicum mattamuskeetense Ashe
Panicum yadkinense Ashe
Panicum meridionale Ashe
Panicum meridionale Ashe var.
  albermarlense (Ashe) Fern.
Panicum lanuginosum Ell.
Panicum lanuginosum Ell. var.
  implicatum (Scribn.) Fern.
Panicum lanuginosum Ell. var.
  lindheimeri (Nash) Fern.
Panicum lanuginosum Ell. var.
  septentrionale (Fern.) Fern.
Panicum praecocius Hitchc. & Chase
Panicum subvillosum Ashe
Panicum villosissimum Nash
Panicum villosissimum Nash var.
  pseudopubescens (Nash) Fern.
Panicum scoparioides Ashe

Panicum columbianum Scribn.
Panicum sphaerocarpon Ell.
Panicum polyanthes Schult.
Panicum wilcoxianum Vasey
Panicum malacophyllum Nash
Panicum oligosanthes Schult.
Panicum oligosanthes Schult. var.
  scribnerianum (Nash) Fern.
Panicum oligosanthes Schult, var.
  helleri (Nash) Fern.
Panicum ravenelii Scribn. & Merr.
Panicum leibergii (Vasey) Scribn.
Panicum scoparium Lam.
Panicum commutatum Schult.
Panicum commutatum Schult. var.
  ashei Fern.
Panicum joori Vasey
Panicum clandestinum L.
Panicum latifolium L.
Panicum boscii Poir.
Panicum boscii Poir. var. molle (Vasey)
  Hitchc. & Chase

Echinochloa walteri (Pursh) Heller
Echinochloa colonum (L.) Link
Echinochloa crus-galli (L.) Beauv.
Echinochloa frumentacea (Roxb.) Link
Echinochloa pungens (Poir.) Rydb.
Echinochloa pungens (Poir.) Rydb. var.
  microstachya (Wieg.) Mohl.
Echinochloa pungens (Poir.) Rydb. var.
  wiegandii Fassett

Setaria geniculata (Lam.) Beauv.
Setaria verticillata (L.) Beauv.
Setaria lutescens (Weigel) Hubb.
Setaria faberi Herrm.
Setaria italica (L.) Beauv.
Setaria viridis (L.) Beauv.
Setaria viridis (L.) Beauv. var. major
  (Gaudin) Pospichal

Cenchrus longispinus (Hack.) Fern.

Miscanthus sinensis Anderss.
Miscanthus sacchariflorus (Maxim.)
  Hack.

Erianthus ravennae (L.) Beauv.
Erianthus alopecuroides (L.) Ell.
Erianthus brevibarbis Michx.

Sorghum halepense (L.) Pers.
Sorghum almum Parodi
Sorghum bicolor (L.) Moench
Sorghum sudanense (Piper) Stapf

Sorghastrum nutans (L.) Nash

Andropogon gerardii Vitman
Andropogon virginicus L.
Andropogon elliottii Chapm.
Andropogon hallii Hack.

Microstegium vimineum (Trin.) A.
  Camus

Bothriochloa saccharoides (Swartz)
  Rydb.

Schizachyrium scoparium (Michx.)
  Nash

Tripsacum dactyloides (L.) L.

Zea mays L.

Eragrostis hypnoides (Lam.) BSP.
Eragrostis reptans (Michx.) Nees
Eragrostis trichodes (Nutt.) Wood
Eragrostis trichodes (Nutt.) Wood var.
  pilifera (Scheele) Fern.
Eragrostis cilianensis (All.) Mosher
Eragrostis poaeoides Beauv.
Eragrostis spectabilis (Pursh) Steud.
Eragrostis curvula (Schrad.) Nees
Eragrostis neomexicana Vasey
Eragrostis hirsuta (Michx.) Nees
Eragrostis pectinacea (Michx.) Nees
Eragrostis diffusa Buckl.
Eragrostis capillaris (L.) Nees
Eragrostis pilosa (L.) Beauv.
Eragrostis frankii C. A. Meyer
Eragrostis frankii C. A. Meyer var.
  brevipes Fassett

Tridens flavus (L.) Hitchcock
Tridens strictus (Nutt.) Nash

Triplasis purpurea (Walt.) Chapm.

Redfieldia flexuosa (Thurb.) Vasey

Calamovilfa longifolia (Hook.) Scribn.

Muhlenbergia asperifolia (Nees &
  Meyen) Parodi
Muhlenbergia capillaris (Lam.) Trin.
Muhlenbergia cuspidata (Torr.) Rydb.
Muhlenbergia schreberi J. F. Gmel.
Muhlenbergia X curtisetosa (Scribn.)
  Pohl
Muhlenbergia sobolifera (Muhl.) Trin.
Muhlenbergia bushii Pohl
Muhlenbergia frondosa (Poir.) Fern.
Muhlenbergia racemosa (Michx.) BSP.
Muhlenbergia glabrifloris Scribn.
Muhlenbergia glomerata (Willd.) Trin.
Muhlenbergia tenuiflora (Willd.) BSP.
Muhlenbergia sylvatica (Torr.) Torr.
Muhlenbergia mexicana (L.) Trin.

Sporobolus cryptandrus (Torr.) Gray
Sporobolus heterolepis (Gray) Gray
Sporobolus asper (Michx.) Kunth
Sporobolus clandestinus (Biehler)
  Hitchc.
Sporobolus vaginiflorus (Torr.) Wood
Sporobolus neglectus Nash

Crypsis schoenoides (L.) Lam.

Eleusine indica (L.) Gaertn.

Dactyloctenium aegyptium (L.) Beauv.

Leptochloa filiformis (Lam.) Beauv.
Leptochloa attenuata (Nutt.) Steud.
Leptochloa fascicularis (Lam.) Gray
Leptochloa acuminata (Nash)
  Mohlenbr.
Leptochloa panicoides (Presl) Hitchc.

Gymnopogon ambiguus (Michx.) BSP.

Schedonnardus paniculatus (Nutt.)
  Trel.

Cynodon dactylon (L.) Pers.

Chloris verticillata Nutt.
Chloris gayana Kunth

Bouteloua curtipendula (Michx.) Torr.
Bouteloua hirsuta Lag.
Bouteloua gracilis (HBK.) Lag.

Buchloë dactyloides (Nutt.) Engelm.

Spartina pectinata Lind.

Distichlis stricta (Torr.) Rydb.

Aristida tuberculosa Nutt.
Aristida desmantha Trin. & Rupr.
Aristida oligantha Michx.
Aristida purpurascens Poir.
Aristida intermedia Scribn. & Ball
Aristida necopina Shinners
Aristida ramosissima Engelm.
Aristida longespica Poir.
Aristida longespica Poir. var.
  geniculata (Raf.) Fern.
Aristida dichotoma Michx.
Aristida basiramea Engelm.
Aristida curtissii (Gray) Nash

Arundinaria gigantea (Walt.) Chapm.

Leersia lenticularis Michx.
Leersia oryzoides (L.) Swartz
Leersia virginica Willd.

Zizania aquatica L.
Zizania aquatica L. var. interior Fassett

Zizaniopsis miliacea (Michx.) Doell &
  Aschers.

Phragmites australis Trin.

Arundo donax L.

Danthonia spicata (L.) Beauv.

Chasmanthium latifolium (Michx.)
  Yates

Cyperaceae
Cyperus densicaespitosus Mattf. &
  Kükenth.
Cyperus flavescens L.
Cyperus diandrus Torr.
Cyperus rivularis Kunth
Cyperus filicinus Vahl
Cyperus aristatus Rottb.
Cyperus pseudovegetus Steud.
Cyperus acuminatus Torr. & Hook.
Cyperus ovularis (Michx.) Torr.
Cyperus grayioides Mohlenbr.
Cyperus compressus L.
Cyperus schweinitzii Torr.
Cyperus houghtonii Torr.
Cyperus filiculmis Vahl
Cyperus filiculmis Vahl var. macilentus
  Fern.
Cyperus erythrorhizos Muhl.
Cyperus iria L.
Cyperus engelmannii Steud.
Cyperus lancastriensis Porter
Cyperus esculentus L.
Cyperus esculentus L. var.
  leptostachyus Boeckl.
Cyperus ferruginescens Boeckl.
Cyperus strigosus L.
Cyperus retrorsus Chapm.

Dulichium arundinaceum (L.) Britt.

Eleocharis equisetoides (Ell.) Torr.
Eleocharis quadrangulata (Michx.)
  Roem. & Schultes
Eleocharis olivacea Torr.
Eleocharis smallii Britt.
Eleocharis macrostachya Britt.
Eleocharis erythropoda Steud.
Eleocharis caribaea (Rottb.) Blake
Eleocharis obtusa (Willd.) Schult.
Eleocharis obtusa (Willd.) Schult. var.
  detonsa (Gray) Drap. & Mohlenbr.
Eleocharis obtusa (Willd.) Schult. var.
  ovata (Roth) Drap. & Mohlenbr.
Eleocharis acicularis (L.) Roem. &
  Schultes
Eleocharis wolfii (Gray) Patterson
Eleocharis tenuis (Willd.) Schult. var.
  verrucosa (Svenson) Svenson
Eleocharis elliptica Kunth
Eleocharis elliptica Kunth var.
  compressa (Sull.) Drap. & Mohlenbr.
Eleocharis intermedia (Muhl.) Schult.
Eleocharis rostellata (Torr.) Torr.
Eleocharis pauciflora (Lightf.) Link
Eleocharis parvula (Roem. & Schultes)
  Link

Bulbostylis capillaris (L.) C. B. Clarke

Fimbristylis vahlii (Lam.) Link
Fimbristylis puberula (Michx.) Vahl var.
  drummondii (Boeckl.) Ward
Fimbristylis baldwiniana (Schult.) Torr.
Fimbristylis autumnalis (L.) Roem. &
  Schultes

Lipocarpha maculata (Michx.) Torr.

Fuirena scirpoides Michx.

Scirpus smithii Gray
Scirpus purshianus Fern.
Scirpus hallii Gray
Scirpus torreyi Olney
Scirpus subterminalis Torr.
Scirpus americanus Pers.
Scirpus acutus Muhl.
Scirpus validus Vahl
Scirpus heterochaetus Chase
Scirpus koilolepis (Steud.) Gl.
Scirpus micranthus Vahl
Scirpus micranthus Vahl var.
  drummondii (Nees) Mohlenbr.
Scirpus fluviatilis (Torr.) Gray
Scirpus paludosus A. Nels.
Scirpus polyphyllus Vahl
Scirpus georgianus Harper
Scirpus atrovirens Willd.
Scirpus hattorianus Mak.
Scirpus microcarpus Presl
Scirpus pendulus Muhl.
Scirpus cyperinus (L.) Kunth
Scirpus pedicellatus Fern.
Scirpus atrocinctus Fern.
Scirpus cespitosus L. var. callosus
  Bigel
Scirpus verecundus Fern.

Eriophorum gracile Koch
Eriophorum tenellum Nutt.
Eriophorum angustifolium Honck.
Eriophorum viridi-carinatum (Engelm.)
  Fern.
Eriophorum virginicum L.

Rhynchospora macrostachya Torr.
Rhynchospora corniculata (Lam.) Gray
Rhynchospora globularis (Chapm.)
  Small
Rhynchospora capillacea Torr.
Rhynchospora glomerata (L.) Vahl
Rhynchospora capitellata (Michx.)
  Vahl
Rhynchospora alba (L.) Vahl

Cladium mariscoides (Muhl.) Torr.

Scleria oligantha Michx.
Scleria triglomerata Michx.
Scleria pauciflora Muhl.
Scleria pauciflora Muhl. var.
  caroliniana (Willd.) Wood
Scleria verticillata Muhl.

Carex stenophylla Wahlenb.
Carex praegracilis W. Boott
Carex chordorrhiza L. f.
Carex sartwellii Dew.
Carex foenea Willd.
Carex foenea Willd. var. enervis Evans
  & Mohlenbr.
Carex retroflexa Muhl.

Carex texensis (Torr.) Bailey
Carex convoluta Mack.
Carex rosea Schk.
Carex socialis Mohlenbr. & Schwegm.
Carex cephalophora Muhl.
Carex leavenworthii Dewey
Carex spicata Huds.
Carex muhlenbergii Schk.
Carex muhlenbergii Schk. var enervis
    Boott
Carex austrina (Small) Mack.
Carex gravida Bailey
Carex gravida Bailey var. lunelliana
    (Mack.) F. J. Herm.
Carex aggregata Mack.
Carex cephaloidea Dewey
Carex sparganioides Muhl.
Carex vulpinoidea Michx.
Carex annectens Bickn.
Carex annectens Bickn. var.
    xanthocarpa (Bickn.) Wieg.
Carex decomposita Muhl.
Carex diandra Schrank
Carex prairea Dewey
Carex alopecoidea Tuckerm.
Carex conjuncta Boott
Carex stipata Muhl.
Carex stipata Muhl. var. maxima
    Chapm
Carex laevivaginata (Kükenth.) Mack.
Carex crus-corvi Shuttlew.
Carex disperma Dewey
Carex trisperma Dewey
Carex bromoides Schk.
Carex interior Bailey
Carex sterilis Willd.
Carex incomperta Bickn.
Carex muskingumensis Schwein.
Carex scoparia Schk.
Carex tribuloides Wahlenb.
Carex projecta Mack.
Carex richii (Fern.) Mack.
Carex cristatella Britt.
Carex beobii Olney
Carex normalis Mack.
Carex terera Dewey
Carex festucacea Schk.
Carex albolutescens Schwein.
Carex cumulata (Bailey) Mack.
Carex suberecta (Olney) Britt.
Carex brevior (Dewey) Mack.
Carex molesta Mack.
Carex reniformis (Bailey) Small
Carex bicknellii Britt.
Carex alata Torr. & Gray
Carex leptalea Wahlenb.
Carex jamesii Schwein.
Carex pensylvanica Lam.
Carex pensylvanica Lam. var. distans
    Peck
Carex communis Bailey
Carex emmonsii Dewey
Carex artitecta Mack.
Carex physorhyncha Liebm.
Carex nigromarginata Schwein.
Carex umbellata Schk.

Carex abdita Bickn.
Carex tonsa (Fern.) Bickn.
Carex pedunculata Muhl.
Carex richardsonii R. Br.
Carex eburnea Boott
Carex hirtifolia Mack.
Carex garberi Fern.
Carex aurea Nutt.
Carex crinita Lam.
Carex crinita Lam. var. brevicrinis Fern.
Carex substricta (Kükenth.) Mack.
Carex stricta Lam.
Carex emoryi Dewey
Carex haydenii Dewey
Carex torta Boott
Carex buxbaumii Wahlenb.
Carex limosa L.
Carex shortiana Dewey
Carex lasiocarpa Ehrh.
Carex lanuginosa Michx.
Carex pallescens L.
Carex hirsutella Mack.
Carex caroliniana Schwein.
Carex bushii Mack.
Carex virescens Muhl.
Carex swanii (Fern.) Mack.
Carex prasina Wahlenb.
Carex gracillima Schwein.
Carex oxylepis Torr. & Hook.
Carex davisii Schwein. & Torr.
Carex debilis Michx.
Carex sprengelii Dewey
Carex granularis Muhl.
Carex granularis Muhl. var. haleana
    (Olney) Porter
Carex crawei Dewey
Carex conoidea Schk.
Carex amphibola Steud.
Carex grisea Wahlenb.
Carex flaccosperma Dewey
Carex glaucodea Tuckerm.
Carex oligocarpa Schk.
Carex hitchcockiana Dewey
Carex meadii Dewey
Carex tetanica Schk.
Carex woodii Dewey
Carex plantaginea Lam.
Carex platyphylla Carey
Carex digitalis Willd.
Carex laxiculmis Schwein.
Carex albursina Sheldon
Carex laxiflora Lam.
Carex striatula Michx.
Carex blanda Dewey
Carex gracilescens Steud.
Carex styloflexa Buckl.
Carex cryptolepis Mack.
Carex viridula Michx.
Carex frankii Kunth
Carex squarrosa L.
Carex typhina Michx.
Carex lacustris Willd.
Carex hyalinolepis Steud.
Carex atherodes Spreng.
Carex X fulleri Ahles
Carex laeviconica Dewey

Carex trichocarpa Muhl.
Carex X subimpressa Clokey
Carex comosa Boott
Carex hystricina Muhl.
Carex lurida Wahlenb.
Carex grayi Carey
Carex intumescens Rudge
Carex louisianica Bailey
Carex lupulina Muhl.
Carex lupuliformis Sartwell
Carex gigantea Rudge
Carex retrorsa Schwein.
Carex rostrata Stokes var. utriculata
    (Boott) Bailey
Carex tuckermanii Boott
Carex oligosperma Michx.
Carex vesicaria L.
Carex careyana Torr.

Araceae
Acorus calamus L.

Symplocarpus foetidus (L.) Nutt.

Arum italicum Mill.

Peltandra virginica (L.) Kunth

Calla palustris L.

Arisaema dracontium (L.) Schott
Arisaema triphyllum (L.) Schott
Arisaema triphyllum (L.) Schott var.
    pusillum Peck

Lemnaceae
Spirodela polyrhiza (L.) Schleiden
Spirodela oligorhiza (Kurtz) Hegelm.

Lemna trisulca L.
Lemna minor L.
Lemna gibba L.
Lemna perpusilla Torr.
Lemna trinervis (Austin) Small
Lemna valdiviana Phil.
Lemna minima Phil.
Lemna obscura (Austin) Daubs

Wolffiella floridana (J. D. Smith)
    Thompson

Wolffia papulifera Thompson
Wolffia punctata Griseb.
Wolffia columbiana Karst.

Xyridaceae
Xyris torta Sm.
Xyris jupicai L. Rich

Commelinaceae
Tradescantia subaspera Ker
Tradescantia ohiensis Raf.
Tradescantia virginiana L.
Tradescantia bracteata Small

Commelina communis L.
Commelina diffusa Burm. f.
Commelina virginica L.
Commelina erecta L.

Pontederiaceae

Pontederia cordata L.

Zosterella dubia (Jacq.) Small

Heteranthera limosa (Sw.) Willd.
Heteranthera reniformis R. & P.

Juncaceae

Luzula acuminata Raf.
Luzula multiflora (Retz.) Lejeune
Luzula multiflora (Retz.) Lejeune var.
    echinata (Small) Mohlenbr.

Juncus effusus L. var. solutus Fern. &
    Wieg.
Juncus balticus Willd. var. littoralis
    Engelm.
Juncus biflorus Ell.
Juncus biflorus Ell. f. andinus Fern. &
    Grisc.
Juncus marginatus Rostk.
Juncus canadensis J. Gay
Juncus brachycephalus (Engelm.)
    Buch.
Juncus alpinus Vill. var. rariflorus
    Hartm.
Juncus alpinus Vill. var. fuscescens
    Fern.
Juncus nodosus L.
Juncus torreyi Coville
Juncus diffusissimus Buckl.
Juncus scirpoides Lam.
Juncus nodatus Coville
Juncus acuminatus Michx.
Juncus brachycarpus Engelm.
Juncus bufonius L.
Juncus bufonius L. var. congestus
    Wahlb.
Juncus gerardii Loisel.
Juncus greenei Oakes & Tuckerm.
Juncus vaseyi Engelm.
Juncus secundus Beauv.
Juncus tenuis Willd.
Juncus dudleyi Wieg.
Juncus interior Wieg.

Liliaceae

Tofieldia glutinosa (Michx.) Pers.

Veratrum woodii Robins.

Melanthium virginicum L.

Zigadenus glaucus Nutt.

Stenanthium gramineum (Ker) Morong

Chamaelirium luteum (L.) Gray

Lilium lancifolium Thunb.
Lilium philadelphicum L. var. andinum
    (Nutt.) Ker
Lilium michiganense Farw.
Lilium superbum L.

Hemerocallis fulva L.
Hemerocallis lilio-asphodelus L.

Hosta lancifolia (Thunb.) Engl.

Ornithogalum umbellatum L.
Ornithogalum nutans L.

Aletris farinosa L.

Muscari botryoides (L.) Mill.
Muscari armeniacum Leicht.
Muscari comosum (L.) Mill.
Muscari atlanticum Boiss. & Reut.

Camassia scilloides (Raf.) Cory
Camassia angusta (Engelm. & Gray)
    Blankenship

Erythronium americanum Ker
Erythronium albidum Nutt.

Uvularia grandiflora Sm.
Uvularia sessilifolia L.

Polygonatum pubescens (Willd.) Pursh
Polygonatum commutatum (Schult.) A.
    Dietr.
Polygonatum biflorum (Walt.) Ell.

Smilacina racemosa (L.) Desf.
Smilacina stellata (L.) Desf.

Convallaria majalis L.

Asparagus officinalis L.

Maianthemum canadense Desf.
Maianthemum canadense Desf. var.
    interius Fern.

Lycoris radiata Herb.

Allium tricoccum Ait.
Allium tricoccum Ait. var. burdickii
    Hanes
Allium sativum L.
Allium ampeloprasum L. var.
    atroviolaceum (Boiss.) Regel
Allium canadense L.
Allium porrum L.
Allium cernuum Roth
Allium stellatum Ker
Allium mutabile Michx.
Allium vineale L.
Allium schoenoprasum L.
Allium schoenoprasum L. var.
    sibiricum (L.) Hartm.
Allium cepa L.
Allium fistulosum L.

Nothoscordum bivalve (L.) Britt.

Medeola virginiana L.

Trillium recurvatum Beck
Trillium sessile L.
Trillium viride Beck
Trillium cuneatum Raf.
Trillium nivale Riddell
Trillium erectum L.
Trillium cernuum L. var. macranthum
    Eames & Wieg.
Trillium grandiflorum (Michx.) Salisb.
Trillium flexipes Raf.

Yucca filamentosa L. var. smalliana
    (Fern.) Ahles

Leucojum aestivum L.

Narcissus pseudo-narcissus L.
Narcissus poeticus L.

Hymenocallis occidentalis (LeConte)
    Kunth

Polianthes virginica (L.) Shinners

Hypoxis hirsuta (L.) Coville

Scilla sibirica Andr.

Smilacaceae

Smilax glauca Walt.
Smilax glauca Walt. var. leurophylla
    Blake
Smilax bona-nox L.
Smilax bona-nox L. var. hederaefolia
    (Beyrich) Fern.
Smilax rotundifolia L.
Smilax hispida Muhl.
Smilax herbacea L.
Smilax lasioneuron Hook.
Smilax pulverulenta Michx.
Smilax illincensis Mangaly
Smilax ecirrata (Engelm.) S. Wats.

Dioscoreaceae

Dioscorea villosa L.
Dioscorea quaternata (Walt.) J. F.
    Gmel.
Dioscorea batatas Dcne.

Iridaceae

Iris pumila L.
Iris germanica L.
Iris brevicaulis Raf.
Iris fulva Ker
Iris shrevei Small
Iris pseudacorus L.
Iris cristata Ait.
Iris flavescens DC.

Gladiolus X colvillei Sweet

Belamcanda chinensis (L.) DC.

Sisyrinchium angustifolium Mill.
Sisyrinchium atlanticum Bickn.
Sisyrinchium albidum Raf.
Sisyrinchium campestre Bickn.
Sisyrinchium mucronatum Michx.
Sisyrinchium montanum Greene
Sisyrinchium montanum Greene var.
   crebrum Fern.

Marantaceae
Thalia dealbata Roscoe

Burmanniaceae
Thismia americana N. E. Pfeiffer

Orchidaceae
Cypripedium acaule Ait.
Cypripedium calceolus L. var.
   parviflorum (Salisb.) Fern.
Cypripedium calceolus L. var.
   pubescens (Willd.) Correll

Cypripedium candidum Muhl.
Cypripedium reginae Walt.

Calopogon tuberosus (L.) BSP.

Orchis spectabilis L.

Habenaria orbiculata (Pursh) Torr.
Habenaria hookeri Torr.
Habenaria clavellata (Michx.) Spreng.
Habenaria viridis (L.) R. Br.
Habenaria flava (L.) R. Br.
Habenaria flava (L.) R. Br. var. herbiola
   (R. Br.) Ames & Correll
Habenaria hyperborea (L.) R. Br. var.
   huronensis (Nutt.) Farw.
Habenaria dilatata (Pursh) Hook.
Habenaria ciliaris (L.) R. Br.
Habenaria blephariglottis (Willd.)
   Hook.
Habenaria lacera (Michx.) Lodd.
Habenaria leucophaea (Nutt.) Gray
Habenaria peramoena Gray
Habenaria psycodes (L.) Spreng.

Liparis liliifolia (L.) Rich.
Liparis loeselii (L.) Rich.

Malaxis unifolia Michx.
Malaxis monophylla (L.) Sw. var.
   brachypoca (Gray) F. Morris

Spiranthes lucida (H. H. Eaton) Ames
Spiranthes ovalis Lindl.
Spiranthes cernua (L.) Rich.
Spiranthes magnicamporum Sheviak
Spiranthes vernalis Engelm. & Gray
Spiranthes lacera (Raf.) Raf.
Spiranthes gracilis (Bigel.) Beck
Spiranthes tuberosa Raf.
Spiranthes romanzoffiana Cham.

Goodyera pubescens (Willd.) R. Br.

Pogonia ophioglossoides (L.) Ker

Isotria verticillata (Willd.) Raf.
Isotria medeoloides (Pursh) Raf.

Triphora trianthophora (Sw.) Rydb.

Epipactis helleborine (L.) Crantz

Corallorhiza trifida Chat.
Corallorhiza maculata Raf.
Corallorhiza wisteriana Conrad
Corallorhiza odontorhiza (Willd.) Nutt.

Hexalectris spicata (Walt.) Barnh.

Aplectrum hyemale (Muhl.) Torr.

Tipularia discolor (Pursh) Nutt.

Saururaceae
Saururus cernuus L.

Salicaceae
Salix nigra Marsh.
Salix caroliniana Michx.
Salix amygdaloides Anderss.
Salix pentandra L.
Salix lucida Muhl.
Salix serissima (Bailey) Fern.
Salix fragilis L.
Salix babylonica L.
Salix alba L.
Salix alba L. var. calva G. F. W. Mey.
Salix alba L. var. vitellina (L.) Stokes
Salix interior Rowlee
Salix rigida Muhl.
Salix X myricoides (Muhl.) Carey
Salix glaucophylloides Fern. var.
   glaucophylla (Bebb) Schneid.
Salix eriocephala Michx.
Salix syrticola Fern.
Salix bebbiana Sarg.
Salix pedicellaris Pursh var.
   hypoglauca Fern.
Salix discolor Muhl.
Salix humilis Marsh.
Salix humilis Marsh. var. hyporhysa
   Fern.
Salix humilis Marsh. var. microphylla
   (Anderss.) Fern.
Salix petiolaris Sm.
Salix sericea Marsh.
Salix caprea L.
Salix candida Fluegge
Salix purpurea L.
Salix X subsericea (Anderss.) Schneid.

Populus deltoides Marsh.
Populus nigra L. var. italica Muenchh.
Populus grandidentata Michx.
Populus X smithii Boivin
Populus tremuloides Michx.

Populus alba L.
Populus canescens (Ait.) Sm.
Populus balsamifera L.
Populus X gileadensis Rouleau
Populus heterophylla L.

Myricaceae
Comptonia peregrina (L.) Coult.

Juglandaceae
Juglans cinerea L.
Juglans nigra L.

Carya illinoensis (Wang.) K. Koch
Carya aquatica (Michx. f.) Nutt.
Carya cordiformis (Wang.) K. Koch
Carya texana Buckl.
Carya pallida (Ashe) Engl. & Graebn.
Carya ovalis (Wang.) Sarg.
Carya ovalis (Wang.) Sarg. var.
   obovalis Sarg.
Carya ovalis (Wang.) Sarg. var. odorata
   (Marsh.) Sarg.
Carya glabra (Mill.) Sweet
Carya glabra (Mill.) Sweet var.
   megacarpa Sarg.
Carya tomentosa (Poir.) Nutt.
Carya ovata (Mill.) K. Koch
Carya ovata (Mill.) K. Koch var. nuttallii
   Sm.
Carya ovata (Mill.) K. Koch var.
   fraxinifolia Sarg.
Carya laciniosa (Michx.) Loud.

Betulaceae
Betula lutea Michx. f.
Betula nigra L.
Betula papyrifera Marsh.
Betula populifolia Marsh.
Betula X sandbergii Britt.
Betula X purpusii Schneid.
Betula pumila L.

Alnus glutinosa (L.) Gaertn.
Alnus rugosa (DuRoi) Spreng. var.
   americana (Regel) Fern.
Alnus serrulata (Ait.) Willd.

Corylus americana Walt.

Ostrya virginiana (Mill.) K. Koch

Carpinus caroliniana Walt.

Fagaceae
Fagus grandifolia Ehrh.

Castanea dentata (Marsh.) Borkh.
Castanea mollissima Blume

Quercus imbricaria Michx.
Quercus phellos L.
Quercus marilandica Muenchh.
Quercus falcata Michx.
Quercus pagodaefolia (Ell.) Ashe

Phylogenetic List 266

Quercus velutina Lam.
Quercus rubra L.
Quercus palustris Muenchh.
Quercus shumardii Buckley
Quercus nuttallii E. J. Palmer
Quercus ellipsoidalis E. J. Hill
Quercus coccinea Muenchh.
Quercus bicolor Willd.
Quercus michauxii Nutt.
Quercus muhlenbergii Engelm.
Quercus prinus L.
Quercus alba L.
Quercus stellata Wangh.
Quercus macrocarpa Michx.
Quercus lyrata Walt.

Ulmaceae
Ulmus rubra Muhl.
Ulmus americana L.
Ulmus pumila L.
Ulmus alata Michx.
Ulmus thomasii Sarg.
Ulmus procera Salisb.

Planera aquatica (Walt.) J. F. Gmel.

Celtis occidentalis L.
Celtis occidentalis L. var. pumila
 (Pursh) Gray
Celtis occidentalis L. var. canina (Raf.)
 Sarg.
Celtis laevigata Willd.
Celtis laevigata Willd. var. smallii
 (Beadle) Sarg.
Celtis laevigata Willd. var. texana Sarg.
Celtis tenuifolia Nutt.
Celtis tenuifolia Nutt. var. georgiana
 (Small) Fern. & Schub.

Moraceae
Morus rubra L.
Morus alba L.
Morus alba L. var. tatarica (L.) Loudon

Broussonetia papyrifera (L.) L'Hér.

Maclura pomifera (Raf.) Schneider

Humulus lupulus L.
Humulus japonicus Sieb. & Zucc.

Cannabis sativa L.

Urticaceae
Urtica dioica L.
Urtica urens L.
Urtica chamaedryoides Pursh

Boehmeria cylindrica (L.) Sw.
Boehmeria cylindrica (L.) Sw. var.
 drummondiana Wedd.

Pilea pumila (L.) Gray
Pilea fontana (Lunell) Rydb.
Pilea opaca (Lunell) Rydb.

Laportea canadensis (L.) Wedd.

Parietaria pensylvanica Muhl.

Santalaceae
Comandra richardsiana Fern.

Loranthaceae
Phoradendron flavescens (Pursh) Nutt.

Aristolochiaceae
Asarum canadense L.
Asarum canadense L. var. reflexum
 (Bickn.) Robins.

Aristolochia serpentaria L.
Aristolochia serpentaria L. var. hastata
 (Nutt.) Duchartre
Aristolochia tomentosa Sims

Polygonaceae
Brunnichia cirrhosa Banks

Polygonella articulata (L.) Meisn.

Rumex acetosella L.
Rumex hastatulus Baldw.
Rumex maritimus L. var. fueginus
 (Phil.) Dusén
Rumex obtusifolius L.
Rumex crispus L.
Rumex altissimus Wood
Rumex orbiculatus Gray
Rumex verticillatus L.
Rumex mexicanus Meisn.
Rumex patientia L.

Fagopyrum esculentum Moench

Polygonum sagittatum L.
Polygonum arifolium L. var. pubescens
 (Keller) Fern.
Polygonum convolvulus L.
Polygonum cristatum Engelm. & Gray
Polygonum scandens L.
Polygonum tenue Michx.
Polygonum aviculare L.
Polygonum aviculare L. var. littorale
 (Link) W. D. J. Koch
Polygonum prolificum (Small) Robbins
Polygonum exsertum Small
Polygonum ramosissimum Michx.
Polygonum erectum L.
Polygonum achoreum Blake
Polygonum sachalinense F. Schmidt
Polygonum cuspidatum Sieb. & Zucc.
Polygonum virginianum L.
Polygonum careyi Olney
Polygonum orientale L.
Polygonum punctatum Ell.
Polygonum hydropiper L.
Polygonum persicaria L.
Polygonum cespitosum Blum var.
 longisetum (DeBruyn) Steward

Polygonum setaceum Baldw. var.
 interjectum Fern.
Polygonum hydropiperoides Michx.
Polygonum hydropiperoides Michx.
 var. bushianum Stanford
Polygonum opelousanum Riddell
Polygonum amphibium L. var.
 stipulaceum (Coleman) Fern.
Polygonum coccineum Muhl.
Polygonum longistylum Small
Polygonum scabrum Moench
Polygonum lapathifolium L.
Polygonum pensylvanicum L.
Polygonum pensylvanicum L. var.
 durum Stanford
Polygonum pensylvanicum L. var.
 laevigatum Fern.

Rheum rhaponticum L.

Chenopodiaceae
Salicornia europaea L.

Polycnemum majus A. Br.

Salsola kali L. var. tenuifolia Tausch

Kochia scoparia (L.) Roth

Suaeda depressa (Pursh) S. Wats.

Corispermum hyssopifolium L.
Corispermum orientale L.

Cycloloma atriplicifolia (Spreng.)
 Coult.

Atriplex hortensis L.
Atriplex patula L.
Atriplex rosea L.
Atriplex argentea Nutt.

Chenopodium botrys L.
Chenopodium ambrosioides L.
Chenopodium ambrosioides L. var.
 anthelminticum (L.) Gray
Chenopodium pallescens Standl.
Chenopodium desiccatum A. Nels. var.
 leptophylloides (Murr.) Wahl
Chenopodium glaucum L.
Chenopodium berlandieri Moq. var.
 zschackei (Murr.) Murr.
Chenopodium bushianum Aellen
Chenopodium album L.
Chenopodium album L. var
 lanceolatum (Muhl.) Coss & Germ.
Chenopodium strictum Roth var.
 glaucophyllum (Aellen) Wahl
Chenopodium missouriense Aellen
Chenopodium capitatum (L.) Aschers
Chenopodium rubrum L.
Chenopodium bonus-henricus L.
Chenopodium gigantospermum Aellen
Chenopodium polyspermum L.
Chenopodium standleyanum Aellen
Chenopodium murale L.

Chenopodium urbicum L.

Monolepis nuttalliana (R. & S.) Greene

Amaranthaceae
Amaranthus spinosus L.
Amaranthus albus L.
Amaranthus graecizans L.
Amaranthus powellii S. Wats.
Amaranthus retroflexus L.
Amaranthus hybridus L.
Amaranthus caudatus L.
Amaranthus cruentus L.
Amaranthus palmeri S. Wats.
Amaranthus torreyi (Gray) Benth.
Amaranthus ambigens Standl.
Amaranthus tuberculatus (Moq.) Sauer
Amaranthus tamariscinus Nutt.

Tidestromia lanuginosa (Nutt.) Standl.

Iresine rhizomatosa Standl.

Froelichia floridana (Nutt.) Moq. var.
  campestris (Small) Fern.
Froelichia gracilis (Hook) Moq.

Nyctaginaceae
Mirabilis linearis (Pursh) Heimerl.
Mirabilis nyctaginea (Michx.) MacM.
Mirabilis hirsuta (Pursh) MacM.
Mirabilis albida (Walt.) Heimerl.

Phytolaccaceae
Phytolacca americana L.

Aizoaceae
Mollugo verticillatus L.

Portulacaceae
Portulaca oleracea L.
Portulaca grandiflora Hook.

Talinum parviflorum Nutt.
Talinum rugospermum Holz.
Talinum calycinum Engelm.

Claytonia virginica L.

Caryophyllaceae
Paronychia canadensis (L.) Wood
Paronychia fastigiata (Raf.) Fern.

Spergularia media (L.) C. Presl
Spergularia rubra (L.) J. & C. Presl

Spergula arvensis L.

Scleranthus annuus L.

Sagina decumbens (Ell.) Torr. & Gray

Holosteum umbellatum L.

Stellaria media (L.) Cyrillo
Stellaria pubera Michx.
Stellaria graminea L.
Stellaria longifolia Muhl.

Myosoton aquaticum (L.) Moench

Cerastium vulgatum L.
Cerastium pumilum Curtis
Cerastium velutinum Raf.
Cerastium tetrandrum Curtis
Cerastium nutans Raf.
Cerastium brachypetalum Pers.
Cerastium viscosum L.
Cerastium brachypodum (Engelm.) B.
  L. Robins.

Arenaria stricta Michx.
Arenaria patula Michx.
Arenaria lateriflora L.
Arenaria serpyllifolia L.

Dianthus deltoides L.
Dianthus armeria L.
Dianthus barbatus L.

Tunica saxifraga (L.) Scop.

Agrostemma githago L.

Lychnis alba Mill.
Lychnis dioica L.
Lychnis coronaria (L.) Desv.

Silene regia Sims
Silene virginica L.
Silene stellata (L.) Ait.
Silene antirrhina L.
Silene armeria L.
Silene noctiflora L.
Silene dichotoma Ehrh.
Silene nivea (Nutt.) Otth.
Silene cserei Baumg.
Silene cucubalus Wibel

Gypsophila elegans Bieb.
Gypsophila paniculata L.
Gypsophila perfoliata L.

Saponaria officinalis L.
Saponaria vaccaria L.

Ceratophyllaceae
Ceratophyllum demersum L.
Ceratophyllum echinatum Gray

Nymphaeaceae
Nuphar luteum L. ssp. variegatum
  (Engelm.) Beal
Nuphar luteum L. ssp. macrophyllum
  (Small) Beal

Nymphaea odorata Ait.
Nymphaea tuberosa Paine

Nelumbonaceae
Nelumbo lutea (Willd.) Pers.

Cabombaceae
Brasenia schreberi Gmel.

Cabomba caroliniana Gray

Ranunculaceae
Ranunculus trichophyllus Chaix
Ranunculus longirostris Godr.
Ranunculus cymbalaria Pursh
Ranunculus ambigens Wats.
Ranunculus laxicaulis (Torr. & Gray)
  Darby
Ranunculus pusillus Poir.
Ranunculus harveyi (Gray) Britt.
Ranunculus rhomboideus Goldie
Ranunculus sceleratus L.
Ranunculus abortivus L.
Ranunculus abortivus L. var. acrolasius
  Fern.
Ranunculus micranthus Nutt.
Ranunculus flabellaris Raf.
Ranunculus gmelinii DC. var. hookeri
  (D. Don) L. Benson
Ranunculus parviflorus L.
Ranunculus recurvatus Poir.
Ranunculus pensylvanicus L. f.
Ranunculus acris L.
Ranunculus bulbosus L.
Ranunculus hispidus Michx.
Ranunculus hispidus Michx. var.
  marilandicus (Poir.) L. Benson
Ranunculus septentrionalis Poir.
Ranunculus septentrionalis Poir. var.
  caricetorum (Greene) Fern.
Ranunculus carolinianus DC.
Ranunculus repens L.
Ranunculus repens L. var. pleniflorus
  Fern.
Ranunculus fascicularis Muhl.
Ranunculus sardous Crantz
Ranunculus arvensis L.

Caltha palustris L.

Eranthis hyemalis Salisb.

Delphinium ajacis L.
Delphinium tricorne Mchx.
Delphinium carolinianum Walt.
Delphinium virescens Nutt.

Trautvetteria caroliniensis (Walt.) Vail

Thalictrum revolutum DC.
Thalictrum revolutum DC. f. glabrum
  Pennell
Thalictrum dasycarpum Fisch. & Lall.
Thalictrum dasycarpum Fisch. & Lall.
  var. hypoglaucum (Rydb.) Boivin
Thalictrum dioicum L.

Actaea rubra (Ait.) Willd.

Actaea pachypoda Ell.

Cimicifuga rubifolia Kearney
Cimicifuga racemosa (L.) Nutt.
Cimicifuga americana Michx.

Hepatica nobilis Schreb. var. obtusa
   (Pursh) Steyerm.
Hepatica nobilis Schreb. var. acuta
   (Pursh) Steyerm.

Hydrastis canadensis L.

Isopyrum biternatum (Raf.) Torr. & Gray

Anemonella thalictroides (L.) Spach

Anemone patens L.
Anemone caroliniana Walt.
Anemone canadensis L.
Anemone quinquefolia L.
Anemone cylindrica Gray
Anemone virginiana L.

Myosurus minimus L.

Aquilegia canadensis L.
Aquilegia vulgaris L.

Helleborus viridis L.

Nigella damascena L.

Clematis virginiana L.
Clematis dioscoreifolia Levl. & Vaniot
   var. robusta (Carr.) Rehd.
Clematis viorna L.
Clematis crispa L.
Clematis pitcheri Torr. & Gray

Berberidaceae
Berberis thunbergii DC.
Berberis canadensis Mill.
Berberis vulgaris L.

Podophyllum peltatum L.

Jeffersonia diphylla (L.) Pers.

Caulophyllum thalictroides (L.) Michx.

Menispermaceae
Calycocarpum lyonii (Pursh) Gray

Menispermum canadense L.

Cocculus carolinus (L.) DC.

Magnoliaceae
Magnolia acuminata L.

Liriodendron tulipifera L.

Calycanthaceae
Calycanthus floridus L.

Annonaceae
Asimina triloba (L.) Dunal

Lauraceae
Sassafras albidum (Nutt.) Nees
Sassafras albidum (Nutt.) Nees var.
   molle (Raf.) Fern.

Lindera benzoin (L.) Blume
Lindera benzoin (L.) Blume var.
   pubescens (Palmer & Steyerm.)
   Rehd.

Papaveraceae
Sanguinaria canadensis L.

Argemone albiflora Hornem.
Argemone mexicana L.

Macleaya cordata (Willd.) R. Br.

Papaver somniferum L.
Papaver dubium L.
Papaver rhoeas L.

Stylophorum diphyllum (Michx.) Nutt.

Chelidonium majus L.

Eschscholtzia californica Cham.

Dicentra cucullaria (L.) Bernh.
Dicentra canadensis (Goldie) Walp.
Dicentra eximia (Ker) Torr.

Adlumia fungosa (Ait.) Greene

Corydalis sempervirens (L.) Pers.
Corydalis flavula (Raf.) DC.
Corydalis micrantha (Engelm.) Gray
Corydalis halei (Small) Fern. & Schub.
Corydalis campestris (Britt.) Buchholz
   & Palmer
Corydalis aurea Willd.
Corydalis montana (Engelm.) Gray

Fumaria officinalis L.

Capparidaceae
Polanisia dodecandra (L.) DC.
Polanisia dodecandra (L.) DC. ssp.
   trachysperma (Torr. & Gray) Iltis
Polanisia jamesii (Torr. & Gray) Iltis

Cleome serrulata Pursh
Cleome hassleriana Chod.

Cruciferae
Dentaria laciniata Muhl.

Raphanus raphanistrum L.
Raphanus sativus L.

Hesperis matronalis L.

Iodanthus pinnatifidus (Michx.) Steud.

Cakile edentula (Bigel.) Hook. var.
   lacustris Fern

Eruca sativa Mill.

Capsella bursa-pastoris (L.) Medic.

Arabis lyrata L.
Arabis canadensis L.
Arabis glabra (L.) Bernh.
Arabis drummondii Gray
Arabis laevigata (Muhl.) Poir.
Arabis shortii (Fern.) Gl.
Arabis shortii (Fern.) Gl. var.
   phalacrocarpa (M. Hopkins)
   Steyerm.
Arabis hirsuta (L.) Scop. var.
   pycnocarpa (M. Hopkins) Rollins
Arabis hirsuta (L.) Scop. var.
   adpressipilis (M. Hopkins) Rollins
Arabis hirsuta (L.) Scop. var. glabrata
   Torr. & Gray

Descurainia sophia (L.) Webb
Descurainia pinnata (Walt.) Britt. var.
   brachycarpa (Richards.) Fern.

Cardamine bulbosa (Schreb.) BSP.
Cardamine douglassii (Torr.) Britt.
Cardamine hirsuta L.
Cardamine pensylvanica Muhl.
Cardamine parviflora L. var. arenicola
   (Britt.) C. E. Schulz
Cardamine pratensis L. var. palustris
   Wimm. & Grab.

Sibara virginica (L.) Rollins

Draba verna L.
Draba verna L. var. boerhaavii Van Hall
Draba cuneifolia Nutt.
Draba cuneifolia Nutt. var. foliosa
   Mohlenbr.
Draba reptans (Lam.) Fern.
Draba reptans (Lam.) Fern. var.
   micrantha (Nutt.) Fern.
Draba brachycarpa Nutt.

Arabidopsis thaliana (L.) Heynh.

Lepidium campestre (L.) R. Br.
Lepidium perfoliatum L.
Lepidium sativum L.
Lepidium virginicum L.
Lepidium ruderale L.
Lepidium densiflorum Schrad.

Armoracia aquatica (Eat.) Wieg.
Armoracia lapathifolia Gilib.

Nasturtium officinale R. Br.

Thlaspi arvense L.
Thlaspi perfoliatum L.

Cardaria draba (L.) Desv.

Alliaria officinalis Andrz.

Berteroa incana (L.) DC.

Lobularia maritima (L.) Desv.

Barbarea verna (Mill.) Aschers.
Barbarea vulgaris R. Br.
Barbarea vulgaris R. Br. var. arcuata
 (Opiz.) Fries

Lesquerella ludoviciana (Nutt.) S.
 Wats.
Lesquerella gracilis (Hook.) S. Wats.

Diplotaxis tenuifolia (L.) DC.
Diplotaxis muralis (L.) DC.

Erysimum capitatum (Dougl.) Greene
Erysimum cheiranthoides L.
Erysimum repandrum L.
Erysimum inconspicuum (S. Wats.)
 MacM.

Alyssum alyssoides L.

Isatis tinctoria L.

Conringia orientalis (L.) Dumort.

Camelina sativa (L.) Crantz
Camelina microcarpa Andrz.

Neslia paniculata (L.) Desv.

Brassica hirta Moench
Brassica kaber (DC.) L. C. Wheeler var.
 pinnatifida (Stokes) L. C. Wheeler
Brassica kaber (DC.) L. C. Wheeler var.
 schkuhriana (Reichenb.) L. C.
 Wheeler
Brassica nigra (L.) Koch
Brassica juncea (L.) Coss
Brassica rapa L.
Brassica oleracea L.
Brassica napus L.

Erucastrum gallicum (Willd.) O. E.
 Schulz

Sisymbrium officinale (L.) Scop.
Sisymbrium officinale (L.) Scop. var.
 leiocarpum DC.
Sisymbrium altissimum L.
Sisymbrium loeselii L.

Rorippa sylvestris (L.) Bess.
Rorippa sinuata (Nutt.) Hitchc.
Rorippa sessiliflora (Nutt.) Hitchc.
Rorippa truncata (Jepson) Stuckey
Rorippa islandica (Oeder) Borbas
Rorippa islandica (Oeder) Borbas var.
 fernaldiana Butt. & Abbe

Rorippa islandica (Oeder) Borbas var.
 hispida (Desv.) Butt. & Abbe

Chorispora tenella (Willd.) DC.

Coronopus didymus (L.) Sm.

Matthiola incana (L.) R. Br.

Resedaceae
Reseda alba L.

Sarraceniaceae
Sarracenia purpurea L.

Droseraceae
Drosera intermedia Hayne
Drosera rotundifolia L.

Crassulaceae
Sedum telephium L.
Sedum acre L.
Sedum sarmentosum Bunge
Sedum pulchellum Michx.
Sedum ternatum Michx.
Sedum telephioides Michx.
Sedum purpureum (L.) Link
Sedum albo-roseum Baker
Sedum rupestre L.

Saxifragaceae
Philadelphus inodorus L.
Philadelphus pubescens Loisel.
Philadelphus coronarius L.

Deutzia scabra Thunb.

Hydrangea arborescens L.

Ribes cynosbati L.
Ribes missouriense Nutt.
Ribes hirtellum Michx.
Ribes odoratum Wendl.
Ribes nigrum L.
Ribes americanum Mill.
Ribes sativum (Reichenb.) Syme

Itea virginica L.

Parnassia glauca Raf.

Mitella diphylla L.

Saxifraga virginiensis Michx.
Saxifraga forbesii Vasey
Saxifraga pensylvanica L.

Heuchera parviflora Bartl. var. rugelii
 (Shuttlw.) Rosend., Butt. & Lak.
Heuchera hirsuticaulis (Wheelock)
 Rydb.
Heuchera richardsonii R. Br. var.
 grayana Rosend., Butt. & Lak.

Sullivantia renifolia Rosend.

Penthorum sedoides L.

Hamamelidaceae
Liquidambar styraciflua L.

Hamamelis virginiana L.

Platanaceae
Platanus occidentalis L.

Rosaceae
Physocarpus opulifolius (L.) Maxim.

Spiraea tomentosa L.
Spirea alba DuRoi
Spiraea latifolia (Ait.) Borkh.
Spiraea prunifolia Sieb. & Zucc.

Prunus persica (L.) Batsch.
Prunus armeniaca L.
Prunus nigra Ait.
Prunus hortulana Bailey
Prunus munsoniana Wight & Hedrick
Prunus angustifolia Marsh.
Prunus mexicana S. Wats.
Prunus americana Marsh.
Prunus americana Marsh. var. lanata
 Sudw.
Prunus susquehanae Willd.
Prunus pensylvanica L. f.
Prunus cerasus L.
Prunus avium L.
Prunus mahaleb L.
Prunus virginiana L.
Prunus serotina Ehrh.
Prunus triloba Lindl.

Cydonia oblonga Mill

Amelanchier humilis Wieg.
Amelanchier arborea (Michx. f.) Fern.
Amelanchier interior Nielsen
Amelanchier laevis Wieg.

Chaenomeles japonica L.

Pyrus communis L.
Pyrus pyrifolia (Burm. f.) Nakai

Malus pumila Mill.
Malus angustifolia (Ait.) Michx.
Malus coronaria (L.) Mill.
Malus coronaria (L.) Mill. var. lancifolia
 Rehder
Malus ioensis (Wood) Britt.
Malus X soulardii (Bailey) Britt.

Aronia prunifolia (Marsh.) Rehd.
Aronia melanocarpa (Michx.) Ell.

Sorbus americana Marsh.
Sorbus aucuparia L.

Crataegus phaenopyrum (L. f.) Medic.
Crataegus marshallii Egglest.

Crataegus monogyna Jacq.
Crataegus punctata Jacq.
Crataegus cuneiformis (Marsh.)
  Egglest.
Crataegus margaretta Ashe
Crataegus collina Chapm.
Crataegus hannibalensis Palmer
Crataegus crus-galli L.
Crataegus crus-galli L. var. barrettiana
  (Sarg.) Palmer
Crataegus engelmannii Sarg.
Crataegus fecunda Sarg.
Crataegus acutifolia Sarg.
Crataegus permixta Palmer
Crataegus succulenta Link
Crataegus calpodendron (Ehrh.)
  Medic.
Crataegus neobushii Sarg.
Crataegus viridis L.
Crataegus nitida (Engelm.) Sarg.
Crataegus lucorum Sarg.
Crataegus faxonii Sarg.
Crataegus tortilis Ashe
Crataegus pruinosa (Wendl.) K. Koch
Crataegus mollis (Torr. & Gray)
  Scheele
Crataegus holmesiana Ashe
Crataegus pedicellata Sarg.
Crataegus pringlei Sarg.
Crataegus corusca Sarg.
Crataegus coccinioides Ashe
Crataegus macrosperma Ashe

Rubus odoratus L.
Rubus pubescens Raf.
Rubus occidentalis L.
Rubus phoenicolasius Maxim.
Rubus idaeus L.
Rubus strigosus Michx.
Rubus procerus P. J. Muell.
Rubus laciniatus Willd.
Rubus hispidus L.
Rubus trivialis Michx.
Rubus occidualis Bailey
Rubus flagellaris Willd.
Rubus enslenii Tratt.
Rubus schneideri Bailey
Rubus allegheniensis Porter
Rubus alumnus Bailey
Rubus argutus Link
Rubus avipes Bailey
Rubus pensylvanicus Poir.
Rubus frondosus Bigel.

Rosa multiflora Thunb.
Rosa setigera Michx.
Rosa setigera Michx. var. tomentosa
  Torr. & Gray
Rosa palustris Marsh.
Rosa carolina L.
Rosa carolina L. var. villosa (Best)
  Rehd.
Rosa pimpinellifolia L.
Rosa rugosa Thunb.
Rosa lunellii Greene
Rosa blanda Ait.

Rosa suffulta Greene
Rosa eglanteria L.
Rosa micrantha Sm.
Rosa canina L.
Rosa gallica L.
Rosa wichuriana Crepin
Rosa rubifolia Vill.

Potentilla palustris (L.) Scop.
Potentilla tridentata Ait.
Potentilla arguta Pursh
Potentilla fruticosa L.
Potentilla simplex Michx.
Potentilla simplex Michx. var.
  calvescens Fern.
Potentilla simplex Michx. var.
  argyrisma Fern.
Potentilla anserina L.
Potentilla recta L.
Potentilla argentea L.
Potentilla intermedia L.
Potentilla millegrana Engelm.
Potentilla norvegica L.
Potentilla rivalis Nutt.
Potentilla paradoxa Nutt.
Potentilla canadensis L.
Potentilla canescens Bess.

Waldsteinia fragarioides (Michx.) Tratt.

Fragaria virginiana Duchesne
Fragaria americana (Porter) Britt.
Fragaria X ananassa Duchesne

Aruncus dioicus (Walt.) Fern.

Filipendula rubra (Hill) Robins.
Filipendula ulmaria (L.) Maxim.

Sanguisorba minor Scop.
Sanguisorba canadensis L.

Gillenia stipulata (Muhl.) Baill.

Geum triflorum Pursh
Geum rivale L.
Geum canadense Jacq.
Geum canadense Jacq. var. grimesii
  Fern. & Weath.
Geum laciniatum Murr.
Geum laciniatum Murr. var.
  trichocarpum Fern.
Geum vernum (Raf.) Torr. & Gray
Geum virginianum L.
Geum strictum Ait.

Duchesnea indica (Andr.) Focke

Agrimonia parviflora Ait.
Agrimonia microcarpa Wallr.
Agrimonia pubescens Wallr.
Agrimonia gryposepala Wallr.
Agrimonia rostellata Wallr.

Leguminosae
Cercis canadensis L.

Caragana arborescens Lam.

Gymnocladus dioica (L.) K. Koch

Gleditsia triacanthos L.
Gleditsia aquatica Marsh.

Desmanthus illinoensis (Michx.)
  MacM.

Albizia julibrissin Duraz.

Wisteria floribunda DC.
Wisteria sinensis Sweet
Wisteria macrostachya Nutt.

Ononis spinosa L.

Amorpha nitens Boynton
Amorpha fruticosa L.
Amorpha fruticosa L. var. angustifolia
  Pursh
Amorpha fruticosa L. var. croceolanata
  (P. W. Wats.) Schneid.
Amorpha canescens Pursh

Robinia pseudoacacia L.
Robinia viscosa Vent.
Robinia hispida L.

Cladrastis lutea (Michx. f.) K. Koch

Crotalaria sagittalis L.
Crotalaria spectabilis Roth

Lupinus perennis L.

Psoralea onobrychis Nutt.
Psoralea psoralioides (Walt.) Cory var.
  eglandulosa (Ell.) Freeman
Psoralea tenuiflora Pursh

Vicia villosa Roth
Vicia caroliniana Walt.
Vicia americana Muhl.
Vicia dasycarpa Ten.
Vicia cracca L.
Vicia sativa L.
Vicia angustifolia Reich.

Lathyrus latifolius L.
Lathyrus odoratus L.
Lathyrus pratensis L.
Lathyrus maritimus (L.) Bigel.
Lathyrus ochroleucus Hook.
Lathyrus palustris L.
Lathyrus myrtifolius Muhl.
Lathyrus venosus Muhl. var. intonsus
  Butt. & St. John

Cassia tora L.
Cassia hebecarpa Fern.
Cassia marilandica L.
Cassia occidentalis L.
Cassia fasciculata Michx.

Cassia fasciculata Michx. var. robusta
(Pollard) Macbr.
Cassia nictitans L.

Sesbania exaltata (Raf.) Cory

Pisum sativum L.

Schrankia uncinata Willd.

Apios americana Medic.
Apios priceana Robins.

Lotus corniculatus L.

Petalostemum candidum (Willd.)
Michx.
Petalostemum purpureum (Vent.)
Rydb.
Petalostemum foliosum Gray

Glycyrrhiza lepidota Pursh

Coronilla varia L.

Dalea alopecuroides Willd.

Tephrosia virginiana (L.) Pers.

Onobrychis viciaefolia Scop.

Astragalus goniatus Nutt.
Astragalus distortus Torr. & Gray
Astragalus canadensis L.
Astragalus trichocalyx Nutt.
Astragalus tennesseensis Gray

Melilotus alba Desr.
Melilotus officinalis (L.) Lam.
Melilotus altissima Thuill.

Trifolium campestre Schreb.
Trifolium dubium Sibth.
Trifolium aureum Pollich
Trifolium pratense L.
Trifolium fragiferum L.
Trifolium arvense L.
Trifolium incarnatum L.
Trifolium reflexum L.
Trifolium reflexum L. var. glabrum
Lajacono
Trifolium resupinatum L.
Trifolium repens L.
Trifolium hybridum L.

Medicago sativa L.
Medicago orbicularis (L.) Bartal
Medicago arabica (L.) Huds.
Medicago lupulina L.
Medicago falcata L.

Baptisia tinctoria (L.) R. Br.
Baptisia australis (L.) R. Br.
Baptisia minor Lehm.
Baptisia leucantha Nutt.
Baptisia leucophaea Nutt.

Baptisia leucophaea Nutt. var.
glabrescens Larisey

Clitoria mariana L.

Stylosanthes biflora (L.) BSP.

Lespedeza striata (Thunb.) Hook. & Arn.
Lespedeza stipulacea Maxim.
Lespedeza procumbens Michx.
Lespedeza repens (L.) Bart.
Lespedeza hirta (L.) Hornem.
Lespedeza capitata Michx.
Lespedeza stuevei Nutt.
Lespedeza X simulata Mack. & Bush
Lespedeza thunbergii (DC.) Nakai
Lespedeza leptostachya Engelm.
Lespedeza cuneata (Dum.-Cours.) G.
Don
Lespedeza violacea (L.) Pers.
Lespedeza intermedia (S. Wats.) Britt.
Lespedeza virginica (L.) Britt.

Hosackia americana (Nutt.) Piper

Phaseolus polystachios (L.) BSP.

Dioclea multiflora (Torr. & Gray) C.
Mohr

Desmodium nudiflorum (L.) DC.
Desmodium glutinosum (Muhl.) Wood
Desmodium pauciflorum (Nutt.) DC.
Desmodium sessilifolium (Torr.) Torr &
Gray
Desmodium rotundifolium DC.
Desmodium illinoense Gray
Desmodium canescens (L.) DC.
Desmodium cuspidatum (Muhl.) Loud.
Desmodium cuspidatum (Muhl.) Loud.
var. longifolium (Torr. & Gray) Schub.
Desmodium laevigatum (Nutt.) DC.
Desmodium marilandicum (L.) DC.
Desmodium ciliare (Muhl.) DC.
Desmodium rigidum (Ell.) DC.
Desmodium canadense (L.) DC.
Desmodium nuttallii (Schindl.) Schub.
Desmodium dillenii Darl.
Desmodium paniculatum (L.) DC.

Glycine max (L.) Merr.

Vigna sinensis (L.) Endl.

Pueraria lobata (Willd.) Ohwi

Strophostyles leiosperma (Torr. &
Gray) Piper
Strophostyles helvola (L.) Ell.
Strophostyles helvola (L.) Ell. var.
missouriensis (S. Wats.) Britt.
Strophostyles umbellata (Muhl.) Britt.

Galactia volubilis (L.) Britt. var.
mississippiensis Vail

Amphicarpa bracteata (L.) Fern.
Amphicarpa bracteata (L.) Fern. var.
comosa (L.) Fern.

Linaceae
Linum usitatissimum L.
Linum perenne L.
Linum sulcatum Riddell
Linum medium (Planch.) Britt. var.
texanum (Planch.) Fern.
Linum virginianum L.
Linum striatum Walt.

Oxalidaceae
Oxalis violacea L.
Oxalis corniculata L.
Oxalis dillenii Jacq.
Oxalis stricta L.
Oxalis grandis Small

Geraniaceae
Erodium cicutarium (L.) L'Hér.

Geranium maculatum L.
Geranium robertianum L.
Geranium pusillum Burm. f.
Geranium sibiricum L.
Geranium dissectum L.
Geranium bicknellii Britt.
Geranium carolinianum L.
Geranium sanguineum L.

Zygophyllaceae
Tribulus terrestris L.

Kallstroemia intermedia Rydb.

Rutaceae
Xanthoxylum americanum Mill.

Ptelea trifoliata L.
Ptelea trifoliata L. var. mollis Torr. &
Gray

Ruta graveolens L.

Simaroubaceae
Ailanthus altissima (Mill.) Swingle

Polygalaceae
Polygala paucifolia Willd.
Polygala cruciata L. var. aquilonia
Fern.
Polygala verticillata L.
Polygala verticillata L. var. isocycla
Fern.
Polygala verticillata L. var. ambigua
(Nutt.) Wood
Polygala incarnata L.
Polygala polygama Walt. var. obtusata
Chod.
Polygala senega L.
Polygala sanguinea L.

Euphorbiaceae

Ricinus communis L.

Phyllanthus caroliniensis Walt.

Croton glandulosus L. var.
  septentrionalis Muell.-Arg.
Croton capitatus Michx.
Croton monanthogynus Michx.
Croton texensis (Klotzsch) Muell.-Arg.

Crotonopsis linearis Michx.
Crotonopsis elliptica Willd.

Tragia cordata Michx.

Acalypha ostryaefolia Riddell
Acalypha rhomboidea Raf.
Acalypha deamii (Weatherby) Ahles
Acalypha virginica L.
Acalypha gracilens Gray
Acalypha gracilens Gray ssp.
  monococca (Engelm.) Webster

Euphorbia marginata Pursh
Euphorbia corollata L.
Euphorbia corollata L. var. mollis
  Millsp.
Euphorbia cyparissias L.
Euphorbia esula L.
Euphorbia commutata Engelm.
Euphorbia peplus L.
Euphorbia helioscopia L.
Euphorbia obtusata Pursh
Euphorbia spathulata Lam.

Poinsettia cyanthophora (Murr.) Kl. &
  Garcke
Poinsettia cyanthophora (Murr.) Kl. &
  Garcke var. graminifolia (Michx.)
  Mohlenbr.
Poinsettia dentata (Michx.) Kl. &
  Garcke
Poinsettia dentata (Michx.) Kl. &
  Garcke var. cuphosperma (Engelm.)
  Mohlenbr.

Chamaesyce serpens (HBK.) Small
Chamaesyce polygonifolia (L.) Small
Chamaesyce geyeri (Engelm.) Small
Chamaesyce supina (Raf.) Moldenke
Chamaesyce humistrata (Engelm.)
  Small
Chamaesyce vermiculata (Raf.) House
Chamaesyce maculata (L.) Small
Chamaesyce serpyllifolia (Pers.) Small
Chamaesyce glyptosperma (Engelm.)
  Small

Callitrichaceae

Callitriche heterophylla Pursh
Callitriche palustris L.
Callitriche terrestris Raf.

Limnanthaceae

Floerkea proserpinacoides Willd.

Anacardiaceae

Toxicodendron vernix (L.) Kuntze
Toxicodendron radicans (L.) Kuntze

Rhus copallina L.
Rhus typhina L.
Rhus glabra L.
Rhus aromatica Ait.
Rhus aromatica Ait. var. serotina
  (Greene) Rehder
Rhus aromatica Ait. var. arenaria
  (Greene) Fern.

Aquifoliaceae

Ilex opaca Ait.
Ilex decidua Walt.
Ilex verticillata (L.) Gray

Nemopanthus mucronatus (L.) Trel.

Celastraceae

Euonymus obovatus Nutt.
Euonymus fortunei (Turcz.) Hand.-Maz.
Euonymus atropurpureus Jacq.
Euonymus europaeus L.
Euonymus americanus L.
Euonymus alatus (Thunb.) Sieb.

Celastrus scandens L.
Celastrus orbiculatus Thunb.

Staphyleaceae

Staphylea trifolia L.

Aceraceae

Acer negundo L.
Acer platanoides L.
Acer barbatum Michx.
Acer saccharum Marsh.
Acer saccharum Marsh. var. schneckii
  Rehder
Acer nigrum Michx. f.
Acer saccharinum L.
Acer rubrum L.
Acer rubrum L. var. drummondii
  (H. & A.) Sarg.

Hippocastanaceae

Aesculus discolor Pursh
Aesculus glabra Willd.
Aesculus glabra Willd. var.
  leucodermis Sarg.
Aesculus hippocastanum L.
Aesculus octandra Marsh.

Sapindaceae

Cardiospermum halicacabum L.

Koelreutera paniculata Laxm.

Balsaminaceae

Impatiens biflora Walt.
Impatiens pallida Nutt.

Rhamnaceae

Berchemia scandens (Hill) K. Koch

Ceanothus ovatus Desf.
Ceanothus americanus L.
Ceanothus americanus L. var. pitcheri
  Torr. & Gray

Rhamnus alnifolia L'Hér.
Rhamnus cathartica L.
Rhamnus lanceolata Pursh
Rhamnus caroliniana Walt.
Rhamnus frangula L.
Rhamnus frangula L. var. angustifolia
  Loud.
Rhamnus davuricus Pall.

Vitaceae

Parthenocissus tricuspidata (Sieb. &
  Zucc.) Planch.
Parthenocissus quinquefolia (L.)
  Planch.
Parthenocissus vitacea (Knerr) Hitchc.

Ampelopsis cordata Michx.
Ampelopsis arborea (L.) Koehne

Vitis labruscana Bailey
Vitis aestivalis Michx.
Vitis aestivalis Michx. var. argentifolia
  (Munson) Fern.
Vitis cinerea Engelm.
Vitis rupestris Scheele
Vitis vulpina L
Vitis palmata Vahl.
Vitis riparia Michx.
Vitis riparia Michx. var. praecox
  Engelm.
Vitis riparia Michx. var. syrticola (Fern.
  & Wieg.) Fern.

Tiliaceae

Tilia americana L.
Tilia americana L. var. neglecta
  (Spach.) Fosberg
Tilia heterophylla Vent.

Sterculiaceae

Melochia corchorifolia L.

Malvaceae

Malva sylvestris L.
Malva sylvestris L. var mauritiana (L.)
  Boiss.
Malva moschata L.
Malva neglecta Wallr.
Malva verticillata L. var. crispa L.
Malva rotundifolia L.

Callirhoë triangulata (Leavenw.) Gray

Callirhoë involucrata (Torr. & Gray)
  Gray
Callirhoë alcaeoides (Michx.) Gray
Callirhoë digitata Nutt.

Iliamna remota Greene

Gossypium hirsutum L.

Sphaeralcea angusta (Gray) Fern.

Hibiscus trionum L.
Hibiscus esculentus L.
Hibiscus militaris Cav.
Hibiscus syriacus L.
Hibiscus palustris L.
Hibiscus lasiocarpus Cav.

Althaea rosea (L.) Cav.

Napaea dioica L.

Anoda cristata (L.) Schlecht.

Abutilon theophrastii Medic.

Sida spinosa L.
Sida elliotti Torr. & Gray

Hypericaceae
Ascyrum hypericoides L.
Ascyrum hypericoides L. var.
  multicaule (Michx.) Fern.

Hypericum perforatum L.
Hypericum punctatum Lam.
Hypericum punctatum Lam. var.
  pseudomaculatum (Bush) Fern.
Hypericum pyramidatum Ait.
Hypericum kalmianum L.
Hypericum lobocarpum Gattinger
Hypericum adpressum Bart.
Hypericum ellipticum Hook.
Hypericum spathulatum (Spach.)
  Steud.
Hypericum densiflorum Pursh
Hypericum denticulatum Walt.
Hypericum sphaerocarpum Michx.
Hypericum sphaerocarpum Michx. var.
  turgidum (Small) Svenson
Hypericum boreale (Britt.) Bickn.
Hypericum mutilum L.
Hypericum gymnanthum Engelm. &
  Gray
Hypericum canadense L.
Hypericum majus (Gray) Britt.
Hypericum gentianoides (L.) BSP.
Hypericum drummondii (Grev. &
  Hook.) Torr. & Gray

Triadenum tubulosum (Walt.) Gl.
Triadenum walteri (Gmel.) Gl.
Triadenum fraseri (Spach.) Gl.
Triadenum virginicum (L.) Raf.

Elatinaceae
Elatine brachysperma Gray

Bergia texana (Hook.) Seubert

Tamaricaceae
Tamarix gallica L.

Cistaceae
Helianthemum canadense (L.) Michx.
Helianthemum bicknellii Fern.

Hudsonia tomentosa Nutt.
Hudsonia tomentosa Nutt. var.
  intermedia Peck

Lechea mucronata Raf.
Lechea minor L.
Lechea tenuifolia Michx.
Lechea tenuifolia Michx. var.
  occidentalis Hodgdon
Lechea stricta Leggett
Lechea pulchella Raf.
Lechea intermedia Leggett

Violaceae
Hybanthus concolor (T. F. Forst.)
  Spreng.

Viola pedata L.
Viola cucullata Ait.
Viola pratincola Greene
Viola pratincola Greene var. albiflora
  (Glover) Mohl.
Viola nephrophylla Greene
Viola missouriensis Greene
Viola affinis LeConte
Viola sororia Willd.
Viola septentrionalis Greene
Viola fimbriatula Smith
Viola sagittata Ait.
Viola triloba Schwein.
Viola triloba Schwein. var. dilatata (Ell.)
  Brainerd
Viola viarum Pollard
Viola pedatifida G. Don
Viola macloskeyi Lloyd ssp. pallens
  (Banks) M. S. Baker
Viola incognita Brainerd
Viola lanceolata L.
Viola lanceolata L. var. vittata (Greene)
  Russell
Viola primulifolia L.
Viola odorata L.
Viola pubescens Ait.
Viola pubescens Ait. var. eriocarpa
  (Schwein.) Russell
Viola canadensis L. var. rugulosa
  (Greene) C. L. Hitchcock
Viola striata Ait.
Viola conspersa Reichenb.
Viola X wittrockiana Hort.
Viola arvensis Murr.
Viola rafinesquii Greene
Viola tricolor L.

Passifloraceae
Passiflora lutea L. var. glabriflora Fern.
Passiflora incarnata L.

Loasaceae
Mentzelia oligosperma Nutt.
Mentzelia nuda Torr. & Gray
Mentzelia decapetala (Pursh) Urban &
  Gilg.

Cactaceae
Opuntia compressa (Salisb.) Macbr.
Opuntia macrorhiza Engelm.

Thymelaeaceae
Dirca palustris L.

Thymelaea passerina (L.) Cosson &
  Germain

Elaeagnaceae
Shepherdia canadensis (L.) Nutt.

Elaeagnus angustifolia L.
Elaeagnus umbellata Thunb.

Lythraceae
Decodon verticillatus (L.) Ell.

Cuphea petiolata (L.) Koehne

Lythrum alatum Pursh
Lythrum salicaria L.

Peplis diandra Nutt.

Rotala ramosior (L.) Koehne

Ammannia coccinea Rottb.
Ammannia auriculata Willd.

Nyssaceae
Nyssa aquatica L.
Nyssa sylvatica Marsh.
Nyssa sylvatica Marsh. var. caroliniana
  (Poir.) Fern.

Melastomaceae
Rhexia virginica L.
Rhexia mariana L.

Onagraceae
Circaea quadrisulcata (Maxim.)
  Franch. & Sav. var. canadensis (L.)
  Hara
Circaea alpina L.

Ludwigia palustris (L.) Ell. var.
  americana (DC.) Fern. & Grisc.
Ludwigia polycarpa Short & Peter
Ludwigia glandulosa Walt.
Ludwigia alternifolia L.
Ludwigia alternifolia L. var. pubescens
  Palmer & Steyerm.

Epilobium angustifolium L.
Epilobium hirsutum L.
Epilobium strictum Muhl.
Epilobium leptophyllum Raf.
Epilobium adenocaulon Haussk.
Epilobium coloratum Muhl.

Jussiaea repens L.
Jussiaea decurrens (Walt.) DC.
Jussiaea leptocarpa Nutt.

Oenothera nuttallii Sweet
Oenothera speciosa Nutt.
Oenothera missouriensis Sims
Oenothera laciniata Hill
Oenothera perennis L.
Oenothera tetragona Roth
Oenothera pilosella Raf.
Oenothera fruticosa L. var. linearis
  (Michx.) S. Wats.
Oenothera biennis L.
Oenothera biennis L. var. canescens
  Torr. & Gray
Oenothera cruciata Nutt.
Oenothera rhombipetala Nutt.
Oenothera serrulata Nutt.
Oenothera linifolia Nutt.
Oenothera triloba Nutt.

Guara biennis L.
Gaura longiflora Spach.
Gaura parviflora Dougl.
Gaura filipes Spach

Haloragidaceae
Myriophyllum hippuroides Nutt.
Myriophyllum pinnatum (Walt.) BSP.
Myriophyllum verticillatum L. var.
  pectinatum Wallr.
Myriophyllum heterophyllum Michx.
Myriophyllum exalbescens Fern.

Proserpinaca palustris L.

Hippuridaceae
Hippuris vulgaris L.

Araliaceae
Hedera helix L.

Aralia spinosa L.
Aralia hispida Vent.
Aralia nudicaulis L.
Aralia racemosa L.

Panax quinquefolius L.

Umbelliferae
Eryngium yuccifolium Michx.
Eryngium prostratum Nutt.

Bupleurum rotundifolium L.

Thaspium barbinode (Michx.) Nutt.
Thaspium trifoliatum (L.) Gray

Thaspium trifoliatum (L.) Gray var.
  flavum Blake

Sanicula marilandica L.
Sanicula gregaria Bickn.
Sanicula canadensis L.
Sanicula trifoliata Bickn.

Torilis japonica (Houtt.) DC.

Daucus pusillus Michx.
Daucus carota L.
Daucus carota L. f. epurpuratus Farw.
Daucus carota L. f. roseus Millsp.

Cynosciadium digitatum DC.

Ptilimnium costatum (Ell.) Raf.
Ptilimnium nuttallii (DC.) Britt.

Limnosciadium pinnatum (DC.) Math.
  & Const.

Oxypolis rigidior (L.) Coulter & Rose
Oxypolis rigidior (L.) Coulter & Rose
  var. ambigua (Nutt.) Robins.

Heracleum maximum Bartr.

Cryptotaenia canadensis (L.) DC.

Falcaria sioides (Wibel) Aschers.

Zizia aptera (Gray) Fern.
Zizia aurea (L.) Koch

Pastinaca sativa L.

Sium suave Walt.

Berula pusilla (Nutt.) Fern.

Osmorhiza longistylis (Torr.) DC.
Osmorhiza longistylis (Torr.) DC. var.
  villicaulis Fern.
Osmorhiza claytonii (Michx.) Clarke

Foeniculum vulgare Mill.

Anethum graveolens L.

Erigenia bulbosa (Michx.) Nutt.

Carum carvi L.

Coriandrum sativum L.

Spermolepis inermis (Nutt.) Math. &
  Const.
Spermolepis echinata (Nutt.) Heller

Chaerophyllum procumbens (L.)
  Crantz
Chaerophyllum tainturieri Hook.

Perideridia americana (Nutt.)
  Reichenb.

Taenidia integerrima (L.) Drude

Polytaenia nuttallii DC.

Cicuta bulbifera L.
Cicuta maculata L.

Angelica atropurpurea L.
Angelica venerosa (Greenway) Fern.

Aegopodium podagraria L.

Conioselinum chinense (L.) BSP.

Aethusa cynapium L.

Conium maculatum L.

Hydrocotyle ranunculoides L. f.

Cornaceae
Cornus canadensis L.
Cornus florida L.
Cornus alternifolia L. f.
Cornus rugosa Lam.
Cornus stolonifera Michx.
Cornus stolonifera Michx. var. baileyi
  (Coult. & Evans) Drescher
Cornus drummondii C. A. Mey.
Cornus racemosa Lam.
Cornus obliqua Raf.
Cornus foemina Mill.
Cornus amomum Mill.

Ericaceae
Monotropa hypopithys L.
Monotropa uniflora L.

Chimaphila maculata (L.) Pursh
Chimaphila umbellata (L.) Bart. var.
  cisatlantica Blake

Pyrola secunda L.
Pyrola elliptica Nutt.
Pyrola americana Sweet

Rhododendron prinophyllum (Small)
  Millais
Rhododendron periclymenoides
  (Michx.) Skinners

Andromeda glaucophylla Link

Epigaea repens L.

Arctostaphylos uva-ursi (L.) Spreng.
  var. coactilis Fern. & Macbr.

Oxydendrum arboreum (L.) DC.

Chamaedaphne calyculata (L.)
  Moench var. angustifolia (Ait.)
  Rehder

Gaultheria procumbens L.

Gaylussac a baccata (Wang.) K. Koch

Vaccinium stamineum L. var.
 neglectum (Small) Deam
Vaccinium arboreum Marsh.
Vaccinium arboreum Marsh. var.
 glaucescens (Greene) Sarg.
Vaccinium angustifolium Ait. var.
 laevifolium House
Vaccinium myrtilloides Michx.
Vaccinium vacillans Torr.
Vaccinium corymbosum L.
Vaccinium macrocarpon Ait.

Primulaceae
Dodecatheon amethystinum Fassett
Dodecatheon meadia L.
Dodecatheon frenchii (Vasey) Rydb.

Primula mistassinica Michx.

Androsace occidentalis Pursh

Centunculus minimus L.

Samolus parviflorus Raf.

Anagallis arvensis L.

Lysimachia quadriflora Sims
Lysimachia radicans Hook.
Lysimachia ciliata L.
Lysimachia lanceolata Walt.
Lysimachia hybrida Michx.
Lysimachia thyrsiflora L.
Lysimachia nummularia L.
Lysimachia vulgaris L.
Lysimachia punctata L.
Lysimachia quadrifolia L.
Lysimachia terrestris (L.) BSP.
Lysimachia X commixta Fern.
Lysimachia clethroides Duby
Lysimachia fraseri Duby

Trientalis borealis Raf.

Hottonia inflata Ell.

Sapotaceae
Bumelia lanuginosa (Michx.) Pers. var.
 oblongifolia (Nutt.) R. B. Clarke
Bumelia lycioides (L.) Gaertn. f.

Ebenaceae
Diospyros virginiana L.

Styracaceae
Halesia carolina L.

Styrax americana Lam.
Styrax grandifolia Ait.

Oleaceae
Fraxinus nigra Marsh.
Fraxinus quadrangulata Michx.
Fraxinus pensylvanica Marsh.
Fraxinus pensylvanica Marsh. var.
 austinii Fern.
Fraxinus pensylvanica Marsh. var.
 subintegerrima (Vahl) Fern.
Fraxinus americana L.
Fraxinus americana L. var. biltmoreana
 (Beadle) J. Wright
Fraxinus tomentosa Michx. f.

Forestiera acuminata (Michx.) Poir.

Syringa vulgaris L.

Ligustrum vulgare L.
Ligustrum obtusifolium Sieb. & Zucc.

Loganiaceae
Spigelia marilandica L.

Polyoremum procumbens L.

Gentianaceae
Swertia caroliniensis (Walt.) Kuntze

Bartonia virginica (L.) BSP.
Bartonia paniculata (Michx.) Muhl.

Obolaria virginica L.

Sabatia angularis (L.) Pursh
Sabatia campestris Nutt.

Centaurium pulchellum (Sw.) Druce

Gentiana alba Gray
Gentiana puberulenta Pringle
Gentiana andrewsii Griseb.
Gentiana saponaria L.
Gentiana crinita Froel.
Gentiana procera Holm.
Gentiana quinquefolia L. var.
 occidentalis (Gray) Hitchc.
Gentiana septemfida Pall.

Menyanthaceae
Menyanthes trifoliata L. var. minor Raf.

Nymphoides peltata (Gmel.) Kuntze

Apocynaceae
Amsonia tabernaemontana Walt.
Amsonia tabernaemontana Walt. var.
 salicifolia (Pursh) Woodson
Amsonia tabernaemontana Walt. var.
 gattingeri Woodson

Vinca minor L.
Vinca major L.

Apocynum androsaemifolium L.
Apocynum X medium Greene

Apocynum cannabinum L.
Apocynum cannabinum L. var.
 pubescens (Mitchell) A. DC.
Apocynum sibiricum Jacq.
Apocynum sibiricum Jacq. var.
 cordigerum (Greene) Fern.

Trachelospermum difforme (Walt.)
 Gray

Asclepiadaceae
Asclepias viridis Walt.
Asclepias stenophylla Gray
Asclepias hirtella (Pennell) Woodson
Asclepias tuberosa L. var. interior
 (Woodson) Shinners
Asclepias verticillata L.
Asclepias lanuginosa Nutt.
Asclepias viridiflora Raf.
Asclepias amplexicaulis Sm.
Asclepias meadii Torr.
Asclepias ovalifolia Decne.
Asclepias purpurascens L.
Asclepias syriaca L.
Asclepias syriaca L. var. kansana (Vail)
 Palmer & Steyerm.
Asclepias quadrifolia Jacq.
Asclepias variegata L.
Asclepias tuberosa Torr.
Asclepias sullivantii Engelm.
Asclepias exaltata L.
Asclepias perennis Walt.
Asclepias incarnata L.

Matelea gonocarpa (Walt.) Shinners
Matelea obliqua (Jacq.) Woodson
Matelea decipiens (Alex.) Woodson

Cynanchum laeve (Michx.) Pers.
Cynanchum nigrum (L.) Pers.

Convolvulaceae
Stylisma pickeringii (Torr.) Gray var.
 pattersonii (Fern. & Schub.) Myint

Jacquemontia tamnifolia (L.) Griseb.

Evolvulus nuttallianus Roem. & Schult.

Convolvulus arvensis L

Calystegia pubescens Lindl.
Calystegia spithamaea (L.) Pursh
Calystegia sepium (L.) R. Br. var.
 americana (Sims) Mohlenbr.
Calystegia sepium (L.) R. Br. var.
 repens (L.) Mohlenbr
Calystegia sepium (L.) R. Br. var.
 fraterniflora (Mack. & Bush) Shinners

Ipomoea coccinea L.
Ipomoea pandurata (L.) G. F. W. Mey.
Ipomoea hederacea (L.) Jacq.
Ipomoea lacunosa L.
Ipomoea purpurea (L.) Roth

Cuscuta cuspidata Engelm.
Cuscuta compacta Juss.
Cuscuta glomerata Choisy
Cuscuta polygonorum Engelm.
Cuscuta cephalanthi Engelm.
Cuscuta coryli Engelm.
Cuscuta gronovii Willd.
Cuscuta pentagona Engelm.
Cuscuta campestris Yuncker
Cuscuta indecora Choisy
Cuscuta indecora Choisy var.
  neuropetala (Engelm.) Hitchc.

Polemoniaceae
Polemonium reptans L.

Gilia rubra (L.) Heller

Collomia linearis Nutt.

Phlox bifida Beck
Phlox bifida Beck var. stellaria (Gray)
  Wherry
Phlox divaricata L.
Phlox divaricata L. ssp. laphamii
  (Wood) Wherry
Phlox pilosa L.
Phlox pilosa L. ssp. fulgida (Wherry)
  Wherry
Phlox pilosa L. spp. sangamonensis
  Levin & Smith
Phlox paniculata L.
Phlox maculata L.
Phlox glaberrima L. ssp. interior
  (Wherry) Wherry

Microsteris gracilis (Dougl.) Greene

Hydrophyllaceae
Hydrolea uniflora Raf.

Ellisia nyctelea L.

Hydrophyllum appendiculatum Michx.
Hydrophyllum canadense L.
Hydrophyllum macrophyllum Nutt.
Hydrophyllum virginianum L.

Phacelia purshii Buckley
Phacelia ranunculacea (Nutt.) Const.
Phacelia bipinnatifida Michx.

Boraginaceae
Mertensia virginica (L.) Pers.

Heliotropium curassavicum L.
Heliotropium tenellum (Nutt.) Torr.
Heliotropium indicum L.
Heliotropium europaeum L.

Cynoglossum virginianum L.
Cynoglossum officinale L.

Lappula echinata Gilib.

Lappula redowskii (Hornem.) Greene
  var. occidentalis (Wats.) Rydb.

Hackelia virginiana (L.) I. M. Johnston
Hackelia americana (Gray) Fern.

Borago officinalis L.

Echium vulgare L.

Asperugo procumbens L.

Symphytum officinale L.

Anchusa officinalis L.

Myosotis virginica (L.) BSP.
Myosotis virginica (L.) BSP. var.
  macrosperma (Engelm.) Fern.
Myosotis scorpioides L.
Myosotis sylvatica Hoffm.
Myosotis arvensis (L.) Hill
Myosotis stricta Link

Lithospermum arvense L.
Lithospermum latifolium Michx.
Lithospermum officinale L.
Lithospermum incisum Lehm.
Lithospermum canescens (Michx.)
  Lehm.
Lithospermum caroliniense (Walt.)
  MacM.

Onosmodium hispidissimum Mack.
Onosmodium occidentale Mack.
Onosmodium molle Michx.

Amsinckia spectabilis Fisch. & Mey.

Plagiobothrys hirtus (Greene) I. M.
  Johnston var. figuratus (Piper) I. M.
  Johnston

Verbenaceae
Lippia lanceolata Michx.
Lippia cuneifolia (Torr.) Steud.

Verbena peruviana Britt.
Verbena canadensis Britt.
Verbena bracteata Lag. & Rodr.
Verbena X deamii Moldenke
Verbena X perriana Moldenke
Verbena simplex Lehm.
Verbena X blanchardii Moldenke
Verbena X moechina Moldenke
Verbena stricta Vent.
Verbena hastata L.
Verbena X rydbergii Moldenke
Verbena urticifolia L.
Verbena urticifolia L. var. leiocarpa
  Perry & Fern.
Verbena X illicita Moldenke
Verbena X engelmannii Moldenke

Phrymaceae
Phryma leptostachya L.

Labiatae
Isanthus brachiatus (L.) BSP.

Trichostema dichotomum L.

Mentha X cardiaca Gerarde
Mentha X gentilis L.
Mentha arvensis L.
Mentha arvensis L. var. villosa (Benth.)
  S. R. Steward
Mentha spicata L.
Mentha X alopecuroides Hull
Mentha rotundifolia (L.) Huds.
Mentha X piperita L.
Mentha citrata Ehrh.
Mentha crispa L.

Lycopus europaeus L.
Lycopus americanus Muhl.
Lycopus virginicus L.
Lycopus uniflorus Michx.
Lycopus rubellus Moench
Lycopus rubellus Moench var.
  arkansanus (Fresn.) Benner
Lycopus asper Greene
Lycopus amplectens Raf.

Teucrium canadense L. var. virginicum
  (L.) Eat.
Teucrium canadense L. var.
  occidentale (Gray) McClintock &
  Epling

Ajuga reptans L.
Ajuga genevensis L.

Scutellaria epilobiifolia Muhl.
Scutellaria nervosa Pursh
Scutellaria parvula Michx.
Scutellaria parvula Michx. var.
  leonardii (Epling) Fern.
Scutellaria lateriflora L.
Scutellaria ovata Hill
Scutellaria ovata Hill var. versicolor
  (Nutt.) Fern.
Scutellaria ovata Hill var. rugosa
  (Wood) Fern.
Scutellaria elliptica Muhl.
Scutellaria incana Biehler

Marrubium vulgare L.

Cunila origanoides (L.) Britt.

Ballota nigra L.

Monarda bradburiana Beck
Monarda didyma L.
Monarda clinopodia L.
Monarda fistulosa L.
Monarda punctata L. var. villicaulis
  Pennell
Monarda punctata L. var. lasiodonta
  Gray
Monarda citriodora Cerv.

Blephilia ciliata (L.) Benth.
Blephilia hirsuta (Pursh) Benth.

Collinsonia canadensis L.

Hedeoma hispida Pursh
Hedeoma pulegioides (L.) Pers.

Salvia lyrata L.
Salvia pratensis L.
Salvia azurea Lam. var. grandiflora
  Benth.
Salvia reflexa Hornem.
Salvia sylvestris L.
Salvia verticillata L.

Thymus serpyllum L.

Ocimum basilicum L.

Satureja hortensis L.
Satureja arkansana (Nutt.) Briq.
Satureja vulgaris (L.) Fritsch

Pycnanthemum muticum (Michx.) Pers.
Pycnanthemum pycnanthemoides
  (Leavenw.) Fern.
Pycnanthemum incanum (L.) Michx.
Pycnanthemum albescens Torr. & Gray
Pycnanthemum tenuifolium Schrad.
Pycnanthemum torrei Benth.
Pycnanthemum virginianum (L.) Dur. &
  Jacks.
Pycnanthemum pilosum Nutt.

Agastache nepetoides (L.) Ktze.
Agastache scrophulariaefolia (Willd.)
  Ktze.
Agastache foeniculum (Pursh) Ktze.

Nepeta cataria L.

Synandra hispidula (Michx.) Baill.

Glecoma hederacea L.
Glecoma hederacea L. var. micrantha
  Moricand

Lamium amplexicaule L.
Lamium purpureum L.
Lamium maculatum L.

Stachys riddellii House
Stachys palustris L. var. pilosa (Nutt.)
  Fern.
Stachys palustris L. var. homotricha
  Fern.
Stachys palustris L. var. phaneropoda
  Weath.
Stachys clingmanii Small
Stachys tenuifolia Willd.
Stachys tenuifolia Willd. var. hispida
  (Pursh) Fern.
Stachys hyssopifolia Michx. var.
  ambigua Gray

Melissa officinalis L.

Leonurus cardiaca L.
Leonurus marrubiastrum L.

Galeopsis tetrahit L.
Galeopsis ladanum L.

Perilla frutescens L.

Physostegia intermedia (Nutt.) Engelm.
  & Gray
Physostegia parviflora Nutt.
Physostegia angustifolia Fern.
Physostegia virginiana (L.) Benth.
Physostegia speciosa (Sweet) Sweet

Dracocephalum parviflorum Nutt.

Prunella vulgaris L.
Prunella vulgaris L. var. lanceolata
  (Bart.) Fern.

Solanaceae
Solanum rostratum Dunal
Solanum carolinense L.
Solanum elaeagnifolium Cav.
Solanum torreyi Gray
Solanum dulcamara L.
Solanum tuberosum L.
Solanum triflorum Nutt.
Solanum americanum Mill.

Lycium halimifolium Mill.
Lycium chinense Mill.

Lycopersicum esculentum Mill.

Hyoscyamus niger L.

Nicandra physalodes (L.) Pers.

Datura stramonium L.
Datura stramonium L. var. tatula (L.)
  Torr.
Datura innoxia Mill.

Physalis ixocarpa Brotero
Physalis angulata L.
Physalis pendula Rydb.
Physalis longifolia Nutt.
Physalis subglabrata Mack. & Bush
Physalis macrophysa Rydb.
Physalis barbadensis Jacq.
Physalis pruinosa L.
Physalis virginiana Mill.
Physalis heterophylla Nees
Physalis heterophylla Nees var.
  ambigua (Gray) Rydb.
Physalis heterophylla Nees var.
  nyctaginea (Dunal) Rydb.
Physalis pubescens L.
Physalis pumila Nutt.
Physalis lanceolata Michx.
Physalis alkekengi L.

Nicotiana rustica L.

Petunia axillaris (Lam.) BSP.
Petunia violacea Lindl.
Petunia X hybrida Vilm.

Scrophulariaceae
Paulownia tomentosa (Thunb.) Steud.

Veronicastrum virginicum (L.) Farw.

Veronica peregrina L.
Veronica peregrina L. var. xalapensis
  (HBK.) St. John
Veronica persica Poir.
Veronica hederaefolia L.
Veronica arvensis L.
Veronica polita Fries
Veronica serpyllifolia L.
Veronica scutellata L.
Veronica comosa Richter
Veronica americana (Raf.) Schwein.
Veronica longifolia L.
Veronica officinalis L.
Veronica chamaedrys L.
Veronica agrestis L.

Gratiola virginiana L.
Gratiola neglecta Torr.
Gratiola aurea Muhl.

Lindernia anagallidea (Michx.) Pennell
Lindernia dubia (L.) Pennell
Lindernia dubia (L.) Pennell var. riparia
  (Raf.) Fern.

Chelone glabra L.
Chelone obliqua L. var. speciosa
  Pennell & Wherry

Penstemon grandiflorus Nutt.
Penstemon cobaea Nutt.
Penstemon tubaeflorus Nutt.
Penstemon digitalis Nutt.
Penstemon alluviorum Pennell
Penstemon deamii Pennell
Penstemon calycosus Small
Penstemon pallidus Small
Penstemon hirsutus (L.) Willd.
Penstemon arkansanus Pennell

Conobea multifida (Michx.) Benth.

Pedicularis lanceolata Michx.
Pedicularis canadensis L.

Gerardia flava L.
Gerardia grandiflora Benth. var.
  pulchra (Pennell) Fern.
Gerardia pedicularia L. var. ambigens
  Fern.
Gerardia auriculata Michx.
Gerardia gattingeri Small
Gerardia skinneriana Wood
Gerardia tenuifolia Vahl

Gerardia tenuifolia Vahl var.
   macrophylla Benth.
Gerardia aspera Dougl.
Gerardia fasciculata Ell.
Gerardia paupercula (Gray) Britt.
Gerardia purpurea L.

Seymeria macrophylla Nutt.

Bacopa rotundifolia (Michx.) Wettst.
Bacopa acuminata (Walt.) Robins.

Melampyrum lineare Desr. var.
   latifolium Bart.

Mimulus glabratus HBK. var. fremontii
   (Benth.) Grant
Mimulus alatus Ait.
Mimulus ringens L.
Mimulus ringens L. var. minthodes
   (Greene) Grant

Buchnera americana L.

Mazus japonicus (Thunb.) Kuntze

Scrophularia marilandica L.
Scrophularia lanceolata Pursh

Collinsia verna Nutt.
Collinsia violacea Nutt.

Cymbalaria muralis Gaertn.

Linaria vulgaris Hill
Linaria dalmatica (L.) Mill.
Linaria canadensis (L.) Dumort.
Linaria canadensis (L.) Dumort. var.
   texana (Scheele) Pennell

Kickxia elatine (L.) Dumort.

Verbascum thapsus L.
Verbascum phlomoides L.
Verbascum blattaria L.
Verbascum virgatum Stokes
Verbascum pulverulenta Schrad.

Antirrhinum majus L.
Antirrhinum orontium L.

Wulfenia bullii (Eat.) Barnh.

Castilleja coccinea L.
Castilleja sessiliflora Pursh

Chaenorrhinum minus (L.) Lange

Bignoniaceae
Campsis radicans (L.) Seem.

Bignonia capreolata L.

Catalpa bignonioides Walt.
Catalpa speciosa Warder

Martyniaceae
Proboscidea louisianica (Mill.) Thell.

Orobanchaceae
Conopholis americana (L. f.) Wallr.

Epifagus virginiana (L.) Bart.

Orobanche ramosa L.
Orobanche ludoviciana Nutt.
Orobanche uniflora L.
Orobanche fasciculata Nutt.

Lentibulariaceae
Utricularia cornuta Michx.
Utricularia gibba L.
Utricularia intermedia Hayne
Utricularia vulgaris L.
Utricularia minor L.

Acanthaceae
Dicliptera brachiata (Pursh) Spreng.

Justicia americana (L.) Vahl
Justicia ovata (Walt.) Lindau

Ruellia humilis Nutt.
Ruellia humilis Nutt. var. longiflora
   (Gray) Fern.
Ruellia pedunculata Torr.
Ruellia caroliniensis (Walt.) Steud. var.
   dentata (Nees) Fern.
Ruellia strepens L.
Ruellia strepens L. f. cleistantha (Gray)
   McCoy

Plantaginaceae
Plantago aristata Michx.
Plantago purshii Roem. & Schultes
Plantago pusilla Nutt.
Plantago pusilla Nutt. var. major
   Engelm.
Plantago heterophylla Nutt.
Plantago cordata Lam.
Plantago lanceolata L.
Plantago media L.
Plantago virginica L.
Plantago rhodosperma Dcne.
Plantago rugelii Dcne.
Plantago major L.
Plantago indica L.

Rubiaceae
Cephalanthus occidentalis L.
Cephalanthus occidentalis L. var.
   pubescens Raf.

Sherardia arvensis L.

Galium virgatum Nutt.
Galium boreale L.
Galium circaezans Michx.
Galium lanceolatum Torr.
Galium pilosum Ait.
Galium triflorum Michx.

Galium aparine L.
Galium verum L.
Galium mollugo L.
Galium tinctorium L.
Galium trifidum L.
Galium asprellum Michx.
Galium pedemontanum All.
Galium concinnum Torr. & Gray
Galium labradoricum Wieg.
Galium obtusum Bigel.

Diodia virginica L.
Diodia teres Walt.

Spermacoce glabra Michx.

Mitchella repens L.

Houstonia caerulea L.
Houstonia minima Beck
Houstonia pusilla Schoepf
Houstonia nigricans (Lam.) Fern.
Houstonia purpurea L.
Houstonia purpurea L. var. calycosa
   Gray
Houstonia longifolia Gaertn.
Houstonia longifolia Gaertn. var.
   ciliolata (Torr.) Torr.
Houstonia longifolia Gaertn. var.
   tenuifolia (Nutt.) Wood

Caprifoliaceae
Sambucus canadensis L.
Sambucus pubens Michx.

Lonicera sempervirens L.
Lonicera prolifera (Kirchn.) Rehd.
Lonicera flava Sims
Lonicera dioica L.
Lonicera japonica Thunb.
Lonicera xylosteum L.
Lonicera morrowi Gray
Lonicera tatarica L.
Lonicera X bella Zabel
Lonicera maackii Maxim.
Lonicera standishii Jacq.
Lonicera dioica L. var. glaucescens
   (Rydb.) Butters

Symphoricarpos orbiculatus Moench
Symphoricarpos albus (L.) Blake var.
   laevigatus (Fern.) Blake
Symphoricarpos occidentalis Hook.

Diervilla lonicera Mill.

Viburnum acerifolium L.
Viburnum trilobum Marsh.
Viburnum opulus L.
Viburnum lantana L.
Viburnum rufidulum Raf.
Viburnum prunifolium L.
Viburnum lentago L.
Viburnum molle Michx.
Viburnum rafinesquianum Schultes

Viburnum dentatum L. var. deamii
(Rehd.) Fern.
Viburnum recognitum Fern.

Linnaea americana Forbes

Triosteum angustifolium L.
Triosteum perfoliatum L.
Triosteum aurantiacum Bickn.
Triosteum i linoense (Wieg.) Rydb.

Adoxaceae
Adoxa moschatellina L.

Valerianaceae
Valeriana c liata Torr. & Gray
Valeriana pauciflora Michx.
Valeriana officinalis L.
Valeriana uiginosa (Torr. & Gray)
Rydb.

Valerianella olitoria (L.) Poll.
Valerianella radiata (L.) Dufr.
Valerianella radiata (L.) Dufr. var.
missouriensis Dyal
Valerianella intermedia Dyal
Valerianella patellaria (Sulliv.) Wood
Valerianella umbilicata (Sulliv.) Wood

Dipsacaceae
Dipsacus laciniatus L.
Dipsacus sy vestris Huds.

Knautia arvensis (L.) Duby

Cucurbitaceae
Citrullus vulgaris Schrad.

Lagenaria siceraria (Molina) Standl.

Cucurbita foetidissima HBK.
Cucurbita pepo L. var. ovifera (L.) Alef.

Cucumis sativus L.

Melothria pendula L.

Sicyos angulatus L.

Echinocystis obata (Michx.) Torr. &
Gray

Campanulaceae
Specularia biflora (R. & P.) Fisch. &
Mey.
Specularia perfoliata (L.) A. DC.
Specularia leptocarpa (Nutt.) Gray

Campanula rotundifolia L.
Campanula rotundifolia L. var. velutina
A. DC.
Campanula aparinoides Pursh
Campanula uliginosa Rydb.
Campanula rapunculoides L.
Campanula americana L.

Lobelia cardinalis L.
Lobelia cardinalis X siphilitica
Schneck
Lobelia silphilitica L.
Lobelia puberula Michx.
Lobelia kalmii L.
Lobelia inflata L.
Lobelia spicata Lam.
Lobelia spicata Lam. var. leptostachys
(A. DC.) Mack. & Bush

Compositae
Polymnia canadensis L.
Polymnia uvedalia (L.) L.

Silphium perfoliatum L.
Silphium laciniatum L.
Silphium terebinthinaceum Jacq.
Silphium integrifolium Michx.
Silphium speciosum Nutt.

Parthenium hysterophorus L.
Parthenium integrifolium L.

Iva annua L.
Iva xanthifolia Nutt.

Ambrosia bidentata Michx.
Ambrosia trifida L.
Ambrosia trifida L. var. texana Scheele
Ambrosia psilostachya DC.
Ambrosia artemisiifolia L.
Ambrosia tomentosa Nutt.

Xanthium spinosum L.
Xanthium strumarium L. var.
canadensis (Mill.) Torr. & Gray
Xanthium strumarium L. var. glabratum
(DC.) Cronq.

Heliopsis helianthoides (L.) Sweet

Eclipta alba (L.) Hassk.

Rudbeckia amplexicaulis Vahl
Rudbeckia laciniata L.
Rudbeckia triloba L.
Rudbeckia subtomentosa Pursh
Rudbeckia hirta L.
Rudbeckia fulgida Ait.
Rudbeckia fulgida Ait. var. sullivantii
(Boynt. & Beadle) Cronq.
Rudbeckia fulgida Ait. var.
missouriensis (Engelm.) Cronq.
Rudbeckia grandiflora (D. Don) DC.

Echinacea purpurea (L.) Moench
Echinacea pallida (Nutt.) Nutt.

Ratibida columnifera (Nutt.) Woot. &
Stardl.
Ratibida pinnata (Vent.) Barnh.

Helianthus ciliaris DC.
Helianthus salicifolius A. Dietr.
Helianthus angustifolius L.

Helianthus annuus L.
Helianthus petiolaris Nutt.

Helianthus rigidus (Cass.) Desf.
Helianthus silphioides Nutt.
Helianthus occidentalis Riddell
Helianthus microcephalus Torr. & Gray
Helianthus decapetalus L.
Helianthus divaricatus L.
Helianthus strumosus L.
Helianthus grosseserratus Martens
Helianthus mollis Lam.
Helianthus tuberosus L.
Helianthus tuberosus L. var.
subcanescens Gray
Helianthus hirsutus Raf.
Helianthus giganteus L.
Helianthus maximiliarii Schrader
Helianthus X doroniccides Lam.
Helianthus X luxurians E. E. Wats.

Melanthera nivea (L.) Small

Verbesina helianthoides Michx.
Verbesina alternifolia (L.) Britt.
Verbesina virginica L.
Verbesina encelioides (Cav.) Benth. &
Hook.

Coreopsis lanceolata L.
Coreopsis pubescens Ell.
Coreopsis palmata Nutt.
Coreopsis tinctoria Nutt.
Coreopsis basalis (Otto & Dietr.) Blake
Coreopsis tripteris L.
Coreopsis grandifolia Hogg

Bidens beckii Torr.
Bidens cernua L.
Bidens coronata (L.) Britt.
Bidens aristosa L.
Bidens aristosa L. var. retrorsa (Sherff)
Wunderlin
Bidens connata Muhl.
Bidens comosa (Gray) Wieg.
Bidens bipinnata L.
Bidens frondosa L.
Bidens vulgata Greene
Bidens discoidea (Torr. & Gray) Britt.

Cosmos sulphureus Cav.
Cosmos bipinnatus Cav.

Galinsoga ciliata (Raf.) Blake
Galinsoga parviflora Cav.

Hymenopappus scabiosaeus L'Hér.

Hymenoxys acaulis (Pursh) Parker

Helenium amarum (Raf.) Rock
Helenium autumnale L.
Helenium flexuosum Raf.

Gaillardia aestivalis (Walt.) Rock
Gaillardia pulchella Foug.

Gaillardia aristata Pursh

Dyssodia papposa (Vent.) Hitchc.

Grindelia lanceolata Nutt.
Grindelia squarrosa (Pursh) Dunal
Grindelia squarrosa (Pursh) Dunal var.
  serrulata (Rydb.) Steyerm.

Xanthocephalum texanum (DC.)
  Shinners
Xanthocephalum dracunculoides
  (DC.) Shinners

Heterotheca villosa (Pursh) Shinners
Heterotheca latifolia Buckl.

Solidago graminifolia (L.) Salisb.
Solidago graminifolia (L.) Salisb. var.
  gymnospermoides (Greene) Croat
Solidago graminifolia (L.) Salisb. var.
  remota (Greene) Harris
Solidago rigida L.
Solidago riddellii Frank
Solidago ohioensis Riddell
Solidago sciaphila Steele
Solidago caesia L.
Solidago flexicaulis L.
Solidago bicolor L.
Solidago bicolor L. var. concolor Torr.
  & Gray
Solidago buckleyi Torr. & Gray
Solidago petiolaris Ait.
Solidago sphacelata Raf.
Solidago missouriensis Nutt.
Solidago juncea Ait.
Solidago uliginosa Nutt.
Solidago speciosa Nutt.
Solidago patula Muhl.
Solidago gigantea Ait.
Solidago sempervirens L.
Solidago arguta Ait.
Solidago strigosa Small
Solidago neurolepis Fern.
Solidago boottii Hook.
Solidago ulmifolia Muhl.
Solidago drummondii Torr. & Gray
Solidago radula Nutt.
Solidago canadensis L.
Solidago nemoralis Ait.
Solidago rugosa Mill.

Haplopappus ciliatus (Nutt.) DC.

Bellis perennis L.

Boltonia asteroides (L.) L'Hér.
Boltonia asteroides (L.) L'Hér. var.
  decurrens (Torr. & Gray) Engelm.
Boltonia diffusa Ell.

Aster undulatus L.
Aster macrophyllus L.
Aster furcatus Burgess
Aster schreberi Nees
Aster anomalus Engelm.

Aster shortii Lindl.
Aster azureus Lindl.
Aster cordifolius L.
Aster sagittifolius Wedem.
Aster sagittifolius Wedem. var.
  drummondii (Lindl.) Shinners
Aster novae-angliae L.
Aster oblongifolius Nutt.
Aster patens Ait.
Aster puniceus L.
Aster puniceus L. var. lucidulus Gray
Aster laevis L.
Aster prenanthoides Muhl.
Aster novi-belgii L.
Aster sericeus Vent.
Aster umbellatus Mill.
Aster ptarmicoides (Nees) Torr. & Gray
Aster ptarmicoides (Nees) Torr. & Gray
  var. lutescens (Lindl.) Gray
Aster brachyactis Blake
Aster parviceps (Burgess) Mack. &
  Bush
Aster pilosus Willd.
Aster pilosus Willd. var. pringlei (Gray)
  Blake
Aster ericoides L.
Aster ericoides L. var. prostratus (Ktze.)
  Blake
Aster vimineus Lam.
Aster dumosus L.
Aster praealtus Poir.
Aster linariifolius L.
Aster junciformis Rydb.
Aster turbinellus Lindl.
Aster tataricus L. f.
Aster ontarionis Wieg.
Aster lateriflorus (L.) Britt.
Aster simplex Willd.
Aster X amethystinus Nutt.

Erigeron pulchellus Michx.
Erigeron philadelphicus L.
Erigeron annuus (L.) Pers.
Erigeron strigosus Muhl.
Erigeron divaricatus Michx.
Erigeron canadensis L.

Anthemis tinctoria L.
Anthemis cotula L.
Anthemis nobilis L.
Anthemis arvensis L.

Achillea millefolium L.
Achillea millefolium L. ssp. lanulosa
  (Nutt.) Piper

Matricaria maritima L.
Matricaria chamomilla L.
Matricaria matricarioides (Less.) Porter

Chrysanthemum leucanthemum L.
Chrysanthemum parthenium (L.)
  Bernh.
Chrysanthemum balsamita L.

Tanacetum vulgare L.

Artemisia serrata Nutt.
Artemisia ludoviciana Nutt.
Artemisia ludoviciana Nutt. var.
  gnaphalodes (Nutt.) Torr. & Gray
Artemisia vulgaris L.
Artemisia absinthium L.
Artemisia frigida Willd.
Artemisia dracunculus L.
Artemisia campestris L.
Artemisia abrotanum L.
Artemisia biennis Willd.
Artemisia annua L.

Pluchea camphorata (L.) DC.

Antennaria neglecta Greene
Antennaria neglecta Greene var.
  attenuata (Fern.) Cronq.
Antennaria plantaginifolia (L.)
  Richards.
Antennaria plantaginifolia (L.)
  Richards. var. arnoglossa (Greene)
  Cronq.
Antennaria plantaginifolia (L.)
  Richards. var. ambigens (Greene)
  Cronq.

Gnaphalium purpureum L.
Gnaphalium macounii Greene
Gnaphalium obtusifolium L.
Gnaphalium uliginosum L.

Thelesperma gracile (Torr.) Gray

Inula helenium L.

Erechtites hieracifolia (L.) Raf.

Cacalia suaveolens L.
Cacalia atriplicifolia L.
Cacalia muhlenbergii (Sch.-Bip.) Fern.
Cacalia tuberosa Nutt.

Sanvitalia procumbens Lam.

Petasites hybridus (L.) Gaertn., Meyer,
  & Scherb.

Senecio jacobaea L.
Senecio obovatus Muhl.
Senecio aureus L.
Senecio plattensis Nutt.
Senecio pauperculus Michx.
Senecio vulgaris L.
Senecio glabellus Poir.
Senecio viscosus L.

Eupatorium maculatum L.
Eupatorium purpureum L.
Eupatorium fistulosum Barratt
Eupatorium coelestinum L.
Eupatorium incarnatum Walt.
Eupatorium serotinum Michx.
Eupatorium rugosum Houtt.
Eupatorium altissimum L.
Eupatorium perfoliatum L.

Eupatorium sessilifolium L.
Eupatorium X polyneuron (F. J. Herm.)
  Wunderlin

Mikania scandens (L.) Willd.

Brickellia eupatorioides (L.) Shinners

Liatris scabra (Greene) K. Schum.
Liatris cylindracea Michx.
Liatris squarrosa (L.) Michx.
Liatris punctata Hook.
Liatris pycnostachya Michx.
Liatris spicata (L.) Willd.
Liatris aspera Michx.
Liatris ligulistylis (A. Nels.) K. Schum.
Liatris X ridgewayi Standl.
Liatris X steelei Gaiser
Liatris X gladewitzii (Farwell) Shinners

Vernonia arkansana DC.
Vernonia missurica Raf.
Vernonia baldwinii Torr.
Vernonia fasciculata Michx.
Vernonia gigantea (Walt.) Trel.

Elephantopus carolinianus Willd.

Echinops sphaerocephalus L.

Arctium lappa L.
Arctium tomentosum Mill.
Arctium minus (Hill) Bernh.

Carduus nutans L.
Carduus acanthoides L.

Cirsium vulgare (Savi) Tenore
Cirsium pitcheri (Torr.) Torr. & Gray
Cirsium discolor (Muhl.) Spreng.
Cirsium altissima (L.) Spreng.
Cirsium carolinianum (Walt.) Fern. &
  Schub.

Cirsium pumilum (Nutt.) Spreng.
Cirsium arvense (L.) Scop.
Cirsium arvense (L.) Scop. var.
  horridum Wimm. & Grab.
Cirsium muticum Michx.

Onopordum acanthium L.

Centaurea solstitalis L.
Centaurea calcitrapa L.
Centaurea diffusa Lam.
Centaurea maculosa Lam.
Centaurea repens L.
Centaurea moschata L.
Centaurea americana Nutt.
Centaurea cyanus L.
Centaurea nigra L.
Centaurea jacea L.
Centaurea dubia Suter
Centaurea pulcherrima Willd.

Cnicus benedictus L.

Cichorium intybus L.

Lapsana communis L.

Microseris cuspidata (Pursh) Sch.-Bip.

Krigia dandelion (L.) Nutt.
Krigia biflora (Walt.) Blake
Krigia virginica (L.) Willd.
Krigia oppositifolia Raf.

Hypochaeris radicata L.
Hypochaeris glabra L.

Leontodon autumnalis L.
Leontodon leysseri (Wallr.) G. Beck

Picris echioides L.
Picris hieracioides L.

Tragopogon porrifolius L.
Tragopogon dubius Scop.
Tragopogon pratensis L.

Taraxacum officinale Weber
Taraxacum laevigatum (Willd.) DC.

Sonchus arvensis L.
Sonchus arvensis L. var. glabrescens
  Guenth., Gram., & Winn
Sonchus oleraceus L.
Sonchus asper (L.) Hill

Lactuca sativa L.
Lactuca canadensis L.
Lactuca ludoviciana (Nutt.) DC.
Lactuca serriola L.
Lactuca saligna L.
Lactuca biennis (Moench) Fern.
Lactuca floridana (L.) Gaertn.
Lactuca tatarica (L.) C. A. Mey.
Lactuca X morssii Robins.
Lactuca hirsuta Muhl.

Pyrrhopappus carolinianus (Walt.) DC.

Crepis capillaris (L.) Wallr.

Prenanthes racemosa Michx.
Prenanthes racemosa Michx. ssp.
  multiflora Cronq.
Prenanthes aspera Michx.
Prenanthes crepidinea Michx.
Prenanthes alba L.
Prenanthes altissima L.

Hieracium murorum L.
Hieracium pratense Tausch
Hieracium aurantiacum L.
Hieracium scabrum Michx.
Hieracium canadense Michx.
Hieracium longipilum Torr.
Hieracium gronovii L.

Literature Cited

Fernald, M. L. 1950. Gray's Manual of Botany, ed. 8. The American Book Company, New York. 1632 pp.

Gleason, H. A. 1952. Illustrated Flora of the Northeastern United States, 3 vols. The New York Botanical Garden, New York.

Jones, G. N. 1963. Flora of Illinois, ed. 3. American Midland Naturalist Monograph No. 7. University of Notre Dame, Notre Dame, Indiana. 402 pp.

Mohlenbrock, R. H. 1975. Guide to the Vascular Flora of Illinois. Southern Illinois University Press, Carbondale. 494 pp.

Winterringer, G. S. & R. A. Evers. 1960. New Records for Illinois Vascular Plants. Illinois State Museum, Scientific Papers Series, Vol. XI, Springfield. 135 pp.